# A Course in
# Combinatorics

# A Course in Combinatorics

### J. H. van Lint
*Technical University of Eindhoven*

### and

### R. M. Wilson
*California Institute of Technology*

CAMBRIDGE
UNIVERSITY PRESS

Published by the Press Syndicate of the University of Cambridge
The Pitt Building, Trumpington Street, Cambridge CB2 1RP
40 West 20th Street, New York, NY 10011-4211, USA
10 Stamford Road, Oakleigh, Victoria 3166, Australia

First published 1992
Reprinted 1993

Printed in Great Britain at the University Press, Cambridge

*Library of Congress cataloguing in publication data available*
*A catalogue record for this book is available from the British Library*

ISBN 0 521 41057 6   hardback
ISBN 0 521 42260 4   paperback

# CONTENTS

# Preface

One of the most popular upper level mathematics courses taught at Caltech for very many years was H. J. Ryser's course *Combinatorial Analysis*, Math 121. One of Ryser's main goals was to show elegance and simplicity. Furthermore, in this course that he taught so well, he sought to demonstrate coherence of the subject of combinatorics. We dedicate this book to the memory of Herb Ryser, our friend whom we admired and from whom we learned much.

Work on the present book was started during the academic year 1988–89 when the two authors taught the course Math 121 together. Our aim was not only to continue in the style of Ryser by showing many links between areas of combinatorics that seem unrelated, but also to try to more-or-less survey the subject. We had in mind that after a course like this, students who subsequently attend a conference on "Combinatorics" would hear no talks where they are completely lost because of unfamiliarity with the topic. Well, at least they should have heard many of the words before. We strongly believe that a student studying combinatorics should see as many of its branches as possible.

Of course, none of the chapters could possibly give a complete treatment of the subject indicated in their titles. Instead, we cover some highlights—but we insist on doing something substantial or nontrivial with each topic. It is our opinion that a good way to learn combinatorics is to see subjects repeated at intervals. For this reason, several areas are covered in more than one part of the book. For example, partially ordered sets and codes appear several times. Enumeration problems and graph theory occur throughout the book. A few topics are treated in more detail (because we like them) and some material, like our proof of the Van der Waerden permanent conjecture, appears here in a text book for the first time.

A course in modern algebra is sufficient background for this book, but is not absolutely necessary; a great deal can be understood with only a certain level of maturity. Indeed, combinatorics is well known for being "accessible". But readers should find this book challenging and will be expected to fill in details (that we hope are instructive and not too difficult). We mention in passing that we believe there is no substitute for a human teacher when trying to learn a subject. An acquaintance with calculus, groups, finite fields, elementary number theory, and especially linear algebra will be necessary for some topics. Both undergraduates and graduate students take the course at Caltech. The material in every chapter has been presented in class, but we have never managed to do all the chapters in one year.

The notes at the end of chapters often include biographical remarks on mathematicians. We have chosen to refrain from any mention of living mathematicians unless they have retired (with the exception of P. Erdős).

Exercises vary in difficulty. For some it may be necessary to consult the hints in Appendix 1. We include a short discussion of formal power series in Appendix 2.

This manuscript was typeset by the authors in $\mathcal{AMS}$-TEX.

J. H. v. L., R. M. W.

Eindhoven and Pasadena, 1992

# 1
# Graphs

A *graph* $G$ consists of a set $V$ (or $V(G)$) of *vertices*, a set $E$ (or $E(G)$) of *edges*, and a mapping associating to each edge $e \in E(G)$ an unordered pair $x, y$ of vertices called the *endpoints* (or simply the *ends*) of $e$. We say an edge is *incident* with its ends, and that it *joins* its ends. We allow $x = y$, in which case the edge is called a *loop*. A vertex is *isolated* when it is incident with no edges.

It is common to represent a graph by a *drawing* where we represent each vertex by a point in the plane, and represent edges by line segments or arcs joining some of the pairs of points. One can think e.g. of a network of roads between cities. A graph is called *planar* if it can be drawn in the plane such that no two edges (that is, the line segments or arcs representing the edges) cross. The topic of planarity will be dealt with in Chapter 32; we wish to deal with graphs more purely combinatorially for the present.

| edge | ends |
|:----:|:----:|
| $a$ | $x, z$ |
| $b$ | $y, w$ |
| $c$ | $x, z$ |
| $d$ | $z, w$ |
| $e$ | $z, w$ |
| $f$ | $x, y$ |
| $g$ | $z, w$ |

Figure 1.1

Thus a graph is described by a table such as the one in Fig. 1.1 that lists the ends of each edge. Here the graph we are describing has vertex set $V = \{x, y, z, w\}$ and edge set $E = \{a, b, c, d, e, f, g\}$; a drawing of this graph may be found as Fig. 1.2(iv).

A graph is *simple* when it has no loops and no two distinct edges have exactly the same pair of ends. Two nonloops are *parallel* when they have the same ends; graphs that contain them are called *multigraphs* by some authors, or are said to have "multiple edges".

If an *ordered* pair of vertices is associated to each edge, we have a *directed graph* or *digraph*. In a drawing of a digraph, we use an arrowhead to point from the first vertex (the *tail*) towards the second vertex (the *head*) incident with an edge. For a *simple* digraph, we disallow loops and require that no two distinct edges have the same ordered pair of ends.

When dealing with simple graphs, it is often convenient to identify the edges with the unordered pairs of vertices they join; thus an edge joining $x$ and $y$ can be called $\{x, y\}$. Similarly, the edges of a simple digraph can be identified with ordered pairs of distinct vertices.

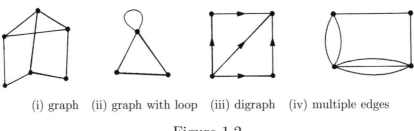

(i) graph    (ii) graph with loop    (iii) digraph    (iv) multiple edges

Figure 1.2

There are several ways to draw the same graph. For example, the two graphs of Fig. 1.3 are essentially the same.

We make this more precise, but to avoid unnecessarily technical definitions at this point, let us assume that all graphs are undirected and simple for the next two definitions.

We say two graphs are *isomorphic* if there is a one-to-one correspondence between the vertex sets such that if two vertices are joined by an edge in one graph, then the corresponding vertices are joined by an edge in the other graph. To show that the two graphs in Fig. 1.3 are the same, find a suitable numbering of the vertices in both graphs (using $1, 2, 3, 4, 5, 6$) and observe that the edge sets are the same sets of unordered pairs.

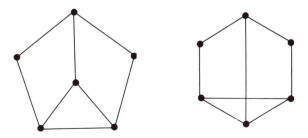

Figure 1.3

A permutation $\sigma$ of the vertex set of a graph $G$ with the property that $\{a, b\}$ is an edge if and only if $\{\sigma(a), \sigma(b)\}$ is an edge, is called an *automorphism* of $G$.

PROBLEM 1A. (i) Show that the drawings in Fig. 1.4 represent the same graph (or isomorphic graphs).

(ii) Find the group of automorphisms of the graph in Fig. 1.4. Remark: There is no quick or easy way to do this unless you are lucky; you will have to experiment and try things.

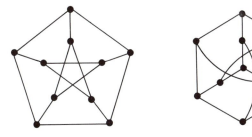

Figure 1.4

The *complete* graph $K_n$ on $n$ vertices is the simple graph that has all $\binom{n}{2}$ possible edges.

Two vertices $a$ and $b$ of a graph $G$ are called *adjacent* if they are distinct and joined by an edge. We will use $\Gamma(x)$ to denote the set of all vertices adjacent to a given vertex $x$; these vertices are also called the *neighbors* of $x$.

The number of edges incident with a vertex $x$ is called the *degree* or the *valency* of $x$. Loops are considered to contribute 2 to the valency, as the pictures we draw suggest. If all the vertices of a graph have the same degree, then the graph is called *regular*.

One of the important tools in combinatorics is the method of *counting* certain objects in two different ways. It is a well known fact that if one makes no mistakes, then the two answers are the same. We give a first elementary example. A graph is *finite* when both $E(G)$ and $V(G)$ are finite sets.

THEOREM 1.1. *A finite graph $G$ has an even number of vertices with odd valency.*

PROOF: Consider a table listing the ends of the edges, as in Fig. 1.1. The number of entries in the right column of the table is twice the number of edges. On the other hand, the degree of a vertex $x$ is, by definition, the number of times it occurs in the table. So the number of entries in the right column is

$$(1.1) \qquad\qquad \sum_{x \in V(G)} \deg(x) = 2|E(G)|.$$

The assertion follows immediately.                                            □

The equation (1.1) is simple but important. It might be called the "first theorem of graph theory", and our Theorem 1.1 is its first corollary.

A *subgraph* of a graph $G$ is a graph $H$ such that $V(H) \subseteq V(G)$, $E(H) \subseteq E(G)$, and the ends of an edge $e \in E(H)$ are the same as its ends in $G$. $H$ is a *spanning* subgraph when $V(H) = V(G)$. The subgraph of $G$ *induced* by a subset $S$ of vertices of $G$ is the subgraph whose vertex set is $S$ and whose edges are *all* the edges of $G$ with both ends in $S$.

A *walk* in a graph $G$ consists of an alternating sequence

$$x_0, e_1, x_1, e_2, x_2, \ldots, x_{k-1}, e_k, x_k$$

of vertices $x_i$, not necessarily distinct, and edges $e_i$ so that the ends of $e_i$ are exactly $x_{i-1}$ and $x_i$, $i = 1, 2, \ldots, k$. If the graph is simple, a walk is determined by its sequence of vertices, any two successive elements of which are adjacent. Such a walk has *length k*.

If the edge terms $e_1, \ldots, e_k$ are distinct, then the walk is called a *path* from $x_0$ to $x_k$. If $x_0 = x_k$, then a walk (or path) is called *closed*. A *simple* path is one in which the vertex terms $x_0, x_1, \ldots, x_k$ are also distinct, although we say we have a *simple closed path* when $k \geq 1$ and all vertex terms are distinct except $x_0 = x_k$.

If a path from $x$ to $y$ exists for every pair of vertices $x, y$ of $G$, then $G$ is called *connected*. Otherwise $G$ consists of a number of connected *components* (maximal connected subgraphs). It will be convenient to agree that the null graph with no vertices and no edges is not connected.

PROBLEM 1B. Suppose $G$ is a simple graph on 10 vertices that is not connected. Prove that $G$ has at most 36 edges. Can equality occur?

The length of the shortest walk from $a$ to $b$, if such walks exist, is called the *distance* $d(a, b)$ between these vertices. Such a shortest walk is necessarily a simple path.

EXAMPLE 1.1. A well known graph has the mathematicians of the world as vertices. Two vertices are adjacent if and only if they have published a joint paper. The distance in this graph from some mathematician to the vertex P. Erdős is known as his or her Erdős-number.

Figure 1.5

A *polygon* is the "graph of" a simple closed path, but more precisely it can be defined as a finite connected graph that is regular of degree 2. There is, up to isomorphism, exactly one polygon $P_n$

with $n$ vertices (often called the *n-gon*) for each positive integer $n$. The sequence of polygons is shown in Fig. 1.5.

A connected graph that contains no simple closed paths, i.e. that has no polygons as subgraphs, is called a *tree*.

PROBLEM 1C. Show that a connected graph on $n$ vertices is a tree if and only if it has $n-1$ edges.

PROBLEM 1D. The *complete bipartite graph* $K_{n,m}$ has $n+m$ vertices $a_1, \ldots, a_n$ and $b_1, \ldots, b_m$, and as edges all $mn$ pairs $\{a_i, b_j\}$. Show that $K_{3,3}$ is not planar.

No introduction to graph theory can omit the problem of the bridges of Königsberg (formerly a city in Prussia). The river Pregel flowed through this city and split into two parts. In the river was the island Kneiphof. There were seven bridges connecting different parts of the city as shown in the diagram of Fig. 1.6.

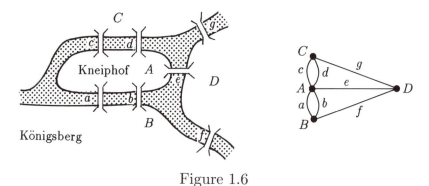

Figure 1.6

In a paper written in 1736 by L. Euler (considered the first paper on graph theory) the author claims that the following question was considered difficult: Is it possible to make a walk through the city, returning to the starting point and crossing each bridge exactly once? This paper has led to the following definition. A closed path through a graph using every edge once is called an *Eulerian circuit* and a graph that has such a path is called an *Eulerian graph*.

THEOREM 1.2. *A finite graph $G$ with no isolated vertices (but possibly with multiple edges) is Eulerian if and only if it is connected and every vertex has even degree.*

PROOF: That $G$ must be connected is obvious. Since the path enters a vertex through some edge and leaves by another edge, it is clear that all degrees must be even. To show that the conditions are sufficient, we start in a vertex $x$ and begin making a path. We keep going, never using the same edge twice, until we cannot go further. Since every vertex has even degree, this can only happen when we return to $x$ and all edges from $x$ have been used. If there are unused edges, then we consider the subgraph formed by these edges. We use the same procedure on a component of this subgraph, producing a second closed path. If we start this second path in a point occurring in the first path, then the two paths can be combined to a longer closed path from $x$ to $x$. Therefore the longest of these paths uses all the edges.  □

The problem of the bridges of Königsberg is described by the graph in Fig. 1.6. No vertex has even degree, so there is no Eulerian circuit.

One can consider a similar problem for digraphs. The necessary and sufficient condition for a directed Eulerian circuit is that the graph is connected and that each vertex has the same "in-degree" as "out-degree".

EXAMPLE 1.2. A puzzle with the name *Instant Insanity* concerns four cubes with faces colored red, blue, green, and yellow, in such a way that each cube has at least one face of each color. The problem is to make a stack of these cubes so that all four colors appear on each of the four sides of the stack. In Fig. 1.7 we describe four possible cubes in flattened form.

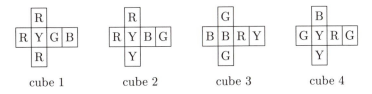

cube 1        cube 2        cube 3        cube 4

Figure 1.7

It is not a very good idea to try all possibilities. A systematic approach is as follows. The essential information about the cubes is given by the four graphs in Fig. 1.8.

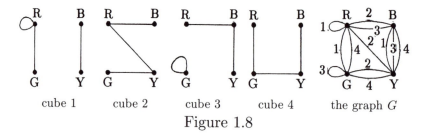

cube 1      cube 2      cube 3      cube 4      the graph $G$

Figure 1.8

An edge indicates that the two adjacent colors occur on opposite faces of the cube. We obtain a graph $G$ by superposition of the four graphs and number the edges according to their origin. It is not difficult to see that we need to find in $G$ two subgraphs that are regular of degree 2, with edges numbered $1, 2, 3, 4$ and such that they have no edge in common. One of the subgraphs tells us which pairs of colors to align on the left side and right side of the stack. The other graph describes the colors on front and back. Of course it is easy to rotate the cubes in such a way that the colors are where we wish them to be. The point of the example is that it takes only a minute to find two subgraphs as described above. In this example the solution is unique.

We mention a concept that seems similar to Eulerian circuits but that is in reality quite different. A *Hamiltonian circuit* in a graph $G$ is a simple closed path that passes through each *vertex* exactly once (rather than each *edge*). So a graph admits a Hamiltonian circuit if and only if it has a polygon as a spanning subgraph. In the mid-19th century, Sir William Rowan Hamilton tried to popularize the exercise of finding such a closed path in the graph of the dodecahedron (Fig. 1.9).

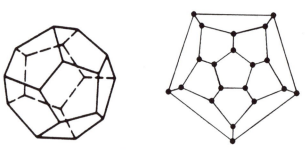

Figure 1.9

The graph in Fig. 1.4 is called the Petersen graph (cf. Chapter 21) and one of the reasons it is famous is that it is *not* "Hamiltonian"; it contains $n$-gons only for $n = 5, 6, 8, 9$, and not when $n = 10$.

By Theorem 1.2, it is easy to decide whether a graph admits an Eulerian circuit. A computer can easily be programmed to check whether the degrees of a graph are even and whether the graph is connected, and even to produce an Eulerian circuit when one exists. In contrast to this, the problem of deciding whether an arbitrary graph admits a Hamiltonian circuit is likely "intractable". To be more precise, it has been proved to be *NP-complete*—see Garey and Johnson (1979).

PROBLEM 1E. Let $A_1, \ldots, A_n$ be $n$ distinct subsets of the $n$-set $N := \{1, \ldots, n\}$. Show that there is an element $x \in N$ such that the sets $A_i \backslash \{x\}$, $1 \le i \le n$, are all distinct. To do this, form a graph $G$ on the vertices $A_i$ with an edge with "color" $x$ between $A_i$ and $A_j$ if and only if the symmetric difference of the sets $A_i$ and $A_j$ is $\{x\}$. Consider the colors occurring on the edges of a polygon. Show that one can delete edges from $G$ in such a way that *no* polygons are left and the number of different colors remains the same. (This idea is due to J. A. Bondy (1972).)

PROBLEM 1F. The *girth* of a graph is the length of the smallest polygon in the graph. Let $G$ be a graph with girth 5 for which all vertices have degree $\ge d$. Show that $G$ has at least $d^2 + 1$ vertices. Can equality hold?

**Notes.**

Paul Erdős (1913– ) (cf. Example 1.1) is probably the most prolific mathematician of this century with well over 1000 papers having been published. His contributions to combinatorics, number theory, set theory, etc., include many important results. He has collaborated with many mathematicians all over the world, all of them proud to have Erdős-number 1, among them the authors of this book.

Leonhard Euler (1707–1783) was a Swiss mathematician who spent most of his life in St. Petersburg. He was probably the most productive mathematician of all times. Even after becoming blind in 1766, his work continued at the same pace. The celebration in 1986 of the 250th birthday of graph theory was based on Euler's paper on the Königsberg bridge problem. Königsberg is now the city of Kaliningrad in Russia.

For an elementary introduction to graph theory, we recommend R. J. Wilson (1979), and J. J. Watkins and R. J. Wilson (1990).

Sir William Rowan Hamilton (1805–1865) was an Irish mathematician. He was considered a genius. He knew 13 languages at the age of 12 and was appointed professor of astronomy at Trinity College Dublin at the age of 22 (before completing his degree). His most important work was in mathematical physics.

## References.

M. Garey and D. S. Johnson (1979), *Computers and Intractability; A Guide to the Theory of NP-completeness*, W. H. Freeman and Co.

J. J. Watkins and R. J. Wilson (1990), *Graphs (An Introductory Approach)*, J. Wiley & Sons.

R. J. Wilson (1979), *Introduction to Graph Theory*, Longman.

# 2
# Trees

We come to the first not so easy theorem. It is due to A. Cayley (1889). We shall give three different proofs here. Two more proofs will occur in later chapters; see Example 14.14 and Example 34.1. The first two proofs illustrate a method that is used very often in combinatorics. In order to count certain objects that seem hard to count, one finds a one-to-one mapping onto a set of other objects whose number is easier to determine.

THEOREM 2.1. *There are $n^{n-2}$ different labeled trees on $n$ vertices.*

The term *labeled* emphasizes that we are not identifying isomorphic graphs. We have fixed the set of vertices, and two trees are counted as the same if and only if exactly the same pairs of vertices are adjacent. A *spanning tree* of a connected graph $G$ is a spanning subgraph of $G$ that is a tree. The theorem could have been stated: the complete graph $K_n$ has $n^{n-2}$ spanning trees.

EXAMPLE 2.1. Here are the 16 labeled trees on four vertices:

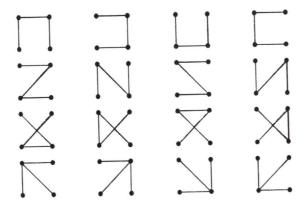

EXAMPLE 2.2. There are three nonisomorphic trees on five vertices:

The number of spanning trees in $K_5$ isomorphic to a specific tree $T$ on five vertices is 5! divided by the order of the automorphism group of $T$ (why?). Thus there are $5!/4! = 5$ trees in $K_5$ isomorphic to the first tree above, and $5!/2 = 60$ trees isomorphic to either of the other two trees, for a total of 125 spanning trees.

Before starting the proofs, we make the following observations. (Probably the reader has already noticed these things in solving Problem 1C.) Firstly, every tree with $n \geq 2$ vertices has at least two monovalent vertices (vertices of degree 1). This is immediate, for example, from Problem 1C and equation (1.1): the sum of the degrees $d_1, d_2, \ldots, d_n$, all of which are at least 1, is $2n - 2$. Secondly, if a monovalent vertex and its incident edge are deleted from a tree, the resulting graph is still a tree. Finally, given a tree $T$, if we introduce a new vertex $x$ and a new edge joining $x$ to *any* vertex of $T$, the new graph is again a tree.

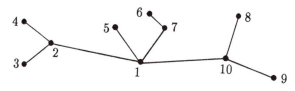

Figure 2.1

PROOF 1: The first proof we present, due to H. Prüfer (1918), uses an algorithm that produces a "code" that characterizes the tree. Any such code uniquely determines a tree.

Let $T$ be a tree with $V(T) = \{1, 2, \ldots, n\}$. We use the order on the vertex set. Let $T_1 := T$. For $i = 1, 2, \ldots, n - 2$, let $b_i$ denote the least monovalent vertex of $T_i$, let $a_i$ be the vertex adjacent to

$b_i$, and let $T_{i+1}$ be the tree obtained by deleting the vertex $b_i$ and the edge $\{a_i, b_i\}$ from $T_i$. The "code" assigned to the tree $T$ is $[a_1, a_2, \ldots, a_{n-2}]$.

As an example, consider the tree of Fig. 2.1. The monovalent vertex with smallest index is 3. It is joined to vertex 2. We define $a_1 = 2$, $b_1 = 3$, then delete vertex 3 and edge $\{3,2\}$ to obtain a tree with one edge less. This procedure is repeated eight times yielding the sequences

$$[a_1, a_2, \ldots, a_8] = [2, 2, 1, 1, 7, 1, 10, 10],$$
$$[b_1, b_2, \ldots, b_8] = [3, 4, 2, 5, 6, 7, 1, 8]$$

and terminating with the edge $\{9,10\}$.

To reverse the procedure start with any code $[a_1, a_2, \ldots, a_{n-2}]$. Write $a_{n-1} := n$. For $i = 1, 2, \ldots, n - 1$, let $b_i$ be the least vertex *not* in

$$\{a_i, a_{i+1}, \ldots, a_{n-1}\} \cup \{b_1, b_2, \ldots, b_{i-1}\}.$$

Then $\{\{b_i, a_i\} : i = 1, \ldots, n - 1\}$ will be the edge set of a spanning tree. (See Problem 2A.)

In general any sequence of $n - 2$ numbers from $\{1, 2, \ldots, n\}$ uniquely determines a tree on this set. There are $n^{n-2}$ such sequences, proving the theorem. $\qquad\square$

PROBLEM 2A. With the sequence $b_i$ defined from the code as indicated above, show that $\{\{b_i, a_i\} : i = 1, \ldots, n - 1\}$ will be the edge set of a tree on $\{1, 2, \ldots, n\}$. Fill in the details of why the mapping associating a code to a tree, and the mapping associating a tree to a code, are inverses.

PROOF 2: We give another proof, again by a reversible algorithm. Consider any mapping $f$ from $\{2, 3, \ldots, n - 1\}$ to $\{1, 2, \ldots, n\}$. There are $n^{n-2}$ such mappings $f$. Construct a digraph $D$ on the vertices 1 to $n$ by defining $(i, f(i))$, $i = 2, \ldots, n-1$, to be the edges. Fig. 2.2 shows an example with $n = 21$.

$D$ consists of two trees "rooted" at 1 and $n$ and a number (say $k$) of circuits (directed polygons) to which trees are attached. (Directed trees with all the edges pointing in the direction of one vertex, called the root, are called *arborescences*.) These circuits are placed as in Fig. 2.2 where the rightmost vertex in the $i$-th component, denoted by $r_i$, is its minimal element (and $l_i$ is the vertex on the left). The circuits are ordered by the condition $r_1 < r_2 < \cdots < r_k$. To $D$ we adjoin the tree obtained by adding the edges $\{1, l_1\}, \{r_1, l_2\}, \ldots, \{r_{k-1}, l_k\}, \{r_k, n\}$ and deleting the edges $\{r_i, l_i\}$ as in Fig. 2.3.

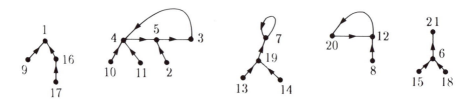

Figure 2.2

If the tree of Fig. 2.3 is given, consider the path from 1 to $n$ (=21). Let $r_0 := 1$. Define $r_1$ to be the minimal number on this path (excluding $r_0 = 1$) and in general $r_i$ as the minimal number on the path from $r_{i-1}$ to $n$. It is easily seen that we recover the function $f$ in this way.                              □

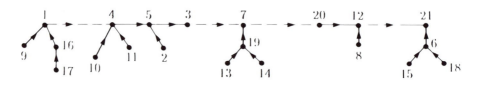

Figure 2.3

Generalizations of this proof may be found in Eğecioğlu and Remmel (1986).

PROBLEM 2B. Let $G$ be a directed graph with vertices $x_1, \ldots, x_n$ for which a (directed) Eulerian circuit exists. A *spanning arborescence* with root $x_i$ is a spanning tree $T$ of $G$, with root $x_i$, such that for all $j \neq i$ there is a directed path from $x_j$ to $x_i$ in $T$. Show that the number of spanning arborescences of $G$ with root $x_i$ does not depend on $i$.

PROOF 3: We now give a proof by counting since it is useful to have seen this method. We remind the reader of the definition of a *multinomial coefficient*. Let $r_1, r_2, \ldots, r_k$ be nonnegative integers with sum $n$. Then $\binom{n}{r_1,\ldots,r_k}$ is defined by

$$(2.1) \qquad (x_1 + x_2 + \cdots + x_k)^n = \sum \binom{n}{r_1, \ldots, r_k} x_1^{r_1} x_2^{r_2} \ldots x_k^{r_k},$$

where the sum is over all $k$-tuples $(r_1, \ldots, r_k)$ with sum $n$.

Since $(x_1 + \cdots + x_k)^n = (x_1 + \cdots + x_k)^{n-1}(x_1 + \cdots + x_k)$, we have

$$(2.2) \qquad \binom{n}{r_1, \ldots, r_k} = \sum_{i=1}^{k} \binom{n-1}{r_1, \ldots, r_i - 1, \ldots, r_k}.$$

We denote the number of labeled trees with $n$ vertices for which the degrees are $d_1, d_2, \ldots, d_n$ by $t(n; d_1, d_2, \ldots, d_n)$. We may assume without loss of generality that $d_1 \geq d_2 \geq \cdots \geq d_n$, so $d_n = 1$. Take the vertex $v_n$ corresponding to $d_n$. It is joined to some vertex $v_i$ of degree $d_i \geq 2$, and any of the remaining vertices is a candidate. Therefore

$$(2.3) \qquad t(n; d_1, \ldots, d_n) = \sum_{i=1}^{n-1} t(n-1; d_1, \ldots, d_i - 1, \ldots, d_{n-1}).$$

It is trivial to check by hand that

$$(2.4) \qquad t(n; d_1, \ldots, d_n) = \binom{n-2}{d_1 - 1, \ldots, d_n - 1}$$

for $n = 3$. Since the numbers on the lefthand side, respectively righthand side, of (2.4) satisfy the same recurrence relation ( (2.3), respectively (2.2) ) it follows by induction that (2.4) is true for all $n$.

In (2.1), we replace $n$ by $n-2$, $k$ by $n$, $r_i$ by $d_i - 1$ and $x_i$ by 1. We find

$$n^{n-2} = \sum t(n; d_1, d_2, \ldots, d_n).$$

$\square$

A spanning tree is easily constructed by starting at any vertex, taking the edges to vertices at distance 1, then one edge to each vertex at distance 2, etc. Several other constructions are possible (e.g. by starting with $G$ and deleting suitable edges).

A graph with no polygons as subgraphs is called a *forest*. Each component $C_1, C_2, \ldots, C_k$ of a forest $G$ is a tree, so if a forest with $n$ vertices has $k$ components, it has

$$(|V(C_1)| - 1) + (|V(C_2)| - 1) + \cdots + (|V(C_k)| - 1) = n - k$$

edges.

A *weighted graph* is a graph $G$ together with a function associating a real number $c(e)$ (usually nonnegative) to each edge $e$, called its *length* or *cost* according to context. Let us use the term "cost" here. Given a weighted connected graph $G$, define the *cost* of a spanning tree $T$ of $G$ as

$$c(T) := \sum_{e \in E(T)} c(e).$$

The graph may represent a network of cities where $c(\{x, y\})$ is the cost of erecting a telephone line joining cities $x$ and $y$, and so it is clear that finding a *cheapest* spanning tree in $G$ is a problem of practical importance.

The following method is often called the *greedy algorithm*. In fact, it is only one of a number of algorithms which can be called greedy algorithms, where one does not plan ahead but takes what seems to be the best alternative at each moment and does not look back. It is surprising that such a simple procedure actually produces a cheapest spanning tree, but this is proved in Theorem 2.2 below. Let us say that a set $S$ of edges of a graph $G$ is *independent* when the spanning subgraph with edge set $S$ (denoted $G : S$) is a forest.

GREEDY ALGORITHM. Let $G$ be a connected weighted graph with $n$ vertices. At each point, we will have a set $\{e_1, e_2, \ldots, e_i\}$ of $i$ independent edges ($i = 0$ to start), so that $G : \{e_1, e_2, \ldots, e_i\}$ has $n - i$ components. If $i < n - 1$, let $e_{i+1}$ be an edge with ends in different components of $G : \{e_1, e_2, \ldots, e_i\}$ and whose cost is minimum with respect to this property. Stop when we have chosen $n - 1$ edges.

THEOREM 2.2. *With $e_1, \ldots, e_{n-1}$ chosen as above, the spanning tree $T_0 := G : \{e_1, \ldots, e_{n-1}\}$ has the property that $c(T_0) \leq c(T)$ for any spanning tree $T$.*

PROOF: Let $\{a_1, a_2, \ldots, a_{n-1}\}$ be the edge set of a tree $T$, numbered so that $c(a_1) \leq c(a_2) \leq \cdots \leq c(a_{n-1})$. We claim something much stronger than $c(T_0) \leq c(T)$; namely, we claim that $c(e_i) \leq c(a_i)$ for each $i = 1, 2 \ldots, n - 1$. If this is false, then

$$c(e_k) > c(a_k) \geq c(a_{k-1}) \geq \cdots \geq c(a_1)$$

for some $k$. Since none of $a_1, a_2, \ldots, a_k$ was chosen at the point when $e_k$ was chosen, each of these $k$ edges has both ends in the same component of $G : \{e_1, e_2, \ldots, e_{k-1}\}$. Then the number of components of $G : \{a_1, a_2, \ldots, a_k\}$ is at least the number $n - k + 1$ of components of $G : \{e_1, e_2, \ldots, e_{k-1}\}$ and this contradicts the fact that $\{a_1, a_2, \ldots, a_k\}$ is independent. $\square$

PROBLEM 2C. Here is a variation on the above greedy algorithm. Let $x_1$ be any vertex of a weighted connected graph $G$ with $n$ vertices and let $T_1$ be the subgraph with the one vertex $x_1$ and no edges. After a tree (subgraph) $T_k$, $k < n$, has been defined, let $e_k$ be a cheapest edge among all edges with one end in $V(T_k)$ and the other end *not* in $V(T_k)$, and let $T_{k+1}$ be the tree obtained by adding that edge and its other end to $T_k$. Prove that $T_n$ is a cheapest spanning tree in $G$.

In many practical situations, it is necessary to search through a tree. There are two well known methods known as *depth-first search* and *breadth-first search*. We explain the terminology by the example of Fig. 2.4.

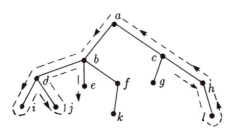

Figure 2.4

In a depth-first search starting at $a$, one essentially considers the tree as a fence and walks around it, keeping the fence on the left, i.e. along the walk $abdidjdbebfk\dots lhca$. If one decides to number the vertices in accordance with the search, one finds the numbering $a = 1$, $b = 2$, $d = 3$, $i = 4$, ... , $l = 12$. Note that this depends on the planar drawing of the tree.

In a breadth-first search, one proceeds as in the construction of a spanning tree mentioned above. The vertices are then numbered in the order of Fig. 2.4.

**Notes.**

A. Cayley (1821–1895), professor at Cambridge from 1863 until his death, was one of the great mathematicians of the 19th century. His work includes important contributions to the theory of elliptic functions, analytic geometry and algebra, e.g. the theory of invariants. His paper on trees appeared in 1889 but it did not contain what we would consider a proof. Of the many proofs (five of which are treated in this book) the one by Prüfer is the best known.

H. Prüfer (1896–1934) was one of I. Schur's many pupils. He was professor at Münster.

**References.**
A. Cayley (1889), A theorem on trees, *Quart. J. Pure and App. Math.* **23**, 376–378.

Ö. Eğecioğlu and J. B. Remmel (1986), Bijections for Cayley trees, spanning trees, and their $q$-analogues, *J. Combinatorial Theory* (A) **42**, 15–30.

H. Prüfer (1918), Neuer Beweis eines Satzes über Permutationen, *Archiv der Math. und Phys.* (3) **27**, 142–144.

# 3
## Colorings of graphs and Ramsey's theorem

We shall first look at a few so-called coloring problems for graphs.

A *proper coloring* of a graph $G$ is a function from the vertices to a set $C$ of "colors" (e.g. $C = \{1, 2, 3, 4\}$) such that any two adjacent vertices have different colors. If $|C| = k$, we say that $G$ is *k-colored*.

The *chromatic number* $\chi(G)$ of a graph $G$ is the minimal number of colors for which a proper coloring exists.

If $\chi(G) = 2$ then $G$ is called *bipartite*. A graph with no odd polygons (equivalently, no closed paths of odd length) is bipartite as the reader should verify.

The famous "four color theorem" (K. Appel and W. Haken, 1977) states that if $G$ is planar, then $\chi(G) \leq 4$.

Clearly $\chi(K_n) = n$. If $k$ is odd then $\chi(P_k) = 3$. In the following theorem, we show that, with the exception of these examples, the chromatic number is at most equal to the maximum degree (R. L. Brooks, 1941).

THEOREM 3.1. *Let $d \geq 3$ and let $G$ be a graph in which all vertices have degree $\leq d$ and such that $K_{d+1}$ is not a subgraph of $G$. Then $\chi(G) \leq d$.*

PROOF: As is the case in many theorems in combinatorial analysis, one can prove the theorem by assuming that it is not true, then considering a *minimal* counterexample (in this case a graph with the minimal number of vertices) and arriving at a contradiction. We shall use the technique of recoloring: it is possible to change the colors of certain vertices to go from one proper coloring to another. For example, let $S$ be a subset of the set $C$ of colors. On any connected component of the subgraph induced by the vertices with colors from $S$, we arbitrarily permute the colors (without changing those of the vertices with colors in $C \backslash S$). Clearly we again have a proper coloring.

So let $G$ be a counterexample with the minimum number of vertices. Let $x \in G$ and let $\Gamma(x) = \{x_1, \ldots, x_l\}$, $l \leq d$. Since $G$ is a minimal counterexample, the graph $H$, obtained by deleting $x$ and the edges incident with $x$, has a $d$-coloring, say with colors $1, 2, \ldots, d$. If one of these colors is not used in the coloring of $\Gamma(x)$, then we can assign this color to $x$ and obtain a $d$-coloring of $G$. It follows that $l = d$ and every $d$-coloring of $H$ must use all the colors on the set $\Gamma(x)$. Let us assume that $x_i$ has color $i$.

Now consider $x_i$ and $x_j$ and the induced subgraph $H_{ij}$ of $H$ with colors $i$ and $j$. If $x_i$ and $x_j$ were in different connected components of $H_{ij}$, then we could interchange the colors in one of these components, after which $x_i$ and $x_j$ would have the same color, which is impossible. So $x_i$ and $x_j$ are in the same component (say $C_{ij}$) of $H_{ij}$. We shall now show that this component is (the graph of) a *simple path* (with alternating colors $i$ and $j$) from $x_i$ to $x_j$. If two neighbors of $x_i$ in $H$ had color $j$, then the neighbors of $x_i$ in $H$ would have at most $d - 2$ different colors. Then we could recolor $x_i$ and that is impossible. Suppose $y$ is the first vertex on a path from $x_i$ to $x_j$ in $C_{ij}$ that has degree $\geq 3$. The neighbors of $y$ in $H$ have at most $d - 2$ colors, so we can recolor $y$ to some color $\notin \{i, j\}$ and then $x_i$ and $x_j$ are no longer connected in $H_{ij}$, which we know to be impossible. So such a $y$ does not exist, proving that $C_{ij}$ is a path.

Suppose that $z$ is a vertex $\neq x_i$ on $C_{ij}$ and on $C_{ik}$. Then $z$ has two neighbors with color $j$ and two with color $k$. Again the neighbors of $z$ in $H$ have at most $d - 2$ colors and $z$ can be recolored to some color $\notin \{i, j, k\}$, again a contradiction. Hence $C_{ij} \cap C_{ik} = \{x_i\}$.

Our assumption that $K^{d+1} \not\subseteq G$ shows that there are two vertices in $\Gamma(x)$, say $x_1$ and $x_2$, that are not connected by an edge. We have the situation of Fig. 3.1. The vertex $a$ is the neighbor of $x_1$ with color 2 on $C_{12}$.

We recolor $H$ by interchanging the colors 1 and 3 on the subgraph $C_{13}$. For the new coloring, we have new paths that we call $C'_{ij}$.

Clearly $a \in C'_{23}$ (since $x_1$ now has color 3). However, on $C_{12}$ no point except $x_1$ has changed color, so $a \in C'_{12}$. Hence $C'_{12} \cap C'_{23} \neq \{x_2\}$, contradicting what we proved above. The contradiction shows that our assumption that a minimal counterexample exists is false.  $\square$

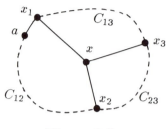

Figure 3.1

We now consider a coloring problem of a completely different nature. It serves as an introduction to a very important theorem of combinatorics, namely Ramsey's theorem. Before reading on, the reader should try the following problem.

PROBLEM 3A. Let the edges of $K_7$ be colored with the colors red and blue. Show that there are at least four subgraphs $K_3$ with all three edges the same color (*monochromatic triangles*). Also show that equality can occur.

The example that is always used to introduce this subject is $K_6$ with its edges colored red or blue. We shall show that there is at least one monochromatic triangle. A proof is as follows. Let $a$ be any vertex. Because $a$ has degree 5, it is incident with at least three edges of the same color, say red edges to the vertices $b$, $c$, $d$. If one of the edges between these three vertices is red, then we have a red triangle; if not, then they form a blue triangle.

The idea of the proof is the same as for the more difficult situation of Ramsey's theorem. However, it does not show as much as the following counting argument.

THEOREM 3.2. *If the edges of $K_n$ are colored red or blue, and $r_i$, $i = 1, 2, \ldots, n$, denotes the number of red edges with vertex $i$ as an endpoint, and if $\Delta$ denotes the number of monochromatic triangles,* *then*

(3.1)         $$\Delta = \binom{n}{3} - \frac{1}{2}\sum_{i=1}^{n} r_i(n-1-r_i).$$

PROOF: Every triangle in $K_n$ that is not monochromatic has exactly two vertices where a red and a blue edge meet. On the $i$-th vertex, two such edges can be chosen in $r_i(n-1-r_i)$ ways. So the sum in (3.1) counts the bichromatic triangles twice.         □

COROLLARY.

(3.2)         $$\Delta \geq \binom{n}{3} - \lfloor\frac{n}{2}\lfloor(\frac{n-1}{2})^2\rfloor\rfloor.$$

PROOF: From (3.1) we see that $\Delta$ is minimized if $r_i = (n-1-r_i)$ for all $i$ when $n$ is odd, or if $r_i = \frac{n}{2}$ or $r_i = \frac{n}{2} - 1$ for all $i$ in the case that $n$ is even. Since $\Delta$ is an integer, the first situation cannot always arise. It is easy to show that (3.2) cannot be improved.         □

Note that this argument shows that a red-blue coloring of $K_6$ must always have at least two monochromatic triangles.

We now treat Ramsey's theorem (Ramsey, 1930).

THEOREM 3.3. *Let $r \geq 1$ and $q_i \geq r$, $i = 1, 2, \ldots, s$ be given. There exists a minimal positive integer $N(q_1, q_2, \ldots, q_s; r)$ with the following property. Let $S$ be a set with $n$ elements. Suppose that all $\binom{n}{r}$ $r$-subsets of $S$ are divided into $s$ mutually exclusive families $T_1, \ldots, T_s$ ("colors"). Then if $n \geq N(q_1, q_2, \ldots, q_s; r)$ there is an $i$, $1 \leq i \leq s$, and some $q_i$-subset of $S$ for which every $r$-subset is in $T_i$.*

(The reader should compare this with our introductory example and show that $N(3, 3; 2) = 6$.)

PROOF: We give the proof only for $s = 2$. The general case only involves a little more bookkeeping.

(a) Trivially, the theorem is true for $r = 1$ and $N(p, q; 1) = p + q - 1$.

(b) For any $r$ and $p \geq r$ it is also obvious that $N(p, r; r) = p$ and similarly $N(r, q; r) = q$ for $q \geq r$.

(c) We proceed by induction on $r$. So assume the theorem is true for $r-1$. We now use induction on $p+q$, using (b). So we can define $p_1 = N(p-1, q; r)$, $q_1 = N(p, q-1; r)$. Let $S$ be a set with $n$ elements, where $n \geq 1 + N(p_1, q_1; r-1)$. Let the $r$-subsets of $S$ be colored with two colors, say red and blue. As in the proof for $K_6$, we pick an arbitrary element $a$ of $S$. We now define a coloring of the $(r-1)$-subsets of $S' := S \backslash \{a\}$ by giving $X \subseteq S'$ the same color as $X \cup \{a\}$. By induction $S'$ either contains a subset $A$ of size $p_1$ such that all its $(r-1)$-subsets are red or a subset $B$ of size $q_1$ such that all its $(r-1)$-subsets are colored blue. Without loss of generality the first situation occurs. Since $A$ has $N(p-1, q; r)$ elements, there are two possibilities. The first is that $A$ has a subset of $q$ elements with all its $r$-subsets blue, in which case we are done. The other possibility is that $A$ has a subset $A'$ of $p-1$ elements with all its $r$-subsets red. The set $A' \cup \{a\}$ also has this property because $A' \subseteq A$. This proves the theorem and furthermore we have shown

$$(3.3) \quad N(p, q; r) \leq N(N(p-1, q; r), N(p, q-1; r); r-1) + 1.$$

$\square$

A special case of (3.3) occurs when we go back to the coloring of edges ($r=2$) of a graph with two colors. Using (a) from the proof, we find

$$(3.4) \quad N(p, q; 2) \leq N(p-1, q; 2) + N(p, q-1; 2).$$

PROBLEM 3B. Show that equality cannot hold in (3.4) if both terms on the righthand side are even.

THEOREM 3.4.
$$N(p, q; 2) \leq \binom{p+q-2}{p-1}.$$

PROOF: Since $N(p, 2; 2) = p$, the result follows from (3.4) because binomial coefficients satisfy the same relation with equality. $\square$

Let us look at what we now know about $N(p, q; 2)$. By Problem 3B, we have $N(3, 4; 2) \leq 9$. To show that equality holds, we have to color $K_8$ such that there is no red triangle and no blue $K_4$. We do this as follows: number the vertices with the elements of $\mathbb{Z}_8$. Let the edge $\{i, j\}$ be red if and only if $i - j \equiv \pm 3$ or $i - j \equiv 4$ (mod 8). One easily checks that this coloring does the job.

PROBLEM 3C. Use the same method to show that $N(4, 4; 2) = 18$ and that $N(3, 5; 2) = 14$.

With a lot more work it has been shown that

$$N(3, 6; 2) = 18, \quad N(3, 7; 2) = 23, \quad N(3, 9; 2) = 36.$$

No other values of $N(p, q; 2)$ are known.

One of the interesting problems in this area that has seen virtually no progress for 20 years is the asymptotic behavior of $N(p, p; 2)$. We know by Theorem 3.4 that

$$(3.5) \qquad N(p, p; 2) \leq \binom{2p - 2}{p - 1} \leq 2^{2p-2}.$$

We now show that $N(p, p; 2)$ grows exponentially, using a method that is used quite often in combinatorics. It is often referred to as "probabilistic" since it estimates the probability that a random coloring has a monochromatic $K_p$. Consider a $K_n$. There are $2^{\binom{n}{2}}$ different ways of coloring the edges red or blue. Now fix a subgraph $K_p$. There are $2^{\binom{n}{2} - \binom{p}{2} + 1}$ colorings for which that $K_p$ is monochromatic. The number of colorings for which some $K_p$ is monochromatic is at most $\binom{n}{p}$ times as large (because we may count some colorings more than once). If this number is less than the total number of colorings, then there exist colorings with no monochromatic $K_p$. Using the fact that $\binom{n}{p} < n^p/p!$, we find that such a coloring certainly exists if $n < 2^{p/2}$ (unless $p = 2$). This proves the following theorem.

THEOREM 3.5. $N(p, p; 2) \geq 2^{p/2}$.

From (3.5) and Theorem 3.5, we know that

$$\sqrt{2} \leq \sqrt[p]{N(p, p; 2)} \leq 4 \qquad (p \geq 2).$$

It would be very nice if one could show that this $p$-th root has a limit for $p \to \infty$.

We have given some examples of an area of combinatorics known as Ramsey theory. We just mention one more example, namely a theorem due to B. L. van der Waerden (1927). It states that there exists a number $N(r)$ such that if $N \geq N(r)$ and the integers from 1 to $N$ are colored red or blue, then there is a monochromatic arithmetic progression of length $r$ in the set. For a short (but not easy) proof see Graham and Rothschild (1974). A general reference for this area is the book *Ramsey Theory* by R. L. Graham, B. L. Rothschild and J. L. Spencer (1980).

An interesting application of Ramsey's theorem is the following theorem due to Erdős and Szekeres (1935).

THEOREM 3.6. *For a given $n$, there is an integer $N(n)$ such that any collection of $N \geq N(n)$ points in the plane, no three on a line, has a subset of $n$ points forming a convex $n$-gon.*

PROOF: (i) First we observe that if we have $n$ points, no three on a line, then they form a convex $n$-gon if and only if every quadrilateral formed by taking four of the points is convex.

(ii) We now claim that $N(n) = N(n, n; 3)$ will do the job. Let $S$ be a set of $N(n)$ points. Number the points and then color triangles red, respectively blue, if the path from the smallest number via the middle one to the largest number is clockwise, respectively counterclockwise. There is an $n$-subset with all its triangles the same color, say red. We shall show that this set cannot contain the configuration of Fig. 3.2.

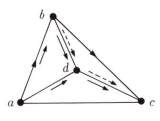

Figure 3.2

Without loss of generality $a < b < c$. From triangle $adc$, we see that $a < d < c$. Then from triangle $abd$ it follows that $a < b < d$. But then triangle $bcd$ is blue, a contradiction. So all quadrilaterals formed from the $n$-subset are convex and by (i), we are done.  □

PROBLEM 3D. A *transitive tournament* is an orientation of a complete graph for which the vertices can be numbered in such a way that $(i, j)$ is an edge if and only if $i < j$.

(a) Show that if $k < \log_2 n$, a directed $K_n$ always has a transitive subtournament on $k$ vertices.

(b) Show that if $k > 1 + 2\log_2 n$, there exists an orientation of $K_n$ with no transitive subtournament on $k$ vertices.

PROBLEM 3E. Prove that for all $r \in \mathbb{N}$ there is a minimal number $N(r)$ with the following property. If $n \geq N(r)$ and the integers in $\{1, 2, \ldots, n\}$ are colored with $r$ colors, then there are three elements $x, y, z$ (not necessarily distinct) with the same color and $x + y = z$. (A result due to I. Schur.) Determine $N(2)$. Show by an elementary argument that $N(3) > 13$.

PROBLEM 3F. Let $m$ be given. Show that if $n$ is large enough, every $n$ by $n$ (0,1)-matrix has a principal submatrix of size $m$, in which all the elements below the diagonal are the same, and all the elements above the diagonal are the same.

## Notes.

The four color conjecture was considered one of the most famous open problems in combinatorics until 1976. Its solution by Appel and Haken has caused much debate because the proof depends on an extensive computer analysis of many cases. The validity of the argument depends on one's trust in computers and programs. (See Chapter 32 for a proof of the Five Color Theorem.)

The theorem by Brooks, Theorem 3.1, which he discovered while an undergraduate at Cambridge, is a typical example of the ingenious arguments that are quite often necessary in that part of graph theory where algebraic methods do not apply.

F. P. Ramsey (1902–1928) died too young to produce the results that he probably would have. He was interested in decision procedures for logical systems and, strangely enough, this led to the theorem that in turn led to so-called Ramsey theory.

Theorem 3.5 is due to P. Erdős (1947). For more on probabilistic methods, see Erdős and Spencer (1974).

Estimates for the numbers $N(p, q; 2)$ can be found in the afore-mentioned book *Ramsey Theory*. The value of $N(3, 9; 2)$ is from Grinstead and Roberts (1982).

The proof of Theorem 3.6 is not the original proof by Erdős and Szekeres that one usually finds in books. This proof was produced by a student in Haifa (M. Tarsy) during an examination! He had missed the class in which the proof had been presented. See Lewin (1976). A proof in a similar vein was given by Johnson (1986).

## References.

K. Appel and W. Haken (1977), Every planar map is four-colorable, *Illinois J. Math.* **21**, 429–567.

R. L. Brooks (1941), On colouring the nodes of a network, *Cambridge Philos. Soc.* **37**, 194–197.

P. Erdős (1947), Some remarks on the theory of graphs, *Bull. Amer. Math. Soc.* **53**, 292–294.

P. Erdős and J. L. Spencer (1974), *Probabilistic Methods in Combinatorics*, Academic Press.

P. Erdős and G. Szekeres (1935), A combinatorial problem in geometry, *Compositio Math.* **2**, 463–470.

R. L. Graham and B. L. Rothschild (1974), A short proof of van der Waerden's theorem on arithmetic progressions, *Proc. Amer. Math. Soc.* **42**, 385–386.

R. L. Graham, B. L. Rothschild, and J. L. Spencer (1980), *Ramsey Theory*, Wiley.

C. M. Grinstead and S. M. Roberts (1982), On the Ramsey numbers $R(3, 8)$ and $R(3, 9)$, *J. Combinatorial Theory* (B) **33**, 27–51.

S. Johnson (1986), A new proof of the Erdős-Szekeres convex $k$-gon result, *J. Combinatorial Theory* (A) **42**, 318–319.

M. Lewin (1976), A new proof of a theorem of Erdős and Szekeres, *The Math. Gazette* **60**, 136–138, 298.

F. P. Ramsey (1930), On a problem of formal logic, *Proc. London Math. Soc.* (2) **30**, 264–286.

B. L. van der Waerden (1927), Beweis einer Baudetschen Vermutung, *Nieuw Archief voor Wiskunde* **15**, 212–216.

# 4
# Turán's theorem and extremal graphs

As an introduction, we first ask the question how many edges a graph must have to guarantee that the graph contains a *triangle*. Since $K_{m,m}$ and $K_{m,m+1}$ do not contain triangles, we see that if the graph has $n$ vertices, then $\lfloor n^2/4 \rfloor$ edges are not enough. We claim that if there are more edges, then the graph contains a triangle (W. Mantel, 1907). The following proof is surprising. Let $G$ have $n$ vertices, numbered from 1 to $n$, and no triangles. We give vertex $i$ a weight $z_i \geq 0$ such that $\sum z_i = 1$ and we wish to maximize $S := \sum z_i z_j$, where the sum is taken over all edges $\{i, j\}$. Suppose that vertex $k$ and vertex $l$ are not joined. Let the neighbors of $k$ have total weight $x$, and those of $l$ total weight $y$, where $x \geq y$. Since $(z_k + \epsilon)x + (z_l - \epsilon)y \geq z_k x + z_l y$, we do not decrease the value of $S$ if we shift some of the weight of vertex $l$ to the vertex $k$. It follows that $S$ is maximal if all of the weight is concentrated on some complete subgraph of $G$, i.e. on *one* edge! Therefore $S \leq \frac{1}{4}$. On the other hand, taking all $z_i$ equal to $n^{-1}$ would yield a value of $n^{-2}|E|$ for $S$. Therefore $|E| \leq \frac{1}{4}n^2$.

Note that Ramsey's theorem states that if a graph on $n$ vertices has $n \geq N(p, q; 2)$, then the graph either has a complete subgraph on $p$ vertices or a set of $q$ vertices with no edges between them (called an *independent* set). We now ask the question whether some condition on the number of edges guarantees a $K_p$ as a subgraph. We saw above what the answer is for $p = 3$. We also already have an idea of how to avoid a $K_p$. Divide the vertices into $p-1$ subsets $S_1, \ldots, S_{p-1}$ of almost equal size, i.e. $r$ subsets of size $t + 1$ and $p - 1 - r$ subsets of size $t$, where $n = t(p-1) + r$, $1 \leq r \leq p - 1$.

Within each $S_i$ there are no edges but every vertex in $S_i$ is joined to every vertex in $S_j$ if $i \neq j$. (This is a *complete multipartite graph*.) The number of edges is

$$M(n, p) := \frac{p-2}{2(p-1)}n^2 - \frac{r(p-1-r)}{2(p-1)}.$$

THEOREM 4.1. *(Turán, 1941) If a simple graph on $n$ vertices has more than $M(n, p)$ edges, then it contains a $K_p$ as a subgraph.*

PROOF: The proof is by induction on $t$. If $t = 0$, the theorem is obvious. Consider a graph $G$ with $n$ vertices, no $K_p$, and the maximum number of edges subject to those properties. Clearly $G$ contains a $K_{p-1}$ (otherwise adding an edge would not produce a $K_p$), say $H$. Each of the remaining vertices is joined to at most $p-2$ vertices of $H$. The remaining $n-p+1$ vertices do not contain a $K_p$ as subgraph. Since $n-p+1 = (t-1)(p-1)+r$, we can apply the induction hypothesis to this set of points. So the number of edges of $G$ is at most

$$M(n-p+1, p) + (n-p+1)(p-2) + \binom{p-1}{2}$$

and this number is equal to $M(n, p)$.                    □

REMARK. The argument that was used for Mantel's theorem would show that if there is no $K_p$, then $|E| \leq \frac{p-2}{2(p-1)}n^2$.

PROBLEM 4A. Let $G$ be a simple graph with 10 vertices and 26 edges. Show that $G$ has at least 5 triangles. Can equality occur?

Turán's paper on graph theory that contains Theorem 4.1 is considered the starting point of what is now known as extremal graph theory—see Bollobás (1978). A simple instance of an extremal problem will ask for the maximum number of edges a graph with a certain property may have. The graphs whose number of edges *is* maximum are called the *extremal graphs* with respect to the property.

The extremal graphs for Turán's problem are *only* the complete multipartite graphs described above. This follows from an analysis of the proof of Theorem 4.1; we ask the reader to do this at least in the case $p = 3$ in the problem below.

PROBLEM 4B. Show that a simple graph on $n$ vertices with $\lfloor n^2/4 \rfloor$ edges and no triangles is a complete bipartite graph $K_{k,k}$ if $n = 2k$, or $K_{k,k+1}$ if $n = 2k+1$.

PROBLEM 4C. If a simple graph on $n$ vertices has $e$ edges, then it has at least $\frac{e}{3n}(4e - n^2)$ triangles.

The *girth* of a graph $G$ is the size of a smallest polygon $P_n$ in $G$. (A forest has infinite girth.) By definition, a graph is simple if and only if it has girth $\geq 3$. By Mantel's theorem, a graph with more than $n^2/4$ edges has girth $\leq 3$.

THEOREM 4.2. *If a graph $G$ on $n$ vertices has more than $\frac{1}{2}n\sqrt{n-1}$ edges, then $G$ has girth $\leq 4$. That is, $G$ is not simple or contains a $P_3$ or a $P_4$ (a triangle or a quadrilateral).*

PROOF: Suppose $G$ has girth $\geq 5$. Let $y_1, y_2, \ldots, y_d$, $d = \deg(x)$, be the vertices adjacent to a vertex $x$. No two of these are adjacent since $G$ has no triangles. Moreover, no vertex (other than $x$) can be adjacent to more than one of $y_1, \ldots, y_d$ since there are no quadrilaterals in $G$. Thus $(\deg(y_1) - 1) + \cdots + (\deg(y_d) - 1) + (d+1)$ cannot exceed the total number $n$ of vertices. That is,

$$\sum_{y \text{ adjacent to } x} \deg(y) \leq n - 1.$$

Then

$$n(n-1) \geq \sum_x \sum_{y \text{ adjacent to } x} \deg(y) = \sum_y \deg(y)^2$$

$$\geq \frac{1}{n}\left(\sum_y \deg(y)\right)^2 = \frac{1}{n}(2|E(G)|)^2.$$

□

The number $\frac{1}{2}n\sqrt{n-1}$ in Theorem 4.2 is only a bound—it is not the exact answer for all $n$. Determination of the extremal graphs for this problem (maximum number of edges subject to girth $\geq 5$) for all values of $n$ is impossibly difficult; a determination of the graphs for which equality holds has, however, been almost possible.

Perhaps surprisingly, there are at most four graphs with $n > 2$ vertices, girth $\geq 5$, and $\frac{1}{2}n\sqrt{n-1}$ edges: The pentagon ($n = 5$), the Petersen graph ($n = 10$), one with $n = 50$, and possibly one with $n = 3250$. See the notes and Chapter 21.

PROBLEM 4D. Suppose $G$ is regular of degree $r$ and has girth $g$ or greater. Find a lower bound for $|V(G)|$. (Consider the cases $g$ even and odd separately.)

It is not so interesting to ask how many edges are required to force a Hamiltonian circuit. But we can ask what bound on the minimum degree will do the job.

THEOREM 4.3. *If a simple graph $G$ on $n$ vertices has all vertices of degree at least $n/2$, then it contains a $P_n$ as a subgraph, i.e. it has a Hamiltonian circuit.*

PROOF: Suppose the theorem is not true and let $G$ be a graph satisfying the hypothesis for some $n$ but having no Hamiltonian circuits. We may take $G$ to be such a counterexample with the maximum number of edges; then the addition of any edge to $G$ (i.e. joining two nonadjacent vertices by an edge) creates a Hamiltonian circuit.

Let $y$ and $z$ be nonadjacent vertices. Since adding $\{y, z\}$ creates a Hamiltonian circuit, there exists a simple path from $y$ to $z$ with vertex terms, $y = x_1, x_2, \ldots, x_n = z$, say. The sets

$$\{i : y \text{ is adjacent to } x_{i+1}\}$$

and

$$\{i : z \text{ is adjacent to } x_i\}$$

each have cardinality $\geq n/2$ and are contained in $\{1, 2, 3, \ldots, n-1\}$, so they must meet; let $i_0$ belong to both. Then

$$y = x_1, x_2, \ldots, x_{i_0}, z = x_n, x_{n-1}, \ldots, x_{i_0+1}, x_1 = y$$

is the vertex sequence of a simple closed path of length $n$ in $G$, contradicting our choice of $G$ as a counterexample.  $\square$

Theorem 4.3 is due to G. A. Dirac and is best possible at least in the sense that it does not remain true if we replace $n/2$ by

$(n-1)/2$. For example, the complete bipartite graphs $K_{k,k+1}$ have no Hamiltonian circuits. But it does admit improvements and generalizations—see e.g. Lovász (1979), Problem 10.21.

PROBLEM 4E. A $3 \times 3 \times 3$ cube of cheese is divided into 27 $1 \times 1 \times 1$ small cubes. A mouse eats one small cube each day and an *adjacent* small cube (sharing a face) the next day. Can the mouse eat the *center* small cube on the last day?

PROBLEM 4F. Let $S$ be a set of $s$ edges in a simple graph $G$ on $n$ vertices. Suppose the edges in $S$ are disjoint, or more generally, form one or more pairwise disjoint paths. Prove: If each vertex of $G$ has degree $\geq (n+s)/2$, then there exists a Hamiltonian circuit in $G$ passing through all edges of $S$.

**Notes.**

P. Turán (1910–1976), one of the famous Hungarian mathematicians of this century, is best known for his work in analytic number theory and real and complex analysis.

For every $r \geq 2$ and $g \geq 2$, there exists a graph that is regular of degree $r$ and has girth $\geq g$. See Lovász (1979), Problem 10.12.

Analysis of the proof of Theorem 4.2 shows that a graph with $n > 2$ vertices, girth $\geq 5$, and $\frac{1}{2}n\sqrt{n-1}$ edges is regular of degree $k := \sqrt{n-1}$ and also that any two vertices are joined by a (unique) path of length 2 if they are not adjacent. With the notation of Chapter 21, such a graph is an $srg(n,k,0,1)$ and the methods of that chapter show that $k = 2, 3, 7,$ or 57. This was first shown in Hoffman and Singleton (1960) where in addition an example with $k = 7$ and $n = 50$ was described (that is now known as the *Hoffman-Singleton graph*). It is not known at this time whether there exists an $srg(3250, 57, 0, 1)$.

**References.**
B. Bollobás (1978), *Extremal Graph Theory*, Academic Press.

A. J. Hoffman and R. R. Singleton (1960), On Moore graphs with diameters two and three, *IBM J. Res. Develop.* **4**, 497–504.

L. Lovász (1979), *Combinatorial Problems and Exercises*, North Holland.

W. Mantel (1907), Problem 28, *Wiskundige Opgaven* **10**, 60–61.

P. Turán (1941), An extremal problem in graph theory (in Hungarian), *Mat. Fiz. Lapok* **48**, 435–452.

# 5
# Systems of distinct representatives

We first give two different formulations of a theorem known as P. Hall's *marriage theorem*. We give a constructive proof and an enumerative one. If $A$ is a subset of the vertices of a graph, then denote by $\Gamma(A)$ the set $\bigcup_{a \in A} \Gamma(a)$. Consider a bipartite graph $G$ with vertex set $X \cup Y$ (every edge has one endpoint in $X$ and one in $Y$). A *matching* in $G$ is a subset $E_1$ of the edge set such that no vertex is incident with more than one edge in $E_1$. A *complete* matching from $X$ to $Y$ is a matching such that every vertex in $X$ is incident with an edge in $E_1$. If the vertices of $X$ and $Y$ are thought of as boys and girls, respectively, or vice versa, and an edge is present when the persons corresponding to its ends have amicable feelings towards one another, then a complete matching represents a possible assignment of marriage partners to the persons in $X$.

THEOREM 5.1. *A necessary and sufficient condition for there to be a complete matching from $X$ to $Y$ in $G$ is that $|\Gamma(A)| \geq |A|$ for every $A \subseteq X$.*

PROOF: (i) It is obvious that the condition is necessary.

(ii) Assume that $|\Gamma(A)| \geq |A|$ for every $A \subseteq X$. Let $|X| = n$, $m < n$, and suppose we have a matching $M$ with $m$ edges. We shall show that a larger matching exists. (We mean larger in *cardinality*; we may not be able to find a complete matching containing these particular $m$ edges.)

Call the edges of $M$ red and all other edges blue. Let $x_0 \in X$ be a vertex not incident with an edge of the matching. We claim that there exists a simple path (of odd length) starting with $x_0$ and a blue edge, using red and blue edges alternately, and terminating with a blue edge and a vertex $y$ not incident with an edge of the matching. If we find such a path $p$, we are done because we obtain a matching with $m + 1$ edges by deleting the red edges of $p$ from $M$ and replacing them with the blue edges of $p$. In other words, we switch the colors on the edges of $p$.

Since $|\Gamma(\{x_0\})| \geq 1$, there is a vertex $y_1$ adjacent to $x_0$ (obviously by a blue edge since $x_0$ is not incident with any red edges). If $y_1$ is also not incident with any red edges, we have the required path (of length one); if $y_1$ is incident with a red edge, let $x_1$ be the other end of that red edge. Recursively, define $x_0, x_1, \ldots$ and $y_1, y_2, \ldots$ as follows. If $x_0, x_1, \ldots, x_k$ and $y_1, \ldots, y_k$ have been defined, then since $|\Gamma(\{x_0, x_1, \ldots, x_k\})| \geq k+1$, there exists a vertex $y_{k+1}$, distinct from $y_1, \ldots, y_k$, that is adjacent to at least one vertex in $\{x_0, x_1, \ldots, x_k\}$. If $y_{k+1}$ is not incident with a red edge, stop; otherwise, let $x_{k+1}$ be the other end of that red edge.

When the procedure terminates, we construct the path $p$ by starting with $y_{k+1}$ and the blue edge joining it to, say, $x_{i_1}$, $i_1 < k + 1$. Then add the red edge $\{x_{i_1}, y_{i_1}\}$. By construction, $y_{i_1}$ is joined by an edge (necessarily blue) to some $x_{i_2}$, $i_2 < i_1$. Then add the red edge $\{x_{i_2}, y_{i_2}\}$. Continue in this way until $x_0$ is reached. $\qquad\square$

PROBLEM 5A. A *perfect* matching in a graph $G$ (not necessarily bipartite) is a matching so that each vertex of $G$ is incident with one edge of the matching. (i) Show that a finite *regular* bipartite graph (regular of degree $d > 0$) has a perfect matching. (ii) Find a trivalent (regular of degree 3) simple graph which does not have a perfect matching.

We now reformulate the theorem in terms of sets and not only prove the theorem but also give a lower bound for the number of matchings. We consider subsets $A_0, A_1, \ldots, A_{n-1}$ of a finite set $S$. We shall say that this collection has property $H$ (Hall's condition) if (for all $k$) the union of any $k$-tuple of subsets $A_i$ has at least $k$ elements. If the union of some $k$-tuple of subsets contains exactly $k$ elements ($0 < k < n$), then we call this $k$-tuple a *critical block*.

We define a *system of distinct representatives* (SDR) of the sets $A_0, \ldots, A_{n-1}$ to be a sequence of $n$ *distinct* elements $a_0, \ldots, a_{n-1}$ with $a_i \in A_i$, $0 \leq i \leq n - 1$.

Let $m_0 \leq m_1 \leq \cdots \leq m_{n-1}$. We define

$$F_n(m_0, m_1, \ldots, m_{n-1}) := \prod_{i=0}^{n-1} (m_i - i)_*,$$

where $(a)_* := \max\{1, a\}$.

From now on, we assume that the sequence $m_i := |A_i|$ is nondecreasing.

For the proof of the main theorem, we need a lemma.

LEMMA 5.2. *For $n \geq 1$ let $f_n : \mathbf{Z}^n \to \mathbf{N}$ be defined by*

$$f_n(a_0, a_1, \ldots, a_{n-1}) := F_n(m_0, m_1, \ldots, m_{n-1})$$

*if $(m_0, \ldots, m_{n-1})$ is a nondecreasing rearrangement of the $n$-tuple $(a_0, \ldots, a_{n-1})$. Then $f_n$ is nondecreasing with respect to each of the variables $a_i$.*

PROOF: Let

$$m_0 \leq \cdots \leq m_{k-1} \leq a_i = m_k \leq m_{k+1} \leq \cdots$$
$$\leq m_l \leq m_{l+1} \leq \cdots \leq m_{n-1}$$

be a nondecreasing rearrangement of $(a_0, \ldots, a_{n-1})$. If $a_i' \geq a_i$ and

$$m_0 \leq \cdots \leq m_{k-1} \leq m_{k+1} \leq \cdots \leq m_l \leq a_i' \leq m_{l+1} \leq \cdots \leq m_{n-1}$$

is a nondecreasing rearrangement of $(a_0, \ldots, a_{i-1}, a_i', a_{i+1}, \ldots, a_{n-1})$ then

$$\frac{f_n(a_0, \ldots, a_{i-1}, a_i', a_{i+1}, \ldots, a_{n-1})}{f_n(a_0, \ldots, a_{n-1})} =$$

$$= \frac{(m_{k+1} - k)_*}{(a_i - k)_*} \cdot \frac{(a_i' - l)_*}{(m_l - l)_*} \prod_{j=k+1}^{l-1} \frac{(m_{j+1} - j)_*}{(m_j - j)_*}$$

and this is $\geq 1$ since $a_i \leq m_{k+1}$, $a_i' \geq m_l$, and $m_{j+1} \geq m_j$ for $j = k+1, \ldots, l-1$. $\qquad\square$

We now come to the second form of Hall's theorem. We denote by $N(A_0, \ldots, A_{n-1})$ the number of SDRs of $(A_0, \ldots, A_{n-1})$.

THEOREM 5.3. *Let* $(A_0, \ldots, A_{n-1})$ *be a sequence of subsets of a set* $S$. *Let* $m_i := |A_i|$ $(i = 0, \ldots, n-1)$ *and let* $m_0 \leq m_1 \leq \cdots \leq m_{n-1}$. *If the sequence has property* $H$, *then*

$$N(A_0, \ldots, A_{n-1}) \geq F_n(m_0, \ldots, m_{n-1}).$$

PROOF: The proof is by induction. Clearly the theorem is true for $n = 1$. We distinguish two cases.

*Case 1.* There is no critical block. In this case, we choose any element $a$ of $A_0$ as its representative and then remove $a$ from all the other sets. This yields sets, that we call $A_1(a), \ldots, A_{n-1}(a)$, and for these sets property $H$ still holds. By the induction hypothesis and by the lemma, we find

$$N(A_0, \ldots, A_{n-1}) \geq \sum_{a \in A_0} f_{n-1}(|A_1(a)|, \ldots, |A_{n-1}(a)|)$$

$$\geq \sum_{a \in A_0} f_{n-1}(m_1 - 1, \ldots, m_{n-1} - 1)$$

$$= m_0 f_{n-1}(m_1 - 1, \ldots, m_{n-1} - 1)$$

$$= F_n(m_0, m_1, \ldots, m_{n-1}).$$

*Case 2.* There is a critical block $(A_{\nu_0}, \ldots, A_{\nu_{k-1}})$ with $\nu_0 < \cdots < \nu_{k-1}$ and $0 < k < n$. In this case, we delete all elements of $A_{\nu_0} \cup \cdots \cup A_{\nu_{k-1}}$ from all the other sets $A_i$ which produces $A'_{\mu_0}, \ldots, A'_{\mu_{l-1}}$, where $\{\nu_0, \ldots, \nu_{k-1}, \mu_0, \ldots, \mu_{l-1}\} = \{0, 1, \ldots, n-1\}$, $k + l = n$.

Now both $(A_{\nu_0}, \ldots, A_{\nu_{k-1}})$ and $(A'_{\mu_0}, \ldots, A'_{\mu_{l-1}})$ satisfy property $H$ and SDRs of the two sequences are always disjoint. Hence by the induction hypothesis and the lemma, we have

(5.1)
$$N(A_0, \ldots, A_{n-1}) = N(A_{\nu_0}, \ldots, A_{\nu_{k-1}}) N(A'_{\mu_0}, \ldots, A'_{\mu_{l-1}})$$

$$\geq f_k(m_{\nu_0}, \ldots, m_{\nu_{k-1}}) f_l(|A'_{\mu_0}|, \ldots, |A'_{\mu_{l-1}}|)$$

$$\geq f_k(m_{\nu_0}, \ldots, m_{\nu_{k-1}}) f_l(m_{\mu_0} - k, \ldots, m_{\mu_{l-1}} - k)$$

$$\geq f_k(m_0, \ldots, m_{k-1}) f_l(m_{\mu_0} - k, \ldots, m_{\mu_{l-1}} - k).$$

Now we remark that

$$m_{\nu_{k-1}} \leq |A_{\nu_0} \cup \cdots \cup A_{\nu_{k-1}}| = k,$$

and therefore we have

$$(m_r - r)_* = 1 \qquad \text{if } k \leq r \leq \nu_{k-1},$$

and

$$(m_{\mu_i} - k - i)_* = 1 \qquad \text{if } \mu_i \leq \nu_{k-1}.$$

This implies that

$$f_k(m_0, \ldots, m_{k-1}) = \prod_{0 \leq i \leq \nu_{k-1}} (m_i - i)_*,$$

$$f_l(m_{\mu_0} - k, \ldots, m_{\mu_{l-1}} - k) = \prod_{\nu_{k-1} < j < n} (m_j - j)_*,$$

i.e. the product (5.1) is equal to $F_n(m_0, \ldots, m_{n-1})$, which proves the theorem. □

PROBLEM 5B. Show that Theorem 5.3 gives the best lower bound for the number of SDRs of the sets $A_i$ that only involves the numbers $|A_i|$.

We now come to a theorem known as König's theorem. It is equivalent (whatever that means) to Hall's theorem. In the theorem, $A$ is a (0,1)-matrix with entries $a_{ij}$. By a *line*, we mean a row or a column of $A$.

THEOREM 5.4. *The minimum number of lines of $A$ that contain all the 1's of $A$ is equal to the maximum number of 1's in $A$, no two on a line.*

PROOF: Let $m$ be the minimum number of lines of $A$ containing all the 1's of $A$ and let $M$ be the maximum number of 1's, no two on a line. Clearly $m \geq M$. Let the minimum covering by lines consist of $r$ rows and $s$ columns ($r + s = m$). Without loss of generality, these are the first $r$ rows and the first $s$ columns. We now define sets $A_i$, $1 \leq i \leq r$, by $A_i := \{j > s : a_{ij} = 1\}$. If some $k$-tuple of the $A_i$'s contained less than $k$ elements, then we could replace the corresponding $k$ rows by $k - 1$ columns, still covering all the 1's. Since this is impossible, we see that the $A_i$'s satisfy property $H$. So the $A_i$'s have an SDR. This means that there are $r$ 1's, no two on a line, in the first $r$ rows and not in the first $s$ columns. By the same argument there are $s$ 1's, no two on a line, in the first $s$ columns and not in the first $r$ rows. This shows that $M \geq r + s = m$ and we are done. $\square$

The following theorem of G. Birkhoff is an application of Hall's theorem.

THEOREM 5.5. *Let $A = (a_{ij})$ be an $n$ by $n$ matrix with nonnegative integers as entries, such that every row and column of $A$ has sum $l$. Then $A$ is the sum of $l$ permutation matrices.*

PROOF: Define $A_i$, $1 \leq i \leq n$, by $A_i := \{j : a_{ij} > 0\}$. For any $k$-tuple of the $A_i$'s, the sum of the corresponding rows of $A$ is $kl$. Since every column of $A$ has sum $l$, the nonzero entries in the chosen $k$ rows must be in at least $k$ columns. Hence the $A_i$'s satisfy property $H$. An SDR of the $A_i$'s corresponds to a permutation matrix $P = (p_{ij})$ such that $a_{ij} > 0$ if $p_{ij} = 1$. The theorem now follows by induction on $l$. $\square$

PROBLEM 5C. In the hypothesis of Theorem 5.5, we replace "integers" by "reals". Show that in this case, $A$ is a nonnegative linear combination of permutation matrices. (Equivalently, every doubly stochastic matrix—see Chapter 11—is a *convex* combination of permutation matrices.)

PROBLEM 5D. Let $S$ be the set $\{1, 2, \ldots, mn\}$. We partition $S$ into $m$ sets $A_1, \ldots, A_m$ of size $n$. Let a second partitioning into $m$ sets of size $n$ be $B_1, \ldots, B_m$. Show that the sets $A_i$ can be renumbered in such a way that $A_i \cap B_i \neq \emptyset$.

**Notes.**

Philip Hall published his result in 1935 (with a rather difficult proof). The proof that we gave is a generalization of ideas of Halmos and Vaughan, Rado, and M. Hall. The proof is due to Ostrand (1970) and Hautus and Van Lint (1972). See Van Lint (1974). The problem of complete matchings is often referred to as the *marriage problem*.

D. König (1884–1944) was professor at Budapest. He wrote the first comprehensive treatise on graph theory (Theorie der endlichen und unendlichen Graphen, 1936). König (1916) contains the first proof of (one of the theorems called) König's theorem.

Just before Theorem 5.4, we referred to the "equivalence" of these theorems. This expression is often used when each of two theorems is more or less an immediate consequence of the other.

The theorem by Birkhoff (1946), i.e. Theorem 5.5, is extremely useful and will be applied a number of times in later chapters.

**References.**

G. Birkhoff (1946), Tres observaciones sobre el algebra lineal, *Univ. Nac. Tucumán, Rev.* Ser. A, **5**, 147–151.

P. Hall (1935), On representatives of subsets, *J. London Math. Soc.* **10**, 26–30.

D. König (1916), Über Graphen und ihre Anwendung auf Determinantentheorie und Mengenlehre, *Math. Annalen* **77**, 453–465.

J. H. van Lint (1974), *Combinatorial Theory Seminar Eindhoven University of Technology*, Lecture Notes in Mathematics **382**, Springer-Verlag.

P. Ostrand (1970), Systems of distinct representatives, *J. of Math. Analysis and Applic.* **32**, 1–4.

# 6

# Dilworth's theorem and extremal set theory

A *partially ordered set* (also *poset*) is a set $S$ with a binary relation $\leq$ (sometimes $\subseteq$ is used) such that:

(i) $a \leq a$ for all $a \in S$ (reflexivity),

(ii) if $a \leq b$ and $b \leq c$ then $a \leq c$ (transitivity),

(iii) if $a \leq b$ and $b \leq a$ then $a = b$ (antisymmetry).

If for any $a$ and $b$ in $S$, either $a \leq b$ or $b \leq a$, then the partial order is called a *total* order, or a *linear* order. If $a \leq b$ and $a \neq b$, then we also write $a < b$. Examples of posets include the integers with the usual order or the subsets of a set, ordered by inclusion. If a subset of $S$ is totally ordered, it is called a *chain*. An *antichain* is a set of elements that are pairwise incomparable.

The following theorem is due to R. Dilworth (1950). This proof is due to H. Tverberg (1967).

THEOREM 6.1. *Let $P$ be a partially ordered finite set. The minimum number $m$ of disjoint chains which together contain all elements of $P$ is equal to the maximum number $M$ of elements in an antichain of $P$.*

PROOF: (i) It is trivial that $m \geq M$.

(ii) We use induction on $|P|$. If $|P| = 0$, there is nothing to prove. Let $C$ be a maximal chain in $P$. If every antichain in $P \backslash C$ contains at most $M - 1$ elements, we are done. So assume that $\{a_1, \ldots, a_M\}$ is an antichain in $P \backslash C$. Now define $S^- := \{x \in P : \exists_i [x \leq a_i]\}$, and define $S^+$ analogously. Since $C$ is a maximal chain, the largest element in $C$ is not in $S^-$ and hence by the induction hypothesis, the theorem holds for $S^-$. Hence $S^-$ is the union of $M$

disjoint chains $S_1^-, \ldots, S_M^-$, where $a_i \in S_i^-$. Suppose $x \in S_i^-$ and $x > a_i$. Since there is a $j$ with $x \leq a_j$, we would have $a_i < a_j$, a contradiction. This shows that $a_i$ is the maximal element of the chain $S_i^-$, $i = 1, \ldots, m$. We do the same for $S^+$. By combining the chains the theorem follows. $\qquad\square$

A "dual" to Dilworth's theorem was given by Mirsky (1971).

THEOREM 6.2. *Let $P$ be a partially ordered set. If $P$ possesses no chain of $m + 1$ elements, then $P$ is the union of $m$ antichains.*

PROOF: For $m = 1$ the theorem is trivial. Let $m \geq 2$ and assume that the theorem is true for $m - 1$. Let $P$ be a partially ordered set that has no chain of $m + 1$ elements. Let $M$ be the set of maximal elements of $P$. $M$ is an antichain. Suppose $x_1 < x_2 < \cdots < x_m$ were a chain in $P \backslash M$. Then this would also be a maximal chain in $P$ and hence we would have $x_m \in M$, a contradiction. Hence $P \backslash M$ has no chain of $m$ elements. By the induction hypothesis, $P \backslash M$ is the union of $m - 1$ antichains. This proves the theorem. $\qquad\square$

The following famous theorem due to Sperner (1928) is of a similar nature. This proof is due to Lubell (1966).

THEOREM 6.3. *If $A_1, A_2, \ldots, A_m$ are subsets of $N := \{1, 2, \ldots, n\}$ such that $A_i$ is not a subset of $A_j$ if $i \neq j$, then $m \leq \binom{n}{\lfloor n/2 \rfloor}$.*

PROOF: Consider the poset of subsets of $N$. $\mathcal{A} := \{A_1, \ldots, A_m\}$ is an antichain in this poset.

A maximal chain $\mathcal{C}$ in this poset will consist of one subset of each cardinality $0, 1, \ldots, n$, and is obtained by starting with the empty set, then any singleton set ($n$ choices), then any 2-subset containing the singleton ($n - 1$ choices), then any 3-subset containing the 2-subset ($n - 2$ choices), etc. Thus there are $n!$ maximal chains. Similarly, there are exactly $k!(n-k)!$ maximal chains which contain a given $k$-subset $A$ of $N$.

Now count the number of ordered pairs $(A, \mathcal{C})$ such that $A \in \mathcal{A}$, $\mathcal{C}$ is a maximal chain, and $A \in \mathcal{C}$. Since each maximal chain $\mathcal{C}$ contains at most one member of an antichain, this number is at most $n!$. If we let $\alpha_k$ denote the number of sets $A \in \mathcal{A}$ with $|A| = k$, then this number is $\sum_{k=0}^{n} \alpha_k k!(n-k)!$. Thus

$$\sum_{k=0}^{n} \alpha_k k!(n-k)! \leq n!, \quad \text{or equivalently,} \quad \sum_{k=0}^{n} \frac{\alpha_k}{\binom{n}{k}} \leq 1.$$

Since $\binom{n}{k}$ is maximal for $k = \lfloor n/2 \rfloor$ and $\sum \alpha_k = m$, the result follows.                                                                              □

Equality holds in Theorem 6.3 if we take all $\lfloor n/2 \rfloor$-subsets of $N$ as the antichain.

We now form a graph $G_n$ (with $2^n$ vertices) by taking the subsets of $N$ as vertices and joining two vertices by an edge if and only if one of the sets is a proper subset of the other (i.e. the sets are comparable in the poset). The set of $i$-subsets of $N$ is denoted by $\mathcal{A}_i$. We define a *symmetric chain* in $G_n$ to be a sequence $P_k, P_{k+1}, \ldots, P_{n-k}$ of vertices such that $P_i \in \mathcal{A}_i$ and $\{P_i, P_{i+1}\}$ is an edge in $G_n$ for $i = k, k+1, \ldots, n-k-1$. We describe an algorithm due to De Bruijn, Van Ebbenhorst Tengbergen and Kruyswijk (1949), that splits $G_n$ into (disjoint) symmetric chains.

Algorithm: Start with $G_1$. Proceed by induction. If $G_n$ has been split into symmetric chains, then for each such symmetric chain $P_k, \ldots, P_{n-k}$ define two symmetric chains in $G_{n+1}$, namely $P_{k+1}, \ldots, P_{n-k}$ and $P_k, P_k \cup \{n+1\}, P_{k+1} \cup \{n+1\}, \ldots, P_{n-k} \cup \{n+1\}$.

It is easy to see that this algorithm does what we claim. Furthermore it provides a natural matching between $k$-subsets and $(n-k)$-subsets in the graph $G_n$ (cf. Theorem 5.1). Also, see Problem 6D below.

PROBLEM 6A. Let $a_1, a_2, \ldots, a_{n^2+1}$ be a permutation of the integers $1, 2, \ldots, n^2+1$. Show that Dilworth's theorem implies that the sequence has a subsequence of length $n+1$ that is monotone.

A nice direct proof of the assertion of Problem 6A is as follows. Suppose there is no increasing subsequence of $n+1$ terms. Define $b_i$ to be the length of the longest increasing subsequence that starts with the term $a_i$. Then by the pigeonhole principle, there are at least $n+1$ terms in the $b_i$-sequence that have the same value. Since $i < j$ and $b_i = b_j$ imply that $a_i > a_j$, we have a decreasing subsequence of $n+1$ terms.

To show a connection between Chapters 5 and 6, we now prove that Theorem 5.1 immediately follows from Theorem 6.1. We consider the bipartite graph $G$ of Theorem 5.1. Let $|X| = n$, $|Y| = n' \geq n$. We introduce a partial order by defining $x_i < y_j$ if and only if there is an edge from vertex $x_i$ to vertex $y_j$. Suppose that the largest antichain contains $s$ elements. Let this antichain be $\{x_1, \ldots, x_h, y_1, \ldots, y_k\}$, where $h + k = s$. Since $\Gamma(\{x_1, \ldots, x_h\}) \subseteq Y \backslash \{y_1, \ldots, y_k\}$, we have $h \leq n' - k$. Hence $s \leq n'$. The partially ordered set is the union of $s$ disjoint chains. This will consist of a matching of size $a$, the remaining $n - a$ elements of $X$, and the remaining $n' - a$ elements of $Y$. Therefore $n + n' - a = s \leq n'$, i.e. $a \geq n$, which means that we have a complete matching.

Theorem 6.3 is a (fairly easy) example of an area known as *extremal set theory* in which the problems are often quite difficult. We first give one more example as an easy exercise.

PROBLEM 6B. Let the sets $A_i$, $1 \leq i \leq k$, be distinct subsets of $\{1, 2, \ldots, n\}$. Suppose $A_i \cap A_j \neq \emptyset$ for all $i$ and $j$. Show that $k \leq 2^{n-1}$ and give an example where equality holds.

We now give one more example of the method that we used to prove Sperner's theorem. We prove the so-called Erdős-Ko-Rado theorem (1961).

THEOREM 6.4. *Let $\mathcal{A} = \{A_1, \ldots, A_m\}$ be a collection of $m$ distinct $k$-subsets of $\{1, 2, \ldots, n\}$, where $k \leq n/2$, with the property that any two of the subsets have a nonempty intersection. Then $m \leq \binom{n-1}{k-1}$.*

PROOF: Place the integers 1 to $n$ on a circle and consider the family $\mathcal{F} := \{F_1, \ldots, F_n\}$ of all consecutive $k$-tuples on the circle, i.e. $F_i$ denotes $\{i, i+1, \ldots, i+k-1\}$ where the integers should be taken mod $n$. We observe that $|\mathcal{A} \cap \mathcal{F}| \leq k$ because if some $F_i$ equals $A_j$, then at most one of the sets $\{l, l+1, \ldots, l+k-1\}$, $\{l-k, \ldots, l-1\}$ $(i < l < i + k)$ is in $\mathcal{A}$. The same assertion holds for the collection $\mathcal{F}^{\pi}$ obtained from $\mathcal{F}$ by applying a permutation $\pi$ to $\{1, \ldots, n\}$. Therefore

$$\Sigma := \sum_{\pi \in S_n} |\mathcal{A} \cap \mathcal{F}^{\pi}| \leq k \cdot n!.$$

We now count this sum by fixing $A_j \in \mathcal{A}$, $F_i \in \mathcal{F}$ and observing that there are $k!(n-k)!$ permutations $\pi$ such that $F_i^\pi = A_j$. Hence $\Sigma = m \cdot n \cdot k!(n-k)!$. This proves the theorem.                    $\square$

By a slight modification of the proof, one can show that the theorem also holds if the sets in $\mathcal{A}$ are assumed to have size at most $k$ and they form an antichain. However we shall give a proof using Theorem 5.1.

THEOREM 6.5. *Let $\mathcal{A} = \{A_1, \ldots, A_m\}$ be a collection of $m$ subsets of $N := \{1, 2, \ldots, n\}$ such that $A_i \not\subseteq A_j$ and $A_i \cap A_j \neq \emptyset$ if $i \neq j$ and $|A_i| \leq k \leq n/2$ for all $i$. Then $m \leq \binom{n-1}{k-1}$.*

PROOF: (i) If all the subsets have size $k$, then we are done by Theorem 6.4.

(ii) Let $A_1, \ldots, A_s$ be the subsets with the smallest cardinality, say $l \leq \frac{n}{2} - 1$. Consider all the $(l+1)$-subsets $B_j$ of $N$ that contain one or more of the sets $A_i$, $1 \leq i \leq s$. Clearly none of these is in $\mathcal{A}$. Each of the sets $A_i$, $1 \leq i \leq s$, is in exactly $n - l$ of the $B_j$'s and each $B_j$ contains at most $l + 1 \leq n - l$ of the $A_i$'s. So by Theorem 5.1, we can pick $s$ distinct sets, say $B_1, \ldots, B_s$, such that $A_i \subseteq B_i$. If we replace $A_1, \ldots, A_s$ by $B_1, \ldots, B_s$, then the new collection $\mathcal{A}'$ satisfies the conditions of the theorem and the subsets of smallest cardinality now all have size $> l$. By induction, we can reduce to case (i).                    $\square$

PROBLEM 6C. Let $\mathcal{A} = \{A_1, \ldots, A_m\}$ be a collection of $m$ distinct subsets of $N := \{1, 2, \ldots, n\}$ such that if $i \neq j$ then $A_i \not\subseteq A_j$, $A_i \cap A_j \neq \emptyset$, $A_i \cup A_j \neq N$. Prove that

$$m \leq \binom{n-1}{\lfloor \frac{n}{2} \rfloor - 1}.$$

PROBLEM 6D. Consider the decomposition of $G_n$ into symmetric chains as described above. Show that Theorem 6.3 is an immediate consequence of this decomposition. Show that Theorem 6.5 reduces to Theorem 6.4 via this decomposition. How many of the chains have their smallest element in $\mathcal{A}_i$?

PROBLEM 6E. Here is an algorithm to construct a symmetric chain in the graph $G_n$ which contains a given vertex $S$ (a subset of

$\{1, 2, \ldots, n\}$). Consider the characteristic vector $x$ of $S$; for example, if $n = 7$ and $S = \{3, 4, 7\}$, then $x = 0011001$. Mark all consecutive pairs 10, temporarily delete these pairs and again mark all consecutive pairs 10, and repeat until only a string of the form $00 \cdots 01 \cdots 11$ remains. In our example, we obtain $00\dot{1}\dot{1}\dot{0}\dot{0}1$, where the $i$-th coordinates are marked for $i = 3, 4, 5, 6$; when these are deleted, the string 001 remains. The characteristic vectors of the subsets in the chain are obtained by fixing all marked coordinates and letting the remaining coordinates range over the strings $0 \cdots 000$, $0 \cdots 001$, $0 \cdots 011$, $\ldots$, $1 \cdots 111$. In our example, these characteristic vectors are

$$00\dot{1}\dot{1}\dot{0}\dot{0}0,$$
$$00\dot{1}\dot{1}\dot{0}\dot{0}1,$$
$$01\dot{1}\dot{1}\dot{0}\dot{0}1,$$
$$11\dot{1}\dot{1}\dot{0}\dot{0}1,$$

which correspond to the subsets

$$\{3, 4\}, \quad \{3, 4, 7\}, \quad \{2, 3, 4, 7\}, \quad \{1, 2, 3, 4, 7\}.$$

Show that this algorithm produces exactly the same symmetric chain containing $S$ as is produced by the inductive algorithm of De Bruijn *et al.* described above.

**Notes.**

We shall return to partially ordered sets in Chapters 23 and 25.

E. Sperner (1905-1980) is best known for a lemma in combinatorial topology known as "Sperner's lemma", which occurred in his thesis (1928). It was used to give a proof of Brouwer's fixed point theorem. (Another connection to combinatorics: his first professorship was in Königsberg!) He was one of the pioneers of the famous Oberwolfach research institute.

For a survey of extremal set theory, we refer to Frankl (1988).

The short proof of the Erdős-Ko-Rado theorem is due to Katona (1974). Theorem 6.5 is due to Kleitman and Spencer (1973) and Schönheim (1971).

**References.**

N. G. de Bruijn, C. van Ebbenhorst Tengbergen and D. Kruyswijk (1949), On the set of divisors of a number, *Nieuw Archief v. Wisk.* (2) **23**, 191–193.

R. P. Dilworth (1950), A decomposition theorem for partially ordered sets, *Annals of Math.* (2) **51**, 161–166.

P. Erdős, Chao Ko, and R. Rado (1961), Extremal problems among subsets of a set, *Quart. J. Math. Oxford* Ser. (2) **12**, 313–318.

P. Frankl (1988), Old and new problems on finite sets, Proc. Nineteenth S. E. Conf. on Combinatorics, Graph Th. and Computing, Baton Rouge, 1988.

G. O. H. Katona (1974), Extremal problems for hypergraphs, in *Combinatorics* (edited by M. Hall, Jr. and J. H. van Lint), Reidel.

D. J. Kleitman and J. Spencer (1973), Families of $k$-independent sets, *Discrete Math.* **6**, 255–262.

D. Lubell (1966), A short proof of Sperner's lemma, *J. Combinatorial Theory* **1**, 299.

L. Mirsky (1971), A dual of Dilworth's decomposition theorem, *Amer. Math. Monthly* **78**, 876–877.

J. Schönheim (1971), A generalization of results of P. Erdős, G. Katona, and D. J. Kleitman concerning Sperner's theorem, *J. Combinatorial Theory* (A) **11**, 111–117.

E. Sperner (1928), Ein Satz über Untermengen einer endlichen Menge, *Math. Zeitschrift* **27**, 544–548.

H. Tverberg (1967), On Dilworth's decomposition theorem for partially ordered sets, *J. Combinatorial Theory* **3**, 305–306.

# 7

# Flows in networks

By a *transportation network*, we will mean a finite directed graph
$D$ together with two distinguished vertices $s$ and $t$ called the *source*
and the *sink*, respectively, and which is provided with a function
$c$ associating to each edge $e$ a nonnegative real number $c(e)$ called
its *capacity*. We may further assume that there are no loops, no
multiple edges, and that no edges enter the source $s$ or leave the sink
$t$ (although there would be no harm in admitting any of these types
of edges other than our having to be more careful in a definition or
two).

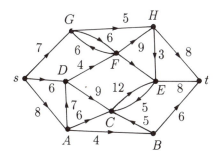

Figure 7.1

In Fig. 7.1 we give an example. We could think of a network
of pipes through which some liquid could flow in the direction of
the arrows. The capacity would indicate the maximal possible flow
(per time unit) in that section of pipe.

A *flow* in a transportation network is a function $f$ assigning a
real number $f(e)$ to each edge $e$ such that:

(a) $0 \le f(e) \le c(e)$ for all edges $e$ (the flow is *feasible*);

(b) for each vertex $x$ (not the source or the sink) the sum of the values of $f$ on incoming edges equals the sum of the values of $f$ on outgoing edges (*conservation of flow*).

The sum of the values of a flow $f$ on the edges leaving the source is called the *strength* of the flow (denoted by $|f|$). It seems obvious that the strength of the flow is also equal to the sum of the values of $f$ on edges entering the sink; the reader might try to verify this formally before reading further.

One of our objectives will be to find a method for constructing a *maximum flow*, that is, a flow with maximum strength. Before we begin, it will be good to have a goal or an upper bound for the strength of a flow; for example, the sum of the capacities of all edges leaving the source is clearly such an upper bound. More generally, by a *cut separating s and t* (or simply a *cut*), we mean here a pair $(X, Y)$ of subsets of the vertex set $V := V(D)$ which partition $V$ and such that $s \in X$ and $t \in Y$. We define the *capacity* $c(X, Y)$ of the cut to be the sum of the capacities of the edges directed from $X$ to $Y$ (that is, edges $e = (x, y)$ with $x \in X$ and $y \in Y$). We claim that the capacity of any cut is an upper bound for the strength of any flow. More strongly, we claim that the conservation law implies (see below) that the strength of a flow $f$ can be computed as

$$(7.1) \qquad |f| = f(X, Y) - f(Y, X),$$

where $f(A, B)$ denotes the sum of the values of $f$ on all edges directed from $A$ to $B$; then the feasibility of $f$ immediately implies that $|f| \leq c(X, Y)$. Thus the minimum capacity of all cuts in a network (e.g. in Fig. 7.1 the minimum cut capacity is 20) is an upper bound for the strength of a flow in that network.

To establish (7.1), we introduce the function $\phi$ by defining for each pair $(x, e)$, where $x$ is a vertex incident with the edge $e$, $\phi(x, e) := -1$ if the edge is incoming, and $\phi(x, e) := +1$ if the edge is outgoing; $\phi(x, e)$ is to be 0 if $x$ is not incident with $e$. (We remark that $\phi$ is essentially the *incidence matrix* of the directed graph—see Chapter 34.) The conservation law is equivalent to $\sum_{e \in E} \phi(x, e) f(e) = 0$ for $x \neq s, t$. Notice that $\sum_{x \in X} \phi(x, e)$ is $+1$ if $e$ is directed from $X$ to $Y$, $-1$ if $e$ is directed from $Y$ to $X$, and

0 if $e$ has both endpoints in $X$ or both in $Y$. Then

$$|f| = \sum_{e \in E} \phi(s,e)f(e) = \sum_{x \in X} \sum_{e \in E} \phi(x,e)f(e)$$
$$= \sum_{e \in E} f(e) \sum_{x \in X} \phi(x,e) = f(X,Y) - f(Y,X).$$

(In the first double sum above, the inner sum is 0 for all terms $x$ other than $s$.)

A special instance of (7.1) is $|f| = f(V\backslash\{t\}, \{t\})$, the assertion which we invited the reader to reflect on earlier.

We now construct flows. Fix a flow $f$, possibly the 0-flow. We shall say that the sequence $x_0, x_1, \ldots, x_{k-1}, x_k$ of distinct vertices is a *special path* from $x_0$ to $x_k$ if for each $i$, $1 \leq i \leq k$, either

(i) $e = (x_{i-1}, x_i)$ is an edge with $c(e) - f(e) > 0$, or

(ii) $e = (x_i, x_{i-1})$ is an edge with $f(e) > 0$.

Edges $e$ with $f(e) = c(e)$ are said to be *saturated* and conditions (i) and (ii) can be stated in words as requiring that "forward" edges of the path are unsaturated while "backward" edges are positive—all with respect to a given flow $f$. Suppose there exists such a special path from $s$ to $t$. Define $\alpha_i$ as $c(e) - f(e)$ in the first case and as $f(e)$ in the second case (picking one of the edges to use if both cases hold) and let $\alpha$ be the minimum of these positive numbers $\alpha_i$. On each edge of type (i) *increase* the flow value by $\alpha$, and on each edge of type (ii) *decrease* the flow by $\alpha$. It is easy to check that the two conditions for a flow (feasibility and conservation of flow) are still satisfied. Clearly the new flow has strength $|f| + \alpha$.

This idea for obtaining a stronger flow becomes an algorithm when we iterate it (starting with the 0-flow) and incorporate a systematic procedure for searching for special paths from $s$ to $t$ with respect to the current flow. We make brief remarks concerning termination in the notes to this chapter. But what happens when we can go no further?

Suppose that *no special path from source to sink exists* with respect to some flow $f_0$. Let $X_0$ be the set of vertices $x$ which can be reached from $s$ by a special path, $Y_0$ the set of remaining vertices.

In this way we produce a cut. If $x \in X_0$, $y \in Y_0$ and $e = (x, y)$ is an edge, then $e$ must be saturated or we could adjoin $y$ to a special path from $s$ to $x$ to get a special path from $s$ to $y$, contradicting the definitions of $X_0$ and $Y_0$. If, on the other hand, $e = (y, x)$ is an edge, then, for a similar reason, $f(e)$ must be 0. In view of (7.1), we have then

$$|f_0| = f_0(X_0, Y_0) - f_0(Y_0, X_0) = c(X_0, Y_0).$$

Now it is clear that not only can no stronger flow be obtained by our method of special paths, but that no stronger flows exist at all because $|f| \leq c(X_0, Y_0)$ for any flow $f$.

If $f_0$ is chosen to be a maximum flow (which exists by continuity reasons in case one is unsure of the termination of the algorithm), then surely no special paths from $s$ to $t$ exist. Note that the constructed cut $(X_0, Y_0)$ is a minimum cut (i.e. a cut of minimum capacity), since $c(X, Y) \geq |f_0|$ for any cut $(X, Y)$. Our observations have combined to prove the following famous theorem of Ford and Fulkerson (1956).

THEOREM 7.1. *In a transportation network, the maximum value of $|f|$ over all flows $f$ is equal to the minimum value of $c(X, Y)$ over all cuts $(X, Y)$.*

This theorem is usually referred to as the "maxflow-mincut" theorem. The procedure for increasing the strength of a flow that we used above shows somewhat more.

THEOREM 7.2. *If all the capacities in a transportation network are integers, then there is a maximum strength flow $f$ for which all values $f(e)$ are integers.*

PROOF: Start with the 0-flow. The argument above provides a way to increase the strength until a maximum flow is reached. At each step $\alpha$ is an integer, so the next flow is integer valued too. □

PROBLEM 7A. Construct a maximum flow for the transportation network of Fig. 7.1.

PROBLEM 7B. An *elementary flow* in a transportation network is a flow $f$ which is obtained by assigning a constant positive value

$\alpha$ to the set of edges traversed by a simple (directed) path from $s$ to $t$, and 0 to all other edges. Show that every flow is the sum of elementary flows and perhaps a flow of strength zero. (This means we can arrive at a maxflow by starting from the 0-flow and using only special paths with "forward" edges.) Give an example of a network and a flow which is not maximum, but with respect to which there are no special paths using only "forward" edges.

PROBLEM 7C. Let $(X_1, Y_1)$ and $(X_2, Y_2)$ be minimum cuts (i.e. cuts of minimum capacity) in a transportation network. Show that $(X_1 \cup X_2, Y_1 \cap Y_2)$ is also a minimum cut. (This can be done either from first principles, or with an argument involving maximum flows.)

PROBLEM 7D. Prove P. Hall's marriage theorem, Theorem 5.1, from Theorems 7.1 and 7.2.

It should be clear that the topic of this chapter is of great practical importance. Routing schemes for all kinds of products depend on algorithms that produce optimal flows through transportation networks. We do not go into the algorithmic aspect of this area. Instead, we shall show a beautiful application of Theorem 7.2 to a problem related to Birkhoff's theorem, Theorem 5.5. Before giving the theorem and its proof, we observe that several attempts were made to prove it by reducing it to Theorem 5.5 but with no success. The proof below is due to A. Schrijver. (If $b = v$ in Theorem 7.3, then we have the situation of Theorem 5.5.)

THEOREM 7.3. *Let $A$ be a $b$ by $v$ $(0, 1)$-matrix with $k$ ones per row and $r$ ones per column (so $bk = vr$). Let $\alpha$ be a rational number, $0 < \alpha < 1$, such that $k' = \alpha k$ and $r' = \alpha r$ are integers. Then there is a $(0, 1)$-matrix $A'$ of size $b$ by $v$ with $k'$ ones per row and $r'$ ones per column such that entries $a'_{ij}$ of $A'$ are 1 only if the corresponding entries of $A$ are 1, i.e. $A'$ can be obtained from $A$ by changing some ones into zeros.*

PROOF: We construct a transportation network with vertices $s$ (the source), $x_1, \ldots, x_b$ (corresponding to the rows of $A$), $y_1, \ldots, y_v$ (corresponding to the columns of $A$), and $t$ (the sink). Edges are $(s, x_i)$ with capacity $k$, $1 \le i \le b$, $(x_i, y_j)$ with capacity 1 if and only if $a_{ij} = 1$, and $(y_j, t)$ with capacity $r$, $1 \le j \le v$. The definition ensures that there is a maximum flow with all edges saturated. We now change the capacities of the edges from the source to $k'$ and those of the edges to the sink to $r'$. Again, all the capacities are integers and clearly a maximum flow exists for which the flows $f((x_i, y_j))$ are equal to $\alpha$. By Theorem 7.2 there is also a maximum flow $f^*$ for which all the flows are integers, i.e. $f^*((x_i, y_j)) = 0$ or 1. From this flow, we immediately find the required matrix $A'$. $\square$

**Notes.**

The term *augmenting path* is often used instead of *special path*.

If the capacities of a transportation network are integers, the special path method for constructing maximum flows will terminate after finitely many iterations, since the strength increases by at least one each time. But Ford and Fulkerson (1962) give an example with irrational capacities where certain contrived choices of special paths lead to an infinite sequence of flows whose strengths converge—but only to one-fourth of the actual maximum flow strength! If one is careful to pick *shortest* special paths, however, then it can be shown that a maximum flow is reached after at most $O(n^3)$ iterations, where $n$ is the number of vertices. See Edmonds and Karp (1972).

The problem of finding a maximum flow is an example of a linear programming problem and can be solved e.g. by the simplex algorithm. The network flow problem is special in that its matrix $\phi$ is totally unimodular, and this is one way of explaining why Theorem 7.2 holds. See the references below for more discussion of linear and integer programming. Graphical methods are usually faster than the simplex algorithm, and add insight.

Theorems 7.1, 7.2, and the algorithm have many further combinatorial applications, since certain combinatorial problems can be phrased in terms of transportation networks. For example, finding a maximum matching in a bipartite graph is equivalent to finding a maximum (integer valued) flow in a certain associated network—see the references—and thus a good algorithm exists to find a maximum matching. We give further applications of Theorems 7.1 and 7.2 in Chapter 16 to an existence problem on (0,1)-matrices, and in Chapter 36 to a problem on partitions of sets.

## References.

J. Edmonds and R. M. Karp (1972), Theoretical improvements in algorithm efficiency for network flow problems, *J. Assn. for Computing Machinery* **19**, 248–264.

L. R. Ford, Jr. and D. R. Fulkerson (1962), *Flows in Networks*, Princeton University Press.

T. C. Hu (1969), *Integer Programming and Network Flows*, Addison-Wesley.

V. Chvátal (1983), *Linear Programming*, W. H. Freeman.

# 8
# De Bruijn sequences

The following problem has a practical origin: the so-called *rotating drum problem*. Consider a rotating drum as in Fig. 8.1.

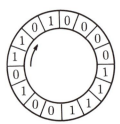

Figure 8.1

Each of the segments is of one of two types, denoted by 0 and 1. We require that any four consecutive segments uniquely determine the position of the drum. This means that the 16 possible quadruples of consecutive 0's and 1's on the drum should be the binary representations of the integers 0 to 15. Can this be done and, if yes, in how many different ways? The first question is easy to answer. Both questions were treated by N. G. de Bruijn (1946) and for this reason the graphs described below and the corresponding circular sequences of 0's and 1's are often called *De Bruijn graphs* and *De Bruijn sequences*, respectively.

We consider a digraph (later to be called $G_4$) by taking all 3-tuples of 0's and 1's (i.e. 3-bit binary words) as vertices and joining the vertex $x_1x_2x_3$ by a directed edge (arc) to $x_2x_30$ and $x_2x_31$. The arc $(x_1x_2x_3, x_2x_3x_4)$ is numbered $e_j$, where $x_1x_2x_3x_4$ is the binary representation of the integer $j$. The graph has a loop at 000 and at 111. As we saw before, the graph has an Eulerian circuit because every vertex has in-degree 2 and out-degree 2. Such a closed path produces the required 16-bit sequence for the drum. Such a (circular) sequence is called a De Bruijn sequence. For example the path $000 \to 000 \to 001 \to 011 \to 111 \to 111 \to 110 \to 100 \to 001 \to 010 \to 101 \to 011 \to 110 \to 101 \to 010 \to 100 \to 000$ corresponds to 0000111100101101 (to be read circularly). We call such a path a *complete cycle*.

We define the graph $G_n$ to be the directed graph on $(n-1)$-tuples of 0's and 1's in a similar way as above. (So $G_n$ has $2^n$ edges.)

The graph $G_4$ is given in Fig. 8.2. In this chapter, we shall call a digraph with in-degree 2 and out-degree 2 for every vertex, a "2-in 2-out graph". For such a graph $G$ we define the "doubled" graph $G^*$ as follows:

(i) to each edge of $G$ there corresponds a vertex of $G^*$;

(ii) if $a$ and $b$ are vertices of $G^*$, then there is an edge from $a$ to $b$ if and only if the edge of $G$ corresponding to $a$ has as terminal end (head) the initial end (tail) of the edge of $G$ corresponding to $b$.

Clearly $G_n^* = G_{n+1}$.

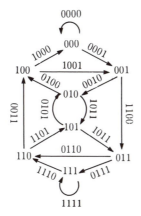

Figure 8.2

THEOREM 8.1. *Let $G$ be a 2-in 2-out graph on $m$ vertices with $M$ complete cycles. Then $G^*$ has $2^{m-1}M$ complete cycles.*

PROOF: The proof is by induction on $m$.

(a) If $m = 1$ then $G$ has one vertex $p$ and two loops from $p$ to $p$. Then $G^* = G_2$ which has one complete cycle.

(b) We may assume that $G$ is connected. If $G$ has $m$ vertices and there is a loop at *every* vertex, then, besides these loops, $G$ is a circuit $p_1 \to p_2 \to \cdots \to p_m \to p_1$. Let $A_i$ be the loop $p_i \to p_i$ and $B_i$ the arc $p_i \to p_{i+1}$. We shall always denote the corresponding vertices in $G^*$ by lower case letters. The situation in $G^*$ is as in Fig. 8.3.

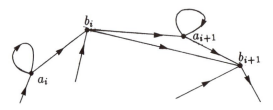

Figure 8.3

Clearly a cycle in $G^*$ has two ways of going from $b_i$ to $b_{i+1}$. So $G^*$ has $2^{m-1}$ complete cycles, whereas $G$ has only one.

(c) We now assume that $G$ has a vertex $x$ that does not have a loop on it. The situation is as in Fig. 8.4, where $P, Q, R, S$ are different edges of $G$ (although some of the vertices $a, b, c, d$ may coincide).

From $G$ we form a new 2-in 2-out graph with one vertex less by deleting the vertex $x$. This can be done in two ways: $G_1$ is obtained by the identification $P = R$, $Q = S$, and $G_2$ is obtained by $P = S$, $Q = R$. By the induction hypothesis, the theorem applies to $G_1$ and to $G_2$.

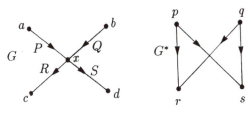

Figure 8.4

There are three different types of complete cycle in $G^*$, depending on whether the two paths leaving $r$ and returning to $p$, respectively $q$, both go to $p$, both to $q$, or one to $p$ and one to $q$. We treat one case; the other two are similar and left to the reader. In Fig. 8.5 we show the situation where path 1 goes from $r$ to $p$, path 2 from $s$ to $q$, path 3 from $s$ to $p$, and path 4 from $r$ to $q$.

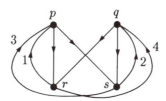

Figure 8.5

These yield the following four complete cycles in $G^*$:

$$
\begin{array}{cccccccc}
1, & pr, & 4, & qs, & 3, & ps, & 2, & qr \\
1, & ps, & 2, & qr, & 4, & qs, & 3, & pr \\
1, & ps, & 3, & pr, & 4, & qs, & 2, & qr \\
1, & ps, & 2, & qs, & 3, & pr, & 4, & qr
\end{array}
$$

In $G_1^*$ and $G_2^*$ the situation reduces to Fig. 8.6.

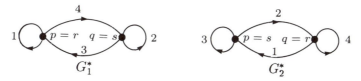

Figure 8.6

In each of $G_1^*$ and $G_2^*$ *one* complete cycle using the paths $1, 2, 3, 4$ is possible. In the remaining two cases, we also find two complete cycles in $G_1^*$ and $G_2^*$ corresponding to four complete cycles in $G^*$. Therefore the number of complete cycles in $G^*$ is twice the sum of the numbers for $G_1^*$ and $G_2^*$. On the other hand, the number of complete cycles in $G$ is clearly equal to the sum of the corresponding numbers for $G_1$ and $G_2$. The theorem then follows from the induction hypothesis. □

We are now able to answer the question how many complete cycles there are in a De Bruijn graph.

THEOREM 8.2. $G_n$ *has exactly* $2^{2^{n-1}-n}$ *complete cycles.*

PROOF: The theorem is true for $n = 1$. Since $G_n^* = G_{n+1}$, the result follows by induction from Theorem 8.1. $\qquad\square$

PROBLEM 8A. Let $\alpha$ be a primitive element in $\mathbb{F}_{2^n}$. For $1 \le i \le m := 2^n - 1$, let

$$\alpha^i = \sum_{j=0}^{n-1} c_{ij}\alpha^j.$$

Show that the sequence

$$0, c_{10}, c_{20}, \ldots, c_{m0}$$

is a De Bruijn sequence.

PROBLEM 8B. Find a circular ternary sequence (with symbols $0, 1, 2$) of length 27 so that each possible ternary ordered triple occurs as three (circularly) consecutive positions of the sequence. First sketch a certain directed graph on 9 vertices so that Eulerian circuits in the graph correspond to such sequences.

**Notes.**

Although the graphs of this chapter are commonly called De Bruijn graphs, Theorem 8.1 was proved in 1894 by C. Flye Sainte-Marie. This went unnoticed for a long time. We refer to De Bruijn (1975).

N. G. de Bruijn (1918–), one of the best-known Dutch mathematicians, worked in many different areas such as analysis, number theory, combinatorics, and also computing science and crystalography.

We mention a peculiarity concerning the spelling of some Dutch names. When omitting the initials of N. G. de Bruijn, one should capitalize the word "de" and furthermore the name should be listed under B. Similarly Van der Waerden is correct when the initials are omitted and he should be listed under W.

For a proof of Theorem 8.1 (and its generalization for arbitrary alphabets) using algebraic methods, we refer to Van Lint (1974).

**References.**

N. G. de Bruijn (1946), A combinatorial problem, *Proc. Kon. Ned. Akad. v. Wetensch.* **49**, 758–764.

N. G. de Bruijn (1975), Acknowledgement of priority to C. Flye Sainte-Marie on the counting of circular arrangements of $2^n$ zeros and ones that show each $n$-letter word exactly once, T. H. report 75-WSK-06, Eindhoven University of Technology.

C. Flye Sainte-Marie (1894), Solution to question nr. 48, *Intermédiaire des Mathématiciens* **1**, 107–110.

J. H. van Lint (1974), *Combinatorial Theory Seminar Eindhoven University of Technology*, Lecture Notes in Math. **382**, Springer-Verlag.

# 9

# The addressing problem for graphs

The following problem originated in communication theory. For a telephone network a connection between terminals $A$ and $B$ is established before messages flow in either direction. For a network of computers it is desirable to be able to send a message from $A$ to $B$ without $B$ knowing that a message is on its way. The idea is to let the message be preceded by some "address" of $B$ such that at each node of the network a decision can be made concerning the direction in which the message should proceed.

A natural thing to try is to give each vertex of a graph $G$ a binary address, say in $\{0,1\}^k$, in such a way that the distance of two vertices in the graph is equal to the so-called *Hamming distance* of the addresses, i.e. the number of places where the addresses differ. This is equivalent to finding $G$ as an *induced subgraph* of the *hypercube* $H_k$, which has $V(H_k) := \{0,1\}^k$ and where $k$-tuples are adjacent when they differ in exactly one coordinate. The example $G = K_3$ already shows that this is impossible. We now introduce a new alphabet $\{0,1,*\}$ and form addresses by taking $n$-tuples from this alphabet. The distance between two addresses is defined to be the number of places where one has a 0 and the other a 1 (so stars do not contribute to the distance). For an addressing of a graph $G$, we require that the distance of any two vertices in $G$ is equal to the distance of their addresses. It is trivial that this can be done if $n$ is large enough. We denote by $N(G)$ the minimum value of $n$ for which there exists an addressing of $G$ with length $n$.

For a tree we can do without the stars as follows. We use induction. For a tree with two vertices, we have a trivial addressing with length 1. Suppose that we can address trees with $k$ vertices. If $x_0, x_1, \ldots, x_k$ are the vertices of the tree $T$ and $x_0$ is a monovalent vertex, then consider an addressing for the tree obtained by removing $x_0$. Let $\mathbf{x}_i$ be the address of $x_i$ and suppose $x_0$ is joined to $x_1$. We change all addresses to $(0, \mathbf{x}_i)$, $1 \leq i \leq k$, and give $x_0$ the address $(1, \mathbf{x}_1)$. Clearly this is now an addressing for $T$. So for a tree, we have $N(T) \leq |V(T)| - 1$.

As a second example, consider $K_m$. In the identity matrix of size $m - 1$, we replace the zeros above the diagonal by stars and add a row of zeros. Any two rows now have distance 1 and hence $N(K_m) \leq m - 1$.

As a third example, we consider the graph of Fig. 9.1.

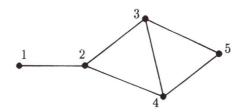

Figure 9.1

A possible (though not optimal) addressing is

$$
\begin{array}{c c c c c c}
1 & \quad 1 & 1 & 1 & * & * \\
2 & \quad 1 & 0 & * & 1 & * \\
3 & \quad * & 0 & 0 & 0 & 1 \\
4 & \quad 0 & 0 & 1 & * & * \\
5 & \quad 0 & 0 & 0 & 0 & 0
\end{array}
$$

We now show a correspondence between addressings of a graph and *quadratic forms* (an idea of Graham and Pollak, 1971). Consider the graph $G$ of Fig. 9.1 and the addressing given above. To the first column of the addressing, we associate the product $(x_1 + x_2)(x_4 + x_5)$. Here $x_i$ is in the first, respectively second, factor if the address of $i$ has a 1, respectively a 0, in the first column. If we do the same thing for each column and then add the terms, we obtain a quadratic form $\sum d_{ij} x_i x_j$, where $d_{ij}$ is the distance of the vertices $i$ and $j$ in $G$. Thus an addressing of $G$ corresponds to writing the quadratic form $\sum d_{ij} x_i x_j$ as a sum of $n$ products $(x_{i_1} + \cdots + x_{i_k})(x_{j_1} + \cdots + x_{j_l})$ such that no $x_i$ occurs in both of the factors. The number of variables is $|V(G)|$.

THEOREM 9.1. *Let* $n_+$, *respectively* $n_-$, *be the number of positive, respectively negative, eigenvalues of the distance matrix* $(d_{ij})$ *of the graph* $G$. *Then* $N(G) \geq \max\{n_+, n_-\}$.

PROOF:

$$(x_1 + \cdots + x_m)(x_{m+1} + \cdots + x_l) =$$
$$\frac{1}{4}\{(x_1 + \cdots + x_l)^2 - (x_1 + \cdots + x_m - x_{m+1} - \cdots - x_l)^2\}.$$

An addressing therefore represents $\sum d_{ij} x_i x_j$ as the difference of two sums of $n$ squares of linear forms. By Sylvester's law (of inertia), $n \geq \max\{n_+, n_-\}$.                                    □

In the proof, we have used "Sylvester's law" which states that if a quadratic form can be transformed into the form

$$x_1^2 + x_2^2 + \cdots + x_i^2 - x_{i+1}^2 - \cdots - x_n^2$$

by a nonsingular transformation, then $i$ is equal to the number of positive eigenvalues of the corresponding matrix and $n - i$ is equal to the number of negative eigenvalues.

THEOREM 9.2. $N(K_m) = m - 1$.

PROOF: We already saw that $N(K_m) \leq m - 1$. Since $J - I$ of size $m$ is the distance matrix of $K_m$ and the eigenvalues of $J - I$ are $m - 1$ with multiplicity 1 and $-1$ with multiplicity $m - 1$, the result follows from Theorem 9.1.                                    □

With slightly more work, we shall now show that the shortest addressing for a tree $T$ has length $|V(T)| - 1$.

THEOREM 9.3. *If* $T$ *is a tree on* $n$ *vertices, then* $N(T) = n - 1$.

PROOF: We first calculate the determinant of the distance matrix $(d_{ij})$ of $T$. We number the vertices $p_1, \ldots, p_n$ in such a way that $p_n$ is an endpoint adjacent to $p_{n-1}$. In the distance matrix, we subtract row $n - 1$ from row $n$, and similarly for the columns. Then all the entries in the new last row and column are 1 except for the diagonal element which is equal to $-2$. Now renumber the vertices $p_1, \ldots, p_{n-1}$ in such a way that the new vertex $p_{n-1}$ is an endpoint of $T \backslash \{p_n\}$ adjacent to $p_{n-2}$. Repeat the procedure for the rows and columns with numbers $n-1$ and $n-2$. After $n-1$ steps, we have the determinant

$$
\begin{vmatrix}
0 & 1 & 1 & \cdots & 1 \\
1 & -2 & 0 & \cdots & 0 \\
1 & 0 & -2 & \cdots & 0 \\
\vdots & \vdots & \vdots & \ddots & \vdots \\
1 & 0 & 0 & \cdots & -2
\end{vmatrix}.
$$

From this we find the remarkable result that the determinant $D_n$ of the distance matrix of a tree on $n$ vertices satisfies

$$
D_n = (-1)^{n-1}(n-1)2^{n-2},
$$

i.e. it depends only on $|V(T)|$. If we number the vertices according to the procedure described above, then the $k$ by $k$ principal minor in the upper lefthand corner of the distance matrix is the distance matrix of a subtree on $k$ vertices. Therefore the sequence $1, D_1, D_2, \ldots, D_n$, where $D_k$ is the determinant of the $k$ by $k$ minor, is equal to

$$
1, 0, -1, 4, -12, \ldots, (-1)^{n-1}(n-1)2^{n-2}.
$$

If we consider the sign of 0 to be positive, then this sequence has only one occurrence of two consecutive terms of the same sign. By an elementary theorem on quadratic forms this implies that the corresponding quadratic form has index 1, and hence $(d_{ij})$ has one positive eigenvalue; see B. W. Jones (1950), Theorem 4. Now the result follows from Theorem 9.1. □

The conjecture that in fact $N(G) \leq |V(G)| - 1$ for all (connected) graphs $G$ was proved by P. Winkler in 1983. The proof is constructive. In order to describe the addressing, we need some preparation. Consider the graph of Fig. 9.2.

 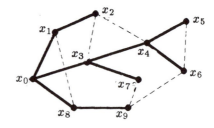

Figure 9.2

We pick a vertex $x_0$, then construct a spanning tree $T$ by a breadth-first search, and then number the vertices by a depth-first search. The result is shown on the righthand side of Fig. 9.2, where edges of $E(G)\backslash E(T)$ are dashed.

Let $n := |V(G)| - 1$. We need several definitions.

For $i \leq n$, we define

$$P(i) := \{j : x_j \text{ is on a path from } x_0 \text{ to } x_i \text{ in } T\}.$$

For example, $P(6) = \{0, 3, 4, 6\}$. Let

$$i\triangle j := \max(P(i) \cap P(j)).$$

We describe the general situation in Fig. 9.3.

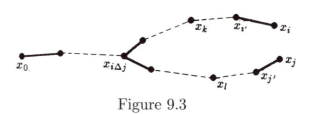

Figure 9.3

Note that in Fig. 9.3, we have $i < j$ if and only if $k < l$.

For $i \leq n$, we define

$$i' := \max(P(i)\backslash\{i\}).$$

For example, $7' = 3$ in Fig. 9.2. Define

$$i \sim j \Leftrightarrow P(i) \subseteq P(j) \text{ or } P(j) \subseteq P(i).$$

We denote distances in $G$, respectively $T$, by $d_G$, respectively $d_T$. The *discrepancy function* $c(i, j)$ is now defined by

$$c(i, j) := d_T(x_i, x_j) - d_G(x_i, x_j).$$

For example, in Fig. 9.2, $c(6, 9) = 4$.

LEMMA 9.4.
   (i) $c(i, j) = c(j, i) \geq 0$ ;
   (ii) if $i \sim j$, then $c(i, j) = 0$ ;
   (iii) if $i \nsim j$, then $c(i, j') \leq c(i, j) \leq c(i, j') + 2$.

PROOF: (i) is trivial; (ii) follows from the definition of $T$ since

$$d_G(x_i, x_j) \geq |d_G(x_j, x_0) - d_G(x_i, x_0)| = d_T(x_i, x_j);$$

(iii) follows from $|d_G(x_i, x_j) - d_G(x_i, x_{j'})| \leq 1$ and the fact that $d_T(x_i, x_j) = 1 + d_T(x_i, x_{j'})$.    □

Now we can define the addressing. For $0 \leq i \leq n$ the vertex $x_i$ is given the address $\mathbf{a}_i \in \{0, 1, *\}^n$, where

$$\mathbf{a}_i = (a_i(1), a_i(2), \ldots, a_i(n))$$

and

$$a_i(j) := \begin{cases} 1 & \text{if } j \in P(i), \\ * & \text{if} \begin{cases} c(i, j) - c(i, j') = 2, \text{ or} \\ c(i, j) - c(i, j') = 1, \ i < j, \ c(i, j) \text{ even, or} \\ c(i, j) - c(i, j') = 1, \ i > j, \ c(i, j) \text{ odd,} \end{cases} \\ 0 & \text{otherwise.} \end{cases}$$

THEOREM 9.5. $d(\mathbf{a}_i, \mathbf{a}_k) = d_G(x_i, x_k)$.

PROOF: We may assume $i < k$.

   (i) Suppose $i \sim k$. Then $d_G(x_i, x_k) = |P(k) \backslash P(i)|$. The values of $j$ such that $j \in P(k) \backslash P(i)$ are exactly the positions where $a_k(j) =$

1, $a_i(j) \neq 1$. For these values of $j$ we see that $c(i,j) = 0$, hence $a_i(j) = 0$ and we are done.

(ii) The hard case is when $i \not\sim k$. The key observation is the following. Let $n_1 \leq n_2 \leq \cdots \leq n_l$ be a nondecreasing sequence of integers such that $|n_{i+1} - n_i| \leq 2$ for all $i$. If $m$ is an even integer between $n_1$ and $n_l$ that does *not* occur in the sequence, then there is an $i$ such that $n_i = m - 1$, $n_{i+1} = m + 1$. Now consider the sequence

$$c(i,k) \geq c(i,k') \geq c(i,k'') \geq \cdots \geq c(i,i\triangle k) = 0.$$

By the definition of $a_i(j)$ and the observation above, $a_i(j) = *$ and $a_k(j) = 1$ exactly as many times as there are even integers between $c(i,i\triangle k)$ and $c(i,k)$. Similarly $a_k(j) = *$ and $a_i(j) = 1$ as many times as there are odd integers between $c(i,i\triangle k)$ and $c(i,k)$. So

$$d(\mathbf{a}_i, \mathbf{a}_k) = |P(k) \backslash P(i)| + |P(i) \backslash P(k)| - c(i,k)$$
$$= d_T(x_i, x_k) - c(i,k) = d_G(x_i, x_k).$$

$\square$

Therefore we have proved the following theorem.

THEOREM 9.6.  $N(G) \leq |V(G)| - 1$.

PROBLEM 9A. If we use the scheme defined above, what are the addresses of $x_2$ and $x_6$ in Fig. 9.2 ?

PROBLEM 9B. Let $G$ be a cycle (polygon) on $2n$ vertices. Determine $N(G)$.

PROBLEM 9C. Let $G$ be a cycle (polygon) on $2n+1$ vertices. Prove that $N(G) = 2n$. Hint: if $C_k$ is the permutation matrix with entries $c_{ij} = 1$ if and only if $j - i \equiv 1 \pmod k$ and $\zeta^k = 1$, then $(1, \zeta, \zeta^2, \ldots, \zeta^{k-1})$ is an eigenvector of $C_k$.

**Notes.**

The problem considered in this chapter was introduced by J. R. Pierce at Bell Laboratories as the *loop switching problem*. Several people (including one of the present authors) tried in vain to solve it. Shortly after R. L. Graham raised the reward for the solution to $200, it was solved by P. Winkler. It is worth noting that Winkler stated that the idea of numbering the vertices as was done in the proof was a regular habit due to his background in computer science. Going over the proof, one sees that this numbering indeed played a crucial role.

**References.**

R. L. Graham and H. O. Pollak (1971), On the addressing problem for loop switching, *Bell System Tech. J.* **50**, 2495–2519.

B. W. Jones (1950), *The Theory of Quadratic Forms*, Carus Math. Monogr. **10**, Math. Assoc. of America.

P. Winkler (1983), Proof of the squashed cube conjecture, *Combinatorica* **3**, 135–139.

# 10

## The principle of inclusion and exclusion; inversion formulae

As we have seen in several of the previous chapters, many problems of combinatorial analysis involve the *counting* of certain objects. We now treat one of the most useful methods for counting. It is known as the *principle of inclusion and exclusion*. The idea is as follows. If $A$ and $B$ are subsets of $S$ and we wish to count the elements of $S \backslash \{A \cup B\}$, then the answer is not $|S| - |A| - |B|$ because the elements of $A \cap B$ have been subtracted twice. However $|S| - |A| - |B| + |A \cap B|$ is correct. The following theorem generalizes this idea.

THEOREM 10.1. *Let $S$ be an $N$-set; $E_1, \ldots, E_r$ not necessarily distinct subsets of $S$. For any subset $M$ of $\{1, \ldots, r\}$, we define $N(M)$ to be the number of elements of $S$ in $\bigcap_{i \in M} E_i$ and for $0 \leq j \leq r$, we define $N_j := \sum_{|M|=j} N(M)$. Then the number of elements of $S$ not in any of the subsets $E_i$, $1 \leq i \leq r$, is*

$$(10.1) \qquad N - N_1 + N_2 - N_3 + \cdots + (-1)^r N_r.$$

PROOF: (i) If $x \in S$ and $x$ is in none of the $E_i$, then $x$ contributes 1 to the expression (10.1).

(ii) If $x \in S$ and $x$ is in exactly $k$ of the sets $E_i$, then the contribution to (10.1) equals

$$1 - \binom{k}{1} + \binom{k}{2} - \cdots + (-1)^k \binom{k}{k} = (1-1)^k = 0.$$

$\square$

REMARK. If we truncate the sum in (10.1) after a positive (respectively, negative) term, then we have an upper (respectively, lower) bound for the number of elements of $S$ not in any of the $E_i$.

Because this method is of great importance, we shall give several examples as illustration.

EXAMPLE 10.1. Let $d_n$ denote the number of permutations $\pi$ of $1, 2, \ldots, n$ such that $\pi(i) \neq i$ for all $i$ (these are called *derangements*). Let $S := S_n$, and let $E_i$ be the subset of those permutations $\pi$ with $\pi(i) = i$. By (10.1) we find

$$(10.2) \qquad d_n = \sum_{i=0}^{n} (-1)^i \binom{n}{i} (n-i)! = n! \sum_{i=0}^{n} \frac{(-1)^i}{i!}.$$

From this formula, we see that for large values of $n$ the probability that a permutation is a derangement is nearly $e^{-1}$. From (10.2) for $n$ and $n-1$, we find a recursion formula for $d_n$:

$$(10.3) \qquad d_n = n d_{n-1} + (-1)^n.$$

The formula (10.2) can also be obtained by inversion as follows. Consider the power series $D(x) := \sum_{n=0}^{\infty} d_n \frac{x^n}{n!}$ ($d_0 = 1$). Now if $F(x) := e^x D(x)$, then

$$F(x) = \sum_{m=0}^{\infty} \left( \sum_{r=0}^{m} \binom{m}{r} d_{m-r} \right) \frac{x^m}{m!}$$

and since $\sum_{r=0}^{m} \binom{m}{r} d_{m-r} = m!$, we find $F(x) = (1-x)^{-1}$. It follows that $D(x) = e^{-x}(1-x)^{-1}$ and by multiplying the power series for the two factors, we find (10.2) again.

EXAMPLE 10.2. Let $X$ be an $n$-set, $Y = \{y_1, \ldots, y_k\}$ a $k$-set. We count the surjections of $X$ to $Y$. Let $S$ be the set of all mappings from $X$ to $Y$, $E_i$ the subset of mappings for which $y_i$ is not in the image of $X$. By (10.1) we find the number of surjections to be $\sum_{i=0}^{k} (-1)^i \binom{k}{i} (k-i)^n$. Now this number is trivially 0 if $k > n$ and clearly $n!$ if $k = n$. So we have proved:

$$(10.4) \qquad \sum_{i=0}^{k} (-1)^i \binom{k}{i} (k-i)^n = \begin{cases} n! & \text{if } k = n, \\ 0 & \text{if } k > n. \end{cases}$$

There are many formulae like (10.4) that are often quite hard to prove directly. The occurrence of $(-1)^i$ is usually a sign that counting the right kind of objects using the principle of inclusion and exclusion can produce the formula, as in this example. Nevertheless it is useful in this case to see another proof.

Let $P(x)$ be a polynomial of degree $n$, with highest coefficient $a_n$. We denote the sequence of values $P(0), P(1), \ldots$ by $\mathbf{P}$. We now consider the sequence of differences $P(1) - P(0), P(2) - P(1), \ldots$. This is $\mathbf{Q}_1$, where $Q_1(x) := P(x+1) - P(x)$, a polynomial of degree $n - 1$ with highest coefficient $na_n$. By repeating this procedure a number of times, we find a sequence $\mathbf{Q}_k$ whose terms are $\sum_{i=0}^{k}(-1)^i \binom{k}{i} P(x+k-i)$, corresponding to the polynomial $Q_k(x)$ of degree $n - k$ with highest coefficient $n(n-1) \ldots (n-k+1)a_n$. If $k = n$, then all the terms of $\mathbf{Q}_k$ are $n!a_n$ and if $k > n$, then they are all 0. Take $P(x) = x^n$. We again find (10.4).

EXAMPLE 10.3. The following identity is a well known relation between binomial coefficients:

$$(10.5) \qquad \sum_{i=0}^{n}(-1)^i \binom{n}{i}\binom{m+n-i}{k-i} = \begin{cases} \binom{m}{k} & \text{if } m \geq k, \\ 0 & \text{if } m < k. \end{cases}$$

We see that if we wish to prove this using inclusion-exclusion, then the sets $E_i$ that we wish to exclude involve choosing from an $n$-set, and after choosing $i$ of them, we must choose $k - i$ elements from some set of size $m + n - i$. This shows us that the following combinatorial problem will lead us to the result (10.5). Consider a set $Z = X \cup Y$, where $X = \{x_1, \ldots, x_n\}$ is an $n$-set of blue points and $Y$ is an $m$-set of red points. How many $k$-subsets consist of red points only? The answer is trivially the righthand side of (10.5). If we take $S$ to be all the $k$-subsets of $Z$ and $E_i$ those $k$-subsets that contain $x_i$, then (10.1) gives us the lefthand side of (10.5).

Again we can ask whether this result can be proved directly. The answer is yes. To do this, we use the following expansion:

$$(10.6) \qquad \sum_{j=0}^{\infty}\binom{a+j}{j}x^j = (1-x)^{-a-1}.$$

Note that $(-1)^i \binom{n}{i}$ is the coefficient of $x^i$ in the expansion of $(1 - x)^n$. From (10.6) we find that $\binom{m+n-i}{k-i}$ is the coefficient of $x^{k-i}$ in the expansion of $(1 - x)^{k-m-n-1}$. So the lefthand side of (10.5) is the coefficient of $x^k$ in the expansion of $(1 - x)^{k-m-1}$. If $m \leq k - 1$, this is obviously 0 and if $m \geq k$, it is $\binom{m}{k}$, again by (10.6).

EXAMPLE 10.4. (The *Euler function*) Let $n = p_1^{a_1} p_2^{a_2} \ldots p_r^{a_r}$ be a positive integer. We denote by $\phi(n)$ the number of integers $k$ with $1 \leq k \leq n$ such that the g.c.d. $(n, k) = 1$. We apply Theorem 10.1 with $S := \{1, 2, \ldots, n\}$ and $E_i$ the set of integers divisible by $p_i$, $1 \leq i \leq r$. Then (10.1) yields

$$(10.7) \quad \phi(n) = n - \sum_{i=1}^{r} \frac{n}{p_i} + \sum_{1 \leq i < j \leq r} \frac{n}{p_i p_j} - \cdots = n \prod_{i=1}^{r} (1 - \frac{1}{p_i}).$$

The next theorem is used quite often.

THEOREM 10.2. $\sum_{d|n} \phi(d) = n$.

PROOF: Consider $\{1, 2, \ldots, n\} = N$. For each $m \in N$, we have $(m, n)|n$. The number of integers $m$ with $(m, n) = d$, i.e. $m = m_1 d$, $n = n_1 d$ and $(m_1, n_1) = 1$ clearly equals $\phi(n_1) = \phi(n/d)$. So $n = \sum_{d|n} \phi(n/d)$ which is equivalent to the assertion. $\square$

At this point, it is useful to introduce the so-called *Möbius function*:

$$(10.8) \quad \mu(d) :=$$
$$\begin{cases} 1 & \text{if } d = \text{product of an even number of distinct primes,} \\ -1 & \text{if } d = \text{product of an odd number of distinct primes,} \\ 0 & \text{otherwise, i.e. } d \text{ not squarefree.} \end{cases}$$

THEOREM 10.3.

$$\sum_{d|n} \mu(d) = \begin{cases} 1 & \text{if } n = 1, \\ 0 & \text{otherwise.} \end{cases}$$

PROOF: If $n = 1$, there is nothing to prove. If $n = p_1^{a_1} \ldots p_r^{a_r}$, then by (10.8) we have

$$\sum_{d|n} \mu(d) = \sum_{i=0}^{r} \binom{r}{i}(-1)^i = (1-1)^r = 0.$$

$\square$

Note how similar the proofs of Theorems 10.1 and 10.3 are. Using the Möbius function, we can reformulate (10.7) as follows:

$$(10.9) \qquad \frac{\phi(n)}{n} = \sum_{d|n} \frac{\mu(d)}{d}.$$

PROBLEM 10A. How many positive integers less than 1000 have no factor between 1 and 10?

PROBLEM 10B. How many monic polynomials of degree $n$ are there in $\mathbb{F}_p[x]$ that do not take on the value 0 for $x \in \mathbb{F}_p$?

PROBLEM 10C. Determine $\sum_{n \le x} \mu(n) \lfloor \frac{x}{n} \rfloor$.

PROBLEM 10D. One of the most famous functions in complex analysis is the so-called *Riemann $\zeta$-function* $\zeta(s) := \sum_{n=1}^{\infty} n^{-s}$, defined in the complex plane for $\text{Re}(s) > 1$. Prove that $1/\zeta(s) = \sum_{n=1}^{\infty} \mu(n) n^{-s}$.

PROBLEM 10E. Let $f_n(z)$ be the function that has as its zeros all numbers $\eta$ for which $\eta^n = 1$ but $\eta^k \ne 1$ for $1 \le k < n$. Prove that

$$f_n(z) = \prod_{k|n} (z^k - 1)^{\mu(n/k)}.$$

Theorem 10.3 makes it possible to derive a very useful inversion formula known as the *Möbius inversion formula.*

THEOREM 10.4. *Let $f(n)$ and $g(n)$ be functions defined for every positive integer $n$ satisfying*

$$(10.10) \qquad f(n) = \sum_{d|n} g(d).$$

*Then g satisfies*

$$(10.11) \qquad g(n) = \sum_{d|n} \mu(d) f\Big(\frac{n}{d}\Big).$$

PROOF: By (10.10) we have

$$\sum_{d|n} \mu(d) f\Big(\frac{n}{d}\Big) = \sum_{d|n} \mu\Big(\frac{n}{d}\Big) f(d)$$

$$= \sum_{d|n} \mu\Big(\frac{n}{d}\Big) \sum_{d'|d} g(d') = \sum_{d'|n} g(d') \sum_{m|(n/d')} \mu(m).$$

By Theorem 10.3 the inner sum on the righthand side is 0 unless $d' = n$. $\qquad\square$

REMARK. The equation (10.11) also implies (10.10).

EXAMPLE 10.5. We shall count the number $N_n$ of circular sequences of 0's and 1's, where two sequences obtained by a rotation are considered the same. Let $M(d)$ be the number of circular sequences of length $d$ that are not periodic. Then $N_n = \sum_{d|n} M(d)$. We observe that $\sum_{d|n} dM(d) = 2^n$ since this counts all possible circular sequences. By Theorem 10.4 we find from this equation that $nM(n) = \sum_{d|n} \mu(d) 2^{n/d}$ and therefore

$$(10.12) \quad N_n = \sum_{d|n} M(d) = \sum_{d|n} \frac{1}{d} \sum_{l|d} \mu\Big(\frac{d}{l}\Big) 2^l$$

$$= \sum_{l|n} \frac{2^l}{l} \sum_{k|\frac{n}{l}} \frac{\mu(k)}{k} = \frac{1}{n} \sum_{l|n} \phi\Big(\frac{n}{l}\Big) 2^l.$$

The final expression has the advantage that all the terms are positive. This raises the question whether we could have obtained that expression by some other counting technique. We shall see that the following theorem, known as *Burnside's lemma* (although the theorem is actually due to Cauchy and Frobenius; see the notes), provides the answer.

THEOREM 10.5. *Let $G$ be a permutation group acting on a set $X$. For $g \in G$ let $\psi(g)$ denote the number of points of $X$ fixed by $g$. Then the number of orbits of $G$ is equal to $\frac{1}{|G|} \sum_{g \in G} \psi(g)$.*

PROOF: Count pairs $(g, x)$, where $g \in G$, $x \in X$, $x^g = x$. Starting with $g$, we find $\sum_{g \in G} \psi(g)$. For each $x \in X$ there are $|G|/|O_x|$ such pairs, where $O_x$ is the orbit of $x$. So the total number equals $|G| \sum_{x \in X} 1/|O_x|$. The orbits of $G$ partition $X$, and if we sum the terms $1/|O_x|$ over all $x$ in a particular orbit, we obtain 1. Thus $\sum_{x \in X} 1/|O_x|$ is the number of orbits.        □

EXAMPLE 10.5. (continued) Let $G$ be the cyclic group of order $n$, i.e. the group of rotations of a circular sequence of 0's and 1's. If $d|n$ there are $\phi(n/d)$ integers $g$ such that $(n, g) = d$ and for each such $g$ there are $2^d$ circular sequences that are fixed by the rotation over $g$ positions. So Theorem 10.5 immediately yields the result (10.12).

EXAMPLE 10.6. The following problem, introduced by Lucas in 1891, is known as the *"problème des ménages"*. We wish to seat $n$ couples at a circular table so that men and women are in alternate places and no husband will sit on either side of his wife. In how many ways can this be done? We assume that the women have been seated at alternate places. Call the ladies 1 to $n$ and the corresponding men also 1 to $n$. The problem amounts to placing the integers 1 to $n$ on a circle with positions numbered 1 to $n$ such that for all $i$ the integer $i$ is not in position $i$ or position $i + 1$ (mod $n$). Let $E_i$ be the set of seatings in which husband $i$ is sitting next to his wife. We now wish to use inclusion-exclusion and we must therefore calculate in how many ways it is possible to seat $r$ husbands incorrectly. Call this number $A_r$. We do this as follows. Consider a circular sequence of $2n$ positions. Put a 1 in position $2i - 1$ if husband $i$ is sitting to the right of his wife; put a 1 in position $2i$ if he is sitting to the left of his wife. Put zeros in the remaining positions. The configurations that we wish to count therefore are circular sequences of $2n$ zeros and ones, with exactly $r$ ones, no two adjacent. Let $A'_r$ be the number of sequences starting with a 1 (followed by a 0). By considering 10 as one symbol, we see that we must choose $r - 1$ out of $2n - r - 1$ positions. To count the number $A_r''$ of sequences starting with a 0, we place the 0 at the end, and then it amounts to choosing $r$ out of $2n - r$ places. Hence

$$A_r = A_r' + A_r'' = \binom{2n - r - 1}{r - 1} + \binom{2n - r}{r} = \frac{2n}{2n - r}\binom{2n - r}{r}.$$

By (10.1) we find that the number of ways to seat the men is

(10.13) $$\sum_{r=0}^{n}(-1)^r(n - r)!\binom{2n - r}{r}\frac{2n}{2n - r}.$$

PROBLEM 10F. We color the integers 1 to $2n$ red or blue in such a way that if $i$ is red then $i - 1$ is not blue. Prove that

$$\sum_{k=0}^{n}(-1)^k\binom{2n - k}{k}2^{2n-2k} = 2n + 1.$$

Can you prove this directly?

PROBLEM 10G. Let $A_i := \{i-1, i, i+1\} \cap \{1, 2, \ldots, n\}, i = 1, \ldots, n$. Let $\theta_n$ denote the number of SDRs of the collection $\{A_1, \ldots, A_n\}$. Determine $\lim_{n\to\infty} \theta_n^{1/n}$.

**Notes.**

The principle of inclusion and exclusion occurred as early as 1854 in a paper by Da Silva and later in a paper by Sylvester in 1883. For this reason (10.1) and similar formulae are sometimes called the formula of Da Silva, respectively Sylvester. A better name that is also often used is "sieve formula". The formula is indeed an example of a principle that is used extensively in number theory, referred to as "sieve methods". An example that is probably familiar to most readers is the sieve of Eratosthenes: to find the primes $\leq n^2$, take the integers $\leq n^2$ and sieve out all the multiples of primes $\leq n$.

The derangements treated in Example 10.1 occur again in Example 14.1 and Example 14.10. The first occurrence of this question is in one of the early books on games of chance: *Essai d'analyse sur les jeux de hazard* by P. R. de Montmort (1678–1719). It is still often referred to by the name that he gave it: "problème des rencontres". Formula (10.2) is sometimes stated as follows. If $n$ persons check their umbrellas (a typical Dutch example; it's always raining in Holland) and subsequently pick one at random in the dark after a power failure, then the probability that nobody gets his own umbrella is roughly $e^{-1}$ (if $n$ is large).

The second proof in Example 10.2 is an example of the use of "calculus of finite differences", used extensively in numerical analysis.

A. F. Möbius (1790–1868) was an astronomer (and before that an assistant to Gauss) who made important contributions to geometry and topology (e.g. the Möbius band).

G. F. B. Riemann (1826–1866) was professor in Göttingen, where he also obtained his doctorate under Gauss. He is famous for many of his ideas, which include the Riemann integral, Riemann surfaces and manifolds, and of course the so-called *Riemann hypothesis* on the location of the zeros of the $\zeta$-function. One wonders what he would have left us if he had not died so young.

In most books in which it occurs, Theorem 10.5 is called Burnside's lemma. This is just one of many examples of theorems, etc. attributed to the wrong person. For a history of this misnomer, we refer to Neumann (1979).

F. E. A. Lucas (1842–1891) was a French number theorist. He is known for his books on number theory and mathematical recreations. The former book contained the problem of Example 10.6. The *Fibonacci numbers* were given this name by Lucas.

**References.**

F. E. A. Lucas (1891), *Théorie des nombres*, Gauthier-Villars, Paris.

P. M. Neumann (1979), A lemma that is not Burnside's, *Math. Scientist*, **4**, 133–141.

# 11
## Permanents

Before introducing the main topic of this chapter, we present a generalization of Theorem 10.1. As in Theorem 10.1, let $S$ be an $n$-set, $E_1, \ldots, E_r$ (not necessarily distinct) subsets of $S$. Let $\mathbb{F}$ be any field. To each element $a \in S$, we assign a *weight* $w(a)$ in $\mathbb{F}$. For any subset $M$ of $\{1, 2, \ldots, r\}$, we define $W(M)$ to be the sum of the weights of the elements of $S$ in $\bigcap_{i \in M} E_i$. For $0 \leq j \leq r$, we define $W_j := \sum_{|M|=j} W(M)$ (so $W_0 = \sum_{a \in S} w(a)$).

THEOREM 11.1. *If $E(m)$ denotes the sum of the weights of the elements of $S$ that are contained in exactly $m$ of the subsets $E_i$, $1 \leq i \leq r$, then*

$$(11.1) \qquad E(m) = \sum_{i=0}^{r-m} (-1)^i \binom{m+i}{i} W_{m+i}.$$

PROOF: The proof is nearly the same as for Theorem 10.1. If $x \in S$ and $x$ is contained in exactly $m$ of the subsets $E_i$, then the contribution of $x$ to the sum in (11.1) is $w(x)$. If $x \in S$ and $x$ is contained in exactly $m + k$ of the subsets $E_i$, then the contribution to the sum equals

$$w(x) \sum_{i=0}^{k} (-1)^i \binom{m+i}{i} \binom{m+k}{m+i}$$

$$= w(x) \binom{m+k}{k} \sum_{i=0}^{k} (-1)^i \binom{k}{i} = 0.$$

$\square$

We now give the definition of a *permanent*. Let $A = (\mathbf{a}_1, \ldots, \mathbf{a}_n)$ be an $n$ by $n$ matrix with columns $\mathbf{a}_j = (a_{1j}, \ldots, a_{nj})^\top$. Then per $A$, the permanent of $A$, is defined by

$$(11.2) \qquad \operatorname{per} A := \sum_{\pi \in S_n} a_{1\pi(1)} \cdots a_{n\pi(n)}.$$

So the permanent is defined in the same way as the determinant but without the signs depending on whether the permutation $\pi$ is even or odd.

From the definition, the following properties of the permanent are obvious.

$$(11.3) \qquad \operatorname{per} A = \operatorname{per} A^\top;$$

(11.4)
   if $P$ and $Q$ are permutation matrices, then per $A = $ per $PAQ$;

$$(11.5) \qquad \operatorname{per} A \text{ is a linear function of } \mathbf{a}_j, \quad 1 \le j \le r.$$

Of course per $A$ is also a linear function of each of the rows of $A$. The permanent of $A$ is much more difficult to calculate than its determinant. However, it is clear from (11.2) that expansion by rows or columns is possible. So define $A_{ij}$ to be the matrix obtained from $A$ by deleting row $i$ and column $j$. Then

$$(11.6) \qquad \operatorname{per} A = \begin{cases} \sum_{i=1}^n a_{ij} \operatorname{per} A_{ij}, & 1 \le j \le r, \\ \sum_{j=1}^n a_{ij} \operatorname{per} A_{ij}, & 1 \le i \le r. \end{cases}$$

The following method of calculating a permanent (due to Ryser) is an application of Theorem 11.1.

THEOREM 11.2. *Let $A$ be an $n$ by $n$ matrix. If $A_r$ is obtained from $A$ by deleting $r$ columns, then $S(A_r)$ denotes the product of the rowsums of $A_r$. We define $\Sigma_r$ to be the sum of the values of $S(A_r)$ for all possible choices of $A_r$. Then*

$$(11.7) \qquad \operatorname{per} A = \sum_{r=0}^{n-1} (-1)^r \Sigma_r.$$

PROOF: Let $S$ be the set of all products $p = a_{1i_1} \ldots a_{ni_n}$ and define $w(p) := p$. Define $E_j$ to be the set of products $p$ for which $j \notin \{i_1, \ldots, i_n\}$. Then the permanent of $A$ is the sum of the weights of the elements of $S$ that are not in any of the subsets $E_j$. So (11.7) is an immediate consequence of (11.1). $\square$

PROBLEM 11A. Prove (10.4) using Theorem 11.2.

REMARK. If $A_1, \ldots, A_n$ are subsets of $\{1, \ldots, n\}$ and $a_{ij} = 1$ if $j \in A_i$, 0 otherwise, then per $A$ counts the number of SDRs of the sets $A_1, \ldots, A_n$.

EXAMPLE 11.1. We find another formula for the number of derangements of $1, 2, \ldots, n$. The permanent of the matrix $J - I$ of size $n$ is clearly $d_n$. From (11.7) we find

$$(11.8) \qquad d_n = \sum_{r=0}^{n-1} (-1)^r \binom{n}{r} (n-r)^r (n-r-1)^{n-r}.$$

By expanding the term $(n - 1 - r)^{n-r}$ and applying (10.4) after changing the order of summation, we find a complicated proof of (10.2).

In the last ten years or so, several well known conjectures on permanents of (0,1)-matrices have been proved, often by ingenious arguments. In fact, much of the research on permanents was motivated by these conjectures. Therefore we will devote attention to a number of these results in this and the next chapter. As an introduction, we consider (0,1)-matrices with two ones in each row and column.

THEOREM 11.3. *If $A$ is a $(0, 1)$-matrix in which all rowsums and columnsums are 2, then*

$$\text{per } A \leq 2^{\lfloor \frac{1}{2} n \rfloor}.$$

PROOF: Consider the graph $G$ whose vertices correspond to the rows of $A$, whose edges correspond to the columns of $A$, and where vertex $i$ and edge $j$ are incident exactly when $A(i, j) = 1$. This graph is regular of degree 2, and hence is the disjoint union of polygons. The submatrix corresponding to the vertices and edges of a polygon is, after reordering rows and columns if necessary, a circulant

$$\begin{pmatrix} 1\ 1\ 0\ 0\ \cdots\ 0\ 0 \\ 0\ 1\ 1\ 0\ \cdots\ 0\ 0 \\ 0\ 0\ 1\ 1\ \cdots\ 0\ 0 \\ \vdots\ \vdots\ \vdots\ \vdots\quad\ \vdots\ \vdots \\ 0\ 0\ 0\ 0\ \cdots\ 1\ 1 \\ 1\ 0\ 0\ 0\ \cdots\ 0\ 1 \end{pmatrix}.$$

(This may degenerate into a 2 by 2 matrix of 1's.) The matrix $A$ is the direct sum of such matrices, each of which has permanent 2. The number of factors is at most $\lfloor \frac{1}{2}n \rfloor$ and we see that equality holds in the theorem if $A$ is the direct sum of $\lfloor \frac{1}{2}n \rfloor$ matrices $J$ of size 2.                                                     □

This elementary theorem is concerned with the relationship between the rowsums of a matrix and the permanent of that matrix, and the same is true for many of the following theorems. This brings us to the first difficult question. It was conjectured by H. Minc in 1967 that if $A$ is a (0,1)-matrix with rowsums $r_1, \ldots, r_n$, then

(11.9)                     $$\text{per } A \le \prod_{j=1}^{n} (r_j!)^{1/r_j}.$$

Observe that Theorem 11.2 shows that equality can hold in (11.9) and in fact, we have equality if $A$ is the direct sum of matrices $J_m$. Several results that were weaker than (11.9) were proved, often by intricate and long arguments. The conjecture was finally proved in 1973 by L. M. Brégman. All the more surprising is the fact that A. Schrijver came up with an extremely elegant and very short proof of Minc's conjecture in 1977. The proof depends on the following lemma.

LEMMA 11.4. *If* $t_1, t_2, \ldots, t_r$ *are nonnegative real numbers, then*

$$\left( \frac{t_1 + \cdots + t_r}{r} \right)^{t_1 + \cdots + t_r} \le t_1^{t_1} \ldots t_r^{t_r}.$$

PROOF: Since $x \log x$ is a convex function, we have

$$\frac{t_1 + \cdots + t_r}{r} \log \left( \frac{t_1 + \cdots + t_r}{r} \right) \leq \frac{t_1 \log \, t_1 + \ldots t_r \log t_r}{r},$$

which proves the assertion. $\square$

In the following, we use (11.6) in the following form:

$$\text{per } A = \sum_{k, a_{ik}=1} \text{per } A_{ik}.$$

THEOREM 11.5. *Let $A$ be an $n$ by $n$ $(0, 1)$-matrix with $r_i$ ones in row $i$, $1 \leq i \leq n$. Then*

$$\text{per } A \leq \prod_{i=1}^{n} (r_i)!^{1/r_i}.$$

PROOF: The proof is by induction on $n$. For $n = 1$, the theorem is trivial. We assume that the theorem is true for matrices of size $n-1$. The idea is to estimate $(\text{per } A)^{n \, \text{per } A}$ and to split this expression into several products. Now note that $r_i$ is the number of values of $k$ for which $a_{ik} = 1$ and apply the lemma. We find:

$$(11.10) \quad (\text{per } A)^{n \, \text{per } A} = \prod_{i=1}^{n} (\text{per } A)^{\text{per } A}$$

$$\leq \prod_{i=1}^{n} \left( r_i^{\text{per } A} \prod_{k, a_{ik}=1} \text{per } A_{ik}^{\text{per } A_{ik}} \right).$$

Now, let $S$ denote the set of all permutations $\nu$ of $\{1, \ldots, n\}$ for which $a_{i\nu_i} = 1$ for $i = 1, \ldots, n$. So $|S| = \text{per } A$. Furthermore, the number of $\nu \in S$ such that $\nu_i = k$ is per $A_{ik}$ if $a_{ik} = 1$ and 0 otherwise. So the righthand side of (11.10) is equal to

$$(11.11) \qquad \prod_{\nu \in S} \left\{ \left( \prod_{i=1}^{n} r_i \right) \cdot \left( \prod_{i=1}^{n} \text{per } A_{i\nu_i} \right) \right\}.$$

We now apply the induction hypothesis to each $A_{i\nu_i}$. This yields

(11.12) $(\text{per } A)^{n \, \text{per} A} \le$

$$\prod_{\nu \in S} \left\{ \left( \prod_{i=1}^n r_i \right) \cdot \prod_{i=1}^n \left[ \prod_{\substack{j \ne i, \\ a_{j\nu_i}=0}} (r_j!)^{1/r_j} \prod_{\substack{j \ne i, \\ a_{j\nu_i}=1}} ((r_j-1)!)^{1/(r_j-1)} \right] \right\}.$$

Since the number of $i$ such that $i \ne j$ and $a_{j\nu_i} = 0$ is $n - r_j$, and the number of $i$ such that $i \ne j$ and $a_{j\nu_i} = 1$ is $r_j - 1$, we can replace the righthand side of (11.12) by

$$\prod_{\nu \in S} \left\{ \left( \prod_{i=1}^n r_i \right) \cdot \left[ \prod_{j=1}^n (r_j!)^{(n-r_j)/r_j} (r_j-1)! \right] \right\} =$$

$$\prod_{\nu \in S} \prod_{i=1}^n (r_i)!^{n/r_i} = \left( \prod_{i=1}^n (r_i!)^{1/r_i} \right)^{n \, \text{per } A}$$

and the assertion is proved. $\square$

We now shall consider a special class of (0,1)-matrices, namely the (0,1)-matrices that have exactly $k$ ones in each row and column. We denote this class by $\mathcal{A}(n,k)$. We define:

(11.13) $\qquad M(n,k) := \max\{\text{per } A : A \in \mathcal{A}(n,k)\},$

(11.14) $\qquad m(n,k) := \min\{\text{per } A : A \in \mathcal{A}(n,k)\}.$

By taking direct sums, we find the following inequalities:

(11.15) $\qquad M(n_1 + n_2, k) \ge M(n_1, k) M(n_2, k),$

(11.16) $\qquad m(n_1 + n_2, k) \le m(n_1, k) m(n_2, k).$

These two inequalities allow us to introduce two more functions using the following result, known as *Fekete's lemma*.

LEMMA 11.6. *Let $f : \mathbf{N} \to \mathbf{N}$ be a function for which $f(m+n) \geq f(m)f(n)$ for all $m, n \in \mathbf{N}$. Then $\lim_{n\to\infty} f(n)^{1/n}$ exists (possibly $\infty$).*

PROOF: Fix $m$ and fix $l$, $l \leq m$. By induction we find from the inequality for $f$ that $f(l+km) \geq f(l)[f(m)]^k$. Therefore

$$\liminf f(l+km)^{1/(l+km)} \geq f(m)^{1/m}$$

and since there are $m$ possible values for $l$, we in fact have

$$\liminf f(n)^{1/n} \geq f(m)^{1/m}.$$

Now let $m \to \infty$. We find that

$$\liminf f(n)^{1/n} \geq \limsup f(m)^{1/m},$$

but then these are equal. □

The assertion of the lemma is also true if in the inequality for $f$ we replace $\geq$ by $\leq$. By applying the lemma to (11.15) and (11.16), we can define:

$$(11.17) \qquad M(k) := \lim_{n\to\infty} \{M(n,k)\}^{1/n},$$

$$(11.18) \qquad m(k) := \lim_{n\to\infty} \{m(n,k)\}^{1/n}.$$

PROBLEM 11B. Prove that $M(n,k) \geq k!$. Prove that $M(k) \leq (k!)^{1/k}$. Show by example that $M(k) \geq (k!)^{1/k}$. This shows that $M(k) = (k!)^{1/k}$.

The function $m(n,k)$ is much more difficult to handle. What we should expect is based on a famous problem still referred to as the *Van der Waerden conjecture*, although in 1981 two different proofs of the conjecture appeared (after nearly 50 years of research on this question!). We formulate the conjecture below, and the proof will be given in the next chapter.

CONJECTURE. *If $A$ is an $n$ by $n$ matrix with nonnegative entries in which all rowsums and columnsums are 1, then*

$$(11.19) \qquad \text{per } A \geq n! \, n^{-n}.$$

The matrices considered in the conjecture are usually called *doubly stochastic matrices*. If $A \in \mathcal{A}(n, k)$ then dividing all the elements of $A$ by $k$ yields a doubly stochastic matrix. Therefore the conjecture (now a theorem) shows that $m(k) \geq k/e$. This is remarkable because the value of $M(k)$ given in Problem 11B tends to $k/e$ for $k \to \infty$ (see the notes). This means that for large $n$, and $A$ an arbitrary element of $\mathcal{A}(n, k)$, the value of $(\text{per } A)^{1/n}$ is nearly $k/e$. For a long time the best lower bound for $m(n, 3)$ was $n + 3$ and even that was not easy to prove. Once again, the next improvement was both considerable and elementary. We now give the proof of that result, due to Voorhoeve (1979).

THEOREM 11.7. $m(n, 3) \geq 6 \cdot (\frac{4}{3})^{n-3}$.

PROOF: Let $U_n$ denote the set of $n$ by $n$ matrices with nonnegative integers as entries and all rowsums and columnsums 3; $u(n) := \min\{\text{per } A : A \in U_n\}$. Denote by $V_n$ the set of all matrices obtained from elements of $U_n$ by decreasing one positive entry by 1; $v(n) := \min\{\text{per } A : A \in V_n\}$. We first show that

$$(11.20) \qquad u(n) \geq \left\lceil \frac{3}{2} v(n) \right\rceil.$$

Let $A$ be an element of $U_n$ with first row $\mathbf{a} = (\alpha_1, \alpha_2, \alpha_3, 0, \ldots, 0)$, where $\alpha_i \geq 0$ for $i = 1, 2, 3$. Since

$$2\mathbf{a} = \alpha_1(\alpha_1 - 1, \alpha_2, \alpha_3, 0, \ldots, 0) + \alpha_2(\alpha_1, \alpha_2 - 1, \alpha_3, 0, \ldots, 0)$$
$$+ \alpha_3(\alpha_1, \alpha_2, \alpha_3 - 1, 0, \ldots, 0),$$

we find from (11.5) that $2u(n) \geq (\alpha_1 + \alpha_2 + \alpha_3)v(n) = 3v(n)$, proving the assertion.

Next, we show that

$$(11.21) \qquad v(n) \geq \left\lceil \frac{4}{3} v(n - 1) \right\rceil.$$

We must distinguish between two cases. In the first one, $A$ is an element of $V_n$ with first row $(1, 1, 0, \ldots, 0)$ and the matrix obtained from $A$ by deleting the first row has the form $(\mathbf{c}_1, \mathbf{c}_2, B)$. The columnsum of $\mathbf{c}_3 := \mathbf{c}_1 + \mathbf{c}_2$ is either 3 or 4. By (11.6), we have

$$\text{per } A = \text{per } (\mathbf{c}_1, B) + \text{per } (\mathbf{c}_2, B) = \text{per } (\mathbf{c}_3, B).$$

If the columnsum of $\mathbf{c}_3$ is 3, then the matrix $(\mathbf{c}_3, B)$ is in $U_{n-1}$ and we are done by (11.20). If the sum is 4, then we use the same trick as above: write $3\mathbf{c}_3$ as a linear combination of four vectors $\mathbf{d}_i$ such that each matrix $(\mathbf{d}_i, B)$ is in $V_{n-1}$ and we find that $3 \text{ per } A \geq 4v(n-1)$. The second case that we have to consider is that $A$ has $(2, 0, \ldots, 0)$ as first row. If we delete the first row and column of $A$, then there are again two possibilities. We obtain a matrix $B$ that is either in $U_{n-1}$ or in $V_{n-1}$. So we have per $A \geq 2 \min\{u(n-1), v(n-1)\}$ and we are done by (11.20). By combining (11.20) and (11.21) with the trivial value $v(1) = 2$, the assertion of the theorem follows. $\qquad \square$

We now consider a larger class of $n$ by $n$ matrices, namely those with nonnegative integers as entries and all rowsums and columnsums equal to $k$. We denote this class by $\Lambda(n, k)$ and the minimal permanent within the class by $\lambda(n, k)$. Again we have $\lambda(m+n, k) \leq \lambda(m, k)\lambda(n, k)$ and by Fekete's lemma, we can define

$$(11.22) \qquad \theta(k) := \lim_{n \to \infty} (\lambda(n, k))^{1/n}.$$

From Theorem 11.3 and Theorem 11.7, we know that $\lambda(n, 2) = 2$ and $\lambda(n, 3) \geq 6 \cdot \left(\frac{4}{3}\right)^{n-3}$. From (11.19), we have seen above that $\lambda(n, k) \geq n!\left(\frac{k}{n}\right)^n$. We have also seen that there is a connection between permanents and SDRs. We now show a proof in which this connection is exploited.

THEOREM 11.8. $\lambda(n, k) \leq k^{2n}/\binom{nk}{n}$.

PROOF: We denote by $P_{n,k}$ the collection of all ordered partitions of the set $\{1, 2, \ldots, nk\}$ into classes of size $k$. We have

$$(11.23) \qquad p_{n,k} := |P_{n,k}| = \frac{(nk)!}{(k!)^n}.$$

Now let $\mathcal{A} := (A_1, \ldots, A_n)$ be such a partition. The number of SDRs of the subsets $A_1, \ldots, A_n$ is $k^n$. Consider a second partition $\mathcal{B} := (B_1, \ldots, B_n)$. We denote by $s(\mathcal{A}, \mathcal{B})$ the number of common SDRs of $\mathcal{A}$ and $\mathcal{B}$. We define an $n$ by $n$ matrix $A$ with entries $\alpha_{ij}$ by $\alpha_{ij} := |A_i \cap B_j|$. The point of the proof is the fact that per $A$ counts the number of common SDRs of $\mathcal{A}$ and $\mathcal{B}$. Furthermore, by definition of the partitions, the matrix $A$ is in $\Lambda(n, k)$. Therefore

$$s(\mathcal{A}, \mathcal{B}) = \text{per } A \geq \lambda(n, k).$$

If $\mathcal{A} \in P_{n,k}$ is given and some SDR of $\mathcal{A}$ is given, then there are $n! p_{n,k-1}$ ordered partitions $\mathcal{B}$ that have this same SDR. Hence we have

$$\sum_{\mathcal{B} \in P_{n,k}} s(\mathcal{A}, \mathcal{B}) = k^n \cdot n! \, p_{n,k-1}.$$

Combining this with (11.23) and the inequality for $\lambda(n, k)$, we find

$$\lambda(n, k) \leq \frac{k^n \cdot n! p_{n,k-1}}{p_{n,k}} = \frac{k^{2n}}{\binom{nk}{n}}.$$

$\square$

This proof is due to Schrijver and Valiant (1980) who also gave the following corollary.

COROLLARY. $\theta(k) \leq \frac{(k-1)^{k-1}}{k^{k-2}}$.

PROOF: This follows in the usual way from the previous theorem by using Stirling's formula: $n! \sim n^n e^{-n} (2\pi n)^{1/2}$. $\square$

The corollary combined with Theorem 11.7 gives us one more value of $\theta(k)$, namely $\theta(3) = \frac{4}{3}$.

PROBLEM 11C. Consider the set of integers $1, 2, \ldots, 64$. We first remove the integers $\equiv 1 \pmod 9$, i.e. $x_1 = 1, \ldots, x_8 = 64$. Then we remove the integers $x_i + 8$, where 72 is to be interpreted as 8. This leaves us with a set $S$ of 48 elements. We partition $S$ into subsets $A_1, \ldots, A_8$ and also into subsets $B_1, \ldots, B_8$, where $A_i$ contains integers in the interval $(8(i-1), 8i]$ and $B_i$ contains the integers $\equiv i$ (mod 8). How many common SDRs are there for the systems $A_1, \ldots, A_8$ and $B_1, \ldots, B_8$?

## Notes.

In his book *Permanents*, H. Minc (1978) mentions that the name permanent is essentially due to Cauchy (1812) although the word as such was first used by Muir in 1882. Nevertheless, a referee of one of Minc's earlier papers admonished him for inventing this ludicrous name! For an extensive treatment of permanents, we refer to Minc's book. There, one can find much of the theory that was developed mainly to solve the Van der Waerden conjecture (without success at the time of writing of the book).

Theorem 11.2 is from Ryser (1963).

For a number of results related to the Minc conjecture, we refer to Van Lint (1974).

The lemma known as *Fekete's lemma* occurs in Fekete (1923). For another application, we refer to J. W. Moon (1968).

The term *doubly stochastic matrix* can be motivated by considering the entries to be conditional probabilities. However, permanents do not seem to play a role of importance in probability theory.

The remarks concerning $m(k)$ and $M(k)$ preceding Theorem 11.7 are based on Stirling's formula and the related inequality $n! \geq n^n e^{-n}$. This inequality is easily proved by induction, using the fact that $(1+n^{-1})^n$ is increasing with limit $e$. Actually Stirling's formula was first given by de Moivre. Stirling derived an asymptotic series for the gamma function which leads to the estimate

$$\Gamma(x) = x^{x-\frac{1}{2}} e^{-x} (2\pi)^{\frac{1}{2}} e^{\theta/(12x)},$$

where $0 < \theta < 1$. $(n! = \Gamma(n+1).)$

It seems to be impossible to generalize the "divide and conquer" method used by Voorhoeve to prove Theorem 11.7. The value of $\theta(k)$ is not known for any $k > 3$.

## References.

L. M. Brégman (1973), Certain properties of nonnegative matrices and their permanents, *Dokl. Akad. Nauk SSSR* **211**, 27–30 (*Soviet Math. Dokl.* **14**, 945–949).

M. Fekete (1923), Über die Verteilung der Wurzeln bei gewissen algebraischen Gleichungen mit ganzzahligen Koeffizienten, *Math. Zeitschr.* **17**, 228–249.

J. H. van Lint (1974), *Combinatorial Theory Seminar Eindhoven University of Technology*, Lecture Notes in Mathematics **382**, Springer-Verlag.

H. Minc (1967), An inequality for permanents of (0,1) matrices, *J. Combinatorial Theory* **2**, 321–326.

H. Minc (1978), *Permanents*, Encyclopedia of Mathematics and its Applications, vol. 6, Addison-Wesley, reissued by Cambridge University Press.

J. W. Moon (1968), *Topics on Tournaments*, Holt, Rinehart and Winston.

H. J. Ryser (1963), *Combinatorial Mathematics*, Carus Math. Monograph **14**.

A. Schrijver (1978), A short proof of Minc's conjecture, *J. Combinatorial Theory* (A) **25**, 80–83.

A. Schrijver and W. G. Valiant (1980), On lower bounds for permanents, *Proc. Kon. Ned. Akad. v. Wetensch.* A **83**, 425–427.

M. Voorhoeve (1979), A lower bound for the permanents of certain (0,1)-matrices, *Proc. Kon. Ned. Akad. v. Wetensch.* A **82**, 83–86.

# 12
## The Van der Waerden conjecture

In this chapter, we denote the set of all doubly stochastic matrices of size $n$ by $\Omega_n$. The subset consisting of matrices for which all entries are *positive* is denoted by $\Omega_n^*$. We define $J_n := n^{-1}J$, where $J$ denotes the $n$ by $n$ matrix for which all entries are 1. The vector $(1, 1, \ldots, 1)^\top$ is denoted by $\mathbf{j}$.

In 1926, B. L. van der Waerden proposed as a problem to determine the minimal permanent among all doubly stochastic matrices. It was natural to assume that this minimum is per $J_n = n!\, n^{-n}$ (as stated in (11.19)). The assertion

$$(12.1) \qquad (A \in \Omega_n \text{ and } A \neq J_n) \Rightarrow (\text{per } A > \text{per } J_n)$$

became known as the "Van der Waerden conjecture" (although in 1969 he told one of the present authors that he had not heard this name before and that he had made no such conjecture). In 1981 two different proofs of the conjecture appeared, one by D. I. Falikman, submitted in 1979, and one by G. P. Egoritsjev, submitted in 1980. We shall give our version of Egoritsjev's proof which had a slightly stronger result than Falikman's, cf. Van Lint (1981).

In the following, we shall use the term *minimizing matrix* for a matrix $A \in \Omega_n$ such that per $A = \min\{\text{per } S : S \in \Omega_n\}$. As usual, the matrix obtained from $A$ by deleting row $i$ and column $j$ is denoted by $A_{ij}$. We often consider $A$ as a sequence of $n$ columns and write $A = (\mathbf{a}_1, \ldots, \mathbf{a}_n)$. Later on, we shall consider permanents of matrices of size $n-1$ but we wish to use the notation for matrices of size $n$. The trick is to write per $(\mathbf{a}_1, \ldots, \mathbf{a}_{n-1}, \mathbf{e}_j)$, where $\mathbf{e}_j$ denotes the $j$-th standard basis vector. This permanent does not change value if the $j$-th row and $n$-th column are deleted. We remind the reader that by Problem 5C (Birkhoff), the set $\Omega_n$ is a convex set with the permutation matrices as vertices.

We need a few elementary results on matrices in $\Omega_n$. The first statement is the same as Theorem 5.4.

THEOREM 12.1. *If $A$ is an $n$ by $n$ matrix with nonnegative entries, then* per $A = 0$ *if and only if $A$ contains an $s$ by $t$ zero submatrix such that $s + t = n + 1$.*

We shall call an $n$ by $n$ matrix *partly decomposable* if it contains a $k$ by $n - k$ zero submatrix. So $A$ is partly decomposable if there exist permutation matrices $P$ and $Q$ such that

$$PAQ = \begin{pmatrix} B & C \\ O & D \end{pmatrix},$$

where $B$ and $D$ are square matrices. If a matrix is not partly decomposable, then we shall say that it is *fully indecomposable*. If $A \in \Omega_n$ and $A$ is partly decomposable, then, in the representation given above, we must have $C = O$, because the sum of the entries of $B$ equals the number of columns of $B$ and the sum of the entries of $B$ and $C$ is equal to the number of rows of $B$. So in that case, $A$ is the direct sum $B \dotplus D$ of an element of $\Omega_k$ and an element of $\Omega_{n-k}$.

PROBLEM 12A. Let $A$ be an $n$ by $n$ matrix with nonnegative entries $(n \geq 2)$. Prove that $A$ is fully indecomposable if and only if per $A_{ij} > 0$ for all $i$ and $j$.

PROBLEM 12B. Let $A$ be an $n$ by $n$ matrix with nonnegative entries. Prove that if $A$ is fully indecomposable, then $AA^\top$ and $A^\top A$ are also fully indecomposable.

THEOREM 12.2. *A minimizing matrix is fully indecomposable.*

PROOF: Let $A \in \Omega_n$ be a minimizing matrix and suppose that $A$ is partly decomposable. Then, as we saw above, $A = B \dotplus C$, where $B \in \Omega_k$ and $C \in \Omega_{n-k}$. By Theorem 12.1, we have per $A_{k,k+1} = 0$ and per $A_{k+1,k} = 0$. By Birkhoff's theorem, we may assume that $B$

and $C$ have positive elements on their diagonals. In $A$ we replace $b_{kk}$ by $b_{kk} - \epsilon$ and $c_{11}$ by $c_{11} - \epsilon$ and we put an $\epsilon$ in the positions $k, k+1$ and $k+1, k$. The new matrix is again in $\Omega_n$, if $\epsilon$ is small enough. The permanent of the new matrix is equal to

$$\operatorname{per} A - \epsilon \operatorname{per} A_{kk} - \epsilon \operatorname{per} A_{k+1,k+1} + O(\epsilon^2).$$

Since per $A_{kk}$ and per $A_{k+1,k+1}$ are both positive, this new permanent is smaller than per $A$ if $\epsilon$ is sufficiently small. This contradiction proves the assertion. □

COROLLARY. *(i) A row of a minimizing matrix has at least two positive entries.*

*(ii) For any $a_{ij}$ in a minimizing matrix, there is a permutation $\sigma$ such that $\sigma(i) = j$ and $a_{s,\sigma(s)} > 0$ for $1 \le s \le n$, $s \ne i$.*

PROOF: Clearly (i) is trivial, and (ii) follows from Problem 12A.

□

Let us now look at how far we can get with *calculus*. A very important step in the direction of a proof of (12.1) is the following surprising result due to Marcus and Newman (1959).

THEOREM 12.3. *If $A \in \Omega_n$ is a minimizing matrix and $a_{hk} > 0$, then per $A_{hk} = $ per $A$.*

PROOF: Let $S$ be the subset of $\Omega_n$ consisting of the doubly stochastic matrices $X$ for which $x_{ij} = 0$ if $a_{ij} = 0$. Then $A$ is an interior point of the set $S$, which is a subset of $\mathbb{R}^m$ for some $m$. If we denote the set of pairs $(i, j)$ for which $a_{ij} = 0$ by $Z$, we can describe $S$ by the relations:

$$\sum_{i=1}^{n} x_{ij} = 1, \quad j = 1, \dots, n;$$

$$\sum_{j=1}^{n} x_{ij} = 1, \quad i = 1, \dots, n;$$

$$x_{ij} \ge 0, \quad i, j = 1, \dots, n;$$

$$x_{ij} = 0, \quad (i, j) \in Z.$$

Since $A$ is minimizing, the permanent function has a relative minimum in the interior point $A$ of the set $S$ and we can use Lagrange multipliers to describe the situation. So we define:

$$F(X) := \text{per } X - \sum_{i=1}^{n} \lambda_i \left( \sum_{k=1}^{n} x_{ik} - 1 \right) - \sum_{j=1}^{n} \mu_j \left( \sum_{k=1}^{n} x_{kj} - 1 \right).$$

For $(i,j) \notin Z$, we have:

$$\partial F(X)/\partial x_{ij} = \text{per } X_{ij} - \lambda_i - \mu_j.$$

It follows that per $A_{ij} = \lambda_i + \mu_j$ and from this we find that for $1 \leq i \leq n$

$$(12.2) \quad \text{per } A = \sum_{j=1}^{n} a_{ij}\text{per } A_{ij} = \sum_{j=1}^{n} a_{ij}(\lambda_i + \mu_j) = \lambda_i + \sum_{j=1}^{n} a_{ij}\mu_j,$$

and similarly for $1 \leq j \leq n$,

$$(12.3) \qquad\qquad \text{per } A = \mu_j + \sum_{i=1}^{n} a_{ij}\lambda_i.$$

We introduce the vectors $\lambda = (\lambda_1, \ldots, \lambda_n)^\top$ and $\mu = (\mu_1, \ldots, \mu_n)^\top$. From (12.2) and (12.3), we find

$$(12.4) \qquad\qquad (\text{per } A)\mathbf{j} = \lambda + A\mu = \mu + A^\top \lambda.$$

Multiplying by $A^\top$ gives us

$$(\text{per } A)\mathbf{j} = A^\top \lambda + A^\top A\mu,$$

and hence $\mu = A^\top A\mu$, and similarly $\lambda = AA^\top \lambda$. The matrices $AA^\top$ and $A^\top A$ are both in $\Omega_n$ and by Problem 12B and Theorem 12.2, they have eigenvalue 1 with multiplicity one corresponding to the eigenvector $\mathbf{j}$. So we see that both $\lambda$ and $\mu$ are multiples of $\mathbf{j}$. By (12.4), we have $\lambda_i + \mu_j = \text{per } A$ and since per $A_{ij} = \lambda_i + \mu_j$, we are finished. $\qquad\square$

REMARK. It was shown by Marcus and Newman that Theorem 12.3 implies that a minimizing matrix in $\Omega_n^*$ must be $J_n$. The proof depends on the following idea. Let $A$ be an element of $\Omega_n$ with the property that per $A_{hk} =$ per $A$ for all $h, k$. If we replace any column of $A$ by a vector $\mathbf{x}$ for which $\sum_{i=1}^n x_{ij} = 1$, then the value of the permanent does not change (by (11.6)). We shall refer to this idea as the *substitution principle*. If $A$ is a minimizing matrix in $\Omega_n^*$, then the substitution principle allows us to replace any two columns of $A$ by their average and thus obtain a new minimizing matrix. In this way, one constructs a sequence of minimizing matrices which tends to $J_n$. The uniqueness of the minimum takes a little extra work.

A final result that uses ideas from calculus is the following generalization of Theorem 12.3 due to London (1971).

THEOREM 12.4. *If $A \in \Omega_n$ is a minimizing matrix, then* per $A_{ij} \geq$ per $A$ *for all $i$ and $j$.*

PROOF: Let $i$ and $j$ be given. By Corollary (ii) of Theorem 12.2, there is a permutation $\sigma$ such that $\sigma(i) = j$ and $a_{s,\sigma(s)} > 0$ for $1 \leq s \leq n$, $s \neq i$. Let $P$ be the corresponding permutation matrix. For $0 \leq \theta \leq 1$ we define $f(\theta) :=$ per $((1 - \theta)A + \theta P)$. Since $A$ is a minimizing matrix, $f'(0) \geq 0$, i.e.

$$0 \leq \sum_{i=1}^n \sum_{j=1}^n (-a_{ij} + p_{ij}) \text{per } A_{ij} = -n \text{ per } A + \sum_{s=1}^n \text{per } A_{s,\sigma(s)}.$$

By Theorem 12.3 we have per $A_{s,\sigma(s)} =$ per $A$ for $s \neq i$ and therefore per $A_{ij} \geq$ per $A$. $\qquad\square$

PROBLEM 12C. Show that Theorem 12.3 implies that if $A \in \Omega_5^*$ is a minimizing matrix, then there is a minimizing matrix $B \in \Omega_5^*$ that has $aJ$ of size 4 as a principal submatrix. Then show that $a$ must be $\frac{1}{5}$.

We now come to the main tool in the proof of the Van der Waerden conjecture. This time we need *linear algebra*. We shall give a direct proof of a theorem on symmetric bilinear forms which leads to the inequality that was derived by Egoritsjev from the so-called Alexandroff-Fenchel inequalities (which we do not treat).

Consider the space $\mathbb{R}^n$ with a symmetric inner product $\langle \mathbf{x}, \mathbf{y} \rangle = \mathbf{x}^\top Q\mathbf{y}$. If $Q$ has one positive eigenvalue and $n - 1$ negative eigenvalues, we shall speak of a *Lorentz space*. We use the following standard terminology: a nonzero vector $\mathbf{x}$ is *isotropic* if $\langle \mathbf{x}, \mathbf{x} \rangle = 0$, *positive*, respectively *negative*, if $\langle \mathbf{x}, \mathbf{x} \rangle$ is positive, respectively negative.

If $\mathbf{a}$ is positive and $\mathbf{b}$ is not a scalar multiple of $\mathbf{a}$, then by Sylvester's law, the plane spanned by $\mathbf{a}$ and $\mathbf{b}$ must contain a negative vector. Therefore the quadratic form in $\lambda$ given by $\langle \mathbf{a} + \lambda\mathbf{b}, \mathbf{a} + \lambda\mathbf{b} \rangle$ must have a positive discriminant. Thus we have the following inequality, which is like the Cauchy inequality but the other way around.

THEOREM 12.5. *If $\mathbf{a}$ is a positive vector in a Lorentz space and $\mathbf{b}$ is arbitrary, then*

$$\langle \mathbf{a}, \mathbf{b} \rangle^2 \geq \langle \mathbf{a}, \mathbf{a} \rangle \langle \mathbf{b}, \mathbf{b} \rangle$$

*and equality holds if and only if $\mathbf{b} = \lambda\mathbf{a}$ for some constant $\lambda$.*

The connection with permanents is provided by the following definition. Consider vectors $\mathbf{a}_1, \ldots, \mathbf{a}_{n-2}$ in $\mathbb{R}^n$ with *positive* coordinates. As usual, let $\mathbf{e}_1, \ldots, \mathbf{e}_n$ be the standard basis of $\mathbb{R}^n$. We define an inner product on $\mathbb{R}^n$ by

$$(12.5) \qquad \langle \mathbf{x}, \mathbf{y} \rangle := \operatorname{per}(\mathbf{a}_1, \mathbf{a}_2, \ldots, \mathbf{a}_{n-2}, \mathbf{x}, \mathbf{y}),$$

i.e.

$$\langle \mathbf{x}, \mathbf{y} \rangle = \mathbf{x}^\top Q\mathbf{y},$$

where $Q$ is given by

$$(12.6) \qquad q_{ij} := \operatorname{per}(\mathbf{a}_1, \mathbf{a}_2, \ldots, \mathbf{a}_{n-2}, \mathbf{e}_i, \mathbf{e}_j).$$

Note that if $A$ is a matrix with columns $\mathbf{a}_1, \ldots, \mathbf{a}_n$ and we delete the last two columns and the rows with index $i$ and $j$, then the reduced matrix has permanent equal to $q_{ij}$.

THEOREM 12.6. *The space* $\mathbb{R}^n$ *with the inner product defined by (12.5) is a Lorentz space.*

PROOF: The proof is by induction. For $n = 2$, we have $Q = \begin{pmatrix} 0 & 1 \\ 1 & 0 \end{pmatrix}$ and the assertion is true. Now assume the theorem is true for $\mathbb{R}^{n-1}$. In the first step of the proof, we show that $Q$ does not have the eigenvalue 0. Suppose $Q\mathbf{c} = 0$, i.e.

$$(12.7) \qquad \text{per } (\mathbf{a}_1, \ldots, \mathbf{a}_{n-2}, \mathbf{c}, \mathbf{e}_j) = 0 \qquad \text{for } 1 \le j \le n.$$

By deleting the last column and the $j$-th row, we can consider (12.7) as a relation for vectors in $\mathbb{R}^{n-1}$. We consider the inner product given by

$$(12.8) \qquad \text{per } (\mathbf{a}_1, \ldots, \mathbf{a}_{n-3}, \mathbf{x}, \mathbf{y}, \mathbf{e}_j)_{jn}$$

and apply the induction hypothesis, (12.7) and Theorem 12.5. Substitution of $\mathbf{x} = \mathbf{a}_{n-2}$, $\mathbf{y} = \mathbf{a}_{n-2}$ in (12.8) gives a positive value, and $\mathbf{x} = \mathbf{a}_{n-2}$, $\mathbf{y} = \mathbf{c}$ gives the value 0. Therefore

$$(12.9) \qquad \text{per } (\mathbf{a}_1, \ldots, \mathbf{a}_{n-3}, \mathbf{c}, \mathbf{c}, \mathbf{e}_j) \le 0 \qquad \text{for } 1 \le j \le n$$

and for each $j$ equality holds if and only if all coordinates of $\mathbf{c}$ except $c_j$ are 0. If we multiply the lefthand side of (12.9) by the $j$-th coordinate of $\mathbf{a}_{n-2}$ and sum over $j$, we find $\mathbf{c}^\top Q\mathbf{c}$. Therefore the assumption $Q\mathbf{c} = 0$ implies that $\mathbf{c} = 0$.

For $0 \le \theta \le 1$, we define a matrix $Q_\theta$ by taking (12.5) and replacing every $\mathbf{a}_i$ by $\theta\mathbf{a}_i + (1 - \theta)\mathbf{j}$. From what we have shown above, it follows that for every $\theta$ in [0,1] the matrix $Q_\theta$ does not have the eigenvalue 0. Therefore the number of positive eigenvalues is constant. Since this number is one for $\theta = 0$, it is also one for $\theta = 1$, which proves our assertion. $\square$

We formulate the combination of Theorem 12.5 and Theorem 12.6 as a corollary. (The final assertion follows by continuity.)

COROLLARY. *If* $\mathbf{a}_1, \ldots, \mathbf{a}_{n-1}$ *are vectors in* $\mathbb{R}^n$ *with positive coordinates and* $\mathbf{b} \in \mathbb{R}^n$, *then*

$$(\text{per } (\mathbf{a}_1, \ldots, \mathbf{a}_{n-1}, \mathbf{b}))^2$$
$$\ge \text{per } (\mathbf{a}_1, \ldots, \mathbf{a}_{n-1}, \mathbf{a}_{n-1}) \cdot \text{per } (\mathbf{a}_1, \ldots, \mathbf{a}_{n-2}, \mathbf{b}, \mathbf{b})$$

and equality holds if and only if $\mathbf{b} = \lambda \mathbf{a}_{n-1}$ for some constant $\lambda$. Furthermore, the inequality also holds if some of the coordinates of the $\mathbf{a}_i$ are 0, but the assertion about the consequence of equality then cannot be made.

We are now able to generalize Theorem 12.3.

THEOREM 12.7. *If $A \in \Omega_n$ is a minimizing matrix, then per $A_{ij} =$ per $A$ for all $i$ and $j$.*

PROOF: Suppose that the statement is false. Then by Theorem 12.4, there is a pair $r, s$ such that per $A_{rs} >$ per $A$.

Choose $t$ such that $a_{rt} > 0$. Consider the product of two factors per $A$. In the first of these we replace $\mathbf{a}_s$ by $\mathbf{a}_t$, and in the second, we replace $\mathbf{a}_t$ by $\mathbf{a}_s$. Subsequently, we develop the first permanent by column $s$ and the second permanent by column $t$. By Theorem 12.5 and the Corollary to Theorem 12.6, we have

$$(\text{per } A)^2 \geq \left( \sum_{k=1}^n a_{kt} \text{ per } A_{ks} \right) \left( \sum_{k=1}^n a_{ks} \text{ per } A_{kt} \right).$$

By Theorem 12.4, every subpermanent on the righthand side is at least per $A$ and per $A_{rs} >$ per $A$. Since per $A_{rs}$ is multiplied by $a_{rt}$ which is positive, we see that the righthand side is larger than $(\text{per } A)^2$, a contradiction. □

We now use the substitution principle as follows. Take a minimizing matrix $A$ and let $\mathbf{u}$ and $\mathbf{v}$ be two columns of $A$. Replace $\mathbf{u}$ and $\mathbf{v}$ by $\frac{1}{2}(\mathbf{u} + \mathbf{v})$. The new matrix is again a minimizing matrix by Theorem 12.7.

Let $A$ be any minimizing matrix and let $\mathbf{b}$ be any column of $A$, say the last column. From Corollary (i) to Theorem 12.2, we know that in every row of $A$ there are at least two positive elements. We now apply the substitution principle (as sketched above) a number of times but we never change the last column. In this way, we can find a minimizing matrix $A' = (\mathbf{a}'_1, \ldots, \mathbf{a}'_{n-1}, \mathbf{b})$ for which $\mathbf{a}'_1, \ldots, \mathbf{a}'_{n-1}$ all have *positive* coordinates. Now apply the Corollary to Theorem 12.6. By the substitution principle, equality must hold. Hence $\mathbf{b}$ is a multiple of $\mathbf{a}'_i$ for any $i$ with $1 \leq i \leq n-1$. This implies that $\mathbf{b} = n^{-1}\mathbf{j}$ and therefore $A = J_n$, which completes the proof of the Van der Waerden conjecture.

THEOREM 12.8. *The implication (12.1) is true.*

**Notes.**

For a survey of the two proofs of the Van der Waerden conjecture, we refer to Van Lint (1982). There one also finds historical comments and a nice anecdote concerning the conjecture.

B. L. van der Waerden, a Dutch mathematician, is known mostly for his work in algebra, although he published in several fields. His work *Moderne Algebra* (1931) set the trend for many decades.

Minc's book on permanents is the best reference for all the work that was done in relation to the Van der Waerden conjecture up to 1978.

The name *Lorentz space* is related to relativity theory and the group of transformations that leave the quadratic form $x^2 + y^2 + z^2 - t^2$ invariant. H. A. Lorentz was a Dutch physicist who won the Nobel prize for his work.

**References.**

J. H. van Lint (1981), Notes on Egoritsjev's proof of the Van der Waerden conjecture, *Linear Algebra and its Applications* **39**, 1–8.

J. H. van Lint (1982), The van der Waerden Conjecture: Two proofs in one year, *The Math. Intelligencer* **39**, 72–77.

D. London (1971), Some notes on the van der Waerden conjecture, *Linear Algebra and its Applications* **4**, 155–160.

M. Marcus and M. Newman (1959), On the minimum of the permanent of a doubly stochastic matrix, *Duke Math. J.* **26**, 61–72.

H. Minc (1978), *Permanents*, Encyclopedia of Mathematics and its Applications, vol. 6, Addison-Wesley, reissued by Cambridge University Press (1984).

B. L. van der Waerden (1926), *Jber. D. M. V.* **35**.

# 13

# Elementary counting; Stirling numbers

The next few chapters will be devoted to counting techniques and some special combinatorial counting problems. We start with a number of elementary methods that are used quite often. Consider mappings from $\{1, 2, \ldots, n\}$ to $\{1, 2, \ldots, k\}$. Their total number is $k^n$. In Example 10.2, we studied the case where the mappings were required to be surjective. We return to this question in Theorem 13.5. If the mappings are injections, then their number is the *falling factorial*

$$(13.1) \qquad (k)_n := k(k-1)\ldots(k-n+1) = k!/(k-n)!.$$

We now consider a similar problem. The $n$ objects to be mapped are no longer distinguishable but the images are. We formulate this as follows. We have $n$ indistinguishable balls that are to be placed in $k$ boxes, marked $1, 2, \ldots, k$. In how many different ways can this be done? The solution is found by using the following trick. Think of the balls as being colored blue and line them up in front of the boxes that they will go into. Then insert a red ball between two consecutive boxes. We end up with a line of $n+k-1$ balls, $k-1$ of them red, describing the situation. So the answer to the problem is $\binom{n+k-1}{k-1}$. We formulate this as a theorem.

THEOREM 13.1. *The number of solutions of the equation*

$$(13.2) \qquad\qquad x_1 + x_2 + \cdots + x_k = n$$

*in nonnegative integers is $\binom{n+k-1}{k-1}$.*

PROOF: Interpret $x_i$ as the number of balls in box $i$. $\qquad\qquad\square$

COROLLARY. *The number of solutions of the equation (13.2) in positive integers is $\binom{n-1}{k-1}$.*

PROOF: Replace $x_i$ by $y_i := x_i - 1$. Then $\sum y_i = n - k$. Apply Theorem 13.1. □

EXAMPLE 13.1. By analogy with the question we encountered in Example 10.6, we consider the problem of selecting $r$ of the integers $1, 2, \ldots, n$ such that no two selected integers are consecutive. Let $x_1 < x_2 < \cdots < x_r$ be such a sequence. Then $x_1 \geq 1$, $x_2 - x_1 \geq 2, \ldots, x_r - x_{r-1} \geq 2$. Define

$$y_1 := x_1, \ y_i := x_i - x_{i-1} - 1, \ 2 \leq i \leq r, \ y_{r+1} := n - x_r + 1.$$

Then the $y_i$ are positive integers and $\sum_{i=1}^{r+1} y_i = n - r + 2$. By the Corollary to Theorem 13.1, we see that there are $\binom{n-r+1}{r}$ solutions.

PROBLEM 13A. Use the result of Example 13.1 to find the number of circular arrangements of $r$ ones and $n - r$ zeros with no two adjacent ones. (Configurations obtained by rotation are considered different.) Compare with Example 10.6.

EXAMPLE 13.2. In how many ways can we arrange $r_1$ balls of color 1, $r_2$ balls of color 2, $\ldots, r_k$ balls of color $k$ in a sequence of length $n := r_1 + r_2 + \cdots + r_k$? If we number the balls 1 to $n$, then there are $n!$ arrangements. Since we ignore the numbering, any permutation of the set of $r_i$ balls of color $i$, $1 \leq i \leq k$, produces the same arrangement. So the answer to the question is the multinomial coefficient $\binom{n}{r_1, \ldots, r_k}$; see (2.1).

EXAMPLE 13.3. We wish to split $\{1, 2, \ldots, n\}$ into $b_1$ subsets of size 1, $b_2$ subsets of size 2, ..., $b_k$ subsets of size $k$. Here $\sum_{i=1}^{k} i b_i = n$. The same argument as used in Example 13.2 applies. Furthermore, the subsets of the same cardinality can be permuted among themselves without changing the configuration. So the solution is

(13.3)
$$\frac{n!}{b_1! \ldots b_k! (1!)^{b_1} (2!)^{b_2} \ldots (k!)^{b_k}}.$$

Several counting problems (often involving binomial coefficients) can be done in a more or less obvious way that leads to a (sometimes difficult) calculation. Often there is a less obvious "combinatorial" way to do the counting that produces an immediate answer. We give a few examples.

EXAMPLE 13.4. Let $A$ run through all subsets of $\{1, 2, \ldots, n\}$. Calculate $S = \sum |A|$. Since there are $\binom{n}{i}$ subsets of size $i$, we apparently must calculate $\sum_{i=0}^{n} i \binom{n}{i}$. By differentiating $(1 + x)^n$, we find

$$\sum_{i=1}^{n} i \binom{n}{i} x^{i-1} = n(1 + x)^{n-1}$$

and substitution of $x = 1$ yields the answer $S = n \cdot 2^{n-1}$. If we had spent a little more time thinking, then this answer would have been obvious! A set $A$ and its complement together contain $n$ elements and there are exactly $2^{n-1}$ such pairs.

EXAMPLE 13.5. In Chapter 10 we saw some examples of formulae involving binomial coefficients for which a combinatorial proof was easier than a direct proof. The familiar relation

(13.4)
$$\sum_{k=0}^{n} \binom{n}{k}^2 = \binom{2n}{n}$$

is another example. Of course one can calculate this sum by determining the coefficient of $x^n$ in $(1 + x)^n (1 + x)^n$ and using the binomial formula. However, each side of (13.4) just counts (in two ways) the number of ways of selecting $n$ balls from a set consisting of $n$ red balls and $n$ blue balls.

PROBLEM 13B. Show that the following formula for binomial coefficients is a direct consequence of (10.6):

$$\binom{n+1}{a+b+1} = \sum_{k=0}^{n} \binom{k}{a} \binom{n-k}{b}.$$

Give a combinatorial proof by considering $(a + b + 1)$-subsets of the set $\{0, 1, \ldots, n\}$, ordering them in increasing order, and then looking at the value of the integer in position $a + 1$.

EXAMPLE 13.6. We consider a slightly more complicated example where our previous knowledge could lead us to an involved solution. How many sequences $A_1, \ldots, A_k$ are there for which $A_i \subseteq$

$\{1, 2, \ldots, n\}$, $1 \leq i \leq k$, and $\bigcup_{i=1}^{k} A_i = \{1, 2, \ldots, n\}$? Since we wish to avoid that $j$, $1 \leq j \leq n$, is not an element of the union of the $A_i$'s, we are tempted to use inclusion-exclusion. If we choose $i$ elements from $\{1, 2, \ldots, n\}$ and consider all sequences $A_1, \ldots, A_k$ not containing any of these $i$ elements, then we find $(2^{n-i})^k$ sequences. So by Theorem 10.1, the solution to the problem is

$$\sum_{i=0}^{n} (-1)^i \binom{n}{i} 2^{(n-i)k} = (2^k - 1)^n.$$

This answer shows that another approach would have been better. If we describe a sequence $A_1, \ldots, A_k$ by a (0,1)-matrix $A$ of size $k$ by $n$, with the characteristic functions of the subsets as its rows, then the condition on the sequences states that $A$ has no column of zeros. So there are $(2^k - 1)^n$ such matrices!

PROBLEM 13C. Give a solution involving binomial coefficients and a combinatorial solution to the following question. How many pairs $(A_1, A_2)$ of subsets of $\{1, 2, \ldots, n\}$ are there such that $A_1 \cap A_2 = \emptyset$?

PROBLEM 13D. Consider the set $S$ of all ordered $k$-tuples $\mathcal{A} = (A_1, \ldots, A_k)$ of subsets of $\{1, 2, \ldots, n\}$. Determine

$$\sum_{\mathcal{A} \in S} |A_1 \cup A_2 \cup \cdots \cup A_k|.$$

PROBLEM 13E. The familiar relation

$$\sum_{m=k}^{l} \binom{m}{k} = \binom{l+1}{k+1}$$

is easily proved by induction. The reader who wishes to can find a more complicated proof by using (10.6). Find a combinatorial proof by counting paths from $(0,0)$ to $(l+1, k+1)$ in the $X$-$Y$ plane where each step is of type $(x, y) \rightarrow (x + 1, y)$ or $(x, y) \rightarrow (x + 1, y + 1)$. Then use the formula to show that the number of solutions of

$$x_1 + x_2 + \cdots + x_k \leq n$$

in nonnegative integers is $\binom{n+k}{k}$. Can you prove this result combinatorially?

Two kinds of numbers that come up in many combinatorial problems are the so-called *Stirling numbers of the first and second kind.* The numbers are often defined by the formulae (13.8) and (13.12) given below. We prefer a combinatorial definition.

Let $c(n, k)$ denote the number of permutations $\pi \in S_n$ with exactly $k$ cycles. (This number is called a *signless* Stirling number of the first kind.) Furthermore define $c(0, 0) = 1$ and $c(n, k) = 0$ if $n \leq 0$ or $k \leq 0$, $(n, k) \neq (0, 0)$. The Stirling numbers of the first kind $s(n, k)$ are defined by

$$(13.5) \qquad s(n, k) := (-1)^{n-k} c(n, k).$$

THEOREM 13.2. *The numbers $c(n, k)$ satisfy the recurrence relation*

$$(13.6) \qquad c(n, k) = (n - 1)c(n - 1, k) + c(n - 1, k - 1).$$

PROOF: If $\pi$ is a permutation in $S_{n-1}$ with $k$ cycles, then there are $n - 1$ positions where we can insert the integer $n$ to produce a permutation $\pi' \in S_n$ with $k$ cycles. We can also adjoin $(n)$ as a cycle to any permutation in $S_{n-1}$ with $k - 1$ cycles. This accounts for the two terms on the righthand side of (13.6).   □

THEOREM 13.3. *For $n \geq 0$ we have*

$$(13.7) \qquad \sum_{k=0}^{n} c(n, k)x^k = x(x + 1) \ldots (x + n - 1)$$

*and*

$$(13.8) \qquad \sum_{k=0}^{n} s(n, k)x^k = (x)_n,$$

*where $(x)_n$ is defined as in (13.1).*

PROOF: Write the righthand side of (13.7) as

$$F_n(x) = \sum_{k=0}^{n} b(n, k)x^k.$$

Clearly $b(0,0) = 1$. Define $b(n,k) = 0$ if $n \leq 0$ or $k \leq 0$, $(n,k) \neq (0,0)$. Since

$$F_n(x) = (x + n - 1)F_{n-1}(x)$$

$$= \sum_{k=1}^{n} b(n-1, k-1)x^k + (n-1)\sum_{k=0}^{n-1} b(n-1,k)x^k,$$

we see that the numbers $b(n,k)$ satisfy the same recurrence relation as the $c(n,k)$, namely (13.6). Since the numbers are equal if $n \leq 0$ or $k \leq 0$, they are equal for all $n$ and $k$.

To prove (13.8) replace $x$ by $-x$ and use (13.5). $\square$

We remark that it is possible to give a combinatorial proof of (13.7) by showing that both sides of the equation count the same objects.

We now define the Stirling numbers of the second kind: denote by $P(n,k)$ the set of all partitions of an $n$-set into $k$ nonempty subsets (blocks). Then

(13.9) $$S(n,k) := |P(n,k)|.$$

Again we have $S(0,0) = 1$ and take the numbers to be 0 for all values of the parameters not covered by the previous definition. Again we have an easy recurrence relation.

THEOREM 13.4. *The Stirling numbers of the second kind satisfy the relation*

(13.10) $$S(n,k) = kS(n-1,k) + S(n-1,k-1).$$

PROOF: The proof is nearly the same as for Theorem 13.3. A partition of the set $\{1, 2, \ldots, n-1\}$ can be made into a partition of $\{1, 2, \ldots, n\}$ by adjoining $n$ to one of the blocks or by increasing the number of blocks by one by making $\{n\}$ a block. $\square$

We define the *Bell number* $B(n)$ to be the *total* number of partitions of an $n$-set, i.e.

(13.11) $$B(n) := \sum_{k=1}^{n} S(n,k), \qquad (n \geq 1).$$

For the Stirling numbers of the second kind there is a formula similar to (13.8).

THEOREM 13.5. *For $n \geq 0$ we have*

(13.12) $$x^n = \sum_{k=0}^{n} S(n,k)(x)_k.$$

PROOF: We first remark that by (13.9) the number of surjective mappings from an $n$-set to a $k$-set is $k!S(n,k)$ (a block of the partition is the inverse image of an element of the $k$-set). So by Example 10.2, we have

(13.13) $$S(n,k) = \frac{1}{k!} \sum_{i=0}^{k} (-1)^i \binom{k}{i}(k-i)^n = \frac{1}{k!} \sum_{i=0}^{k} (-1)^{k-i} \binom{k}{i} i^n.$$

Now let $x$ be an integer. There are $x^n$ mappings from the $n$-set $N := \{1, 2, \ldots, n\}$ to the $x$-set $\{1, 2, \ldots, x\}$. For any $k$-subset $Y$ of $\{1, 2, \ldots, x\}$, there are $k!S(n,k)$ surjections from $N$ to $Y$. So we find

$$x^n = \sum_{k=0}^{n} \binom{x}{k} k! S(n,k) = \sum_{k=0}^{n} S(n,k)(x)_k.$$

$\square$

In Example 10.1, we saw that it can be useful to associate a so-called *generating function* with a sequence $a_1, a_2, \ldots$ of numbers. In Chapter 14, we will see many uses of the generating function. At this point we treat the generating functions for the Stirling numbers.

THEOREM 13.6. $\sum_{n \geq k} S(n,k)\frac{x^n}{n!} = \frac{1}{k!}(e^x - 1)^k$    $(k \geq 0)$.

PROOF: Let $F_k(x)$ denote the sum on the lefthand side. By (13.10) we have

$$F_k'(x) = kF_k(x) + F_{k-1}(x).$$

The result now follows by induction. Since $S(n,1) = 1$ the assertion is true for $k = 1$ and the induction hypothesis yields a differential equation for $F_k$, which with the condition $S(k,k) = 1$ has the righthand side of the assertion as unique solution.    $\square$

For the Stirling numbers of the first kind, it is slightly more difficult to find the generating function.

THEOREM 13.7. $\sum_{n=k}^{\infty} s(n,k)\frac{z^n}{n!} = \frac{1}{k!}(\log(1+z))^k$.

PROOF: Since

$$(1+z)^x = e^{x\log(1+z)} = \sum_{k=0}^{\infty} \frac{1}{k!}(\log(1+z))^k x^k,$$

the righthand side in the assertion is the coefficient of $x^k$ in the expansion of $(1+z)^x$. On the other hand, we have for $|z| < 1$,

$$(1+z)^x = \sum_{n=0}^{\infty} \binom{x}{n} z^n = \sum_{n=0}^{\infty} \frac{1}{n!}(x)_n z^n$$

$$= \sum_{n=0}^{\infty} \frac{z^n}{n!} \sum_{r=0}^{n} s(n,r) x^r = \sum_{r=0}^{\infty} x^r \sum_{n=r}^{\infty} s(n,r) \frac{z^n}{n!}.$$

This completes the proof. $\qquad\qquad\qquad\qquad\qquad\qquad\square$

PROBLEM 13F. Show directly that the number of permutations of the integers 1 to $n$ with an even number of cycles is equal to the number of permutations with an odd number of cycles ($n > 1$). Also show that this is a consequence of Theorem 13.7.

Finally we mention that the Stirling numbers of the first and second kind are related by

$$(13.14) \qquad\qquad \sum_{k=m}^{n} S(n,k)s(k,m) = \delta_{mn}.$$

This follows immediately if we substitute (13.8) in (13.12). Since the functions $x^n$, respectively $(x)_n$, with $n \geq 0$ both form a basis of the vector space $\mathbb{C}[x]$, the formula (13.14) is just the standard relation between the matrices for basis transformation.

PROBLEM 13G. Show that (13.12) leads to $B(n) = \frac{1}{e}\sum_{k=0}^{\infty} \frac{k^n}{k!}$.

**Notes.**
Stirling numbers of the first and second kind appear in many areas of mathematics; for example, they play a role in a number

of interpolation formulae and in the calculus of finite differences. There are tables of the numbers in several books on tables of mathematical functions.

Later, in Chapter 35, these numbers will reappear.

James Stirling (1692–1770), a Scottish mathematician, studied at Oxford. He taught mathematics at Venice and London but switched to a career in business at the age of 43.

# 14

# Recursions and generating functions

Many combinatorial counting problems with a solution $a_n$ depending on a parameter $n$, can be solved by finding a recursion relation for $a_n$ and then solving that recursion. Sometimes this is done by introducing an *ordinary generating function*

$$f(x) := \sum_{n \geq 0} a_n x^n,$$

or an *exponential generating function*

$$f(x) := \sum_{n \geq 0} a_n \frac{x^n}{n!},$$

and using the recursion to find an equation or a differential equation for $f(x)$, and solving that equation. We shall demonstrate several of the techniques involved.

EXAMPLE 14.1. As an introduction, consider once again Example 10.1. Let $\pi$ be a derangement of $\{1, 2, \ldots, n+1\}$. There are $n$ choices for $\pi(n+1)$. If $\pi(n+1) = i$ and $\pi(i) = n+1$, then $\pi$ is also a derangement on the set $\{1, 2, \ldots, n\} \setminus \{i\}$. If $\pi(n+1) = i$ and $\pi(i) \neq n+1 = \pi(j)$, then replacing $\pi(j)$ by $i$ yields a derangement on the set $\{1, 2, \ldots, n\}$. Therefore

(14.1) $$d_{n+1} = n(d_n + d_{n-1}),$$

which is also an immediate consequence of (10.3). Let $D(x)$ be the *exponential* generating function for the sequence $d_0 = 1, d_1 =$

$0, d_2, \ldots$. From (14.1) we immediately find

$$(1-x)D'(x) = xD(x),$$

and from this we find $D(x) = e^{-x}/(1-x)$ and (10.2).

In many cases, we use the generating functions only as a book-keeping device, and our operations of addition, multiplication (and even substitution and derivation, as we shall see below) are to be interpreted formally. It is possible to give a completely rigorous theory of formal power series (as algebraic objects) and we give an introduction to this theory in Appendix 2. In most cases, it is intuitively clear and easy to check that the operations are legiti-mate. If the series that we use actually converge, then we can use all appropriate knowledge from analysis concerning these series, as we did in Example 14.1. We give another elementary example.

EXAMPLE 14.2. Suppose that we have $k$ boxes numbered 1 to $k$ and suppose that box $i$ contains $r_i$ balls, $1 \leq i \leq k$. A formal bookkeeping device to list all possible configurations is to let the named one correspond to the term $x_1^{r_1} x_2^{r_2} \cdots x_k^{r_k}$ in the product

$$(1 + x_1 + x_1^2 + \cdots)(1 + x_2 + x_2^2 + \cdots) \cdots (1 + x_k + x_k^2 + \cdots).$$

We can collect all the terms involving exactly $n$ balls by taking $x_i = x$ for all $i$, and considering the terms equal to $x^n$. Therefore we find that the number of ways to divide $n$ balls over $k$ distinguishable boxes is the coefficient of $x^n$ in the expansion of $(1-x)^{-k}$, and by (10.6) this is $\binom{k-1+n}{n}$, giving a second proof of Theorem 13.1.

In many cases, the combinatorial problem that we are interested in leads to a linear recursion relation with constant coefficients, which is easily solved by standard methods.

EXAMPLE 14.3. We consider paths of length $n$ in the $X$-$Y$ plane starting from (0,0) with steps $R : (x,y) \to (x+1,y)$, $L : (x,y) \to (x-1,y)$, and $U : (x,y) \to (x,y+1)$ (i.e. to the right, to the left, or up). We require that a step $R$ is not followed by a step $L$ and vice versa. Let $a_n$ denote the number of such paths. First observe that if we denote by $b_n$ the number of paths of length $n$ starting with a step $U$, then $b_n = a_{n-1}$ and furthermore trivially $b_{n+m} \geq b_n b_m$ and $b_n \leq 3^{n-1}$. So by Fekete's lemma, Lemma 11.6, $\lim_{n \to \infty} b_n^{1/n}$ exists and is at most 3. Next, note that $a_0 = 1$ and $a_1 = 3$. We split the set of paths of length $n$ into subsets depending on the last one or two steps. Clearly there are $a_{n-1}$ paths ending with the step $U$. Take a path of length $n-1$ and repeat the last step if it is $L$ or $R$, and adjoin a step $L$ if the last step was $U$. In this way, we obtain all the paths of length $n$ that end in $LL$, $RR$, or $UL$. So there are $a_{n-1}$ of these. It remains to count the paths ending with $UR$ and again it is trivial that there are $a_{n-2}$ of these. We have shown that

$$a_n = 2a_{n-1} + a_{n-2} \qquad (n \geq 2).$$

Let $f(x) = \sum_{n=0}^{\infty} a_n x^n$. Then the recursion implies that

$$f(x) = 1 + 3x + 2x\left(f(x) - 1\right) + x^2 f(x),$$

i.e.

$$f(x) = \frac{1+x}{1 - 2x - x^2} = \frac{\frac{1}{2}\alpha}{1 - \alpha x} + \frac{\frac{1}{2}\beta}{1 - \beta x},$$

where $\alpha = 1 + \sqrt{2}$, $\beta = 1 - \sqrt{2}$. Therefore

$$a_n = \frac{1}{2}(\alpha^{n+1} + \beta^{n+1})$$

and we find $\lim_{n \to \infty} a_n^{1/n} = 1 + \sqrt{2}$.

PROBLEM 14A. Let $a_n$ denote the number of sequences of 0's and 1's that do not contain two consecutive 0's. Determine $a_n$.

EXAMPLE 14.4. Let $a(r, n)$, where $0 \leq r \leq n$, denote the number of solutions of the problem of Example 13.1 ($a(0, 0) = 1$). We divide the set of possible sequences into two subsets: those with $x_1 = 1$ and those with $x_1 > 1$. The first subset clearly contains $a(r-1, n-2)$ elements, the second one $a(r, n-1)$ elements. So

(14.2)      $a(r, n) = a(r, n - 1) + a(r - 1, n - 2)$      $(n > 1)$.

From this recursion, we can prove the result $a(r,n) = \binom{n-r+1}{r}$ by induction. Using the generating function is more difficult. Try

$$f(x,y) := \sum_{n=0}^{\infty} \sum_{r=0}^{\infty} a(r,n) x^n y^r.$$

From (14.2) we find

$$f(x,y) = 1 + x + xy + x\left(-1 + f(x,y)\right) + x^2 y f(x,y),$$

i.e.

$$f(x,y) = \frac{1 + xy}{1 - x - x^2 y} = \frac{1}{1-x} + \sum_{a=1}^{\infty} \frac{x^{2a-1} y^a}{(1-x)^{a+1}}.$$

Substitution of (10.6) for $(1-x)^{-a-1}$ produces the required binomial coefficient for $a(r,n)$.

As we saw in Example 14.3 (and Problem 14A) a linear recursion with constant coefficients leads to a rational function as generating function (and vice versa). Indeed, if $a_n = \sum_{k=1}^{l} \alpha_k a_{n-k}$ $(n > l)$ and $f(x) = \sum_{n=0}^{\infty} a_n x^n$, then $(1 - \sum_{k=1}^{l} \alpha_k x^k) f(x)$ is a power series for which the coefficient of $x^n$ is 0 if $n > l$.

The following example due to Klarner (1967) shows an interesting case of a linear recursion that is found with the help of the generating function and that is not at all obvious in a combinatorial way.

EXAMPLE 14.5. Consider configurations in the plane, called *polyominoes*, as in Fig. 14.1. The configuration consists of layers, each consisting of consecutive squares. Two consecutive layers meet along the edges of a number ($\geq 1$) of squares. (More precisely, we are considering *horizontally convex polyominoes*.)

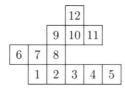

Figure 14.1

Let $a_n$ denote the number of polyominoes with $n$ squares and define $f(x) := \sum_{n=1}^{\infty} a_n x^n$. To find $f$, we introduce $a(m,n)$ for the number of polyominoes for which the bottom layer has $m$ squares (and a total of $n$). We define $a(m,n) := 0$ if $m > n$. Clearly

$$(14.3) \qquad a(m,n) = \sum_{l=1}^{\infty} (m + l - 1) a(l, n - m).$$

We define

$$(14.4) \qquad F(x,y) := \sum_{n=1}^{\infty} \sum_{m=1}^{\infty} a(m,n) x^n y^m.$$

Then $f(x) = F(x,1)$. Because the series will turn up below, we also define

$$g(x) := \sum_{n=1}^{\infty} \sum_{m=1}^{\infty} m a(m,n) x^n.$$

We would like to write

$$(14.5) \qquad g(x) = \left( \frac{\partial F}{\partial y} \right)_{y=1}.$$

Even though we have a theory of formal power series, it may be instructive to show that the righthand side of (14.4) converges in a sufficiently large region. This gives us the opportunity to show a quick way of getting a rough estimate for $a_n$. Number the squares of the polyomino in the obvious way as in Fig. 14.1. To each square associate a quadruple $(x_0, x_1, x_2, x_3)$ of 0's and 1's, where $x_0 = 1$ means that there is a square of the polyomino below this square, $x_1 = 1$ means that there is a square of the polyomino to the left, $x_2 = 1$ means that there is a square above, and $x_3 = 1$ means that there is a square to the right. For example, in Fig. 14.1 the first quadruple is (0,0,1,1). The sequence of quadruples uniquely determines the polyomino (e.g. the fifth quadruple is the first one in the sequence that ends in a 0, showing that $m = 5$, etc.). This shows that $a_n \leq 15^n$. From this and (14.3) we find $a(m,n) \leq$

$n \cdot 15^{n-m}$ which is enough to justify (14.5). From (14.4) we find by substituting (14.3) and a straightforward calculation

$$(14.6) \qquad F(x,y) = \frac{xy}{1-xy} + \frac{(xy)^2}{(1-xy)^2}f(x) + \frac{xy}{1-xy}g(x).$$

Differentiation of both sides of (14.6) with respect to $y$ and taking $y = 1$ yields (using (14.5)):

$$(14.7) \qquad g(x) = \frac{x}{(1-x)^2} + \frac{2x^2}{(1-x)^3}f(x) + \frac{x}{(1-x)^2}g(x).$$

From (14.7) we can find $g(x)$ and substitute this in (14.6); then take $y = 1$, which yields

$$(14.8) \qquad f(x) = \frac{x(1-x)^3}{1-5x+7x^2-4x^3}.$$

From (14.8) we see that $a_n$ satisfies the recurrence relation

$$(14.9) \qquad a_n = 5a_{n-1} - 7a_{n-2} + 4a_{n-3} \qquad (n \geq 5).$$

As we remarked above, it is not at all clear how one could prove this directly; it has been done however.

REMARK. From (14.9) we find that $\lim_{n \to \infty} a_n^{1/n} = \theta$, where $\theta$ is the zero with largest absolute value of the polynomial $x^3 - 5x^2 + 7x - 4$ ($\theta \approx 3.2$).

The following example produces a result that is important for the theory of finite fields. In this example, we combine a generalization of the idea of Example 14.2 with the method of formal operations with power series. The reader should convince herself that the operations with logarithms are correct without using convergence.

EXAMPLE 14.6. We shall count the number of irreducible polynomials of degree $n$ over a field of $q$ elements.

Number all the monic irreducible polynomials of degree at least one over a field of $q$ elements:

$$f_1(x), f_2(x), f_3(x), \ldots$$

with, say, respective degrees $d_1, d_2, d_3, \ldots$ . Let $N_d$ denote the number of degree $d$, $d = 1, 2, 3, \ldots$ .

Now for any sequence $i_1, i_2, i_3, \ldots$ of nonnegative integers (all but finitely many of which are zero), we get a monic polynomial

$$f(x) = (f_1(x))^{i_1}(f_2(x))^{i_2}(f_3(x))^{i_3} \cdots$$

whose degree is $n = i_1 d_1 + i_2 d_2 + i_3 d_3 + \cdots$ . By unique factorization, every monic polynomial of degree $n$ arises exactly once in this way. To repeat, there is a one-to-one correspondence between monic polynomials of degree $n$ and sequences $i_1, i_2, \ldots$ of nonnegative integers satisfying $n = i_1 d_1 + i_2 d_2 + i_3 d_3 + \cdots$ .

Of course, the number of monic polynomials of degree $n$ is $q^n$, i.e. the coefficient of $x^n$ in

$$\frac{1}{1 - qx} = 1 + qx + (qx)^2 + (qx)^3 + \cdots .$$

Clearly (cf. Example 14.2), the number of sequences $i_1, i_2, \ldots$ with $n = i_1 d_1 + i_2 d_2 + \ldots$ is the coefficient of $x^n$ in the formal power series

$$(1 + x^{d_1} + x^{2d_1} + x^{3d_1} + \cdots)(1 + x^{d_2} + x^{2d_2} + x^{3d_2} + \cdots) \cdots .$$

Thus we conclude

$$\frac{1}{1 - qx} = \prod_{i=1}^{\infty} \frac{1}{1 - x^{d_i}} = \prod_{d=1}^{\infty} \left(\frac{1}{1 - x^d}\right)^{N_d} .$$

Recalling that $\log \frac{1}{1-z} = z + \frac{1}{2}z^2 + \frac{1}{3}z^3 + \cdots$ , we take (formal) logarithms of both extremes of the above displayed equation to find

$$\sum_{n=1}^{\infty} \frac{(qx)^n}{n} = \sum_{d=1}^{\infty} N_d \sum_{j=1}^{\infty} \frac{x^{jd}}{j} .$$

Then, comparing coefficients of $x^n$ on both sides, we get

$$\frac{q^n}{n} = \sum_{d|n} N_d \frac{1}{n/d} ,$$

i.e.

$$q^n = \sum_{d|n} d N_d.$$

This last equation was our goal. We have derived it combinatorially from unique factorization, but we remark that it has in fact an elegant interpretation in terms of the factorization of $x^{q^n} - x$ over the field of $q$ elements. Applying Möbius inversion (cf. Theorem 10.4), we obtain the following theorem.

THEOREM 14.1. *For a prime power $q$, the number of monic irreducible polynomials of degree $n$ over the field of $q$ elements is given by*

$$N_n = \frac{1}{n} \sum_{d|n} \mu(\frac{n}{d}) q^d.$$

We note that a direct consequence of Theorem 14.1 is that $N_d > 0$, that is, there exist irreducible polynomials of every degree $d$. This immediately leads to a proof of the existence of fields of $p^d$ elements for every prime $p$, without the usual expedient of reference to the algebraic closures of the prime fields.

$$* * *$$

We now come to a large class of counting problems, all with the same solution. The class is known as the *Catalan family* because Catalan (1838) treated Example 14.7. Actually, the equivalent problem called Question 3 in Example 14.9 below was treated by von Segner and Euler in the 18th century. We denote the solutions to the problems by $u_n$. We shall show that $u_n = \frac{1}{n}\binom{2n-2}{n-1}$. These numbers are called *Catalan numbers*.

EXAMPLE 14.7. Suppose that we have a set $S$ with a *nonassociative* product operation. An expression $x_1 x_2 \ldots x_n$ with $x_i \in S$ does not make sense, and brackets are needed to indicate the order in which the operations are to be carried out. Let $u_n$ denote the number of ways to do this if there are $n$ factors $x_i$. For example, $u_4 = \frac{1}{4}\binom{6}{3} = 5$ corresponding to the products $(a(b(cd)))$, $(a((bc)d))$, $((ab)(cd))$, $((a(bc))d)$, and $(((ab)c)d)$. Each product contains within the outer brackets two expressions, the first a product of $m$ factors and the second a product of $n - m$ factors, where $1 \leq m \leq n - 1$. It follows that the numbers $u_n$ satisfy the recurrence relation

$$(14.10) \qquad u_n = \sum_{m=1}^{n-1} u_m u_{n-m} \qquad (n \geq 2).$$

From (14.10) and $u_1 = 1$, we find that the generating function $f(x) := \sum_{n=1}^{\infty} u_n x^n$ satisfies the equation

$$(14.11) \qquad f(x) = x + \sum_{n=2}^{\infty} \left( \sum_{m=1}^{n-1} u_m u_{n-m} \right) x^n = x + (f(x))^2 .$$

Solving the quadratic equation and taking into account that $f(0) = 0$, we find

$$f(x) = \frac{1 - \sqrt{1 - 4x}}{2}.$$

From the binomial series we then obtain

$$(14.12) \qquad u_n = \frac{1}{n} \binom{2n - 2}{n - 1}.$$

The operations carried out above can all be justified as formal operations on power series, and the binomial series can also be treated formally without being concerned about convergence. If one feels uneasy about this approach, there are two remedies. The first is to find the solution as we did above and afterwards prove that it is correct by using (14.10) and induction. If one really wishes to be sure that the generating function is defined by a convergent power series, then use the result of the calculation to find some rough estimate for $u_n$ and prove that the estimate is correct using (14.11), and then show that the series converges. For example, we could try $u_n \leq c^{n-1}/n^2$. This is true for $n = 1$ and any positive $c$ and it is true for $n = 2$ if $c \geq 4$. The induction step with (14.11) yields

$$u_n \leq c^{n-2} \sum_{m=1}^{n-1} \frac{1}{m^2(n - m)^2} \leq c^{n-2} \frac{2}{(n/2)^2} \sum_{m=1}^{\infty} \frac{1}{m^2} < \frac{c^{n-1}}{n^2}$$

if $c > \frac{4}{3}\pi^2$. So the radius of convergence of $\sum_{n=1}^{\infty} u_n x^n$ is positive.

REMARK. Formula (14.10) shows that $u_n \geq u_m u_{n-m}$, and then Fekete's lemma implies that $\lim_{n \to \infty} u_n^{1/n}$ exists. In fact, (14.12) now shows that this limit is 4.

As we remarked above, the Catalan numbers reappear regularly in combinatorial counting problems. This has led to the question of proving the other results combinatorially, i.e. by showing that there is a one-to-one correspondence between the objects to be counted and those of Example 14.7. Some of these will be treated in Example 14.9. First we look at another famous counting problem that has the Catalan numbers as solution.

EXAMPLE 14.8. Consider walks in the $X$-$Y$ plane where each step is $U : (x, y) \to (x+1, y+1)$ or $D : (x, y) \to (x+1, y-1)$. We start at (0,0) and ask in how many ways we can reach $(2n, 0)$ without *crossing* the $X$-axis. The solution of this problem uses an elegant trick known as André's *reflection principle* (1887). In Fig. 14.2 we consider two points $A$ and $B$ in the upper halfplane and a possible path between them which meets and/or crosses the $X$-axis.

By reflecting the part of the path between $A$ and the first meeting with the $X$-axis (i.e. $C$ in Fig. 14.2) with respect to the $X$-axis, we find a path from the reflected point $A'$ to $B$. This establishes a one-to-one correspondence between paths from $A'$ to $B$ and paths from $A$ to $B$ that meet or cross the $X$-axis.

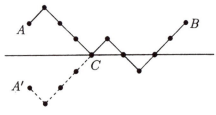

Figure 14.2

It follows that if $A = (0, k)$ and $B = (n, m)$, then there are $\binom{n}{l_1}$ paths from $A$ to $B$ that cross or meet the $X$-axis, where $2l_1 := n -$

$k-m$. Since there are $\binom{n}{l_2}$ paths from $A$ to $B$, where $2l_2 := n-m+k$, we find $\binom{n}{l_2} - \binom{n}{l_1}$ paths from $A$ to $B$ that do not meet the $X$-axis. Any path from $(0,0)$ to $(2n,0)$ in the upper halfplane that does not meet the $X$-axis between these points goes from $(0,0)$ to $(1,1):=A$, from $A$ to $B := (2n-1,1)$ without meeting the $X$-axis, and then from $(2n-1,1)$ to $(2n,0)$. By the argument above, we find that there are $u_n$ such paths. If we allow the paths to meet the $X$-axis without crossing, then there are $u_{n+1}$ such paths.

We remark that the number of paths from $(0,0)$ to $(2n,0)$ in the upper halfplane that do not meet the $X$-axis between these points is equal to the number of sequences of zeros and ones $(x_1, x_2, \ldots, x_{2n})$ with

$$(14.13) \qquad x_1 + x_2 + \cdots + x_j \begin{cases} < \frac{1}{2}j & \text{for } 1 \leq j \leq 2n-1, \\ = n & \text{for } j = 2n. \end{cases}$$

The correspondence is given by letting a 1 correspond to a step $D$ of the path.

We now show by combinatorial arguments that several counting problems lead to the Catalan numbers.

EXAMPLE 14.9. Consider the following three questions.

Question 1. A tree that is drawn in the plane is called a *plane tree*. How many rooted plane trees are there with $n$ vertices and a root with degree 1 (so-called *planted* plane trees)?

Question 2. A planted plane tree is called *trivalent* or a *binary tree* if every vertex has degree 1 or 3. How many trivalent planted plane trees are there with $n$ vertices of degree 1?

Question 3. In how many ways can one decompose a convex $n$-gon into triangles by $n-3$ nonintersecting diagonals?

The first correspondence we shall show is between Question 2 and Example 14.7.

For this, Fig. 14.3 suffices.

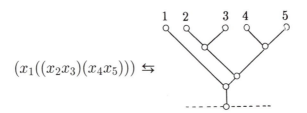

$$(x_1((x_2x_3)(x_4x_5))) \leftrightarrows$$

Figure 14.3

It follows that the solution to Question 2 is $u_{n-1}$.

The correspondence between Question 1 and Example 14.8 is also seen from a figure, namely Fig. 14.4.

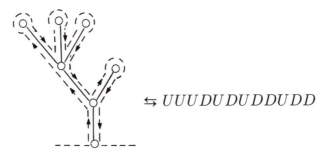

$\leftrightarrows UUUDUDUDDUDD$

Figure 14.4

In the figure, the tree can be described by a walk around it as shown by the dotted line. This walk can be described by calling the steps $U$ for a step going up, and $D$ for a step going down. This yields a sequence of twelve steps. Replacing $U$ by 0 and $D$ by 1 yields a sequence that obviously satisfies (14.13). This shows that the solution to Question 1 is also $u_{n-1}$.

Finally, we show a correspondence between Question 3 and Question 2. Consider an $n$-gon decomposed into triangles and distinguish some edge. We now construct a tree as in Fig. 14.5.

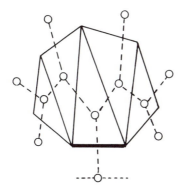

Figure 14.5

The tree is a binary tree with vertices of degree 1 corresponding to the sides of the $n$-gon, and vertices of degree 3 corresponding to the triangles. The tree is planted at the vertex corresponding to the special edge. So again we find that the number of solutions is $u_{n-1}$.

PROBLEM 14B. Describe (with a figure) a correspondence between the graphs of Question 1 and those of Question 2.

PROBLEM 14C. Find a direct one-to-one mapping from the $n$-gons of Question 3 to the products of Example 14.7.

We now turn to several problems in which the exponential generating function turns out to be useful. We shall give a somewhat more systematic treatment for this case. For a complete treatment of the methods, we refer to A. Joyal (1981).

Let $M$ denote a "type" of combinatorial structure. For example, trees, polygons, sets (the "uniform" structures), permutations, etc. Let $m_k$ be the number of ways of giving a labeled $k$-set such a structure. In each separate case we shall specify whether we take $m_0 = 0$ or $m_0 = 1$. We use capital letters for the structure and lower case letters for the counting sequence. We define

$$(14.14) \qquad M(x) := \sum_{k=0}^{\infty} m_k \frac{x^k}{k!}.$$

So if $T$ denotes the structure (labeled) tree, then we know from

Theorem 2.1 that

$$T(x) = \sum_{k=0}^{\infty} k^{k-2} \frac{x^k}{k!}.$$

If $S$ denotes the uniform structure (a set) then $s_k = 1$ for all $k$ and therefore $S(x) = e^x$. If $C$ denotes "oriented circuit", then we can start at vertex 1 and there are clearly $(k-1)!$ ways to proceed. Therefore $C(x) = -\log(1-x)$.

Suppose we wish to consider the number of ways a labeled $n$-set can be partitioned into two parts, one with a structure of type $A$ and the other with a structure of type $B$. It is easy to see that the number of ways that we can do this is $\sum_{k=0}^{n} \binom{n}{k} a_k b_{n-k}$. It follows that if we call this a structure of type $A \cdot B$, then

$$(14.15) \quad (A \cdot B)(x) = \sum_{n=0}^{\infty} \left( \sum_{k=0}^{n} \binom{n}{k} a_k b_{n-k} \right) \frac{x^n}{n!} = A(x) \cdot B(x).$$

EXAMPLE 14.10. Once again we consider derangements. Call this a structure of type $D$ and let $\Pi$ denote the structure "permutation". Clearly $\Pi(x) = (1-x)^{-1}$. Any permutation consists of a set of fixed points (that we interpret as just a set) and a derangement on the remaining points. So by (14.15) we have

$$(1-x)^{-1} = D(x) \cdot S(x), \text{ i.e. } D(x) = e^{-x}(1-x)^{-1}$$

as was shown in Example 10.1.

EXAMPLE 14.11. In how many ways can a labeled $n$-set be split into a number of pairs ($= P$) and a number of singletons ($= S$)? First, observe that if we wish to split $2k$ points into pairs, we must choose a point $x_1$ to go with point 1 and then split the remaining $2k-2$ points into pairs. Therefore $p_{2k} = (2k-1)!!$ and we find that

$$P(x) = \sum_{k=0}^{\infty} (2k-1)!! \frac{x^{2k}}{(2k)!} = \exp(\frac{1}{2}x^2).$$

It follows that

$$(14.16) \qquad (P \cdot S)(x) = \exp(x + \frac{1}{2}x^2).$$

Let us now try to find the same result using a recursion relation. We denote the structure $P \cdot S$ by $B$. In the set $\{1, 2, \ldots, n\}$ we can either let $n$ be a singleton or make a pair $\{x, n\}$ with $1 \le x \le n-1$. So

$$b_n = b_{n-1} + (n-1)b_{n-2} \qquad (n \ge 1).$$

It follows that

$$B'(x) = (1+x)B(x),$$

and since $B(0) = 1$, we again find (14.16) as solution.

We return to the recurrence of Example 14.11 in Example 14.15.

PROBLEM 14D. Consider the structures $M_0 :=$ mapping of a set to itself with no fixed point, $M_1 :=$ mapping of a set to itself with exactly one fixed point, and $A :=$ arborescence. Find a relation between $M_0(x)$, $M_1(x)$, and $A(x)$ and check the first few terms to see if the result is correct.

We make things slightly more difficult. We wish to partition an $n$-set into parts and then impose structure $N$ on each of the parts, where $n_0 = 0$. We claim that the exponential generating function for the compound structure is $\exp(N(x))$. A simple example is the case where $N$ is the uniform structure (with the convention $n_0 = 0$) and hence $N(x) = e^x - 1$. The compound structure is of course "partition" for which we know from Theorem 13.6 that the exponential generating function is equal to

$$\sum_{n=0}^{\infty} \left( \sum_{k=0}^{n} S(n,k) \right) \frac{x^n}{n!} = \sum_{k=0}^{\infty} \frac{(e^x - 1)^k}{k!} = \exp(e^x - 1) = \exp(N(x)).$$

We formulate this method as a theorem.

THEOREM 14.2. *If the compound structure $S(N)$ is obtained by splitting a set into parts, each of which gets a structure of type $N$, then*

$$S(N)(x) = \exp(N(x)).$$

PROOF: By a slight generalization of (13.3) we see that if the partition of the $n$-set consists of $b_1$ parts of size 1, $b_2$ parts of size 2,

..., $b_k$ parts of size $k$, where $b_1 + 2b_2 + \cdots + kb_k = n$, then there are

$$\left(\frac{n_1}{1!}\right)^{b_1} \cdots \left(\frac{n_k}{k!}\right)^{b_k} \cdot \frac{n!}{b_1! \ldots b_k!}$$

ways to make the compound structure, and this has to be divided by $n!$ to obtain the contribution to the coefficient of $x^n$ in the exponential generating function. If $b_1 + \cdots + b_k = m$, then this same contribution is found in $\exp(N(x))$ from the term $(N(x))^m/m!$, namely as

$$\frac{1}{m!}\binom{m}{b_1, \ldots, b_k}\left(n_1 \frac{x}{1!}\right)^{b_1} \cdots \left(n_k \frac{x^k}{k!}\right)^{b_k}.$$

This proves the assertion.                                               □

   In fact, it is not difficult to see that this theorem is a special case of a more general method. Interpret the previous situation as follows. We have a uniform structure on a $k$-set of points and we replace each point by some structure of type $N$. The resulting configuration is an element of the compound structure that we just discussed. If the first of the structures is not uniform but, say, of type $R$, then the exponential generating function for the compound structure will be $R(N(x))$ by the same argument as we used above. We sometimes call this procedure the substitution of $N$ into $R$, and it is nice that we then must do the same for the generating functions!

EXAMPLE 14.12. If we substitute the structure "oriented cycle" into the uniform structure, then we are considering the compound structure consisting of a partition of an $n$-set into oriented cycles, i.e. the structure $\Pi$ with as usual $\pi_0 = 1$. So we must have $\Pi(x) = \exp(C(x))$ and, indeed, $\Pi(x) = (1-x)^{-1}$ and $C(x) = -\log(1-x)$.

EXAMPLE 14.13. Let us look at Fig. 2.2 again, adding loops at 1 and 21. We then have a description of a mapping of an $n$-set (with $n = 21$ in this case) to itself. But we can also interpret the figure as an element of the structure $\Pi$, namely $(1)(4, 5, 3)(7)(20, 12)(21)$ in which each point has been replaced by an arborescence with that point as root. Letting $A$ denote arborescence, we find from Cayley's theorem, Theorem 2.1, that

$$(14.17) \qquad A(x) = \sum_{n=1}^{\infty} n^{n-1} \frac{x^n}{n!}.$$

(The extra factor $n$ comes from the choice of the root.) Since there are $n^n$ mappings from an $n$-set to itself, the method of Theorem 14.2 shows that the following relation must hold:

$$(14.18) \qquad \Pi(A(x)) = \frac{1}{1 - A(x)} = \sum_{n=0}^{\infty} n^n \frac{x^n}{n!}.$$

We shall verify this after the next example.

The following well known result from complex analysis often plays a role in combinatorial problems involving generating functions. It is known as the *Lagrange inversion formula*. It may be found in textbooks on analysis with analytic proofs; see the notes. It is possible to prove the theorem within the theory of formal power series (using formal derivation, etc.) and we give such a formal proof in Appendix 2.

THEOREM 14.3. *Let $f$ be analytic in a neighborhood of $z = 0$ and $f(0) \neq 0$. Then, if $w = z/f(z)$, $z$ can be expressed as a power series $z = \sum_{k=1}^{\infty} c_k w^k$ with a positive radius of convergence, in which*

$$(14.19) \qquad c_k = \frac{1}{k!} \left\{ \left( \frac{d}{dz} \right)^{k-1} (f(z))^k \right\}_{z=0}.$$

EXAMPLE 14.14. We shall find a fourth proof of Theorem 2.1. Let $T$ be the structure "labeled tree". We wish to show that $t_n = n^{n-2}$ for $n \geq 1$. If $A$ denotes arborescences as before, then obviously $a_n = n t_n$, i.e. $A(x) = xT'(x)$. Furthermore, from Theorem 14.2 we see that $\exp(A(x))$ is the exponential generating function for the structure "rooted forest" ($= F$). Consider a labeled tree on $n + 1$ vertices as an arborescence with vertex $n + 1$ as its root and then delete the root and all incident edges. The result is a rooted forest on $n$ vertices. Since we can reverse this process, we have a one-to-one correspondence. Therefore

$$(14.20) \quad e^{A(x)} = 1 + \sum_{n=1}^{\infty} f_n \frac{x^n}{n!} = \sum_{n=0}^{\infty} t_{n+1} \frac{x^n}{n!} = T'(x) = x^{-1} A(x),$$

i.e.

$$(14.21) \qquad \frac{A(x)}{e^{A(x)}} = x.$$

We apply Theorem 14.3 to (14.21) with $z = A(x)$, $f(z) = e^z = e^{A(x)}$, $w = x$. We find $A(x) = \sum_{k=1}^{\infty} c_k x^k$ with

$$c_k = \frac{1}{k!} \left\{ \left( \frac{d}{dz} \right)^{k-1} e^{kz} \right\}_{z=0} = \frac{k^{k-1}}{k!}$$

and it follows that $t_n = n^{n-2}$. Furthermore we see that the number of labeled rooted forests on $n$ vertices is $(n+1)^{n-1}$.

REMARK. We now have verification of (14.18) because (14.20) implies that $A'(x) = e^{A(x)} + x e^{A(x)} A'(x)$, i.e.

$$\sum_{n=1}^{\infty} n^n \frac{x^n}{n!} = x A'(x) = \frac{x e^{A(x)}}{1 - x e^{A(x)}} = \frac{A(x)}{1 - A(x)}.$$

REMARK. The procedure of removing a point from a combinatorial structure with $n+1$ points that we used in Example 14.14 is called *derivation* and it indeed corresponds to the derivation of generating functions. Another example is removing a vertex from an oriented cycle on $n+1$ vertices. This yields an oriented path on $n$ vertices, of which there are $n!$. So their exponential generating function is $(1-x)^{-1}$ and that is $C'(x)$.

PROBLEM 14E. Let $a_n$ denote the number of ways of decomposing a convex $n+1$-gon into quadrilaterals by inserting a number of nonintersecting chords. By convention $a_0 = 0$, $a_1 = 1$. Show that $\sum_{k+l+m=n} a_k a_l a_m = a_n$ for $n \geq 3$. If $f(x)$ is the ordinary generating function for the sequence $a_n$, then find a functional equation for

$f(x)$ and solve this equation using Theorem 14.3. (We remark that the result that is obtained can also be proved combinatorially.)

Although we have stressed that in most cases the power series can be considered as formal power series, it has been clear in a number of examples that analysis can be an important tool in many combinatorial problems. We give one more example that shows a method that can be used for many recurrences.

EXAMPLE 14.15. In Example 14.11 we encountered the recursion

$$(14.22) \qquad a_n = a_{n-1} + (n-1)a_{n-2}$$

and the corresponding exponential generating function. In previous examples we showed that the counting function grows more or less as $c^n$ for some constant $c$. Clearly $a_n$ grows more rapidly, but what is a good approximation for its asymptotic behavior? The first step in obtaining the answer to this question involves a method that is applicable to many recurrences. In (14.22) we substitute

$$(14.23) \qquad a_n = \int_C \psi(z) z^n \, dz,$$

where $C$ is a path in $\mathbb{C}$ that we can still choose and $\psi$ is a function also still to be determined. We substitute (14.23) into (14.22). The term $(n-1)a_{n-2}$ yields $\int_C \psi(z)(n-1)z^{n-2} \, dz$, and we require that integration by parts yields only the integral involving $\psi'(z)$, the other term being 0 by a suitable choice of $C$. Then (14.22) becomes

$$(14.24) \qquad \int_C \left\{ \psi(z)[z^n - z^{n-1}] + \psi'(z)z^{n-1} \right\} \, dz = 0,$$

which is true for all $n \in \mathbb{N}$ if $\psi(z)(1-z) = \psi'(z)$, i.e. $\psi(z) = \alpha e^{z-\frac{1}{2}z^2}$. Once we know $\psi$, the requirement on $C$ shows that the real axis from $-\infty$ to $\infty$ is a good choice for the path of integration. From $a_0 = 1$ we find that $\alpha = (2\pi e)^{-\frac{1}{2}}$.

To find the asymptotic behavior of $a_n$, we must now analyze the behavior of

$$(14.25) \qquad I := \int_{-\infty}^{\infty} e^{x - \frac{1}{2}x^2} x^n \, dx.$$

Since the integrand is maximal near $x = \sqrt{n}$, we substitute $x = y + \sqrt{n}$ and find

(14.26)
$$I = e^{-\frac{1}{2}n+\sqrt{n}}n^{\frac{1}{2}n}\int_{-\infty}^{\infty}e^{y-\frac{1}{2}y^2}\exp\left(-y\sqrt{n}+n\log(1+\frac{y}{\sqrt{n}})\right)dy.$$

Now note that if $u$ and $v$ are negative, then $|e^u - e^v| < |u - v|$ and use the representation $\log(1+t) = \int_0^t \frac{ds}{1+s} = s - \frac{s^2}{2} + \int_0^t \frac{s^2}{1+s}ds$ to obtain

$$\left|-y\sqrt{n}+n\log(1+\frac{y}{\sqrt{n}})+\frac{y^2}{2}\right| \le \frac{|y|^3}{\sqrt{n}}.$$

Substitution in (14.26) shows that the integral tends to

$$\int_{-\infty}^{\infty}e^{y-y^2}\,dy = \sqrt{\pi}e^{\frac{1}{4}}.$$

We have proved that

(14.27)
$$a_n \sim \frac{e^{-\frac{1}{4}}}{\sqrt{2}}n^{\frac{1}{2}n}e^{-\frac{1}{2}n+\sqrt{n}}.$$

It is clear that this result cannot be found in an easy way!

PROBLEM 14F. Let $F_n(x)$ denote the expansion of $(1-x^n)^{-\mu(n)/n}$ in a power series. Also consider the expansion of $e^x$ as a formal power series. Prove that

$$e^x = \prod_{n=1}^{\infty}F_n(x)$$

is true as a relation between formal power series.

PROBLEM 14G. Find the exponential generating function for the number of symmetric $n$ by $n$ permutation matrices.

PROBLEM 14H. On a circle we place $n$ symbols 0 and $n$ symbols 1 in an arbitrary order. Show that it is possible to number the positions on the circle consecutively from 1 to $2n$ such that the sequence satisfies (14.13) if we replace the strict inequality '$<$' with '$\leq$'.

PROBLEM 14I. Let the points $1, \ldots, 2n$ be on a circle (consecutively). We wish to join them in pairs by $n$ nonintersecting chords. In how many ways can this be done?

PROBLEM 14J. Find the exponential generating function for the number of labeled regular graphs of valency 2, with $n$ vertices.

**Notes.**

For an extensive treatment of the material in this chapter, we refer to Goulden and Jackson (1983) and Stanley (1986). A very readable survey is Stanley (1978).

The first of the references mentioned above has much on the theory of formal power series. For this topic also see Niven (1969) and Appendix 2. Generally speaking, operations with power series are "formal" and require no discussion of convergence when the coefficient of any monomial $x^n$, say, in a sum or product is determined by a finite number of operations (i.e. no limits). For example, $\sum_{n=0}^{\infty}(\frac{1}{2} + x)^n$ is not allowed in formal operations since even the constant term is not a finite sum; but $\sum_{n=0}^{\infty}(n!x + x^2)^n$ *is* allowed (even though it does not converge for $x \neq 0$) since the coefficient of any power of $x$ requires only a finite computation (cf. Problem 14F).

The first linear recurrence that one usually learns to solve is $a_{n+1} = a_n + a_{n-1}$. It leads to the famous *Fibonacci sequence*; cf. Problem 14A. As we observed in Chapter 10, this name is due to Lucas. The sequence is related to a problem occurring in the book *Liber abaci* (1203) by Leonardo of Pisa (known as Fibonacci).

The number $a_n$ of polyominoes of size $n$, treated in Example 14.5, was shown to be less than $c^n$ (where we used $c = 15$). Note that this assertion also follows from Fekete's lemma. We derived the inequality by using a suitable coding for the cells. The reader may wish to try to prove that $a(m, n) < m \cdot c^n$ for a suitable $c$ by using (14.3) and induction. The remark following (14.9) is based on a statement in Stanley (1986), attributing an unpublished combinatorial proof to D. Hickerson.

The result of Example 14.6 is related to the fact that $\mathbb{F}_{p^n}$ has as subfields the fields $\mathbb{F}_{p^d}$, where $d$ divides $n$. From this, one finds that $x^{p^n} - x$ is the product of all the irreducible polynomials over $\mathbb{F}_p$ with a degree $d$ dividing $n$. This yields the formula proved in the example.

Despite the fact that there is extensive literature on all the problems with the Catalan numbers as solution, the problems reappear regularly (e.g. in problem sections of journals). As stated earlier, the Belgian mathematician E. Catalan (1814–1894) studied well-formed bracketings (as in Example 14.7).

The reflection principle of Fig. 14.2 was used by the French combinatorialist D. André (1840–1917) in his solution of Bertrand's famous *ballot problem*: if, at the end of an election, candidate $P$ has $p$ votes and $Q$ has $q$ votes, $p < q$, then the probability that $Q$ was ahead all the time during the election is $(q - p)/(q + p)$.

The Lagrange inversion formula is one of the many contributions to analysis by J. L. Lagrange (see notes to Chapter 19). For a proof see §7.32 in Whittaker and Watson (1927). The theorem was published in 1770. We also refer the reader to G. N. Raney (1960).

**References.**

E. Catalan (1838), Note sur une équation aux différences finies, *J. M. Pures Appl.* **3**, 508–516.

I. P. Goulden and D. M. Jackson (1983), *Combinatorial Enumeration*, Wiley-Interscience.

A. Joyal (1981), Une théorie combinatoire des séries formelles, *Advances in Mathematics* **42**, 1–82.

D. A. Klarner (1967), Cell growth problem, *Canad. J. Math.* **19**, 851–863.

I. Niven (1969), Formal power series, *Amer. Math. Monthly* **76**, 871–889.

G. N. Raney (1960), Functional composition patterns and power series reversion, *Trans. Amer. Math. Soc.* **94**, 441–451.

R. P. Stanley (1978), Generating functions, pp. 100–141 in *Studies in Combinatorics* (G.-C. Rota, ed.), Studies in Math. **17**, Math. Assoc. of America.

R. P. Stanley (1986), *Enumerative Combinatorics*, Vol. I, Wadsworth and Brooks/Cole.

E. T. Whittaker and G. N. Watson (1927), *A Course in Modern Analysis*, Cambridge University Press.

# 15
## Partitions

We have considered several partition problems in the previous chapters. We now come to the most difficult one, namely the problem of *unordered* partitions of $n$ into $k$ parts, respectively any number of parts. We define $p_k(n)$ as the number of solutions of

$$(15.1) \qquad n = x_1 + x_2 + \cdots + x_k, \qquad x_1 \geq x_2 \geq \cdots \geq x_k \geq 1.$$

For example, $7 = 5 + 1 + 1 = 4 + 2 + 1 = 3 + 3 + 1 = 3 + 2 + 2$, so $p_3(7) = 4$.

Using the same idea as in the Corollary to Theorem 13.1, we see that $p_k(n)$ equals the number of solutions of $n - k = y_1 + \cdots + y_k$ with $y_1 \geq \cdots \geq y_k \geq 0$. If exactly $s$ of the integers $y_i$ are positive, then by (15.1) there are $p_s(n - k)$ solutions $(y_1, \ldots, y_k)$. Therefore

$$(15.2) \qquad\qquad p_k(n) = \sum_{s=1}^{k} p_s(n - k).$$

PROBLEM 15A. Show that $p_k(n) = p_{k-1}(n - 1) + p_k(n - k)$ and use this to prove (15.2).

Since we have the trivial initial conditions $p_k(n) = 0$ for $n < k$ and $p_k(k) = 1$, we can recursively calculate the numbers $p_k(n)$. Clearly $p_1(n) = 1$ and $p_2(n) = \lfloor n/2 \rfloor$.

EXAMPLE 15.1. We shall show that $p_3(n) = \{\frac{n^2}{12}\}$, i.e. the number nearest to $\frac{n^2}{12}$, using a method that can be used for other values of $k$. Let $a_3(n)$ denote the number of solutions of $n = x_1 + x_2 + x_3$, $x_1 \geq x_2 \geq x_3 \geq 0$. Then $a_3(n) = p_3(n + 3)$ and writing $y_3 = x_3$,

$y_2 = x_2 - x_3$, $y_1 = x_1 - x_2$, we see that $a_3(n)$ is the number of solutions of $n = y_1 + 2y_2 + 3y_3$, $y_i \geq 0$, $i = 1, 2, 3$. Therefore (cf. Example 14.2)

$$(15.3) \qquad \sum_{n=0}^{\infty} a_3(n)x^n = (1-x)^{-1}(1-x^2)^{-1}(1-x^3)^{-1}.$$

Let $\omega = e^{2\pi i/3}$. The partial fraction decomposition of (15.3) yields

$$(15.4) \quad \sum_{n=0}^{\infty} a_3(n)x^n = \frac{1}{6}(1-x)^{-3} + \frac{1}{4}(1-x)^{-2} + \frac{17}{72}(1-x)^{-1}$$

$$+ \frac{1}{8}(1+x)^{-1} + \frac{1}{9}(1-\omega x)^{-1} + \frac{1}{9}(1-\omega^2 x)^{-1}.$$

Using (10.6), we find from (15.4)

$$a_3(n) = \frac{1}{12}(n+3)^2 - \frac{7}{72} + \frac{(-1)^n}{8} + \frac{1}{9}(\omega^n + \omega^{2n}),$$

and this implies that

$$|a_3(n) - \frac{1}{12}(n+3)^2| \leq \frac{7}{72} + \frac{1}{8} + \frac{2}{9} < \frac{1}{2}.$$

Therefore

$$(15.5) \qquad\qquad p_3(n) = \{\frac{1}{12}n^2\}.$$

PROBLEM 15B. We choose three vertices of a regular $n$-gon and consider the triangle they form. Prove directly that there are $\{\frac{1}{12}n^2\}$ mutually incongruent triangles that can be formed in this way, thereby providing a second proof of (15.5).

We shall now show that (15.5) is an example of a more general result.

THEOREM 15.1. *If $k$ is fixed then*

$$p_k(n) \sim \frac{n^{k-1}}{k!(k-1)!} \qquad (n \to \infty).$$

PROOF: (i) If $n = x_1 + \cdots + x_k$, $x_1 \geq \cdots \geq x_k \geq 1$, then the $k!$ permutations of $(x_1, \ldots, x_k)$ yield solutions of (13.2) in positive integers, not necessarily all different. So

$$(15.6) \qquad\qquad k!\, p_k(n) \geq \binom{n-1}{k-1}.$$

If $n = x_1 + \cdots + x_k$, $x_1 \geq \cdots \geq x_k \geq 1$, then if $y_i := x_i + (k-i)$, $1 \leq i \leq k$, the integers $y_i$ are distinct and $y_1 + \cdots + y_k = n + \frac{k(k-1)}{2}$. Therefore

$$(15.7) \qquad\qquad k!\, p_k(n) \leq \binom{n + \frac{k(k-1)}{2} - 1}{k-1}.$$

The result follows from (15.6) and (15.7). $\qquad\qquad\qquad\qquad\square$

PROBLEM 15C. Let $a_1, a_2, \ldots, a_t$ be positive integers (not necessarily distinct) with greatest common divisor 1. Let $f(n)$ denote the number of solutions of

$$n = a_1 x_1 + a_2 x_2 + \cdots + a_t x_t$$

in nonnegative integers $x_1, \ldots, x_t$. What is the generating function $F(x) := \sum f(n) x^n$? Show that $f(n) \sim c n^{t-1}$ for some constant $c$ and explicitly give $c$ as a function of $a_1, \ldots, a_t$.

The problem that we considered in the Corollary to Theorem 13.1 concerns what is called the number of *compositions* of $n$ into $k$ parts. By this corollary the total number of compositions of $n$ equals $\sum_{k=1}^{n} \binom{n-1}{k-1} = 2^{n-1}$. The method that we used in Theorem 13.1 shows this directly as follows: consider $n$ blue balls with a space between any two of them in which we can place a red ball if we wish. The total number of configurations that can be made in this way is clearly $2^{n-1}$ and they represent all compositions of $n$. We could have proved the same result without our knowledge of Chapter 13 as follows. Let $c_{nk}$ denote the number of compositions of $n$ into $k$ parts. Define

$$c_k(x) := \sum_{n=k}^{\infty} c_{nk} x^n.$$

Then, using a method that we have used several times before, we find

$$c_k(x) = (x + x^2 + x^3 + \cdots)^k = x^k(1-x)^{-k} = \sum_{n=k}^{\infty} \binom{n-1}{k-1} x^n$$

and

$$\sum_{k=1}^{\infty} c_k(x) = \sum_{k=1}^{\infty} x^k(1-x)^{-k} = \frac{x}{1-2x},$$

i.e. $n$ has $2^{n-1}$ compositions.

PROBLEM 15D. Show that in a list of all $2^{n-1}$ compositions of $n$, the integer 3 occurs exactly $n \cdot 2^{n-5}$ times.

The most difficult and mathematically the most interesting and important function of all in this area of combinatorics is defined by

$$(15.8) \qquad p(n) := \text{number of unordered partitions of } n.$$

For example, $p(5) = 7$, the partitions of 5 being

$$1+1+1+1+1 = 2+1+1+1 = 3+1+1 = 2+2+1 = 4+1 = 3+2 = 5.$$

Clearly, $p(n)$ is the number of solutions of

$$(15.9) \qquad n = x_1 + x_2 + \cdots + x_n \qquad (x_1 \geq x_2 \geq \ldots x_n \geq 0).$$

As we saw above, this can also be formulated as the number of solutions of $n = y_1 + 2y_2 + \cdots + ny_n$, where $y_i \geq 0$, $1 \leq i \leq n$. (Here $y_i$ denotes the number of terms $x_k$ equal to $i$ in the first definition.)

The study of the so-called *partition function* $p(n)$ has led to some of the most fascinating areas of analytic number theory, analysis, algebraic geometry, etc. We shall obviously only be able to look at properties that are proved by combinatorial arguments.

THEOREM 15.2. *The generating function for the partition function is*

$$P(x) := \sum_{n=0}^{\infty} p(n)x^n = \prod_{k=1}^{\infty} (1 - x^k)^{-1}.$$

PROOF: We again use the idea of Example 14.2. The partition of $n = \sum_{i=1}^{m} i r_i$ with $r_i$ terms equal to $i$ (for $1 \leq i \leq m$) corresponds to the term $x_1^{r_1} \cdots x_m^{r_m}$ in the (formal) infinite product $\prod (1 + x_k + x_k^2 + \cdots)$. Replacing $x_k$ by $x^k$, we find the result. □

Many theorems about partitions can be proved easily by representing each partition by a diagram of dots, known as a *Ferrers diagram*. Here we represent each term of the partition by a row of dots, the terms in descending order, with the largest at the top. Sometimes it is more convenient to use squares instead of dots (in this case the diagram is called a *Young diagram* by some authors but we shall use Ferrers diagram as the name for the figure). For example, the partition (5,4,2,1) of 12 is represented by each of the diagrams of Fig. 15.1.

Figure 15.1

The partition we get by reading the Ferrers diagram by columns instead of rows is called the *conjugate* of the original partition. So the conjugate of $12 = 5 + 4 + 2 + 1$ is $12 = 4 + 3 + 2 + 2 + 1$. The relationship is symmetric. We first show a few easy examples of the use of a Ferrers diagram.

THEOREM 15.3. *The number of partitions of $n$ into parts, the largest of which is $k$, is $p_k(n)$.*

PROOF: For each partition for which the largest part is $k$, the conjugate partition has $k$ parts (and vice versa). □

PROBLEM 15E. Show with a Ferrers diagram that the number of partitions of $n + k$ into $k$ parts equals the number of partitions of $n$ into at most $k$ parts. (This is (15.2).)

In some cases a proof using generating functions requires less ingenuity than using a Ferrers diagram would. The next theorem can be proved with a diagram similar to a Ferrers diagram though we shall use generating functions instead.

THEOREM 15.4. *The number of partitions of $n$ into odd parts equals the number of partitions of $n$ into unequal parts.*

PROOF: The generating function for the number of partitions of $n$ into odd parts is (by an obvious generalization of Theorem 15.2) equal to $\prod_{m=1}^{\infty}(1 - x^{2m-1})^{-1}$, and the generating function for the number of partitions of $n$ into unequal parts is (by the same argument as used in the proof of Theorem 15.2) $\prod_{k=1}^{\infty}(1 + x^k)$. Since

$$\prod_{k=1}^{\infty}(1 + x^k) = \prod_{k=1}^{\infty} \frac{(1 - x^{2k})}{(1 - x^k)}$$

$$= \prod_{k=1}^{\infty}(1 - x^{2k}) \prod_{l=1}^{\infty}(1 - x^l)^{-1} = \prod_{m=1}^{\infty}(1 - x^{2m-1})^{-1},$$

the proof is complete.                                                    □

Now consider the function $(P(x))^{-1} = \prod_{k=1}^{\infty}(1 - x^k)$. In the expansion of this product, a partition of $n$ into unequal parts contributes $+1$ to the coefficient of $x^n$ if the number of parts is even, and $-1$ if the number of parts is odd. So the coefficient of $x^n$ is $p_e(n) - p_o(n)$, where $p_e(n)$, respectively $p_o(n)$, denotes the number of partitions of $n$ into an even, respectively odd, number of unequal parts. It was shown by Euler that $p_e(n) = p_o(n)$ unless $n$ has the form $n = \omega(m) := (3m^2 - m)/2$ or $n = \omega(-m) = (3m^2 + m)/2$, in which case $p_e(n) - p_o(n) = (-1)^m$. The numbers $\omega(m)$ and $\omega(-m)$ are called *pentagonal numbers*. This name is explained by the relation $\omega(m) = \sum_{k=0}^{m-1}(3k + 1)$ and Fig. 15.2.

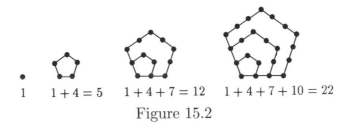

$$1 \qquad 1 + 4 = 5 \qquad 1 + 4 + 7 = 12 \qquad 1 + 4 + 7 + 10 = 22$$

Figure 15.2

The following extremely elegant pictorial proof of *Euler's identity* is due to Franklin (1881).

THEOREM 15.5. *We have*

$$\prod_{k=1}^{\infty}(1 - x^k) = 1 + \sum_{m=1}^{\infty}(-1)^m \left( x^{\omega(m)} + x^{\omega(-m)} \right).$$

PROOF: Consider a Ferrers diagram of a partition of $n$ into unequal parts as for $23 = 7 + 6 + 5 + 3 + 2$ in Fig. 15.3.

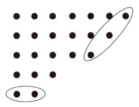

Figure 15.3

The last row is called the *base* of the diagram. The number of dots in the base is denoted by $b$. In Fig. 15.3 the base is indicated by a line segment with $b = 2$ dots. The longest 45° line segment joining the last point in the top row with other points of the diagram is called the *slope* of the diagram. Its number of points is denoted by $s$. In Fig. 15.3 the slope is indicated by a segment with $s = 3$ points. We now define two operations on the diagram, which we call $A$ and $B$.

$A$: If $b \leq s$, then remove the base and adjoin it to the diagram at the right to form a new slope parallel to the original one, unless $b = s$ and the base and the slope have a point in common. This exception is considered below.

$B$: If $b > s$, remove the slope and adjoin it at the bottom of the diagram as a new base, except if $b = s + 1$ and the base and the slope have a point in common.

In Fig. 15.3 operation $A$ can be carried out and it yields the partition $23 = 8+7+5+3$. An example of the exceptional situation for operation $B$ is shown in Fig. 15.4.

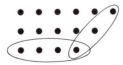

Figure 15.4

The exception for operation $A$ can occur only if

$$n = b + (b+1) + \cdots + (2b - 1),$$

i.e. $n = \omega(b)$. The exception for operation $B$ occurs if $n = (s + 1) + \cdots + 2s$, i.e. $n = \omega(-s)$. In all other cases exactly one of the operations can be carried out and we thus have a one-to-one correspondence between partitions of $n$ into an even number of unequal parts and partitions of $n$ into an odd number of unequal parts; and hence for these values of $n$, we have $p_e(n) - p_o(n) = 0$. In the exceptional cases the difference is $+1$ or $-1$.  □

Euler used Theorem 15.5 to find the following recursion formula for $p(n)$.

THEOREM 15.6. *Let $p(n) = 0$ for $n < 0$. Then for $n \geq 1$,*

$$(15.10) \quad p(n) = \sum_{m=1}^{\infty} (-1)^{m+1} \left\{ p(n - \omega(m)) + p(n - \omega(-m)) \right\}.$$

PROOF: This is an immediate consequence of Theorem 15.2 and Theorem 15.5.  □

Note that the sum in (15.10) is finite. With this recursion one can rapidly generate a table for $p(n)$ for small values of $n$.

The first two terms of the recursion formula are the same as in the famous Fibonacci recursion (cf. Problem 10G). Although these are followed by two negative terms, one might think that $p(n)$ will increase in a way similar to the Fibonacci numbers, i.e. as $c^n$ for some constant $c$. This is not the case. We shall show that $p(n)$ grows much more slowly. The actual asymptotics of $p(n)$ involve complicated methods from analytic number theory. We mention only the main term of the asymptotic formula:

$$\lim_{n \to \infty} n^{-\frac{1}{2}} \log p(n) = \pi \sqrt{\frac{2}{3}}.$$

One can show that in fact $p(n)$ is much smaller than $\exp\left(\pi\sqrt{\frac{2}{3}n}\right)$.

THEOREM 15.7. *For $n > 2$ we have*

$$p(n) < \frac{\pi}{\sqrt{6(n-1)}} e^{\pi\sqrt{\frac{2}{3}n}}.$$

PROOF: Let $f(t) := \log P(t)$. From Theorem 15.2 we find

$$f(t) = -\sum_{k=1}^{\infty} \log(1 - t^k) = \sum_{k=1}^{\infty}\sum_{j=1}^{\infty} \frac{t^{kj}}{j} = \sum_{j=1}^{\infty} \frac{j^{-1}t^j}{1 - t^j}.$$

From now on let $0 < t < 1$. Then from

$$(1-t)^{-1}(1-t^j) = 1 + t + \cdots + t^{j-1} > jt^{j-1},$$

we find

$$f(t) < \frac{t}{1-t}\sum_{j=1}^{\infty} j^{-2} = \frac{1}{6}\pi^2\frac{t}{1-t}.$$

Since $p(n)$ is increasing, we have $P(t) > p(n)t^n(1-t)^{-1}$. By combining the two inequalities and then substituting $t = (1+u)^{-1}$, we find

$$\log p(n) < f(t) - n\log t + \log(1-t)$$
$$< \frac{\pi^2}{6}\cdot\frac{t}{1-t} - n\log t + \log(1-t)$$
$$= \frac{\pi^2}{6}u^{-1} + n\log(1+u) + \log\frac{u}{1+u}.$$

Therefore

$$\log p(n) < \frac{1}{6}\pi^2 u^{-1} + (n-1)u + \log u.$$

We get the required inequality by substituting $u = \pi\{6(n-1)\}^{-\frac{1}{2}}$. □

Euler's identity is a special case of a famous identity of Jacobi that is important in the theory of theta functions. We would not mention this so-called *Jacobi triple product identity* if we had not known of the beautiful combinatorial proof using a Ferrers diagram, due to Wright (1965).

THEOREM 15.8. *We have*

$$\prod_{n=1}^{\infty}(1 - q^{2n})(1 + q^{2n-1}t)(1 + q^{2n-1}t^{-1}) = \sum_{r=-\infty}^{\infty} q^{r^2}t^r.$$

PROOF: We rewrite the assertion as

$$\prod_{n=1}^{\infty}(1 + q^{2n-1}t)(1 + q^{2n-1}t^{-1}) = \sum_{r=-\infty}^{\infty} q^{r^2}t^r \prod_{n=1}^{\infty}(1 - q^{2n})^{-1},$$

and substitute $x = qt$, $y = qt^{-1}$. This yields the relation

$$(15.11) \quad \prod_{n=1}^{\infty}\left\{(1 + x^n y^{n-1})(1 + x^{n-1}y^n)\right\}$$

$$= \sum_{r=-\infty}^{\infty} x^{\frac{1}{2}r(r+1)}y^{\frac{1}{2}r(r-1)} \prod_{n=1}^{\infty}(1 - x^n y^n)^{-1}.$$

We shall prove this relation by interpreting both sides as generating functions corresponding to the counting of appropriate combinatorial objects, and then showing a one-to-one mapping between these objects. For the lefthand side, we find as a combinatorial interpretation the generating function for the number of partitions of the Gaussian integer $n + mi$ into parts $a + (a - 1)i$ and $(b - 1) + bi$ ($a \geq 1, b \geq 1$), with no two parts equal. Call this number $\alpha(n, m)$. So the lefthand side of (15.11) is $\sum_{n=1}^{\infty}\sum_{m=1}^{\infty} \alpha(n, m)x^n y^m$. On the righthand side, we use Theorem 15.2 and replace the product by $\sum_{k=1}^{\infty} p(k)x^k y^k$. We must therefore prove that $\alpha(n, m) = p(k)$, where $n = k + \frac{1}{2}r(r + 1)$ and $m = k + \frac{1}{2}r(r - 1)$. Without loss of generality, we may assume that $n \geq m$, i.e. $r \geq 0$. A partition of $n + mi$ must have $v \geq 0$ terms of type $(b - 1) + bi$ and $v + r$ terms of type $a + (a - 1)i$, and therefore $n \geq \frac{1}{2}r(r + 1)$, so $k \geq 0$. In Fig. 15.5 below, we consider the example $(n, m) = (47, 44)$, so $r = 3$ and $k = 41$. We have taken the Ferrers diagram of the partition $41 = 12 + 10 + 8 + 5 + 2 + 2 + 2$ and above the top row we have added rows of length $r, r - 1, \ldots, 1$.

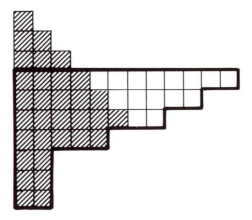

Figure 15.5

The resulting diagram is split into two parts. The shaded part on the left is determined by the "staircase" starting with the part that was added at the top. We read the shaded part by columns. By the construction, the sequence is decreasing. These numbers are the $a$'s for terms of type $a + (a - 1)i$. The unshaded part is read by rows and again the numbers are decreasing. These are the $b - 1$'s for terms of type $(b-1)+bi$ (the last $b$ could be 1, corresponding to one empty unshaded row). The number of $a$'s exceeds the number of $b-1$'s by $r$ and the sum of the $a$'s and $b-1$'s is clearly $k+\frac{1}{2}r(r+1)$. So we have indeed produced a partition of $n + mi$ of the required type. It is easy to see that the procedure can be reversed, i.e. we have defined the one-to-one mapping we were looking for.          □

PROBLEM 15F. Prove that the number of self-conjugate partitions of $n$ equals the number of partitions of $n$ into unequal odd parts.

As a final problem related to partitions we shall consider objects known as *Young tableaux* or *standard tableaux*. A Young tableau of shape $(n_1, n_2, \ldots, n_m)$ is a Ferrers diagram (or Young diagram) of squares in which the integers 1 to $n$ have been inserted (one in each square) in such a way that all rows and columns are increasing. For example, in Fig. 15.6 we display a Young tableau of shape (5,4,2,1).

| 1 | 3 | 4 | 7 | 11 |
|---|---|---|---|---|
| 2 | 5 | 10 | 12 | |
| 6 | 9 | | | |
| 8 | | | | |

Figure 15.6

We are interested in determining the number of Young tableaux of a given shape. At first, this may seem like a somewhat unnatural problem. However, Young tableaux play an important role in the theory of group representations (and other areas). One of the interesting facts concerning these tableaux is that there is a one-to-one correspondence between the Young tableaux with $n$ squares (that we shall call "*cells*") and involutions of 1 to $n$, where we include the identity as an involution. (Note that we can therefore count the total number of tableaux using Problem 14G.) For a treatment of this one-to-one correspondence and several related problems we refer to D. Knuth (1973).

In order to count the Young tableaux of a given shape, we need to introduce the function $\Delta(x_1, \ldots, x_m)$ defined by

$$(15.12) \qquad \Delta(x_1, \ldots, x_m) := \prod_{1 \le i < j \le m} (x_i - x_j).$$

(Note that $\Delta$ is the value of the Vandermonde determinant.)

LEMMA 15.9. *Let*

$$g(x_1, \ldots, x_m; y) := x_1 \, \Delta(x_1 + y, x_2, \ldots, x_m)$$
$$+ x_2 \, \Delta(x_1, x_2 + y, \ldots, x_m) + \cdots + x_m \, \Delta(x_1, x_2, \ldots, x_m + y).$$

*Then*

$$g(x_1, \ldots, x_m; y) = \left( x_1 + \cdots + x_m + \binom{m}{2} y \right) \Delta(x_1, \ldots, x_m).$$

PROOF: Clearly the function $g$ is a homogeneous polynomial of degree $1 + \deg \Delta(x_1, \ldots, x_m)$ in the variables $x_1, \ldots, x_m, y$. If we interchange $x_i$ and $x_j$, then $g$ changes sign. So if $x_i = x_j$, then

$g$ must be 0, i.e. $g$ is divisible by $x_i - x_j$ and hence by the polynomial $\Delta(x_1, \ldots, x_m)$. If $y = 0$, the assertion is obvious. Therefore we only have to prove the coefficient $\binom{m}{2}$ for $y$. If we expand $g$, then the terms of degree 1 in $y$ are $\frac{x_i y}{x_i - x_j} \Delta(x_1, \ldots, x_m)$ and $\frac{-x_j y}{x_i - x_j} \Delta(x_1, \ldots, x_m)$ for all pairs $(i, j)$ with $1 \le i < j \le m$. The sum of these terms is clearly $\binom{m}{2} \Delta(x_1, \ldots, x_m)$.   □

We introduce a function $f$ defined on all $m$-tuples $(n_1, \ldots, n_m)$, $m \ge 1$, with the following properties:

(15.13)
$$f(n_1, \ldots, n_m) = 0 \qquad \text{unless } n_1 \ge n_2 \ge \cdots \ge 0;$$

(15.14)
$$f(n_1, \ldots, n_m, 0) = f(n_1, \ldots, n_m);$$

(15.15)
$$f(n_1, \ldots, n_m) = f(n_1 - 1, n_2, \ldots, n_m) +$$
$$f(n_1, n_2 - 1, \ldots, n_m) + \cdots + f(n_1, n_2, \ldots, n_m - 1),$$
$$\text{if } n_1 \ge n_2 \ge \cdots \ge n_m \ge 0;$$

(15.16)
$$f(n) = 1 \qquad \text{if } n \ge 0.$$

Clearly $f$ is well defined. We claim that $f(n_1, \ldots, n_m)$ counts the number of Young tableaux of shape $(n_1, \ldots, n_m)$. Condition (15.13) is obvious and so are (15.14) and (15.16). To see that the number of Young tableaux of shape $(n_1, \ldots, n_m)$ satisfies (15.15), we consider the entry $n$. It must be the last entry in one of the rows, and if we remove the square with $n$, then we obtain a Young tableau for $n - 1$. In fact, if two or more rows of the tableau have the same length, then $n$ can only be the last entry in the lowest of these; but including all terms in (15.15) does not matter since if, say, $n_1 = n_2$, then $f(n_1 - 1, n_2, \ldots, n_m) = 0$.

THEOREM 15.10. *The number of Young tableaux that have shape* $(n_1, \ldots, n_m)$ *satisfies*

(15.17)   $$f(n_1, \ldots, n_m) = \frac{\Delta(n_1 + m - 1, n_2 + m - 2, \ldots, n_m) \, n!}{(n_1 + m - 1)!(n_2 + m - 2)! \ldots n_m!},$$

*and in fact this formula for $f$ is correct if* $n_1 + m - 1 \ge n_2 + m - 2 \ge \cdots \ge n_m$.

PROOF: We first observe that, if for some $i$ we have $n_i + m - i = n_{i+1} + m - i - 1$, then the expression $\Delta$ on the righthand side of (15.17) is 0, in accordance with the fact that the shape is not allowed. We must show that the righthand side of (15.17) satisfies (15.13) to (15.16). All but (15.15) are trivial. We find the relation (15.15) by substituting $x_i = n_i + m - i$ and $y = -1$ in Lemma 15.9. $\qquad\square$

There is a formulation of this theorem that makes it more interesting. We introduce the concept of *"hook"* in a Young tableau. The hook corresponding to a cell in a tableau is the union of the cell and all cells to the right in the same row and all cells below it in the same column. The *"hooklength"* is the number of cells in the hook. In Fig. 15.7 we show a Young diagram with in each cell a number equal to the length of the corresponding hook. The shaded area is the hook corresponding to the cell in row 2 and column 3.

| 12 | 11 | 9 | 7 | 5 | 2 | 1 |
|----|----|---|---|---|---|---|
| 9  | 8  | 6 | 4 | 2 | ● |   |
| 8  | 7  | 5 | 3 | 1 | ● |   |
| 6  | 5  | 3 | 1 | ● |   |   |
| 4  | 3  | 1 | ● |   |   |   |
| 2  | 1  | ● |   |   |   |   |

Figure 15.7

We now have the following remarkable theorem due to J. S. Frame, G. de Beauregard Robinson and R. M. Thrall (1954).

THEOREM 15.11. *The number of Young tableaux of a given shape and a total of $n$ cells is $n!$ divided by the product of all hooklengths.*

PROOF: If we have a Young tableau of shape $(n_1, \ldots, n_m)$, then the hooklength for the cell in row 1 and column 1 is $n_1 + m - 1$. Consider the hooklengths for the cells in row 1. In Fig. 15.7 the second column has the same length as column 1 and hence the hooklength for the next cell is $n_1 + m - 2$. For the next cell the length drops by 2 because of the "missing cell" indicated by a dot in the last row. Similarly, there is a jump of 3 for the column with two dots. So in row 1 we find as hooklengths the numbers from 1 to $n_1 + m - 1$ with the exception of the numbers $(n_1+m-1)-(n_j+m-j)$, $2 \le j \le m$. Similarly row $i$ contains the integers from 1 to $n_i+m-i$ except for $(n_i+m-i)-(n_j+m-j)$, $i+1 \le j \le m$. It follows that the product of the hooklengths is $\{\prod_{i=1}^{m}(n_i + m - i)!\}/\Delta(n_1 + m - 1, \ldots, n_m)$. By Lemma 15.9 we are done. $\qquad\qquad\qquad\qquad\qquad\qquad\qquad\qquad\square$

PROBLEM 15G. Consider a Young tableau of shape $(n, n)$. Define a sequence $a_k$, $1 \le k \le 2n$, by $a_k := i$ if $k$ is in row $i$ of the tableau, $i = 1, 2$. Use this to show that the number of Young tableaux of this shape is the Catalan number $u_{n+1}$ (in accordance with Theorem 15.11).

**Notes.**

The use of Ferrers diagrams to prove theorems such as Theorem 15.3 was introduced by Sylvester in 1853. He wrote that the proof had been communicated to him by N. M. Ferrers.

For other proofs involving Ferrers diagrams and analogues (e.g. a proof of Theorem 15.4) we refer to Hardy and Wright (1954), MacMahon (1916), and Van Lint (1974).

Euler's proof of Theorem 15.5 was by induction. Polygonal numbers occurred in ancient mathematics, e.g. triangular numbers were considered by Pythagoras before 500 BC.

Theorem 15.7 is due to Van Lint (1974).

The asymptotic behavior of the partition function $p(n)$ was given by G. H. Hardy and S. Ramanujan (1918) in the paper in which they developed the famous "circle method" that has had so many applications in number theory. A further asymptotic result was given by H. Rademacher (1937). The proofs depend on the remarkable functional equation for the famous Dedekind $\eta$-function

$$\eta(z) := e^{\frac{\pi i z}{12}} \prod_{n=1}^{\infty} \left(1 - e^{2\pi i n z}\right).$$

For an exposition of these proofs see Chandrasekharan (1970).

C. G. J. Jacobi (1804–1851) is best known for developing the theory of elliptic functions. He became professor at Königsberg at the age of 23. Two years later he published his book *Fundamenta*

*Nova Theoriae Functionum Ellipticarum.* In §64 we find what is now known as the triple product identity.

The first to use tableaux such as in Fig. 15.6 was Frobenius! A year later (1901) A. Young independently introduced them in his work on matrix representations of permutation groups. The name *Young tableaux* has become standard terminology.

**References.**

K. Chandrasekharan (1970), *Arithmetical Functions*, Springer-Verlag.

J. S. Frame, G. de Beauregard Robinson, and R. M. Thrall (1954), The hook graphs of $S_n$, *Canad. J. Math.* **6**, 316–324.

F. Franklin (1881), Sur le développement du produit infini $(1 - x)(1 - x^2)(1 - x^3)(1 - x^4) \cdots$, *Comptes Rendus Acad. Sci.* (Paris) **92**, 448–450.

G. H. Hardy and S. Ramanujan (1918), Asymptotic formulae in combinatory analysis, *Proc. London Math. Soc.* (2) **17**, 75–115.

G. H. Hardy and E. M. Wright (1954), *An Introduction to the Theory of Numbers*, 3d edn., Clarendon Press.

D. Knuth (1973), *The Art of Computer Programming*, Vol. 3, Addison-Wesley .

J. H. van Lint (1974), *Combinatorial Theory Seminar Eindhoven University of Technology*, Lecture Notes in Math. **382**, Springer-Verlag.

P. A. MacMahon (1916), *Combinatory Analysis* Vol. II, Cambridge University Press.

H. Rademacher (1937), On the partition function $p(n)$, *Proc. London Math. Soc.* **43**, 241–254.

E. M. Wright (1965), An enumerative proof of an identity of Jacobi, *J. London Math. Soc.* **40**, 55–57.

A. Young (1901), On quantitative substitutional analysis, *Proc. London Math. Soc.* **33**, 97–146.

# 16
# (0,1)-Matrices

In the previous chapters (0,1)-matrices have turned up a number of times; and especially those with constant linesums led to interesting problems. In this chapter we shall consider the existence problem for (0,1)-matrices with given linesums, and try to count or estimate how many there are in the case of constant linesums. In the first problem, the matrices are not necessarily square. If the rowsums of a matrix $A$ are $r_1, r_2, \ldots, r_k$, then we shall call the vector $\mathbf{r} := (r_1, r_2, \ldots, r_k)$ the *rowsum* of $A$, and similarly for the columnsums.

We consider the problem of the existence of a $(0, 1)$-matrix with given rowsum $\mathbf{r}$ and columnsum $\mathbf{s}$. For convenience, we shall assume that the coordinates of $\mathbf{r}$ and $\mathbf{s}$ are nonincreasing; that is, $\mathbf{r}$ and $\mathbf{s}$ are *partitions* (we may allow trailing zero coordinates in partitions for the purposes of this chapter).

Given two partitions $\mathbf{r} = (r_1, r_2, \ldots, r_n)$ and $\mathbf{s} = (s_1, s_2, \ldots, s_m)$ of the same integer $N$, we say that $\mathbf{r}$ *majorizes* $\mathbf{s}$ when $r_1 + r_2 + \cdots + r_k \geq s_1 + s_2 + \cdots + s_k$ for all $k$ (interpreting $r_k$ or $s_k$ as 0 when $k$ exceeds $n$ or $m$, respectively). Recall that the *conjugate* of a partition $\mathbf{r}$ is the partition $\mathbf{r}^*$ where $r_i^*$ is the number of $j$ such that $r_j \geq i$.

PROBLEM 16A. Prove the following assertion: $\mathbf{r}$ majorizes $\mathbf{s}$ if and only if the latter can be obtained from the former by a sequence of operations "pick two coordinates $a$ and $b$ so that $a \geq b + 2$ and replace them with $a - 1$ and $b + 1$". (That is, $\mathbf{s}$ is "more average" than $\mathbf{r}$.)

THEOREM 16.1. *Let $r_1, r_2, \ldots, r_n$ and $s_1, s_2, \ldots, s_m$ be two nonincreasing sequences of nonnegative integers each summing to a com-*

*mon value $N$. There exists an $n$ by $m$ $(0,1)$-matrix with rowsum $\mathbf{r}$ and columnsum $\mathbf{s}$ if and only if $\mathbf{r}^*$ majorizes $\mathbf{s}$.*

PROOF: To prove necessity, let $k$ be given and consider the first $k$ columns of a hypothetical $(0,1)$-matrix with rowsum $\mathbf{r}$ and columnsum $\mathbf{s}$. The number of 1's in these columns is

$$s_1 + s_2 + \cdots + s_k \le \sum_{i=1}^{n} \min(k, r_i) = \sum_{j=0}^{k} r_j^*.$$

The latter equality is most evident from the Ferrers diagram of the partition $\mathbf{r}$ (see Fig. 16.1) where we see both expressions count the number of cells in the first $k$ columns.

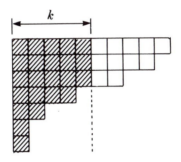

Figure 16.1

We now introduce a transportation network with a source $S$, a sink $T$, one vertex $x_i$ for each row, and one vertex $y_j$ for each column. There is to be an edge of capacity $r_i$ directed from $S$ to $x_i$, $1 \le i \le n$, an edge of capacity $s_j$ from $y_j$ to $T$, $1 \le j \le m$, and an edge of capacity 1 from $x_i$ to $y_j$, $1 \le i \le n, 1 \le j \le m$. We claim that a $(0,1)$-matrix $M = (a_{ij})$ with rowsum $\mathbf{r}$ and columnsum $\mathbf{s}$ exists if and only if this network admits a flow of strength $N$ (there are at least two cuts of capacity $N$, so this would be a maxflow). Given such a matrix, we get a flow of strength $N$ by saturating the edges incident with $S$ and $T$ and assigning flow $a_{ij}$ to the edge from $x_i$ to $y_j$. Conversely, if the network admits a flow of strength $N$, then there exists an *integral* flow of that strength (cf. Theorem 7.2); clearly, the edges incident with either $S$ or $T$ must be saturated, and on the other edges the flow must be either 0 or 1.

Consider a cut $A, B$ in this network which separates $S$ and $T$. Say $A$ consists of $S$, $n_0$ vertices of $X := \{x_i : i = 1, \ldots, n\}$, and $m_0$ vertices of $Y := \{y_i : i = 1, \ldots, m\}$. The edges crossing from $A$ to $B$ include $n - n_0$ edges leaving $S$, $m_0$ edges into $T$, and $n_0(m - m_0)$ edges from $X$ to $Y$. The capacity of this cut is at least

$$r_{n_0+1} + r_{n_0+2} + \cdots + r_n + s_{m-m_0+1} + s_{m-m_0+2} + \cdots + s_m + n_0(m - m_0).$$

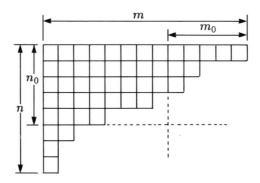

Figure 16.2

It is again convenient to refer to the Ferrers diagram of the partition $\mathbf{r}$ (see Fig. 16.2). The number of cells in the Ferrers diagram is $N$. So clearly $N$ is at most $n_0(m - m_0)$ plus the number of cells in the last $n - n_0$ rows plus the number of cells in the last $m_0$ columns. The number of cells in the last $m_0$ columns is the sum of the last $m_0$ parts of the conjugate $\mathbf{r}^*$. But under the assumption that $\mathbf{r}^*$ majorizes $\mathbf{s}$,

$$r^*_{m-m_0+1} + r^*_{m-m_0+2} + \cdots + r^*_m \leq s_{m-m_0+1} + s_{m-m_0+2} + \cdots + s_m,$$

and we conclude that the capacity of any cut is at least $N$. By the maxflow-mincut theorem, Theorem 7.1, the required flow of strength $N$ exists.     $\square$

We shall now try to find a (rough) estimate for the number of $(0, 1)$-matrices with given linesums.

THEOREM 16.2. *Given partitions* **r** *and* **s** *of an integer* $N$, *let* $M(\mathbf{r}, \mathbf{s})$ *denote the number of* $(0, 1)$-*matrices* $A$ *with rowsum* **r** *and columnsum* **s**. *If* **r** *majorizes* $\mathbf{r}_0$ *and* **s** *majorizes* $\mathbf{s}_0$, *then*

$$M(\mathbf{r}_0, \mathbf{s}_0) \geq M(\mathbf{r}, \mathbf{s}).$$

PROOF: We will establish the following simple observation. Let us fix a rowsum vector $\mathbf{r} = (r_1, r_2, \ldots, r_n)$ of length $n$ and a columnsum vector $\mathbf{s} = (s_1, s_2, \ldots, s_m)$ of length $m$ that have equal sums. (We do not require them to be nonincreasing.) If $s_1 > s_2$, then we claim that

$$M(\mathbf{r}, (s_1 - 1, s_2 + 1, s_3, \ldots, s_m)) \geq M(\mathbf{r}, (s_1, s_2, s_3, \ldots, s_m)).$$

Of course, the same will hold for any two columns (we have used the first two only for notational convenience). The same idea applies to the transpose, where keeping **s** constant and replacing $r_i$ and $r_j$ $(r_i > r_j)$ with $r_i - 1$ and $r_j + 1$ will not decrease the number of associated $(0, 1)$-matrices. The inequality of the theorem then follows from this observation and the result of Problem 16A.

To prove the observation, consider $(0, 1)$-matrices $A$ of size $n$ by $(m - 2)$ with columnsum $(s_3, s_4, \ldots, s_m)$. For a given matrix $A$, it may or may not be possible to prepend two columns to $A$ to get rowsum **r**; if it is possible, we need to add two 1's to $a$ rows, one 1 to $b$ rows, and no 1's to $c$ rows, say, where $a + b + c = n$ and $2a + b = s_1 + s_2$. But the number of ways to prepend the two columns to get new columnsums $s_1$ and $s_2$ is

$$\binom{b}{s_1 - a} = \binom{b}{s_2 - a},$$

and this is at most the number

$$\binom{b}{s_1 - 1 - a} = \binom{b}{s_2 + 1 - a}$$

of ways to prepend the two columns to get new columnsums $s_1 - 1$ and $s_2 + 1$ since the latter sums are closer to $b/2$. $\qquad\square$

COROLLARY. *The number of $n$ by $n$ $(0,1)$-matrices with all line-sums $k$ is at least*

$$\left\{ \binom{n}{k} / 2^{k+1} \right\}^n.$$

PROOF: The number of $(0,1)$-matrices with rowsum $(k, k, \ldots, k)$ is $\binom{n}{k}^n$. Each has a columnsum $\mathbf{s} = (s_1, s_2, s_3, \ldots, s_n)$ satisfying $0 \le s_i \le n$, $s_1 + \cdots + s_n = nk$. Ignoring the restriction $s_i \le n$, we see that by Theorem 13.1 there are at most $\binom{nk+n-1}{n-1} \le 2^{n(k+1)}$ such columnsums. Since the greatest number of associated $(0,1)$-matrices occurs for columnsum $(k, k, \ldots, k)$, the number of such matrices is at least the average number. $\qquad\square$

PROBLEM 16B. Prove the following theorem.

THEOREM 16.3. *Let $\mathbf{d}$ and $\mathbf{d}'$ be two partitions of an (even) integer $N$. If $\mathbf{d}$ majorizes $\mathbf{d}'$, then there are at least as many labeled simple graphs with degree sequence $\mathbf{d}'$ as with degree sequence $\mathbf{d}$.*

PROBLEM 16C. If $n$ is even, then

$$A\left(n, \frac{1}{2}n\right) \ge \frac{2^{n^2}}{n^{2n}}.$$

Show this by modifying the proof of the Corollary to Theorem 16.2. (Substituting $k = \frac{1}{2}n$ in the corollary would give a poor result.)

As in Chapter 11, let $\mathcal{A}(n, 2)$ denote the set of $(0,1)$-matrices of size $n$ with all linesums equal to 2. A method that we learned in Chapter 14 allows us to count the number of elements of $\mathcal{A}(n, 2)$. Call this number $A(n, 2)$.

Let $\mathcal{A}(n, 2)^*$ denote the subset of $\mathcal{A}(n, 2)$ consisting of indecomposable matrices (cf. Chapter 11). Define $a_n := |\mathcal{A}^*(n, 2)|$. It is easy to see that $a_n = \frac{1}{2}n! \, (n-1)!$ as follows. We have $\binom{n}{2}$ choices for the first row. If we choose $(1\ 1\ 0\ \ldots 0)$ as first row, then there is one more row (out of $n-1$) with a 1 in the first column and a second 1 not in the second column (so $n-2$ choices). This gives us $(n-1)(n-2)$ possibilities, etc.

Now define

$$m_n := A(n, 2)/(n!)^2,$$
$$b_k := a_k/(k!)^2.$$

Then by the same argument as was used in Theorem 14.2, we have

$$(16.1) \qquad 1 + \sum_{n=2}^{\infty} m_n x^n = \exp\left(\sum_{k=2}^{\infty} b_k x^k\right).$$

Therefore we find

$$(16.2) \quad 1 + \sum_{n=2}^{\infty} m_n x^n = \exp\left(\frac{-x - \log(1-x)}{2}\right) = e^{-\frac{1}{2}x}(1-x)^{-\frac{1}{2}}.$$

It is an easy exercise to show that from the expansion

$$(1-x)^{-\frac{1}{2}} = \sum_{n=1}^{\infty} \binom{2n}{n}\left(\frac{x}{4}\right)^n,$$

we can conclude that

$$m_n \sim e^{-\frac{1}{2}} \binom{2n}{n} 4^{-n} \qquad \text{for } n \to \infty.$$

We have thus proved the following theorem.

THEOREM 16.4.
$$A(n, 2) \sim e^{-\frac{1}{2}} \frac{(2n)!}{(2!)^{2n}}.$$

This theorem is a special case of the following theorem (that we shall only prove for $k = 3$, using a method that can be applied for other values of $k$).

THEOREM 16.5. *If $A(n, k) := |\mathcal{A}(n, k)|$, then*

$$A(n, k) = \frac{(nk)!}{(k!)^{2n}} \exp\left[-\frac{(k-1)^2}{2}\right]\left\{1 + O\left(\frac{1}{n^{\frac{3}{4}}}\right)\right\} \qquad (n \to \infty),$$

*uniformly in $k$ for $1 \le k < \log n$.*

For more information on this question and generalizations to the case where the linesums are not constant we refer to B. D. McKay (1984).

For the case $k = 3$, we now prove a slightly weaker result. We shall use the truncated form of the principle of inclusion and exclusion (cf. Theorem 10.1, Remark).

Consider $N := 3n$ elements, numbered

$$1_a, 1_b, 1_c, 2_a, 2_b, 2_c, \ldots, n_a, n_b, n_c.$$

Form a permutation of these $N$ elements and subsequently form the corresponding ordered partition into triples: $(x, y, z)(u, v, w) \ldots$ . We shall say that a *repetition* occurs if, in at least one of these triples, some number occurs more than once, as in $(5_a, 3_b, 5_c)$. Assume that the chosen permutation results in a sequence of $n$ triples with *no* repetition. In that case, we associate an $n$ by $n$ $(0, 1)$-matrix with the permutation as follows. If the $i$-th triple is $(x_\alpha, y_\beta, z_\gamma)$ (where $\{\alpha, \beta, \gamma\} \subseteq \{a, b, c\}$), then the matrix has ones in row $i$ in the columns numbered $x, y, z$, and zeros elsewhere. Since the indices $a, b, c$ do not influence the matrix, and the order of the three elements within one triple is also irrelevant, each such matrix corresponds to $(3!)^{2n}$ different permutations. Clearly the matrix is in $\mathcal{A}(n, 3)$. We must find an estimate for the number $P$ of permutations with no repetition among all $N_0 := N!$ permutations of the $3n$ elements.

Let $1 \leq r \leq n$. Specify $r$ triples and count the permutations that have a repetition in these triples. Then sum this over all choices of the $r$ triples. As in Theorem 10.1, we call this number $N_r$. Then if $R$ is *even*, we have

(16.3)          $$\sum_{r=0}^{R+1} (-1)^r N_r \leq P \leq \sum_{r=0}^{R} (-1)^r N_r.$$

The difficult part of the proof is finding a suitable upper estimate and lower estimate for $N_r$.

We start with the upper estimate. The $r$ triples can be chosen in $\binom{n}{r}$ ways. Within each triple we specify two positions, where we require a repetition. This can be done in $3^r$ ways. Now we choose $r$ numbers that will occur in the repetitions, and the indices $a, b, c$. This can also be done in $\binom{n}{r} \cdot 3^r$ ways. Subsequently we distribute the chosen numbers over the chosen positions, and that can be done in $2^r \cdot r!$ ways. Finally, we distribute the remaining $N - 2r$ numbers arbitrarily. Clearly, a permutation with a repetition of type $(5_a, 5_b, 5_c)$ has been counted more than once. So we have

(16.4)

$$N_r \leq \binom{n}{r}^2 \cdot 3^{2r} \cdot 2^r \cdot r! \cdot (N - 2r)!$$

$$\leq \frac{2^r}{r!}(3^2 \cdot n^2)^r (N - 2r)! \leq N! \frac{2^r}{r!} \left(\frac{N - 2r}{N}\right)^{-2r}$$

$$\leq N! \frac{2^r}{r!} \left(1 + \frac{8r^2}{N}\right),$$

if $r < \frac{1}{2}\sqrt{n}$. Here the last step is based on the power series expansion of $(1 - \frac{2r}{N})^{-2r}$, and choosing $r$ in such a way that each term is at most half of the previous term.

For the lower estimate, we start in the same way. After the $r$ pairs have been chosen and distributed over the chosen positions, we first complete the $r$ triples that have a repetition, using elements with a number that differs from the two repeated numbers for each triple. This can clearly be done in more than $(N - 3r)^r$ ways. Then we distribute the remaining numbers. Some permutations that should be counted are not, i.e.

(16.5)

$$N_r \geq \binom{n}{r}^2 \cdot 3^{2r} \cdot 2^r \cdot r! \cdot (N - 3r)^r (N - 3r)!$$

$$\geq N! \frac{2^r}{r!}(3^2 \cdot n^2)^r \left[\frac{n(n-1)\ldots(n-r+1)}{n^r}\right]^2 \frac{(N-3r)^r}{N^{3r}}$$

$$\geq N! \frac{2^r}{r!}\left(1 - \frac{r}{n}\right)^{2r} \left(1 - \frac{3r}{N}\right)^r$$

$$\geq N! \frac{2^r}{r!}\left(1 - \frac{3r^2}{n}\right)$$

if $r < \sqrt{n}$. Here in the last step, we used the well known inequality

$$\left(1 - \frac{r}{n}\right)^r \geq 1 - \frac{r^2}{n} \quad \text{if} \quad 1 \leq r \leq \sqrt{n}.$$

We now combine (16.3), (16.4) and (16.5). We take $R = \lfloor \frac{1}{2}\sqrt{n} \rfloor$. We find

$$\frac{P}{N!} = e^{-2} + \Delta, \qquad \text{where} \qquad |\Delta| < \frac{2^{R+1}}{(R+1)!} + \frac{3}{n}\sum_{r=0}^{\infty}\frac{2^r r^2}{r!},$$

and hence

$$|\Delta| < \frac{48}{n} + \frac{18e^2}{n} < \frac{200}{n}.$$

We have therefore found the following estimate (in accordance with Theorem 16.5).

THEOREM 16.6. *We have*

$$A(n,3) = \frac{(3n)!}{(3!)^{2n}}e^{-2}\left(1 + O(\frac{1}{n})\right) \qquad (n \to \infty).$$

REMARK. Note that for $k = 3$, we find from the Corollary to Theorem 16.2 that $A(n,3)$ grows at least as fast as $(cn^3)^n$, for a suitable constant $c$, and this differs from Theorem 16.6 only in the constant.

PROBLEM 16D. Determine $A(5,3)$.


**Notes.**

Theorem 16.1 is due to D. Gale (1957) and to H. J. Ryser (1957). For a more extensive treatment of the classes of (0,1)-matrices that we have considered, we refer to Ryser (1963).

**References.**
D. Gale (1957), A theorem on flows in networks, *Pacific J. Math.* **7**, 1073–1082.
B. D. McKay (1984), Asymptotics for $(0,1)$-matrices with prescribed line sums, in: *Enumeration and Design* (D. M. Jackson and S. A. Vanstone, eds.), Academic Press.
H. J. Ryser (1957), Combinatorial properties of matrices of zeros and ones, *Canad. J. Math.* **9**, 371–377.
H. J. Ryser (1963), *Combinatorial Mathematics*, Carus Math. Monograph **14**.

# 17
# Latin squares

A *Latin square of order* $n$ is a quadruple $(R, C, S; L)$ where $R$, $C$, and $S$ are sets of cardinality $n$ and $L$ is a mapping $L : R \times C \to S$ such that for any $i \in R$ and $x \in S$, the equation

$$L(i, j) = x$$

has a unique solution $j \in C$, and for any $j \in C$, $x \in S$, the same equation has a unique solution $i \in R$. That is, any two of $i \in R$, $j \in C$, $x \in S$ uniquely determine the third so that $L(i, j) = x$. Elements of $R$ are called *rows*, elements of $C$ are called *columns*, and elements of $S$ are called the *symbols* or *entries* of the Latin square. A Latin square is usually written as an $n$ by $n$ array for which the cell in row $i$ and column $j$ contains the symbol $L(i, j)$. In Fig. 17.1 we have an example of a Latin square of order 5.

| a | b | c | d | e |
|---|---|---|---|---|
| b | a | e | c | d |
| c | d | b | e | a |
| d | e | a | b | c |
| e | c | d | a | b |

Figure 17.1

The terminology "Latin square" originated with Euler who used a set of Latin letters for $S$.

A *quasigroup* is a Latin square $(X, X, X; \circ)$ with a common row, column, and symbol set $X$. Here we abbreviate the quadruple and simply denote the quasigroup by $(X, \circ)$. The mapping $\circ$ is now a binary operation on $X$ and the image of $(x, y)$ under $\circ$ is denoted by $x \circ y$. So as a special case, we obtain Latin squares that are the multiplication tables of groups.

PROBLEM 17A. If we fix the first two rows in Fig. 17.1, then there are many ways to fill in the remaining three rows to obtain a Latin square (in fact 24 ways). Show that none of these Latin squares is the multiplication table of a group.

Note that if $(R, C, S; L)$ is a Latin square and the mappings $\sigma : R \to R'$, $\tau : C \to C'$, $\pi : S \to S'$ are bijections, and if we define $L'(\sigma(i), \tau(j)) := \pi(L(i, j))$, then $(R', C', S'; L')$ is a Latin square. The two Latin squares are called *equivalent*. So with this notion of equivalence, we may assume that a Latin square of order $n$ on the set $S := \{1, 2, \ldots, n\}$ has the integers from 1 to $n$ in that order as its first row and as its first column. Note that two Latin squares that are normalized in this way and different, can still be equivalent.

Sometimes we denote a Latin square as just "$L : R \times C \to S$".

An *orthogonal array* $OA(n, 3)$ of *order* $n$ and *depth* 3 is a 3 by $n^2$ array with the integers 1 to $n$ as entries, such that for any two rows of the array, the $n^2$ vertical pairs occurring in these rows are different. Suppose that we have such an array. Call the rows $\mathbf{r}$, $\mathbf{c}$, and $\mathbf{s}$, in any order. For any pair $(i, j)$, there is a $k$ such that $r_k = i$, $c_k = j$. We make a square with entry $s_k$ in the $i$-th row and $j$-th column (for all $i, j$). The definition of orthogonal array ensures that this is a Latin square and we can reverse the procedure. Therefore the concepts of Latin square and orthogonal array are equivalent. Fig. 17.2 shows an $OA(4, 3)$ and two corresponding Latin squares.

Two Latin squares for which the corresponding orthogonal arrays have the same three rows (possibly in different order) are called *conjugates*. For example, one of the conjugates of a Latin square (with $R = C$) is its transpose. As an exercise the reader should write out the six conjugates of the Latin square in Fig. 17.2. Two orthogonal arrays are called *isomorphic* if one can be obtained from the other by permutations of the elements in each of the rows and permutations of the rows and columns of the array. Two Latin squares for which the corresponding orthogonal arrays are isomorphic are also called *isomorphic*. This means that one is equivalent to a conjugate of the other.

|       |   |   |   |   |
|-------|---|---|---|---|
| rows    | 1 1 1 1 2 2 2 2 3 3 3 3 4 4 4 4 |
| columns | 1 2 3 4 1 2 3 4 1 2 3 4 1 2 3 4 |
| symbols | 3 2 4 1 1 4 2 3 4 3 1 2 2 1 3 4 |

| 3 | 2 | 4 | 1 |
|---|---|---|---|
| 1 | 4 | 2 | 3 |
| 4 | 3 | 1 | 2 |
| 2 | 1 | 3 | 4 |

|       |   |   |   |   |
|-------|---|---|---|---|
| symbols | 1 1 1 1 2 2 2 2 3 3 3 3 4 4 4 4 |
| columns | 1 2 3 4 1 2 3 4 1 2 3 4 1 2 3 4 |
| rows    | 3 2 4 1 1 4 2 3 4 3 1 2 2 1 3 4 |

| 2 | 4 | 3 | 1 |
|---|---|---|---|
| 4 | 1 | 2 | 3 |
| 1 | 3 | 4 | 2 |
| 3 | 2 | 1 | 4 |

Figure 17.2

PROBLEM 17B. Consider the two Latin squares obtained from Fig. 17.3 by setting $a = 1$, $b = 2$, respectively $a = 2$, $b = 1$.

| 1 | 2 | 3 | 4 | 5 |
|---|---|---|---|---|
| 2 | 1 | 4 | 5 | 3 |
| 3 | 5 | a | b | 4 |
| 4 | 3 | 5 | a | b |
| 5 | 4 | b | 3 | a |

Figure 17.3

Show that these two squares are equivalent. Show that a square that is not equivalent to these corresponds to the cyclic group of order 5.

One can also interpret a Latin square as a three-dimensional array of 0's and 1's with exactly one 1 in each line of the array that is parallel to one of the sides. For example, in Fig. 17.3 the entry 4 in row 2 and column 3 should then be interpreted as an entry 1 in position $(2, 3, 4)$ of the array.

In a later chapter we shall consider orthogonal arrays with depth greater than 3. These are often much more difficult to construct than the arrays $OA(n, 3)$.

PROBLEM 17C. Construct an $OA(4, 4)$, i.e. a 4 by 16 matrix $A$ with entries 1,2,3,4, such that for any two rows of $A$, say row $i$ and row $j$, the 16 pairs $(a_{ik}, a_{jk})$, $1 \le k \le 16$, are all different.

A *sub-Latin square* (or *subsquare*) of a Latin square $(R, C, S; L)$ is a Latin square $(R_1, C_1, S_1; L_1)$ with $R_1 \subseteq R$, $C_1 \subseteq C$, $S_1 \subseteq S$, and $L_1(i, j) = L(i, j)$ for $(i, j) \in R_1 \times C_1$.

PROBLEM 17D. Let $m$ and $n$ be positive integers, $m < n$. Show that $m \leq \frac{1}{2}n$ is a necessary and sufficient condition for the existence of a Latin square of order $n$ containing a subsquare of order $m$.

The following problems are all of a similar nature. They concern a problem similar to the situation of Problems 17B and 17D, namely the completion of a square that has been partly filled. An $n$ by $n$ array $A$ with cells which are either empty or contain exactly one symbol is called a *partial* Latin square if no symbol occurs more than once in any row or column. (It might be better to say that the array "has the Latin property", but we will use the former terminology.) We are interested in conditions that ensure that a partial Latin square can be *completed* to a Latin square of order $n$, i.e. when the empty cells can be filled with symbols so that a Latin square is obtained. For example, if $A$ is a partial Latin square with filled (i.e. nonempty) cells everywhere except in the last row, there is clearly a unique way to complete $A$ to a Latin square. On the other hand, the two partial Latin squares of Fig. 17.4 are clearly not completable.

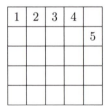 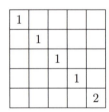

Figure 17.4

To a partial Latin square we can associate an array in the same way that we did for Latin squares by only considering filled cells. This allows us to define the conjugate of a partial Latin square. The two partial Latin squares in Fig. 17.4 are not really essentialy different but are conjugates, as is evident from their "partial

$OA(5,3)$'s" shown below:

$$
\begin{array}{cc}
\text{rows} \\
\text{columns} \\
\text{symbols}
\end{array}
\begin{pmatrix} 1\ 1\ 1\ 1\ 2 \\ 1\ 2\ 3\ 4\ 5 \\ 1\ 2\ 3\ 4\ 5 \end{pmatrix}
\qquad
\begin{pmatrix} 1\ 2\ 3\ 4\ 5 \\ 1\ 2\ 3\ 4\ 5 \\ 1\ 1\ 1\ 1\ 2 \end{pmatrix}
$$

If the first $k$ rows of a partial Latin square ($k \leq n$) are filled and the remaining cells are empty, then $A$ is called a $k$ by $n$ *Latin rectangle*.

THEOREM 17.1. *A $k$ by $n$ Latin rectangle, $k < n$, can be extended to a $k+1$ by $n$ Latin rectangle (and hence it can be completed).*

PROOF: Let $B_j$ denote the set of positive integers that do *not* occur in column $j$ of $A$. Each of the numbers 1 to $n$ occurs $k$ times in $A$ and therefore $n - k$ times in the sets $B_j$. Any $l$ of the sets $B_j$ together contain $l(n - k)$ elements, and therefore at least $l$ different ones. So the sets $B_j$, $1 \leq j \leq n$, have property $H$ of Chapter 5, and by Theorem 5.1 they therefore have an SDR. This SDR can be adjoined as the $(k+1)$-st row of the Latin rectangle. $\qquad\square$

We denote by $L(n)$ the total number of different Latin squares of order $n$. The exact value of $L(n)$ is known only for $n \leq 9$. The numbers grow very rapidly, e.g. although there are only two inequivalent Latin squares of order 5, there are $5! \cdot 4! \cdot 56$ different ones.

THEOREM 17.2. $L(n) \geq (n!)^{2n} / n^{n^2}$.

PROOF: To construct a Latin square of order $n$, we can take any permutation of 1 to $n$ as the first row. If we have constructed a Latin rectangle with $k$ rows, then by the proof of Theorem 17.1, the number of choices for the next row is per $B$ where $b_{ij} = 1$ if $i \in B_j$. By Theorem 12.8 (the Van der Waerden conjecture) this permanent is at least $(n - k)^n \cdot n! / n^n$. So we find

$$
(17.1) \qquad L(n) \geq n! \prod_{k=1}^{n-1} \{(n - k)^n n! / n^n\} = (n!)^{2n} / n^{n^2}.
$$

$\qquad\square$

REMARK. If we had applied Theorem 5.3, we would have found the estimate $L(n) \geq n!(n-1)! \cdots 1!$ (H. J. Ryser, 1969), which is better for small $n$ but asymptotically much smaller than (17.1).

For $n = 8$ the estimate (17.1) is less than $10^{-4}$ times the true value but asymptotically (17.1) is best possible as the next theorem shows.

THEOREM 17.3. *If we define* $\mathcal{L}(n) := \{L(n)\}^{1/n^2}$, *then*

$$\mathcal{L}(n) \sim e^{-2}n \quad \text{for } n \to \infty.$$

PROOF: From (17.1) and Stirling's formula we find $n^{-1}\mathcal{L}(n) \geq e^{-2}$. By the same argument as in the proof of Theorem 17.2, and by estimating per $B$ using Theorem 11.5, we find

$$L(n) \leq \prod_{k=1}^{n} M(n, k) \leq \prod_{k=1}^{n} (k!)^{n/k}.$$

Hence, again using Stirling's formula with $C$ some constant $> \sqrt{2\pi}$, we have

$$
\begin{aligned}
\log \mathcal{L}(n) &\leq \frac{1}{n} \sum_{k=1}^{n} \frac{1}{k} \log k! \\
&\leq \frac{1}{n} \sum_{k=1}^{n} \left\{ \log k - 1 + \frac{1}{2k} \log k + \frac{1}{k} \log C \right\} \\
&= \frac{1}{n} \sum_{k=1}^{n} \log k - 1 + o(1) \\
&= -2 + \log n + o(1) \text{ for } n \to \infty.
\end{aligned}
$$

Combining this with the lower bound, we find the result. □

The following theorem due to H. J. Ryser (1951) is a generalization of Problem 17D.

THEOREM 17.4. *Let $A$ be a partial Latin square of order $n$ in which cell $(i, j)$ is filled if and only if $i \leq r$ and $j \leq s$. Then $A$ can* *be completed if and only if $N(i) \geq r + s - n$ for $i = 1, \ldots, n$, where $N(i)$ denotes the number of elements of $A$ that are equal to $i$.*

PROOF: First, observe that in a Latin square of order $n$, the first $r$ rows contain exactly $r$ elements equal to $i$ of which at most $n - s$ are in the last $n - s$ columns. So the condition on $N(i)$ is trivially necessary. To show that the condition is also sufficient, we again use Hall's theorem. Let $B$ be the $(0, 1)$-matrix of size $r$ by $n$ with $b_{ij} = 1$ if and only if the element $j$ does not occur in row $i$ of $A$. Clearly every row of $B$ has sum $n - s$. The $j$-th column of $B$ has sum $r - N(j) \leq n - s$. Therefore it is possible to add $n - r$ rows of nonnegative integers to $B$ such that the resulting matrix $B^*$ has all rowsums and columnsums equal to $n - s$ (this is an easy exercise). By Theorem 5.5, $B^*$ is the sum of $n - s$ permutation matrices. We ignore the last $n - r$ rows to find

$$B = L^{(s+1)} + \cdots + L^{(n)}$$

where each $L^{(t)}$ is an $r$ by $n$ $(0,1)$-matrix with one 1 in each row and at most one 1 in each column.

As an example, suppose $r = s = 4$, $n = 7$, and the first four rows of $A$ are

| 1 | 2 | 3 | 4 |  |  |  |
|---|---|---|---|---|---|---|
| 5 | 3 | 1 | 6 |  |  |  |
| 3 | 1 | 5 | 2 |  |  |  |
| 7 | 4 | 2 | 5 |  |  |  |

Then e.g.

$$B := \begin{pmatrix} 0\,0\,0\,0\,1\,1\,1 \\ 0\,1\,0\,1\,0\,0\,1 \\ 0\,0\,0\,1\,0\,1\,1 \\ 1\,0\,1\,0\,0\,1\,0 \end{pmatrix} = L^{(5)} + L^{(6)} + L^{(7)} =$$

$$\begin{pmatrix} 0\,0\,0\,0\,0\,0\,1 \\ 0\,0\,0\,1\,0\,0\,0 \\ 0\,0\,0\,0\,0\,1\,0 \\ 1\,0\,0\,0\,0\,0\,0 \end{pmatrix} + \begin{pmatrix} 0\,0\,0\,0\,0\,1\,0 \\ 0\,0\,0\,0\,0\,0\,1 \\ 0\,0\,0\,1\,0\,0\,0 \\ 0\,0\,1\,0\,0\,0\,0 \end{pmatrix} + \begin{pmatrix} 0\,0\,0\,0\,1\,0\,0 \\ 0\,1\,0\,0\,0\,0\,0 \\ 0\,0\,0\,0\,0\,0\,1 \\ 0\,0\,0\,0\,0\,1\,0 \end{pmatrix}$$

(where we have not bothered to indicate any of the possible extensions of $B$ to a square matrix $B^*$).

Say $L^{(t)} = [l_{ij}^{(t)}]$. Then we fill the cell in position $(i,j)$ of $A$, $i = 1, \ldots, r$, $j = s+1, \ldots, n$, by $k$ if $l_{ik}^{(j)} = 1$. In our example, we would fill in the last three columns with

| 7 |
|---|
| 4 |
| 6 |
| 1 |

,

| 6 |
|---|
| 7 |
| 4 |
| 3 |

,

| 5 |
|---|
| 2 |
| 7 |
| 6 |

.

Thus $A$ is changed into a partial Latin square of order $n$ with $r$ complete rows, i.e. a Latin rectangle. By Theorem 17.1 this can be completed to a Latin square of order $n$. □

The examples of partial Latin squares that cannot be completed that were given in Fig. 17.4 both have $n$ filled cells. The conjecture that a partial Latin square with less than $n$ filled cells is completable to a Latin square was known as the *Evans conjecture* until it was finally proved by B. Smetaniuk (1981). We shall follow a slight modification of his proof, due to J. Brandt. The proof uses an earlier partial result due to C. C. Lindner (1970), that we show first. The proof is another application of Hall's theorem but this time it is rather tricky to show that property $H$ is satisfied.

THEOREM 17.5. *If $A$ is a partial Latin square of order $n$ with at most $n - 1$ filled cells and at most $\lfloor n/2 \rfloor$ distinct symbols, then $A$ is completable to a Latin square of order $n$.*

PROOF: We first tranform the problem. Since $A$ is a partial Latin square if and only if a conjugate is a partial Latin square, we may replace "at most $\lfloor \frac{n}{2} \rfloor$ distinct symbols" by the condition that $A$ has its entries in at most $\lfloor \frac{n}{2} \rfloor$ rows and we can also take these rows to be the top rows. For example, given the partial square

|   | 1 |   |   |   |   |
|---|---|---|---|---|---|
|   |   |   |   |   |   |
|   | 1 |   |   |   |   |
|   | 3 |   |   |   |   |
|   |   | 2 |   |   |   |
|   |   |   |   | 3 |   |

we consider the corresponding partial $OA(6,3)$ and interchange the role of rows and symbols as indicated below:

$$
\begin{array}{l}
\text{rows} \\
\text{columns} \\
\text{symbols}
\end{array}
\begin{pmatrix}
1\ 3\ 4\ 5\ 6 \\
2\ 3\ 3\ 4\ 6 \\
1\ 1\ 3\ 2\ 3
\end{pmatrix}
\quad \longrightarrow \quad
\begin{pmatrix}
1\ 1\ 3\ 2\ 3 \\
2\ 3\ 3\ 4\ 6 \\
1\ 3\ 4\ 5\ 6
\end{pmatrix}
$$

The latter partial $OA(6,3)$ corresponds to the partial Latin square

|   | 1 | 3 |   |   |   |
|---|---|---|---|---|---|
|   |   |   | 5 |   |   |
|   |   | 4 |   | 6 |   |
|   |   |   |   |   |   |
|   |   |   |   |   |   |
|   |   |   |   |   |   |

and it should be clear that this latter partial Latin square is completable if and only if the original Latin square is completable.

So in the following we assume without loss of generality that $A$ has $n-1$ filled cells, all of them in rows $1, \ldots, m$, where $m \leq \frac{1}{2}n$. Let the number of filled cells in row $i$ be $n_i$. We assume that $n_1 \geq n_2 \geq \cdots \geq n_m$. Suppose that for some $r \leq m$ we have completed $r-1$ rows of $A$ and still have a partial Latin square. We shall try to complete the next row. There are already $n_r$ filled cells in this row; we assume they are at the end of the row. The diagram in Fig. 17.5 below is suggestive; here the shading indicates filled cells.

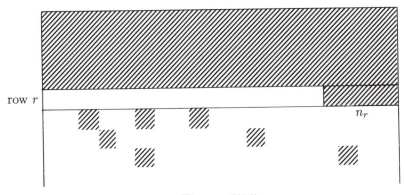

row $r$

$n_r$

Figure 17.5

Let $S_0$ denote the set of $n - n_r$ symbols not in row $r$. For $j = 1, \ldots, n - n_r$, let $B_j$ consist of those symbols in $S_0$ which *do not* occur in column $j$, either in the $r - 1$ cells above cell $(r, j)$ or in any of the filled cells which may lie below that cell.

In order to complete row $r$, we need an SDR of the $n - n_r$ sets $B_1, B_2, \ldots, B_{n-n_r}$. Do these sets have property $H$? Consider any $\ell$ of the $B_j$'s, $1 \leq \ell \leq n - n_r$, and let $W$ be their union. We claim that $|W| \geq \ell$.

First note that no symbol in $S_0$ can occur in more than $r - 1 + n_{r+1} + n_{r+2} + \cdots + n_m$ rows and hence each $x \in S_0$ appears in at most that many columns. Thus if $\ell > r - 1 + n_{r+1} + n_{r+2} + \cdots + n_m$, each symbol in $S_0$ is missing from at least one of a set of $\ell$ columns. So $W = S_0$ and $|W| = n - n_r \geq \ell$ in this case.

Now consider the number $\alpha$ of cells in the union of the $\ell$ columns corresponding to the chosen $\ell$ $B_j$'s that contain a member of $S_0$. There are at most $(r - 1)\ell$ such cells above row $r$ and at most $n_{r+1} + n_{r+2} + \cdots + n_m$ below. That is,

$$\alpha \leq (r - 1)\ell + n_{r+1} + n_{r+2} + \cdots + n_m.$$

On the other hand, each $x \in S_0 \setminus W$ appears in each of the $\ell$ columns, so $\alpha \geq \ell |S_0 \setminus W|$. It follows that

$$|W| \geq n - n_r - r + 1 - \frac{1}{\ell}(n_{r+1} + n_{r+2} + \cdots + n_m).$$

Thus $|W| > \ell - 1$ will follow from

$$\ell(n - n_r - r + 2 - \ell) > n_{r+1} + n_{r+2} + \cdots + n_m.$$

This inequality is easily seen to be true for $\ell = 1$ and for $\ell = n - n_r - r + 1$, and hence for all values of $\ell$ between 1 and $n - n_r - r + 1$ (the righthand side is a quadratic in $\ell$).

Thus $|W| \geq \ell$ has been shown unless

(17.2)     $n - n_r - r + 1 < r - 1 + n_{r+1} + n_{r+2} + \cdots + n_m.$

This inequality implies that $n_1 + \cdots + n_{r-1} < 2r - 2$, which in turn implies that $n_{r-1} = 1$. Then $n_i = 1$ for $i \geq r - 1$ and substitution in

(17.2) yields $n < 2r - 2 + (m - r + 1) = m + r - 1$. This contradicts our hypothesis that $m$ and $r$ do not exceed $\frac{1}{2}n$.

So indeed, an SDR for the $B_j$'s exists and row $r$ can be completed. Therefore the first $m$ rows can be completed, and, by Theorem 17.1, $A$ can be completed. □

The proof of the Evans conjecture depends on an algorithm that constructs Latin squares of order $n + 1$ by taking "one half" of a known Latin square of order $n$ and completing this as shown in the next two lemmas; see Fig. 17.7. We suggest that the reader go through the proofs of Lemmas 17.6, 17.7, and Theorem 17.8 in reverse order.

Let $A$ be a Latin rectangle of size $a$ by $n$. In Fig. 17.6 we show what we call the *cutoff* $C(A)$ of $A$.

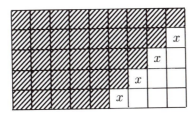

Figure 17.6

Here the shaded part has the same entries as $A$, $x$ denotes $n + 1$, and the remaining cells are empty.

LEMMA 17.6. *$C(A)$ can be completed to a Latin rectangle of size $a$ by $n$ on the symbols $1, \ldots, n+1$ such that each column contains the same entries as there were in $A$, except for one so-called "missing" element in each of the columns containing $n + 1$, and such that the missing elements $x_j$, $j \geq n - a + 2$, are all distinct.*

PROOF: The proof is by induction. The cases $a = 1$ and $a = 2$ are trivial. For the induction step we consider the completion of the cutoff of the first $a-1$ rows of $A$, with missing elements $x_j$, which for convenience are numbered from 1 to $a - 2$. In the cutoff $C(A)$, the elements of $A$ in row $a$ that are removed are called $y_0, y_1, \ldots, y_{a-2}$. Here $y_0$ is the element replaced by $n + 1$ and it is therefore the new missing element in that column. Note that all $x_i$ are different by the induction hypothesis and all $y_i$ are different by definition and also $x_i \neq y_i$.

We first consider completing row $a$ with symbols $y_1, y_2, \ldots, y_{a-2}$ so that the new missing symbols will be $y_0, x_1, x_2, \ldots, x_{a-2}$. This will work unless $y_0$ is equal to one of the $x_i$'s, say $y_0 = x_1$. In that case, we consider completing row $a$ with symbols $x_1, y_2, \ldots, y_{a-2}$ so that the new missing symbols will be $y_0, y_1, x_2, \ldots, x_{a-2}$. Since $x_1 = y_0$, $x_1$ does not occur elsewhere in row $a$ and thus this will work unless $y_1$ is equal to one of $x_2, \ldots, x_{a-2}$, say $y_1 = x_2$. In that case, we consider completing row $a$ with symbols $x_1, x_2, y_3, \ldots, y_{a-2}$ so that the new missing symbols will be $y_0, y_1, y_2, x_3 \ldots, x_{a-2}$. This procedure clearly terminates with the desired result. $\qquad\square$

LEMMA 17.7. *Let $A$ be a Latin square of order $n$ on the symbols 1 to $n$. Define $A^*$ to be the partial Latin square of order $n+1$ that has $n+1$ on all the cells of the back diagonal and the entries of the cutoff of $A$ above this back diagonal. Then $A^*$ can be completed to a Latin square of order $n+1$.*

(Fig. 17.7 shows a Latin square $A$ and the corresponding partial square $A^*$.)

| 1 | 2 | 3 | 4 | 5 |
|---|---|---|---|---|
| 4 | 3 | 5 | 1 | 2 |
| 2 | 5 | 1 | 3 | 4 |
| 5 | 1 | 4 | 2 | 3 |
| 3 | 4 | 2 | 5 | 1 |

| 1 | 2 | 3 | 4 | 5 | 6 |
|---|---|---|---|---|---|
| 4 | 3 | 5 | 1 | 6 |   |
| 2 | 5 | 1 | 6 |   |   |
| 5 | 1 | 6 |   |   |   |
| 3 | 6 |   |   |   |   |
| 6 |   |   |   |   |   |

Figure 17.7

PROOF: By Lemma 17.6, the cutoff of $A$ can be completed and we can insert the missing elements in row $n+1$ to obtain a Latin rectangle of size $n+1$ by $n$. Then add the unique last column that is possible. $\qquad\square$

THEOREM 17.8. *A partial Latin square of order $n$ with at most $n-1$ filled cells can be completed to a Latin square of order $n$.*

PROOF: The proof is by induction. Suppose the theorem is true for $n$ and consider a partial Latin square $A$ of order $n+1$ with $n$

filled cells. By permuting rows and columns we move the filled cells above the back diagonal as follows. If there are $n_i$ filled cells in row $r_i$, $i = 1, \ldots, j$, then move row $r_1$ to position $n + 1 - n_1$, the filled cells all to the left, row $r_2$ to position $n + 1 - n_1 - n_2$, entries as far to the left as possible, etc. If the number of distinct symbols is $\leq \lfloor n/2 \rfloor$ then we are done by Theorem 17.5. If this is not the case, then there is a symbol that occurs only once. We call this symbol $n + 1$ and alter the moving strategy described above such that this single entry $n + 1$ ends up on the back diagonal. By the induction hypothesis the part above the back diagonal can be completed to a Latin square of order $n$. We then fill the back diagonal with entries $n + 1$ and apply Lemma 17.7, completing the induction step. □

PROBLEM 17E. Let $A$ denote the Latin square of order 10 on the left below, and let $B_0 = C(A)$ be the partial 10 by 10 Latin rectangle on the 11 symbols $0, 1, 2, \ldots, 9, x$ on the right (not including the last row and column). Use the algorithm of the proof of Lemma 17.6 to complete $B_0$ to a 10 by 10 Latin rectangle $B$ on the symbols $0, 1, 2, \ldots, 9, x$. Finally, complete $B$ to a Latin square of order 11.

| 4 | 1 | 3 | 5 | 9 | 6 | 0 | 2 | 8 | 7 |
|---|---|---|---|---|---|---|---|---|---|
| 5 | 8 | 9 | 2 | 6 | 7 | 3 | 4 | 1 | 0 |
| 9 | 4 | 5 | 6 | 0 | 3 | 7 | 8 | 2 | 1 |
| 1 | 3 | 4 | 7 | 5 | 2 | 8 | 9 | 0 | 6 |
| 2 | 6 | 1 | 3 | 7 | 4 | 5 | 0 | 9 | 8 |
| 7 | 9 | 0 | 1 | 4 | 8 | 2 | 3 | 6 | 5 |
| 6 | 7 | 8 | 9 | 1 | 0 | 4 | 5 | 3 | 2 |
| 8 | 5 | 6 | 0 | 2 | 9 | 1 | 7 | 4 | 3 |
| 3 | 0 | 2 | 4 | 8 | 5 | 6 | 1 | 7 | 9 |
| 0 | 2 | 7 | 8 | 3 | 1 | 9 | 6 | 5 | 4 |

| 4 | 1 | 3 | 5 | 9 | 6 | 0 | 2 | 8 | 7 | $x$ |
|---|---|---|---|---|---|---|---|---|---|-----|
| 5 | 8 | 9 | 2 | 6 | 7 | 3 | 4 | 1 | $x$ | |
| 9 | 4 | 5 | 6 | 0 | 3 | 7 | 8 | $x$ | | |
| 1 | 3 | 4 | 7 | 5 | 2 | 8 | $x$ | | | |
| 2 | 6 | 1 | 3 | 7 | 4 | $x$ | | | | |
| 7 | 9 | 0 | 1 | 4 | $x$ | | | | | |
| 6 | 7 | 8 | 9 | $x$ | | | | | | |
| 8 | 5 | 6 | $x$ | | | | | | | |
| 3 | 0 | $x$ | | | | | | | | |
| 0 | $x$ | | | | | | | | | |
| $x$ | | | | | | | | | | |

(For example, the algorithm fills in the first five rows of $B_0$ as indicated on the right below. The elements "missing" from the seventh through tenth columns at this point are, respectively, $5, 9, 2, 0$. Now consider the sixth row. The $x$ in column six forces 8 to be the missing element from column six. But 8 is distinct from the elements currently missing from the other columns. So there is nothing more to do; we just fill in row six with the entries $2, 3, 6, 5$ from the sixth row of $A$.)

$$
\begin{array}{|cccccccccc|}
\hline
4 & 1 & 3 & 5 & 9 & 6 & 0 & 2 & 8 & 7 \\
5 & 8 & 9 & 2 & 6 & 7 & 3 & 4 & 1 & 0 \\
9 & 4 & 5 & 6 & 0 & 3 & 7 & 8 & 2 & 1 \\
1 & 3 & 4 & 7 & 5 & 2 & 8 & 9 & 0 & 6 \\
2 & 6 & 1 & 3 & 7 & 4 & 5 & 0 & 9 & 8 \\
7 & 9 & 0 & 1 & 4 & 8 & 2 & 3 & 6 & 5 \\
\hline
\end{array}
\qquad
\begin{array}{|cccccccccc|}
\hline
4 & 1 & 3 & 5 & 9 & 6 & 0 & 2 & 8 & 7 \\
5 & 8 & 9 & 2 & 6 & 7 & 3 & 4 & 1 & x \\
9 & 4 & 5 & 6 & 0 & 3 & 7 & 8 & x & 1 \\
1 & 3 & 4 & 7 & 5 & 2 & 8 & x & 0 & 6 \\
2 & 6 & 1 & 3 & 7 & 4 & x & 0 & 9 & 8 \\
7 & 9 & 0 & 1 & 4 & x & & & & \\
\hline
\end{array}
$$

PROBLEM 17F.

(i) Let $A$ and $B$ be Latin squares of order $n$ on symbols $1, 2, \ldots, n$. Let $A'$ and $B'$ be the Latin squares on symbols $1, 2, \ldots, n+1$ (with $n+1$'s on their back diagonals) that agree with $A$ and $B$, respectively, above their back diagonals as constructed in the proof of Lemma 17.7 via the algorithm of Lemma 17.6. Show that if $A$ and $B$ are distinct, then $A'$ and $B'$ are distinct. This proves that the number of Latin squares of order $n+1$ with $n+1$ on the back diagonal is greater than or equal to the number $N(n)$ of Latin squares of order $n$.

(ii) Explain how part (i) leads to the result

$$
L(n) \geq n!(n-1)! \cdots 2!1!
$$

mentioned in the Remark following Theorem 17.2.

PROBLEM 17G.

(a) Let $n$ be even. Find a permutation $x_1, x_2, \ldots, x_n$ of the elements of $\mathbb{Z}_n$ such that the differences $x_{i+1} - x_i$, $1 \leq i < n$, are all different.

(b) Show that this is not possible if $n$ is odd.

(c) Consider the permutation of (a). Define $a_{ij} := x_i + x_j$. Show that the square with entries $a_{ij}$, $1 \leq i, j \leq n$, is a Latin square that has the following property : The $n(n-1)$ adjacent pairs $(a_{ij}, a_{i,j+1})$ are different. Such a square is called *rowcomplete*. This square is also columncomplete.

**Notes.**

Although Latin squares have been studied for more than 200 years, much of the material in this chapter is recent. Theorem 17.1 is the oldest; it is due to M. Hall, Jr. (1945).

Theorem 17.2 and the remark following it can be found in H. J. Ryser (1969). At the time that paper was written, the Van der Waerden conjecture was still open.

Theorem 17.3 is new.

Theorem 17.4 is again due to H. J. Ryser (1951).

The so-called Evans conjecture occurs as a question (not as a conjecture) in a paper by T. Evans (1960). Before the proof by B. Smetaniuk appeared, several partial results were published. Of these, we have only treated the result by C. C. Lindner (1970) because it was used in the proof.

The most extensive treatment of Latin squares can be found in the book *Latin Squares and their Applications* by J. Dénes and A. D. Keedwell (1974).

**References.**

J. Dénes and A. D. Keedwell (1974), *Latin Squares and their Applications*, Academic Press.

T. Evans (1960), Embedding incomplete Latin squares, *Amer. Math. Monthly* **67**, 958–961.

M. Hall, Jr. (1945), An existence theorem for Latin squares, *Bull. Amer. Math. Soc.* **51**, 387–388.

C. C. Lindner (1970), On completing Latin rectangles, *Canad. Math. Bull.* **13**, 65–68.

H. J. Ryser (1951), A combinatorial theorem with an application to Latin rectangles, *Proc. Amer. Math. Soc.* **2**, 550–552.

H. J. Ryser (1969), Permanents and systems of distinct representatives, in: *Combinatorial Mathematics and its Applications*, University of North Carolina Press.

B. Smetaniuk (1981), A new construction on Latin squares I: A proof of the Evans conjecture, *Ars Combinatoria* **11**, 155–172.

# 18

# Hadamard matrices, Reed-Muller codes

Hadamard considered the following question. Let $A$ be an $n$ by $n$ matrix with real entries of absolute value at most 1. How large can the determinant of $A$ be (in absolute value)? Since each row of $A$ is a vector with length $\sqrt{n}$, the determinant cannot be larger than $n^{n/2}$. (The absolute value of the determinant is the $n$-dimensional volume of the parallelopiped spanned by the row vectors in Euclidean $n$-space.) Can equality hold? In that case all entries must be $+1$ or $-1$ and furthermore any two rows must be orthogonal, i.e. they have inner product 0. This leads to the following definition.

A *Hadamard matrix* of *order* $n$ is an $n$ by $n$ matrix $H$ with entries $+1$ and $-1$, such that

$$(18.1) \qquad\qquad HH^\top = nI.$$

Of course, any two columns of $H$ are also orthogonal. This property does not change if we permute rows or columns or if we multiply some rows or columns by $-1$. Two such Hadamard matrices are called equivalent. For a given Hadamard matrix we can find an equivalent one for which the first row and the first column consist entirely of $+1$'s. Such a Hadamard matrix is called *normalized*. Clearly the remaining rows (if any) have as many $+1$'s as $-1$'s, i.e. if $n \neq 1$ then $n$ must be even. Some small examples are

$$(1), \quad \begin{pmatrix} 1 & 1 \\ 1 & -1 \end{pmatrix}, \quad \begin{pmatrix} + & + & + & + \\ + & + & - & - \\ + & - & + & - \\ + & - & - & + \end{pmatrix},$$

where, in the last example, we have only indicated the sign of the entry.

PROBLEM 18A. Show that any two Hadamard matrices of order 12 are equivalent.

THEOREM 18.1. *If $H$ is a Hadamard matrix of order $n$, then $n = 1$, $n = 2$, or $n \equiv 0 \pmod 4$.*

PROOF: Let $n > 2$. Normalize $H$. We can permute columns in such a way that the first three rows of $H$ become

$$
\begin{array}{cccc}
+\,+\,\cdots\,+\,+ & +\,+\,\cdots\,+\,+ & +\,+\,\cdots\,+\,+ & +\,+\,\cdots\,+\,+ \\
+\,+\,\cdots\,+\,+ & +\,+\,\cdots\,+\,+ & -\,-\,\cdots\,-\,- & -\,-\,\cdots\,-\,- \\
+\,+\,\cdots\,+\,+ & -\,-\,\cdots\,-\,- & +\,+\,\cdots\,+\,+ & -\,-\,\cdots\,-\,- \\
\underbrace{\phantom{+\,+\,\cdots\,+\,+}}_{a \text{ columns}} & \underbrace{\phantom{+\,+\,\cdots\,+\,+}}_{b \text{ columns}} & \underbrace{\phantom{+\,+\,\cdots\,+\,+}}_{c \text{ columns}} & \underbrace{\phantom{+\,+\,\cdots\,+\,+}}_{d \text{ columns}}
\end{array}
$$

We have $a + b + c + d = n$ and the three inner products formed from these rows yield $a + b - c - d = 0$, $a - b + c - d = 0$, and $a - b - c + d = 0$. If we add these equations, we find $n = 4a$, proving the theorem. (In a similar way we see that $4b = 4c = 4d = n$.) □

One of the famous conjectures in the area of combinatorial designs states that a Hadamard matrix of order $n$ exists for every $n \equiv 0 \pmod 4$. We are still very far from a proof of this conjecture. The smallest $n$ for which a Hadamard matrix could exist but no example is known is presently 428. There are very many construction methods for Hadamard matrices, of which we shall treat a few. First we define a second class of matrices, very similar to Hadamard matrices.

A *conference matrix* $C$ of *order* $n$ is an $n$ by $n$ matrix with 0's on the diagonal, $+1$ or $-1$ in all other positions and with the property

$$(18.2) \qquad CC^{\mathsf{T}} = (n-1)I.$$

The name conference matrix originates from an application to conference telephone circuits. V. Belevitch (1950) studied so-called ideal nondissipative networks consisting of ideal transformers, to be used to set up a conference telephone network. The theory led to a necessary condition for the existence of such a network, namely the existence of a conference matrix of order $n$, where $n$ is the number of terminals of the network. This explains the name of these matrices.

PROBLEM 18B. Let $C$ be a conference matrix of order $n \neq 1$. Show that $n$ is even. Show that by permuting rows and columns, and multiplying certain rows and columns by $-1$, we can find an equivalent conference matrix that is symmetric if $n \equiv 2 \pmod 4$, and antisymmetric if $n \equiv 0 \pmod 4$.

THEOREM 18.2. *If $C$ is an antisymmetric conference matrix, then $I + C$ is a Hadamard matrix.*

PROOF: $(I+C)(I+C)^{\top} = I+C+C^{\top}+CC^{\top} = I+(n-1)I = nI.$ $\square$

THEOREM 18.3. *If $C$ is a symmetric conference matrix of order $n$, then*
$$H = \begin{pmatrix} I+C & -I+C \\ -I+C & -I-C \end{pmatrix}$$
*is a Hadamard matrix of order $2n$.*

PROOF: Again calculate $HH^{\top}$ and the result follows. $\square$

One of the most common construction methods for several combinatorial structures is the so-called *recursive* method in which a large object of the required type is made by applying some procedure to two or more smaller objects of the same type. In the next theorem we demonstrate such a method for the construction of Hadamard matrices.

Let $A$ be an $m$ by $n$ matrix with entries $a_{ij}$ and $B$ another matrix. The matrix
$$\begin{pmatrix} a_{11}B & a_{12}B & \dots & a_{1n}B \\ a_{21}B & a_{22}B & \dots & a_{2n}B \\ \vdots & \vdots & \ddots & \vdots \\ a_{m1}B & a_{m2}B & \dots & a_{mn}B \end{pmatrix}$$
consisting of $mn$ blocks with the size of $B$ is called the *Kronecker product* $A \otimes B$ of the matrices $A$ and $B$.

THEOREM 18.4. *If $H_m$ and $H_n$ are Hadamard matrices of order $m$ and $n$ respectively, then $H_m \otimes H_n$ is a Hadamard matrix of order $mn$.*

PROOF: By straightforward calculation one sees that
$$(A \otimes B)(C \otimes D) = (AC) \otimes (BD)$$

and that
$$(A \otimes B)^\top = A^\top \otimes B^\top.$$

We take $A = C = H_m$ and $B = D = H_n$. The result follows from the definition of a Hadamard matrix and the fact that $I_m \otimes I_n = I_{mn}$. $\square$

By repeatedly applying this theorem to $\mathbf{H}_2 := \begin{pmatrix} + & + \\ + & - \end{pmatrix}$, we find a sequence of Hadamard matrices that we shall denote by $\mathbf{H}_n$, where $n = 2^m$, $m = 1, 2, \dots$ .

We now come to a direct construction method for conference matrices, and these can then be used with Theorems 18.2 and 18.3 to construct Hadamard matrices. In the following $q$ is a power of an odd prime. On the field $\mathbb{F}_q$, we define a function $\chi$ (a so-called *character*) by

$$\chi(x) := \begin{cases} 0 & \text{if } x = 0, \\ 1 & \text{if } x \text{ is a square,} \\ -1 & \text{if } x \text{ is a nonsquare.} \end{cases}$$

For any $x$ and $y$ in $\mathbb{F}_q$, we have $\chi(x)\chi(y) = \chi(xy)$, and since there are as many squares as nonsquares we also have

(18.3)
$$\sum_{x \in \mathbb{F}_q} \chi(x) = 0.$$

Now let $0 \neq c \in \mathbb{F}_q$. Then (18.3) implies that

(18.4)
$$\sum_{b \in \mathbb{F}_q} \chi(b)\chi(b + c) = -1.$$

This is seen by ignoring the term with $b = 0$, which is 0, and then writing $\chi(b + c) = \chi(b)\chi(1 + cb^{-1})$; (note that $\chi(b)^2 = 1$ if $b \neq 0$). If $b$ runs through all nonzero elements of the field, then $1 + cb^{-1}$ takes on every value except 1.

Number the elements of $\mathbb{F}_q$ : $0 = a_0, a_1, \dots, a_{q-1}$. We define a $q$ by $q$ matrix $Q$ by

$$q_{ij} := \chi(a_i - a_j), \quad 0 \leq i, j < q.$$

Note that $Q$ is symmetric if $q \equiv 1 \pmod 4$, and antisymmetric if $q \equiv 3 \pmod 4$. As a direct consequence of the elementary properties of $\chi$ and of (18.4), we find that $QQ^\top = qI - J$ and $QJ = JQ = O$. A matrix $C$ of size $q+1$ by $q+1$ is defined by

(18.5)
$$
C := \begin{pmatrix} 0 & 1 & 1 & \cdots & 1 \\ \pm 1 & & & & \\ \vdots & & & Q & \\ \pm 1 & & & & \end{pmatrix},
$$

where the signs of the terms $\pm 1$ are chosen in such a way that $C$ is symmetric or antisymmetric. From the properties of $Q$, it now follows that $C$ is a conference matrix of order $q + 1$. This construction is due to Paley (1933) and the conference matrices of this type are usually called *Paley matrices*. For the special case that $q$ is a prime, the matrix $Q$ is a circulant. We summarize the constructions as a theorem.

THEOREM 18.5. *If $q$ is a power of an odd prime, then a Hadamard matrix of order $q + 1$ exists if $q \equiv 3 \pmod 4$, and a Hadamard matrix of order $2(q + 1)$ exists if $q \equiv 1 \pmod 4$.*

Figure 18.1

```
+ + + + + + | − + + + + +
+ + + − − + | + − + − − +
+ + + + − − | + + − + − −
+ − + + + − | + − + − + −
+ − − + + + | + − − + − +
+ + − − + + | + + − − + −
------------|------------
− + + + + + | − − − − − −
+ − + − − + | − − − + + −
+ + − + − − | − − − − + +
+ − + − + − | − + − − − +
+ − − + − + | − + + − − −
+ + − − + − | − − + + − −
```

Figure 18.2

Figures 18.1 and 18.2 illustrate Hadamard matrices of order 12 constructed from Paley matrices of orders $11 + 1$ and $5 + 1$, respectively.

PROBLEM 18C. Show that a Hadamard matrix of order $n$ exists if $n \equiv 0 \pmod 4$, $n \le 100$, except possibly for $n = 92$.

There was a period of thirty years between Paley's result and the discovery of a Hadamard matrix of order 92 by L. D. Baumert, S. W. Golomb and M. Hall. The method that they used was developed by Williamson in 1944 but a computer search was necessary to find the actual matrix. Williamson's method is based on the following observation. Let the matrices $A_i$, $1 \le i \le 4$, of order $n$, $n$ odd, be symmetric and assume that they commute with each other. Consider the matrix $H$ defined by

$$(18.6) \qquad \begin{pmatrix} A_1 & A_2 & A_3 & A_4 \\ -A_2 & A_1 & -A_4 & A_3 \\ -A_3 & A_4 & A_1 & -A_2 \\ -A_4 & -A_3 & A_2 & A_1 \end{pmatrix}.$$

We then have

$$(18.7) \qquad HH^\top = I_4 \otimes (A_1^2 + A_2^2 + A_3^2 + A_4^2).$$

To construct a Hadamard matrix in this way, we have to find matrices $A_i$ that satisfy the conditions stated above, have entries $\pm 1$, and furthermore satisfy

$$(18.8) \qquad A_1^2 + A_2^2 + A_3^2 + A_4^2 = 4nI_n.$$

Let $U$ be the permutation matrix of order $n$ corresponding to the permutation $(12\ldots n)$, i.e. $u_{ij} = 1$ if and only if $j - i \equiv 1$ (mod $n$). Then $U^n = I$ and any circulant is a linear combination of powers of $U$. If we assume that the matrices $A_i$ have the form $A_i = \sum_{j=0}^{n-1} a_{ij}U^j$ with $a_{i0} = 1$ and $a_{ij} = a_{i,n-j}$, then the matrices do indeed commute and they are symmetric. From now on, we also assume that all the $a_{ij}$ are $\pm 1$ and that (18.8) is satisfied. A simple example of a Hadamard matrix constructed in this way is found by taking $n = 3$, $A_1 = J$, and $A_i = J - 2I$, $i = 2, 3, 4$. We find a Hadamard matrix of order 12; see Fig. 18.3.

| + + + | − + + | − + + | − + + |
|-------|-------|-------|-------|
| + + + | + − + | + − + | + − + |
| + + + | + + − | + + − | + + − |
| + − − | + + + | + − − | − + + |
| − + − | + + + | − + − | + − + |
| − − + | + + + | − − + | + + − |
| + − − | − + + | + + + | + − − |
| − + − | + − + | + + + | − + − |
| − − + | + + − | + + + | − − + |
| + − − | + − − | − + + | + + + |
| − + − | − + − | + − + | + + + |
| − − + | − − + | + + − | + + + |

Figure 18.3

EXAMPLE 18.1. Note that our construction implies that $A_i$ has constant rowsum $a_i$ (odd) and by (18.8) we have $a_1^2 + \cdots + a_4^2 = 4n$. If we wish to use this method to find a Hadamard matrix of order 20, we first write 20 as a sum of four squares: $20 = 9 + 9 + 1 + 1$. From this we see that two of the matrices $A_i$ must be $2I - J$ and then it is not difficult to see that the other two have as first row $+ - + + -$, respectively $+ + - - +$.

To analyze the situation a little further, we introduce matrices $W_i$ and $P_i$ as follows:

(18.9)
$$2W_i := (A_1 + \cdots + A_4) - 2A_i,$$

(18.10)
$$A_i = 2P_i - J.$$

Our conventions on the coefficients $a_{ij}$ imply that if $p_i$ is defined by $p_i J := P_i J$, then $p_i$ is an odd integer. Furthermore (18.8) and (18.9) imply that

(18.11)
$$W_1^2 + W_2^2 + W_3^2 + W_4^2 = 4nI.$$

By substituting (18.10) in (18.8), we find

(18.12)
$$\sum_{i=1}^{4} P_i^2 = (\sum_{i=1}^{4} p_i - n)J + nI.$$

Suppose the term $U^k$, $k \neq 0$, occurs in $\alpha$ of the matrices $P_i$. Considering the equation (18.12) mod 2, we find $U^{2k}$ on the lefthand side with coefficient $\alpha$ and on the righthand side with coefficient 1. So $\alpha$ is odd and this implies that $U^k$ occurs in exactly *one* of the matrices $W_i$ (with coefficient $\pm 2$). From Example 18.1 and (18.9), we see that the constant rowsums $w_i$ of the matrices $W_i$ satisfy $w_1^2 + \cdots + w_4^2 = 4n$. These facts reduce the number of possibilities for matrices $W_i$ also satisfying (18.11) sufficiently to make a computer search for such matrices feasible. The $A_i$ are then found from (18.9). We list the first rows of the four matrices of order 23 that produced the first Hadamard matrix of order 92:

$A_1$ :  $+ + - - - + - - - + - + + - + - - - + - - - +$

$A_2$ :  $+ - + + - + + - - + + + + + + - - + + - + + -$

$A_3$ :  $+ + + - - - + + - + - + + - + - + + - - - + +$

$A_4$ :  $+ + + - + + + - + - - - - - - + - + + + + - + +$

PROBLEM 18D. Construct a Hadamard matrix of order 28 using Williamson's method.

We look at a different problem concerning Hadamard matrices. If a Hadamard matrix of order $n$ is normalized, then it obviously has exactly $n$ more entries $+1$ than entries $-1$. We define the *excess* of a Hadamard matrix to be the sum of all the entries, and then define $\sigma(n)$ as the maximal value of the excess of all Hadamard matrices of order $n$. The following bound due to Best (1977) shows that $\sigma(n)$ grows as $n^{3/2}$.

**THEOREM 18.6.** $n^2 2^{-n} \binom{n}{\frac{1}{2}n} \le \sigma(n) \le n\sqrt{n}$.

**PROOF:** (a) Let $H$ be a Hadamard matrix of order $n$. Let $s_k$ be the sum of the $k$-th column of $H$. Let $\mathbf{c}_i$ be the $i$-th row of $H$, $1 \le i \le n$. We calculate $\sum_{1 \le i,j \le n} \langle \mathbf{c}_i, \mathbf{c}_j \rangle$ in two ways. By the definition of Hadamard matrix this sum is $n^2$. On the other hand

$$\sum_{1 \le i,j \le n} \sum_{k=1}^{n} c_{ik} c_{jk} = \sum_{k=1}^{n} s_k^2.$$

From this and the Cauchy-Schwarz inequality we find

$$\sigma(n) \le \sum_{k=1}^{n} s_k \le (n \sum_{k=1}^{n} s_k^2)^{1/2} = n\sqrt{n}.$$

(b) Let $\mathbf{x}$ be any vector in $\{+1, -1\}^n$. We multiply column $j$ of $H$ by $x_j$, $1 \le j \le n$. Subsequently, multiply those rows that have more terms $-1$ than $+1$ by $-1$. Call the resulting matrix $H_\mathbf{x}$ and define $\sigma(H_\mathbf{x}) := \sum_{i=1}^{n} |\langle \mathbf{x}, \mathbf{c}_i \rangle|$. Clearly $\sigma(n)$ is at least equal to the average value of $\sigma(H_\mathbf{x})$. So we find

$$\sigma(n) \ge 2^{-n} \sum_{\mathbf{x} \in \{+1,-1\}^n} \sigma(H_\mathbf{x}) = 2^{-n} \sum_{\mathbf{x}} \sum_{i=1}^{n} |\langle \mathbf{x}, \mathbf{c}_i \rangle|$$

$$= 2^{-n} \sum_{i=1}^{n} \sum_{d=0}^{n} \sum_{\mathbf{x}, d(\mathbf{x}, \mathbf{c}_i) = d} |n - 2d|$$

$$= 2^{-n} n \sum_{d=0}^{n} |n - 2d| \binom{n}{d} = n^2 2^{-n} \binom{n}{\frac{1}{2}n}.$$

$\square$

COROLLARY. $2^{-1/2}n^{3/2} \leq \sigma(n) \leq n^{3/2}$.

PROOF: That the lefthand side of the inequality in Theorem 18.6 is asymptotically equal to $2^{1/2}\pi^{-1/2}n^{3/2}$ follows from Stirling's formula. To obtain an inequality that holds for all $n$, we must replace the constant by $2^{-1/2}$. $\qquad\square$

EXAMPLE 18.2. Consider a square of size 4 by 4, divided into 16 cells that we number from 1 to 16. A matrix $A$ with rows $\mathbf{a}_i$, $1 \leq i \leq 16$, is defined by taking $a_{ij} = -1$ if $j$ occurs in the square in the same row or column as $i$, but $j \neq i$; otherwise $a_{ij} = 1$. For any two rows of $A$ there are two positions where they both have entry $-1$. It follows that $A$ is a Hadamard matrix of order 16 with excess 64, i.e. for this matrix we have equality for the upper bound in Theorem 18.5.

This Hadamard matrix can also be constructed as follows. Define $H := J - 2I$, a Hadamard matrix of order 4 with maximal excess, namely 8. Every row of $H$ has the same number of terms $+1$, namely 3. The matrix $H \otimes H$ is a Hadamard matrix of order 16, with 10 terms $+1$ in each row, and the maximal excess 64. If we take the Kronecker product of this matrix with $H$, we again find a Hadamard matrix with constant rowsums and maximal excess, etc. In general, a Hadamard matrix of order $4u^2$, all rowsums equal to $2u$, and hence maximal excess, is called a *regular* Hadamard matrix. Other constructions of such matrices are known.

One of the recent very interesting and successful applications of Hadamard matrices is their use as so-called *error-correcting codes*. Many readers will have seen the excellent pictures that were taken of Mars, Saturn and other planets by satellites such as the Mariners, Voyagers, etc. To transmit such a picture to Earth, it is first divided into very small squares (*pixels*) and for each such square the degree of blackness is measured and expressed, say in a scale of 0 to 63. These numbers are expressed in the binary system, i.e. each pixel produces a string of six 0's and 1's (bits). The bits are transmitted to the receiver station on Earth (the Jet Propulsion Laboratory at Caltech). Due to a number of sources of *noise*, one of which is thermal noise from the amplifier, it happens occasionally that a signal that was transmitted as a 0, respectively a 1, is interpreted by the receiver as a 1, respectively a 0. If each sextuple corresponding to a pixel is transmitted as such, then the *errors* made by the receiver would make the pictures very poor. Let us assume that there is a fixed probability $p$, say $p = 0.01$, that a bit is changed. For each pixel the probability that it is received incorrectly is about 0.06. We could do the following. Instead of sending a bit, say 0, we send five 0's and the receiver translates a received fivetuple into the bit that occurs most. This will take five times as long (or if we take as a model a fixed amount of available energy, then we have only one-fifth of the original amount available per bit to be transmitted). The probability that a transmitted pixel is received correctly now becomes

$$\left[(1-p)^5 + 5p(1-p)^4 + 10p^2(1-p)^3\right]^6$$

and in our example that means a probability $6 \cdot 10^{-5}$ that the pixel is received incorrectly. For this example we say that we have used a *code*, called the *repetition code*, with two *words* of *length n*, namely 00000 and 11111. The number $1/5$ is called the *information rate* of this code. Let us now look at what was actually done in the Mariner 1969 expedition. The 64 possible information strings (corresponding to the possible degrees of blackness of a pixel) were mapped onto the rows of the matrices $\mathbf{H}_{32}$ and $-\mathbf{H}_{32}$. Since we now have codewords of length 32 representing information words of length 6, the information rate is $6/32$, nearly the same as for the repetition code. Note that any two of these codewords either differ in all 32 positions or differ in exactly 16 of the positions. It follows that if a received word contains at most 7 errors, then it resembles the correct word more than any of the 63 other words. (We have changed our symbols to $\pm 1$ instead of 0 and 1.) We say that this code is a 7-error-correcting code. The probability that a received word is *decoded* incorrectly now is

$$\sum_{i=8}^{32} \binom{32}{i} p^i (1-p)^{32-i}$$

and for our example that is $8 \cdot 10^{-10}$, very much smaller than for the repetition code, whereas the transmission of a complete picture would take about the same time for the two codes.

What we have seen is one example of a sequence of (low rate) codes, known as *first order Reed-Muller codes*. Consider once again the construction of the Hadamard matrices $\mathbf{H}_n$, where $n = 2^m$. In Fig. 18.4 we show $\mathbf{H}_8$.

In the matrix $\mathbf{H}_n$ we replace each $+1$ by $0$ and each $-1$ by $1$. Number the rows from $0$ to $n - 1 = 2^m - 1$. What happens in the Kronecker product construction when we go from $m$ to $m + 1$, i.e. double the order of the matrix? The new rows (now interpreted as vectors in $\mathbb{F}_2^{2n}$) have the form $(\mathbf{c}_i, \mathbf{c}_i)$ for $0 \leq i < n$ and they have the form $(\mathbf{c}_i, \mathbf{c}_i + \mathbf{1})$ for $n \leq i < 2n$, where $\mathbf{1}$ denotes the all-one vector.

| + + | + + | + + | + + |
|---|---|---|---|
| + − | + − | + − | + − |
| + + | − − | + + | − − |
| + − | − + | + − | − + |
| + + | + + | − − | − − |
| + − | + − | − + | − + |
| + + | − − | − − | + + |
| + − | − + | − + | + − |

Figure 18.4

THEOREM 18.7. *Let $R'(1, m)$ denote the set of row vectors in $\mathbb{F}_2^n$ obtained from $\mathbf{H}_n$ as described above. Then $R'(1, m)$ is an $m$-dimensional subspace of $\mathbb{F}_2^n$.*

PROOF: For $m = 1$ the assertion is trivial. We proceed by induction. Let $\mathbf{v}_1, \mathbf{v}_2, \ldots, \mathbf{v}_m$ be a basis of $R'(1, m)$. Our observation made above shows that $R'(1, m + 1)$ consists of all linear combinations of the vectors $(\mathbf{v}_i, \mathbf{v}_i)$ and the vector $(\mathbf{0}, \mathbf{1})$. This completes the proof. □

We now define the first order Reed-Muller code $R(1, m)$ of length $n = 2^m$ and dimension $m + 1$ to be the subspace of $\mathbb{F}_2^n$ spanned by the space $R'(1, m)$ and the all-one vector of length $n$. In this terminology, the code used by Mariner 1969 was $R(1, 5)$. From the properties of Hadamard matrices, we immediately see that any two codewords in $R(1, m)$ differ in at least $\frac{1}{2}n$ places. We say that the code has *minimum distance* $d = 2^{m-1}$. Therefore the code can correct up to $2^{m-2} - 1$ errors.

We can give the codewords a nice geometric interpretation. Let us number the points of the space $\mathbb{F}_2^m$ by considering the vector as the binary representation of its number. For example, $(0, 1, 1, 0, 1)$ is considered as the point $P_{22}$ in $\mathbb{F}_2^5$. Let these representations be the columns of an $m$ by $n$ matrix. Let $\mathbf{v}_i$ be the $i$-th row of this matrix, $1 \leq i \leq m$. Then from Fig. 18.4 and the observations above, we see that the vectors $\mathbf{v}_i$ are the natural basis for $R'(1, m)$. The basis vector $\mathbf{v}_i$ is the characteristic function of the hyperplane $\{(x_1, x_2, \ldots, x_m) \in \mathbb{F}_2^m : x_i = 1\}$. By taking linear combinations, we see that the codewords of $R(1, m)$ are exactly all the characteristic functions of the affine hyperplanes in the vector space and the characteristic function of the space itself (for **1**) and of the empty set (for **0**). Any two affine hyperplanes are either parallel or they meet in an affine subspace of dimension $m - 2$, in accordance with the fact that two rows of the Hadamard matrix have the same entries in exactly half of the positions.

As our definition of $R(1, m)$ suggests, Reed-Muller codes of higher order are also defined in the theory of error-correcting codes. They also have a geometric interpretation that we shall not go into here (but see Example 26.4).

PROBLEM 18E. For $n = 2^m$ and $1 \leq i \leq m$ we define the matrix $M_n^{(i)}$ by
$$M_n^{(i)} := I_{2^{m-i}} \otimes H_2 \otimes I_{2^{i-1}}.$$

Prove that
$$H_n = M_n^{(1)} M_n^{(2)} \ldots M_n^{(m)}.$$

A received word $\mathbf{x}$ is decoded by calculating $\mathbf{x} H_n^\top$. If there are not too many errors, then all entries of this product will be nearly 0 except one with an absolute value close to $n$, telling us what the true message was. A multiplication by $\pm 1$ is called an operation. Compare the number of operations that are necessary for decoding when $H$ is used and when the representation of $H$ as a product of matrices $M_n^{(i)}$ is used. (This is an example of what is known as a *Fast Fourier Transform*.)

PROBLEM 18F. Let $\mathbf{v}_i$, $0 \le i \le m$, where $\mathbf{v}_0 = \mathbf{1}$, be the basis of $R(1,m)$ given above. We consider the subspace $R(2,m)$ of $\mathbb{F}_2^n$, where $n = 2^m$, spanned by all the vectors

$$\mathbf{v}_i \cdot \mathbf{v}_j := (v_{i0}v_{j0}, \ldots, v_{i,n-1}v_{j,n-1}).$$

What is the dimension of this space? Show that any two vectors of $R(2,m)$ differ in at least $\frac{1}{4}n$ places.

PROBLEM 18G. Suppose $M$ is an $m$ by $n$ (0,1)-matrix so that the Hamming distance between any two distinct rows is at least $d$. (If $M$ is the result of stacking a Hadamard matrix $H$ on top of $-H$ and changing the symbols to 0's and 1's, we have an example with $m = 2n$ and $d = n/2$.)

(1) Count the number of ordered triples $(i, j, k)$ such that $i$ and $j$ are (indices of) distinct rows and $k$ is a column where $M(i,k) \ne M(j,k)$ in two ways—one yielding an inequality involving $d$ and the other involving the columnsums of $M$. Prove that if $2d > n$, then

$$m \le \frac{2d}{2d - n}.$$

(This is known as *Plotkin's bound* in coding theory.) What conditions ensure equality?

(2) Suppose $d = n/2$. Prove that $m \le 2n$ and show that equality implies the existence of a Hadamard matrix of order $n$.

## Notes.

J. Hadamard (1865–1963) was a leading mathematician around the turn of the century. His most important work was in the theory of analytic functions and in mathematical physics. He is best known for the proof of the so-called *prime number theorem*, jointly with C. J. De La Vallée-Poussin.

R. E. A. C. Paley (1907–1933) died in an avalanche while skiing at the age of 26. In his short life he produced 26 papers of excellent quality (mostly on Fourier Theory).

A long outstanding conjecture of H. J. Ryser asserts that no Hadamard matrix of order $n > 4$ can be a circulant.

J. S. Wallis (1976) proved that for any integer $s$, Hadamard matrices of orders $2^t s$ exist whenever $t > 2 \log_2(s - 3)$. However, it is still not known whether the set of orders of Hadamard matrices has "positive density".

For more on error-correcting codes, see Chapter 20.

The codes that are now called Reed-Muller codes were (surprisingly?) indeed first treated by D. E. Muller (1954) and I. S. Reed (1954).

For a detailed account of the coding and decoding for the Mariner missions, see E. C. Posner (1968).

**References.**

L. D. Baumert, S. W. Golomb, and M. Hall, Jr. (1962), Discovery of a Hadamard matrix of order 92, *Bull. Amer. Math. Soc.* **68**, 237–238.

V. Belevitch (1950), Theory of $2n$-terminal networks with applications to conference telephony, *Electrical Communication* **27**, 231–244.

M. R. Best (1977), The excess of a Hadamard matrix, *Proc. Kon. Ned. Akad. v. Wetensch.* **80**, 357–361.

D. E. Muller (1954), Application of Boolean algebra to switching circuit design and to error detection, *IEEE Trans. Computers* **3**, 6–12.

R. E. A. C. Paley (1933), On orthogonal matrices, *J. Math. Phys.* **12**, 311–320.

E. C. Posner (1968), Combinatorial structures in planetary reconnaissance, in: *Error Correcting Codes* (H. B. Mann, ed.), J. Wiley and Sons.

I. S. Reed (1954), A class of multiple-error-correcting codes and the decoding scheme, *IEEE Trans. Information Theory* **4**, 38–49.

J. S. Wallis (1976), On the existence of Hadamard matrices, *J. Combinatorial Theory* (A) **21**, 188–195.

J. Williamson (1944), Hadamard's determinant theorem and the sum of four squares, *Duke Math. J.* **11**, 65–81.

# 19
# Designs

In this chapter we give an introduction to a large and important area of combinatorial theory which is known as *design theory*. The most general object that is studied in this theory is a so-called *incidence structure*. This is a triple $\mathbf{S} = (\mathcal{P}, \mathcal{B}, \mathbf{I})$, where:

(1) $\mathcal{P}$ is a set, the elements of which are called *points*;

(2) $\mathcal{B}$ is a set, the elements of which are called *blocks*;

(3) $\mathbf{I}$ is an incidence relation between $\mathcal{P}$ and $\mathcal{B}$ (i.e. $\mathbf{I} \subseteq \mathcal{P} \times \mathcal{B}$). The elements of $\mathbf{I}$ are called *flags*.

If $(p, B) \in \mathbf{I}$, then we say that point $p$ and block $B$ are *incident*. We allow two different blocks $B_1$ and $B_2$ to be incident with the same subset of points of $\mathcal{P}$. In this case one speaks of "repeated blocks". If this does not happen, then the design is called a *simple* design and we can then consider blocks as subsets of $\mathcal{P}$. In fact, from now on we shall always do that, taking care to realize that different blocks are possibly the same subset of $\mathcal{P}$. This allows us to replace the notation $(p, B) \in \mathbf{I}$ by $p \in B$, and we shall often say that point $p$ is "in block $B$" instead of incident with $B$.

It has become customary to denote the cardinality of $\mathcal{P}$ by $v$ and the cardinality of $\mathcal{B}$ by $b$. So the incidence structure then is a set of $v$ points and a collection of $b$ not necessarily distinct subsets of the point set. The structure obtained by replacing each block by its complement is, of course called the *complement* of the structure. (This means that we replace $\mathbf{I}$ by its complement in $\mathcal{P} \times \mathcal{B}$.)

To obtain an interesting theory, we must impose some regularity conditions on the structure $\mathbf{S}$. As a first example, we mention incidence structures that have the confusing name "*linear spaces*". Here the blocks are usually called *lines* and the regularity conditions are that every line contains (i.e. is incident with) at least two points and any two points are on exactly one line. Example 19.6 below shows a simple but important linear space. The following theorem is due to De Bruijn and Erdős (1948). The elegant proof is due to Conway.

THEOREM 19.1. *For a linear space we have $b = 1$ or $b \geq v$, and equality implies that for any two lines there is exactly one point incident with both.*

PROOF: For $x \in \mathcal{P}$, denote by $r_x$ the number of lines incident with $x$, and similarly for $B \in \mathcal{B}$, let $k_B$ be the number of points on $B$. Let there be more than one line. If $x \notin L$ then $r_x \geq k_L$ because there are $k_L$ lines "joining" $x$ to the points on $L$. Suppose $b \leq v$. Then $b(v - k_L) \geq v(b - r_x)$ and hence

$$1 = \sum_{x \in \mathcal{P}} \sum_{L \not\ni x} \frac{1}{v(b - r_x)} \geq \sum_{L \in \mathcal{B}} \sum_{x \notin L} \frac{1}{b(v - k_L)} = 1$$

and this implies that in all the inequalities, equality must hold. Therefore $v = b$, and $r_x = k_L$ if $x \notin L$. □

A trivial example of equality in Theorem 19.1 is a so-called *near pencil*, a structure with one line that contains all the points but one, and all pairs containing that point as lines of size two. Much more interesting examples are the projective planes that we shall define later in this chapter.

In the rest of this chapter, we shall be interested in highly regular incidence structures called "$t$-designs". Let $v, k, t$ and $\lambda$ be integers with $v \geq k \geq t \geq 0$ and $\lambda \geq 1$. A $t$-design on $v$ points with *blocksize* $k$ and *index* $\lambda$ is an incidence structure $\mathcal{D} = (\mathcal{P}, \mathcal{B}, \mathbf{I})$ with:

(i) $|\mathcal{P}| = v$,
(ii) $|B| = k$ for all $B \in \mathcal{B}$,
(iii) for any set $T$ of $t$ points, there are exactly $\lambda$ blocks incident with all points in $T$.

So all blocks have the same size and every $t$-subset of the point set is contained in the same number of blocks. Two different notations for such a design are widely used, namely $t$-$(v, k, \lambda)$ design and $S_\lambda(t, k, v)$. We shall use both of them. A *Steiner system* $S(t, k, v)$

is a *t*-design with $\lambda = 1$, and we suppress the index in the notation. Most of the early theory of designs originated in statistics, where 2-designs are used in the design of experiments for statistical analysis. These designs are often called *balanced incomplete block designs* (BIBDs). Usually trivial designs are excluded from the theory: a design with one block that contains all the points or a design that has all the *k*-subsets of the point set as blocks is of course a *t*-design for $t \leq k$, but is not very interesting.

We give a few examples; more will follow further on.

EXAMPLE 19.1. Let the nonzero vectors of $\mathbb{F}_2^4$ be the points. As blocks we take all triples $\{\mathbf{x}, \mathbf{y}, \mathbf{z}\}$ with $\mathbf{x} + \mathbf{y} + \mathbf{z} = \mathbf{0}$. Any pair $\mathbf{x}, \mathbf{y}$ with $\mathbf{x} \neq \mathbf{y}$ uniquely determines a third element $\mathbf{z}$, different from both, satisfying this equation. So we have constructed an $S(2, 3, 15)$. The blocks are the 2-dimensional subspaces of $\mathbb{F}_2^4$ with $\mathbf{0}$ deleted.

We construct a second design by taking all the vectors as point set and defining blocks to be 4-tuples $\{\mathbf{w}, \mathbf{x}, \mathbf{y}, \mathbf{z}\}$ for which $\mathbf{w} + \mathbf{x} + \mathbf{y} + \mathbf{z} = \mathbf{0}$. This defines an $S(3, 4, 16)$. Note that if we take the blocks that contain $\mathbf{0}$ and delete this vector, we find the blocks of the previous design.

EXAMPLE 19.2. We take the ten edges of a $K_5$ as point set. Each of the three kinds of 4-tuples shown in Fig. 19.1 will be a block.

Figure 19.1

There are $5+10+15 = 30$ blocks. No triple (of edges) is contained in more than one block. Therefore the blocks contain 120 different triples, i.e. all the triples. We have constructed an $S(3, 4, 10)$.

EXAMPLE 19.3. Let $H$ be a normalized Hadamard matrix of order $4k$. Delete the first row and the first column. We now identify points with rows of this matrix. Each column defines a subset of the rows, namely those rows for which there is a $+$ in that column. These subsets are the blocks. From the argument in Theorem 18.1, we see that any pair of points is contained in exactly $k-1$ blocks and clearly all blocks have size $2k-1$. We have a 2-$(4k-1, 2k-1, k-1)$ design and such a design is called a *Hadamard 2-design*.

Consider the same matrix $H$ but now delete only the first row. Each of the other rows determines two $2k$-subsets of the set of columns. This partition is unaffected if we change the sign of the row. The argument of Theorem 18.1 now shows that for any three columns, there are exactly $k - 1$ of the subsets that have three elements in these columns. So these $2k$-sets are the blocks of a 3-$(4k, 2k, k - 1)$ design called a *Hadamard 3-design*.

EXAMPLE 19.4. Consider a regular Hadamard matrix of order $4u^2$ (see Example 18.2). If we replace $+$ by 1 and $-$ by 0, we find a $(0,1)$-matrix with $2u^2 + u$ ones in every row and column, and furthermore, any two rows or columns have inner product $u^2 + u$. Let the columns be the characteristic functions of the blocks of a design on $4u^2$ points. The properties of the matrix show that this is a 2-$(4u^2, 2u^2 + u, u^2 + u)$ design. One usually prefers considering the complement of this design, i.e. a 2-$(4u^2, 2u^2 - u, u^2 - u)$ design.

PROBLEM 19A. Here are two more examples in the spirit of Example 19.2.

(i) Take the edges of $K_6$ as points of an incidence structure. The blocks are to be all sets of three edges that either are the edges of a perfect matching, or the edges of a triangle. Show that this is an $S(2, 3, 15)$ and show that it is isomorphic to the design in Example 19.1.

(ii) Take the edges of $K_7$ as points of an incidence structure. The blocks are to be all sets of five edges of these three types: (a) "claws" with five edges incident with a common vertex, (b) edge sets of pentagon subgraphs, and (c) five edges that form a triangle and two disjoint edges. Show that this is an $S_3(3, 5, 21)$.

We now give two elementary theorems on $t$-designs.

THEOREM 19.2. *The number of blocks of an $S_\lambda(t,k,v)$ is*

(19.1)
$$b = \lambda \binom{v}{t} / \binom{k}{t}.$$

PROOF: Count in two ways the number of pairs $(T, B)$, where $T$ is a $t$-subset of $\mathcal{P}$ and $B$ is a block incident with all points of $T$. We find $\lambda\binom{v}{t} = b\binom{k}{t}$. □

THEOREM 19.3. *Given $i$, $0 \le i \le t$, the number of blocks incident with all the points of an $i$-subset $I$ of $\mathcal{P}$ is*

(19.2)
$$b_i = \lambda \binom{v-i}{t-i} / \binom{k-i}{t-i}.$$

*That is, every $S_\lambda(t,k,v)$ is also an $i$-design for $i \le t$.*

PROOF: Count in two ways the number of pairs $(T, B)$, where $T$ is a $t$-subset of $\mathcal{P}$ that contains $I$ and $B$ is a block that is incident with all the points of $T$. □

COROLLARY. *If $\mathcal{D}$ is a $t$-design with point set $\mathcal{P}$ and block set $\mathcal{B}$ and if $I$ is a subset of $\mathcal{P}$ with $|I| \le t$, then the point set $\mathcal{P}\backslash I$ and the blocks $\{B\backslash I : I \subseteq B\}$ form an $S_\lambda(t-i, k-i, v-i)$. This design is called the derived design $\mathcal{D}_I$.*

In Example 19.1 we already saw an example of a derived design. If we take $I = \{0\}$, the derived design for $S(3,4,16)$ is $S(2,3,15)$.

PROBLEM 19B. Show that an $S(3,6,11)$ does not exist.

The number of blocks incident with any point, i.e. $b_1$, is usually denoted by $r$ (*replication number*). Two special cases of Theorem 19.3 are the following relations for the parameters of a 2-design:

(19.3)
$$bk = vr,$$

(19.4)
$$\lambda(v-1) = r(k-1).$$

THEOREM 19.4. *Let $0 \le j \le t$. The number of blocks of an $S_\lambda(t,k,v)$ that are incident with none of the points of a $j$-subset $J$ of $\mathcal{P}$ is*

(19.5)
$$b^j = \lambda \binom{v-j}{k} / \binom{v-t}{k-t}.$$

PROOF: For $x \in \mathcal{P}$, let $\mathcal{B}_x$ be the set of blocks incident with $x$. We use inclusion-exclusion, Theorem 10.1. We find that

$$b^j = \sum_{i=0}^{j} (-1)^i \binom{j}{i} b_i.$$

The result follows by substitution of (19.2) and then using (10.6).

It is quicker to observe that $b^j$ apparently does not depend on the particular set $J$ and then count in two ways the pairs $(J, B)$, where $J$ is a $j$-subset of $\mathcal{P}$ and $J \cap B = \emptyset$. So $\binom{v}{j} b^j = b \binom{v-k}{j}$. Then the result follows from Theorem 19.2. $\qquad\square$

COROLLARY. *If $i + j \le t$, then the number of blocks of an $S_\lambda(t, k, v)$ that are incident with all of a set of $i$ points and none of a disjoint set of $j$ points is a constant*

$$(19.6) \qquad\qquad b_i^j = \lambda \frac{\binom{v-i-j}{k-i}}{\binom{v-t}{k-t}}.$$

PROOF: The result follows upon application of Theorem 19.4 to the $(t-i)$-design $\mathcal{D}_I$, where $I$ is the set of $i$ points. $\qquad\square$

COROLLARY. *If $J$ is a $j$-subset of $\mathcal{P}$, $j \le t$, then the point set $\mathcal{P} \setminus J$ and the blocks $B$ with $B \cap J = \emptyset$ form an $S_\mu(t - j, k, v - j)$ called the residual design $\mathcal{D}^J$.*

PROBLEM 19C. Prove that the complement of an $S_\lambda(t, k, v)$ is a $t$-design and determine its parameters.

EXAMPLE 19.5. Consider a Hadamard 3-design 3-$(4k, 2k, k-1)$ and form the residual with respect to a set with one point. We find a Hadamard 2-design 2-$(4k - 1, 2k, k)$, i.e. the complement of the design of Example 19.3.

An obvious necessary condition for the existence of an $S_\lambda(t, k, v)$ is that the numbers $b_i$ of (19.2) are integers. However, this condition is not sufficient. An $S(10, 16, 72)$ does not exist, as is demonstrated by the following theorem due to Tits (1964).

THEOREM 19.5. *In any nontrivial Steiner system* $S(t, k, v)$,

$$v \geq (t + 1)(k - t + 1).$$

PROOF: In a Steiner system, any two distinct blocks have at most $t-1$ points in common. Choose a set $S$ of $t+1$ points not contained in any block. For each set $T \subseteq S$ with $|T| = t$, there is a unique block $B_T$ containing $T$. Each such $B_T$ is incident with $k - t$ points not in $S$, and any point not in $S$ is incident with at most one such block $B_T$ since two such blocks already have $t - 1$ points of $S$ in common. This shows that the union of all blocks $B_T$ contains $(t + 1) + (t + 1)(k - t)$ points and the result follows. □

Given an incidence structure with $|\mathcal{P}| = v$ and $|\mathcal{B}| = b$, the *incidence matrix* $N$ is the $v$ by $b$ matrix with rows indexed by the elements $p$ of $\mathcal{P}$, columns indexed by the elements $B$ of $\mathcal{B}$, and with the entry $N(p, B) = 1$ if $p$ is incident with $B$, $N(p, B) = 0$ otherwise. Note that the entry in row $p$ and column $q$ of $NN^\top$ is the sum of $N(p, B)N(q, B)$ over all blocks $B$, and this is the number of blocks that contain both $p$ and $q$. Dually, the entry in row $A$ and column $B$ of $N^\top N$ is the cardinality of $A \cap B$.

Two designs $\mathcal{D}$ and $\mathcal{D}'$ with incidence matrices $N$ and $N'$ are called *isomorphic* or *equivalent* if there are permutation matrices $P$ and $Q$ such that $N' = PNQ$.

We shall often identify $N$ with the design, i.e. we refer to the columns as blocks instead of as characteristic functions of blocks.

Now if $N$ is the incidence matrix of a 2-design, then $NN^\top$ has the entry $r$ everywhere on the diagonal and entries $\lambda$ in all other positions, i.e.

$$(19.7) \qquad NN^\top = (r - \lambda)I + \lambda J,$$

where $I$ and $J$ are $v$ by $v$ matrices.

PROBLEM 19D. Let $N$ be an 11 by 11 (0,1)-matrix with the following properties: (i) every row of $N$ has six ones; (ii) the inner product of any two distinct rows of $N$ is at most 3. Show that $N$ is the incidence matrix of a 2-(11,6,3) design. Furthermore show that this design is unique (up to isomorphism).

The following theorem is known as *Fisher's inequality*.

THEOREM 19.6. *For a 2-$(v, k, \lambda)$ design with $b$ blocks and $v > k$ we have*

$$b \geq v.$$

PROOF: Since $v > k$, we have $r > \lambda$ by (19.4). Since $J$ has one eigenvalue $v$ and its other eigenvalues are 0, the matrix on the right-hand side of (19.7) has $v - 1$ eigenvalues $(r - \lambda)$ and one eigenvalue $(r - \lambda) + \lambda v = rk$. So it has determinant $rk(r - \lambda)^{v-1} \neq 0$ and hence $N$ has rank $v$. This implies that $b \geq v$.                              □

From the argument in the preceding proof, we can make a very important conclusion, given in the next theorem.

THEOREM 19.7. *If a 2-$(v, k, \lambda)$ design has $b = v$ blocks and $v$ is even, then $k - \lambda$ must be a square.*

PROOF: Since $b = v$, we have $r = k$. Now $N$ is a $v$ by $v$ matrix and by (19.7)

$$(\det N)^2 = k^2 (k - \lambda)^{v-1}.$$

Since $\det N$ is an integer, we are done.                              □

Theorem 19.6 was generalized by A. Ya. Petrenjuk (1968) to $b \geq \binom{v}{2}$ for any $S_\lambda(4, k, v)$ with $v \geq k + 2$ and finally generalized to arbitrary $t$-designs by Ray-Chaudhuri and Wilson (1975).

THEOREM 19.8. *For an $S_\lambda(t, k, v)$ with $t \geq 2s$ and $v \geq k + s$, we have $b \geq \binom{v}{s}$.*

PROOF: We introduce the *higher incidence matrices* of the $t$-design $\mathcal{D} = S_\lambda(t, k, v)$. For $i = 0, 1, 2, \ldots$, let $N_i$ denote the $\binom{v}{i}$ by $b$ matrix with rows indexed by the $i$-element subsets of points, columns indexed by the blocks, and with entry 1 in row $Y$ and column $B$ if $Y \subseteq B$, 0 otherwise. For $0 \leq i \leq j \leq v$, we use $W_{ij}$ to denote the $i$-th incidence matrix of the incidence structure whose blocks are all the $j$-element subsets of a $v$-set. Thus $W_{ij}$ is a $\binom{v}{i}$ by $\binom{v}{j}$ matrix.

We claim that

$$N_s N_s^\top = \sum_{i=0}^{s} b_{2s-i}^i W_{is}^\top W_{is}.$$

To see this, note that $N_s N_s^\top$ has rows indexed by $s$-element subsets $E$ and columns indexed by $s$-element subsets $F$ of the points, and for given $E$ and $F$, the entry in row $E$ and column $F$ of $N_s N_s^\top$ is the number of blocks that contain both $E$ and $F$. This number is $b_{2s-\mu}$, where $\mu := |E \cap F|$. The entry in row $E$ and column $F$ of $W_{is}^\top W_{is}$ is the number of $i$-subsets of the points contained in both $E$ and $F$, i.e. $\binom{\mu}{i}$. So the $(E, F)$-entry on the righthand side of the equation is $\sum_{i=1}^{s} b_{2s-i}^i \binom{\mu}{i}$, and from (19.6) it follows that this is $b_{2s-\mu}$.

The $\binom{v}{s}$ by $\binom{v}{s}$ matrices $b_{2s-i}^i W_{is}^\top W_{is}$ are all positive semidefinite, and $b_s^s W_{ss}^\top W_{ss} = b_s^s I$ is positive definite since $b_s^s > 0$ ($v \geq k + s$). Therefore $N_s N_s^\top$ is positive definite and hence nonsingular. The rank of $N_s N_s^\top$ is equal to the rank of $N_s$, i.e. $N_s$ has rank $\binom{v}{s}$, and this cannot exceed the number of columns of $N_s$, which is $b$. $\qquad\square$

If equality holds in the Wilson-Petrenjuk inequality, Theorem 19.8, then the $2s$-design is called *tight*. The only known examples with $s > 1$ and $v > k + s$ are the unique Steiner system $S(4, 7, 23)$ that we treat in the next chapter and its complement.

It is useful to give some idea of the history of $t$-designs. Only finitely many Steiner systems $S(t, k, v)$ with $t \geq 4$ are known. The most famous are the designs $S(5, 8, 24)$ and $S(5, 6, 12)$ found by E. Witt (1938) and the derived 4-designs. These will appear in the next chapter. R. H. F. Denniston (1976) constructed $S(5, 6, 24)$, $S(5, 7, 28)$, $S(5, 6, 48)$, and $S(5, 6, 84)$. W. H. Mills (1978) constructed an $S(5, 6, 72)$. Again, the derived designs are Steiner systems. Since then, no others have been found. In 1972, W. O. Alltop constructed the first infinite sequence of 5-designs without repeated blocks. We remark that it is easy to show that $t$-designs with repeated blocks exist for any $t$, but for a long time many design theorists believed that nontrivial $t$-designs without repeated blocks did not exist for $t > 6$. The first simple 6-design was found by D. W. Leavitt and S. S. Magliveras in 1982, and in 1986 D. L. Kreher and S. P. Radziszowski found the smallest possible simple 6-design, an $S_4(6, 7, 14)$. The big sensation in this area was the paper by L. Teirlinck (1987) proving that nontrivial simple $t$-designs exist for all $t$. His construction produces designs with tremendously large parameters and hence the construction of small examples is still an open problem. For a number of special parameter sets, it has been shown that the corresponding designs do not exist.

For the remainder of this chapter we shall mainly be interested in 2-designs. When $t = 2$, we often omit this parameter in the "$t$-$(v, k, \lambda)$" notation and speak of $(v, k, \lambda)$-designs.

A class of designs of special interest are the 2-designs with $b = v$. In this case the incidence matrix $N$ of the design is a square matrix and these designs should be called *square designs*. However, the confusing name *symmetric designs* is standard terminology. (Note that $N$ is *not* necessarily symmetric.) For a symmetric 2-$(v, k, \lambda)$ design (19.4) becomes

$$\lambda(v - 1) = k(k - 1).$$

Some authors use the name *projective design*, a name derived from the fact that a 2-$(v, k, 1)$ design with $b = v$ is called a projective plane (see Example 19.7). Despite the fact that we are not happy with the name, we shall use the terminology *symmetric designs* for these designs.

PROBLEM 19E. Let $\mathcal{D}$ be a 2-$(v, k, \lambda)$ design with $b$ blocks and $r$ blocks through every point. Let $B$ be any block. Show that the number of blocks that meet $B$ is at least

$$k(r - 1)^2/[(k - 1)(\lambda - 1) + (r - 1)].$$

Show that equality holds if and only if any block not disjoint from $B$ meets it in a constant number of points.

EXAMPLE 19.6. Take as points the elements of $\mathbb{Z}_7$ and as blocks all triples $B_x := \{x, x + 1, x + 3\}$ with $x \in \mathbb{Z}_7$. It is easy to check that this yields an $S(2, 3, 7)$. The following Fig. 19.2 is often drawn. The lines repesent blocks, but one block must be represented by the circle.

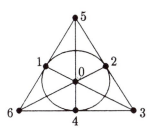

Figure 19.2

This design is known as the *Fano plane*. The idea of the construction will be extended in Chapter 27. It is based on the fact that the six differences among the elements of $\{0,1,3\}$ are exactly all the nonzero elements of $\mathbb{Z}_7$. If we wish to find the block containing say $\{1,6\}$, we observe that $6 - 1 = 1 - 3$ and we therefore take $x = 5$ and find $x + 1 = 6$, $x + 3 = 1$, i.e. the pair is indeed in $B_5$. The reader should have no difficulty finding an $S(2,4,13)$ in the same way, using $\mathbb{Z}_{13}$.

A symmetric design with $\lambda = 1$ is called a *projective plane*. If $k$ is the size of the blocks, then $n = k - 1$ is called the *order* of the plane (why this is done will become clear in Example 19.7). Expressed in $n$, the parameters of a projective plane of order $n$ are:

$$v = n^2 + n + 1, \qquad k = n + 1, \qquad \lambda = 1.$$

The blocks are usually called *lines*. The Fano plane is the (unique) projective plane of order 2.

EXAMPLE 19.7. Consider the vector space $\mathbb{F}_q^3$. This vector space contains $(q^3 - 1)/(q - 1) = q^2 + q + 1$ 1-dimensional subspaces and the same number of 2-dimensional subspaces. We now construct an incidence structure $(\mathcal{P}, \mathcal{B}, \mathbf{I})$, where $\mathcal{P}$ and $\mathcal{B}$ are these two classes of subspaces of $\mathbb{F}_q^3$. If a 1-dimensional subspace is contained in a 2-dimensional subspace, we say they are incident. It is immediately clear that we have thus defined a projective plane of order $q$, i.e. a 2-$(q^2 + q + 1, q + 1, 1)$ design. This design is usually denoted by $PG(2, q)$ or $PG_2(q)$, which stands for projective geometry of dimension 2 and order $q$.

The construction defined above can also be applied if we replace $\mathbb{F}_q$ by $\mathbb{R}$. We then obtain the classical real projective plane, where points are the 1-dimensional subspaces and lines are the 2-dimensional subspaces. This geometry contrasts with classical affine geometry in the fact that no two lines are parallel. When speaking about the designs defined above, we use terminology from geometry.

PROBLEM 19F. Find a subset $S = \{s_1, \ldots, s_5\}$ of $\mathbb{Z}_{21}$ such that the elements of $\mathbb{Z}_{21}$ as points and the 21 blocks $S + x$ $(x \in \mathbb{Z}_{21})$ form a projective plane of order 4. (Hint: there is a solution $S$ for which $2S = S$.)

PROBLEM 19G. Let $(R, C, S; L)$ be a Latin square of order 6. Define $\mathcal{P} := R \times C$. Let $\mathcal{B}$ be the set of blocks

$$B_{ij} :=$$
$$\{(x, y) \in R \times C : x = i \text{ or } y = j \text{ or } L(x, y) = L(i, j)\} \setminus \{(i, j)\}$$

for $(i, j) \in R \times C$.

  (1) Show that this defines a 2-(36,15,6) design.
  (2) Show that a regular Hadamard matrix of order 36 exists.

PROBLEM 19H. Let $\mathcal{D}$ be a 3-$(v, k, \lambda)$ design. Suppose that the derived design of $\mathcal{D}$ with respect to a point $p$ (i.e. the case $i = 1$ in the Corollary to Theorem 19.3) is a symmetric design.

  (1) Show that $\lambda(v - 2) = (k - 1)(k - 2)$.
  (2) Show that any two blocks of $\mathcal{D}$ meet in 0 or $\lambda + 1$ points.
  (3) Show that the set of points not on a block $B$ together with the blocks disjoint from $B$ form a 2-design $\mathcal{D}^B$.
  (4) Apply Fisher's inequality to the design $\mathcal{D}^B$ and deduce that $v = 2k$ or otherwise $k = (\lambda + 1)(\lambda + 2)$ or $k = 2(\lambda + 1)(\lambda + 2)$.

What are the possibilities for the design $\mathcal{D}$? Do we know any designs with these properties?

PROBLEM 19I. Let $O$ be a subset of the points of a projective plane of order $n$ such that no three points of $O$ are on one line. Show that $|O| \leq n + 1$ if $n$ is odd and that $|O| \leq n + 2$ if $n$ is even. A set of $n + 1$ points, no three on a line, is called an *oval*; a set of

$n + 2$ points, no three on a line, is a *hyperoval*. Two constructions of $PG_2(4)$ were given in Example 19.7 and Problem 19F. In each case, construct a hyperoval.

PROBLEM 19J. Let $O$ be a hyperoval (with $q+2$ points) in $PG_2(q)$, $q = 2^m$. Any of the $q^2 - 1$ points $p \notin O$ has the property that there are exactly $\frac{1}{2}(q + 2)$ secants of $O$ through $p$. Take five points on $O$ and split them into

$$\{\{p_1, p_2\}, \{p_3, p_4\}, \{p_5\}\} \, .$$

This can be done in 15 ways. The two pairs determine two secants that meet in a point $p \notin O$. The line through $p$ and $p_5$ meets $O$ in a point, that we call $p_6$. This defines 15 (not necessarily distinct) 6-tuples of points on $O$, containing the given five points. This defines an $S_\lambda(5, 6, q + 2)$ (a construction due to D. Jungnickel and S. A. Vanstone). Construct a hyperoval $O$ in $PG_2(q)$ and then show that the 5-design is not a simple design. (Hint: use coordinates, cf. Example 19.7.)

Any 2-$(n^2, n, 1)$ design is called an *affine plane*. The points and lines of the plane (= 2-dimensional vector space) $\mathbb{F}_q^2$ form an affine plane of order $q$. For such a design we use the notation $AG_2(n)$ (2-dimensional affine geometry of order $n$).

EXAMPLE 19.8. Let $\mathcal{D}$ be a projective plane of order $n$. If we delete one line and all the points on that line, we find an affine plane of order $n$.

PROBLEM 19K. Let $\mathcal{D}$ be any affine plane of order $n$. If $B_1$ and $B_2$ are two blocks, then we write $B_1 \sim B_2$ if the two blocks are the same or if they have no points in common. Show that $\sim$ is an equivalence relation. A class of this relation is called a *parallel class*. Show that there exists a projective plane of order $n$ such that $\mathcal{D}$ can be obtained from that plane by the construction of Example 19.8.

We shall now show that if $N$ is the incidence matrix of a symmetric design $\mathcal{D}$, then $N^\top$ is also the incidence matrix of a symmetric design $\mathcal{D}^\top$, called the *dual* of $\mathcal{D}$.

THEOREM 19.9. *Let $N$ be the incidence matrix of a symmetric 2-$(v, k, \lambda)$ design. Then $N^\top$ is also the incidence matrix of a design.*

PROOF: Consider any block $B$ of the design. For $0 \leq i \leq k$ let $a_i$ be the number of blocks ($\neq B$) that have $i$ points in common with $B$. Then counting blocks, pairs $(p, B')$ with $p \in B \cap B'$ and triples $(p, q, B')$ with $p \neq q$ and $\{p, q\} \subseteq B \cap B'$ we find:

$$\sum_{i=0}^{k} a_i = v - 1, \quad \sum_{i=0}^{k} i a_i = k(k-1), \quad \sum_{i=0}^{k} \binom{i}{2} a_i = \binom{k}{2}(\lambda - 1),$$

from which we find $\sum_{i=0}^{k}(i - \lambda)^2 a_i = 0$. Hence, any block $B' \neq B$ has $\lambda$ points in common with $B$, i.e. $N^\top N = (k - \lambda)I + \lambda J$.  □

Note that in Example 19.7, we did not need to specify whether the set $\mathcal{P}$ was the 1-dimensional subspaces or the 2-dimensional subspaces. In the latter situation, we have the dual of the former.

In many cases the designs $\mathcal{D}$ and $\mathcal{D}^\top$ are not isomorphic.

Let $\mathcal{D}$ be a symmetric 2-$(v, k, \lambda)$ design. There are two other ways to obtain a design from $\mathcal{D}$. These two designs are called the *derived* design and *residual* design of $\mathcal{D}$. This could be somewhat confusing since we have already introduced that terminology. We shall always indicate which of the two we mean. Take any block $B$ of $\mathcal{D}$. The residual of $\mathcal{D}$ with respect to $B$ has $\mathcal{P} \backslash B$ as point set and as blocks all $B' \backslash B$ with $B' \neq B$. It is a 2-$(v - k, k - \lambda, \lambda)$ design. The derived design has $B$ as point set and as blocks all $B' \cap B$ with $B' \neq B$. It is a 2-$(k, \lambda, \lambda - 1)$ design. If a design with parameters $v, k, b, r, \lambda$ is the residual of a symmetric design, then $r = k + \lambda$. Any 2-design for which this equation holds is called a *quasiresidual* design. If such a design is not the residual of a symmetric design, then we say that it is *nonembeddable*. The assertion of Problem 19K is that every affine plane is embeddable in a projective plane. A theorem due to W. S. Connor (1952), that we shall leave until Chapter 21, states that every quasiresidual design with $\lambda = 2$ is embeddable.

EXAMPLE 19.9. Let $C := \begin{pmatrix} 0 & 1 & 0 \\ 0 & 0 & 1 \\ 1 & 0 & 0 \end{pmatrix}$ and let $E_i$ denote a 3 by 3 matrix with ones in column $i$ and zeros elsewhere. Then

$$N := \begin{pmatrix} E_1 & I & I & I \\ E_2 & I & C & C^2 \\ E_3 & I & C^2 & C \end{pmatrix}$$

is the 9 by 12 incidence matrix of $AG_2(3)$. Define

$$A := \begin{pmatrix} 1\,1\,1\,1\,1\,1\,0\,0\,0\,0\,0\,0 \\ 1\,1\,1\,0\,0\,0\,1\,1\,1\,0\,0\,0 \\ 1\,1\,1\,0\,0\,0\,0\,0\,0\,1\,1\,1 \\ 0\,0\,0\,1\,1\,1\,1\,1\,1\,0\,0\,0 \\ 0\,0\,0\,1\,1\,1\,0\,0\,0\,1\,1\,1 \\ 0\,0\,0\,0\,0\,0\,1\,1\,1\,1\,1\,1 \end{pmatrix}, \quad B := \begin{pmatrix} 1\,1\,0\,0 \\ 1\,1\,0\,0 \\ 1\,1\,0\,0 \\ 1\,0\,1\,0 \\ 1\,0\,1\,0 \\ 1\,0\,1\,0 \\ 1\,0\,0\,1 \\ 1\,0\,0\,1 \\ 1\,0\,0\,1 \end{pmatrix}.$$

Form the 24 by 16 matrix

$$D := \begin{pmatrix} A & O \\ N & B \\ N & J - B \end{pmatrix}.$$

One easily checks that $D^\top$ is the 16 by 24 incidence matrix of a 2-(16,6,3) design. This is a quasiresidual design. However, it cannot be the residual of a 2-(25,9,3) symmetric design because the inner product of row $i + 6$ and row $i + 15$ of $D$, $1 \le i \le 9$, equals 4 and, by Theorem 19.9, the inner product of the columns of the incidence matrix of a 2-(25,9,3) design is 3. This shows that nonembeddable designs with $\lambda = 3$ exist.

The combination of a counting argument and a suitable quadratic form that we used to prove Theorem 19.9 is widely used in combinatorics. However, sometimes it is easier to use algebraic methods as we shall demonstrate in the following theorem, due to Ryser. (The reader can try to prove the theorem by using counting arguments.)

THEOREM 19.10. *Let* $\mathcal{D} = (\mathcal{P}, \mathcal{B}, \mathbf{I})$ *be an incidence structure with* $|\mathcal{P}| = |\mathcal{B}| = v$, *blocksize* $k$, *such that any two blocks meet in* $\lambda$ *points. Then* $\mathcal{D}$ *is a symmetric 2-design.*

PROOF: Let $N$ be the incidence matrix of $\mathcal{D}$. Then

(19.8) $$N^\top N = (k - \lambda)I + \lambda J,$$

and

(19.9) $$JN = kJ.$$

By Theorem 19.9, we are done if we can show that $NJ = kJ$. From (19.8) we see that $N$ is nonsingular and hence (19.9) can be read as $J = kJN^{-1}$. From (19.8) we find $JN^\top N = (k - \lambda + \lambda v)J$ and therefore

$$JN^\top = (k - \lambda + \lambda v)JN^{-1} = (k - \lambda + \lambda v)k^{-1}J,$$

i.e. $N$ has constant rowsums. Then these rowsums must be $k$. This proves the theorem and yields $(k - \lambda + \lambda v)k^{-1} = k$ as was to be expected from (19.3) and (19.4). □

As a preparation for the best known nonexistence theorem for designs, we need two results, both due to Lagrange. For the first, consider the matrix $H$ of (18.6) with $n = 1$, i.e. $A_i = (a_i)$. Define $\mathbf{y} = (y_1, y_2, y_3, y_4)$ by $\mathbf{y} := \mathbf{x}H$, where $\mathbf{x} = (x_1, x_2, x_3, x_4)$. Then from (18.7) we find

(19.10) $\quad (a_1^2 + a_2^2 + a_3^2 + a_4^2)(x_1^2 + x_2^2 + x_3^2 + x_4^2) = (y_1^2 + y_2^2 + y_3^2 + y_4^2).$

Using this identity, Lagrange proved that every integer is the sum of four squares. Clearly the identity shows that it is sufficient to prove this for primes. For an elegant proof that a prime is the sum of four squares we refer to Chandrasekharan (1968).

The following nonexistence theorem is known as the Bruck-Ryser-Chowla theorem.

THEOREM 19.11. *If $v, k, \lambda$ are integers such that $\lambda(v - 1) = k(k - 1)$, then for the existence of a symmetric 2-$(v, k, \lambda)$ design it is necessary that:*

(i) *if $v$ is even then $k - \lambda$ is a square;*
(ii) *if $v$ is odd, then the equation $z^2 = (k - \lambda)x^2 + (-1)^{(v-1)/2}\lambda y^2$ has a solution in integers $x, y, z$, not all zero.*

PROOF: Assertion (i) was proved in Theorem 19.7. So assume that $v$ is odd. Let $\mathcal{D}$ be a symmetric 2-$(v, k, \lambda)$ design with incidence matrix $N = (n_{ij})$ and write $n := k - \lambda$. We now introduce $v$ linear forms $L_i$ in the variables $x_1, \ldots, x_v$ by

$$L_i := \sum_{j=1}^{v} n_{ij} x_j, \qquad 1 \leq i \leq v.$$

Then the equation $N^\top N = (k - \lambda)I + \lambda J$ implies that

$$(19.11) \qquad L_1^2 + \cdots + L_v^2 = n(x_1^2 + \cdots + x_v^2) + \lambda(x_1 + \cdots + x_v)^2.$$

By Lagrange's theorem, $n$ can be written as $n = a_1^2 + \cdots + a_4^2$. This and (19.10) allow us to take four of the variables $x_j$ and write

$$(19.12) \qquad n(x_i^2 + x_{i+1}^2 + x_{i+2}^2 + x_{i+3}^2) = (y_i^2 + y_{i+1}^2 + y_{i+2}^2 + y_{i+3}^2),$$

where each $y_j$ is a linear form in the four variables $x_i, \ldots, x_{i+3}$.

We now first assume that $v \equiv 1 \pmod 4$. By applying this to (19.11) four variables at a time and introducing $w$ for $x_1 + \cdots + x_v$, we reduce (19.11) to

$$(19.13) \qquad L_1^2 + \cdots + L_v^2 = y_1^2 + \cdots + y_{v-1}^2 + nx_v^2 + \lambda w^2.$$

Since $H$ in (18.6) is invertible, we can express the variables $x_j$ for $1 \leq j \leq v - 1$ as linear forms in the corresponding $y_j$ and hence $w$ is a linear form in these variables and $x_v$. Next, we reduce the number of variables in the following way. If the linear form $L_1$, expressed in $y_1, \ldots, y_{v-1}, x_v$, does not have coefficient $+1$ for $y_1$, then we set $L_1 = y_1$, and if the coefficient is $+1$, we set $L_1 = -y_1$, and in both cases, we subsequently solve this equation for $y_1$ as a linear expression in the remaining variables $y_j$ and $x_v$. This is substituted in the expression $w$. So (19.11) has been reduced to

$$L_2^2 + \cdots + L_v^2 = y_2^2 + \cdots + y_{v-1}^2 + nx_v^2 + \lambda w^2.$$

We proceed in this way for $y_2, \ldots, y_{v-1}$. In each step, $w$ is replaced by another linear form in the remaining variables, and hence we end up with

$$L_v^2 = nx_v^2 + \lambda w^2,$$

in which both $L_v$ and $w$ are rational multiples of the variable $x_v$. If we multiply this by the common denominator of the factors, we find an equation

$$z^2 = (k - \lambda)x^2 + \lambda y^2$$

in integers. This proves the assertion that if $v \equiv 1 \pmod 4$. If $v \equiv 3 \pmod 4$, the same procedure is applied to (19.13) after adding $nx_{v+1}^2$ to both sides, where $x_{v+1}$ is a new variable. The equation is then finally reduced to $nx_{v+1}^2 = y_{v+1}^2 + \lambda w^2$ and again we multiply by a common denominator to find an equation of type

$$(k - \lambda)x^2 = z^2 + \lambda y^2$$

in accordance with assertion (ii).                                        □

EXAMPLE 19.10. From Example 19.7, we know that a projective plane of order $n$ exists for $2 \leq n \leq 9$, except possibly for $n = 6$. By Theorem 19.11, a necessary condition for the existence of a projective plane of order 6 is that the equation $z^2 = 6x^2 - y^2$ has a nontrivial solution. If such a solution exists, then also one for which $x$, $y$, and $z$ have no prime factor in common, i.e. $z$ and $y$ are both odd. Then $z^2$ and $y^2$ are both $\equiv 1 \pmod 8$. Since $6x^2 \pmod 8$ is either 0 or 6, we see that the equation has only the trivial solution (0,0,0). Therefore a projective plane of order 6 does not exist.

If we try the same thing for a plane of order 10, we find the equation $z^2 = 10x^2 - y^2$, which has the solution $x = 1$, $y = 1$, $z = 3$. In this case Theorem 19.11 tells us nothing. A few years ago it was announced that a computer search involving several hundred hours on a Cray 1, had excluded the existence of a projective plane of order 10. This is the only case where the nonexistence of a symmetric 2-design has been shown using something other than Theorem 19.11.

COROLLARY. *If there exists a projective plane of order $n \equiv 1$ or 2 (mod 4), then $n$ is the sum of two integral squares.*

PROOF: The condition $n \equiv 1$ or 2 (mod 4) implies that $v = n^2 + n + 1 \equiv 3 \pmod 4$. Theorem 19.11 asserts that $n$ is the sum of two rational squares. It is well known that $n$ is the sum of two rational squares if and only if $n$ is the sum of two integral squares. (This follows from the condition that $n$ is the sum of two integral squares if and only if no prime divisor of the square free part of $n$ is $\equiv 3$ (mod 4).) $\qquad\square$

PROBLEM 19L. Show that a symmetric 2-(29,8,2) design does not exist.

PROBLEM 19M. Suppose $M$ is a rational square matrix of order $v$ and that $MM^\top = mI$. Show that if $v$ is odd, then $m$ is a square. Show that if $v \equiv 2$ (mod 4), then $m$ is the sum of two rational squares.

(Note that one consequence of this latter result is that the existence of a conference matrix of order $n \equiv 2$ (mod 4) implies that $n - 1$ is the sum of two squares.)

A great deal of work has been done on the construction of 2-designs. We shall only treat a number of examples that will give some idea of the kind of methods that have been used. The smallest nontrivial pair $(k, \lambda)$ to consider is $(3, 1)$. A 2-$(v, 3, 1)$ design is called a *Steiner triple system*. One uses the notation $STS(v)$ for such a design. By (19.3) and (19.4) a necessary condition for the existence of such a design is that $v \equiv 1$ (mod 6) or $v \equiv 3$ (mod 6). We shall show that this condition is also sufficient. This will be done by direct construction in Examples 19.11 and 19.15. However, it is useful to see a number of examples of a more complicated approach. The methods that we demonstrate can be used for the construction of other designs than Steiner triple systems. Furthermore, they can be used to produce designs with certain subdesigns (see Problem 19N) or prescribed automorphism group. The idea of this approach is to find direct constructions for small examples and some recursive constructions, and subsequently show that, for any $v$ that satisfies the necessary conditions, an $STS(v)$ can be constructed by the recursive methods, using the list of known small examples. We shall see below that this in fact reduces to a (not very difficult) problem in number theory. As stated above, we restrict ourselves to a number of examples. The reader may wish to try to show that our examples suffice to find an $STS(v)$ for all possible values of $v$, without using Examples 19.11 and 19.15.

We consider the trivial design with only one block of size 3 as $STS(3)$. We have already seen constructions of $STS(7) = PG_2(2)$ and $STS(9) = AG_2(3)$. In Example 19.1, we constructed an $STS(15)$.

EXAMPLE 19.11. Let $n = 2t + 1$. We define $\mathcal{P} := \mathbb{Z}_n \times \mathbb{Z}_3$. As blocks we take all triples $\{(x,0),(x,1),(x,2)\}$ with $x \in \mathbb{Z}_n$ and all triples $\{(x,i),(y,i),(\frac{1}{2}(x+y),i+1)\}$ with $x \neq y$ in $\mathbb{Z}_n$ and $i \in \mathbb{Z}_3$. This simple construction provides an $STS(6t + 3)$ for every $t$.

EXAMPLE 19.12. Let $q = 6t + 1$ be a prime power and let $\alpha$ be a primitive element in $\mathbb{F}_q$, i.e. $\mathbb{F}_q^*$ is a cyclic group generated by $\alpha$. We define

$$(19.14) \quad B_{i,\xi} := \{\alpha^i + \xi, \alpha^{2t+i} + \xi, \alpha^{4t+i} + \xi\}, \qquad 0 \leq i < t, \quad \xi \in \mathbb{F}_q.$$

We claim that the elements of $\mathbb{F}_q$ as points and the blocks $B_{i,\xi}$ form an $STS(q)$. The idea of the proof is the same as in Example 19.6. Note that $\alpha^{6t} = 1$, $\alpha^{3t} = -1$ and define $s$ by $\alpha^s = (\alpha^{2t} - 1)$. We consider the six differences of pairs from $B_{0,0}$. These are:

$$\alpha^{2t} - 1 = \alpha^s, \qquad\qquad -(\alpha^{2t} - 1) = \alpha^{s+3t},$$
$$\alpha^{4t} - \alpha^{2t} = \alpha^{s+2t}, \qquad -(\alpha^{4t} - \alpha^{2t}) = \alpha^{s+5t},$$
$$\alpha^{6t} - \alpha^{4t} = \alpha^{s+4t}, \qquad\quad -(1 - \alpha^{4t}) = \alpha^{s+t}.$$

It follows that for any $\eta \neq 0$ in $\mathbb{F}_q$, there is a unique $i$, $0 \leq i < t$, such that $\eta$ occurs as the difference of two elements of $B_{i,0}$. Hence for any $x$ and $y$ in $\mathbb{F}_q$, there is a unique $i$ and a unique $\xi \in \mathbb{F}_q$ such that the pair $x, y$ occurs in the block $B_{i,\xi}$. $\qquad\square$

The method of Examples 19.6 and 19.12 is known as the *method of differences*. Example 19.15 will show a more complicated use of the same idea.

We now know that an $STS(v)$ exists for $v = 13, 19, 25, 31, 37, 43$ and 49 as well as the values mentioned above. This includes all $v \equiv 1 \pmod{6}$ less than 50. In fact, we now know at least one $STS(v)$ for each feasible value of $v$ less than 100, except $v = 55$, $v = 85$, $v = 91$.

EXAMPLE 19.13. Let there be an $STS(v_i)$ on the point set $V_i$ $(i = 1, 2)$. We take $V_1 \times V_2$ as a new point set and define as blocks all triples: $\{(x_1, y_1), (x_2, y_2), (x_3, y_3)\}$ for which

(1) $x_1 = x_2 = x_3$ and $\{y_1, y_2, y_3\}$ is a block of $STS(v_2)$;
(2) $\{x_1, x_2, x_3\}$ is a block of $STS(v_1)$ and $y_1 = y_2 = y_3$;
(3) $\{x_1, x_2, x_3\}$ is a block of $STS(v_1)$ and $\{y_1, y_2, y_3\}$ is a block of $STS(v_2)$.

It is practically obvious that this defines an $STS(v_1 v_2)$. The reader should check that we have defined the correct number of blocks.

This construction provides us with an $STS(91)$.

EXAMPLE 19.14. We show a slightly more complicated construction. Suppose that we have an $STS(v_1)$ on the point set $V_1 = \{1, 2, \ldots, v_1\}$ with block set $S_1$, and furthermore suppose that the blocks that are completely contained in $V = \{s+1, \ldots, v_1\}$, where $s = v_1 - v$, form an $STS(v)$. Let $S_2$ be the set of triples of an $STS(v_2)$ on the point set $V_2 = \{1, 2, \ldots, v_2\}$.

We consider a new point set

$$\mathcal{P} := V \cup \{(x, y) : 1 \leq x \leq s, \ 1 \leq y \leq v_2\}.$$

This set has $v + v_2(v_1 - v)$ points. We introduce a set $\mathcal{B}$ of four kinds of blocks:

(1) those of the subsystem $STS(v)$;
(2) $\{(a, y), (b, y), c\}$ with $c \in V$, $\{a, b, c\} \in S_1$ and $y \in V_2$;
(3) $\{(a, y), (b, y), (c, y)\}$ with $\{a, b, c\}$ a block in $S_1$ with no point in $V$, and $y \in V_2$;
(4) $\{(x_1, y_1), (x_2, y_2), (x_3, y_3)\}$, where $\{y_1, y_2, y_3\}$ is a block in $S_2$ and the integers $x_1, x_2, x_3$ satisfy

$$x_1 + x_2 + x_3 \equiv 0 \pmod{s}.$$

Again, one easily checks that any two points of $\mathcal{P}$ uniquely determine a block in $\mathcal{B}$. Hence $\mathcal{P}$ and $\mathcal{B}$ are the points and blocks of a Steiner triple system on $v + v_2(v_1 - v)$ points. A simple example is obtained by letting the subsystem be just one block, i.e. $v = 3$. Taking $v_1 = 7$, $v_2 = 13$, we find an $STS(55)$.

We have thus constructed an $STS(v)$ for every feasible value of $v$ less than 100, except $v = 85$.

PROBLEM 19N. (a) Show that if an $STS(v_1)$ and an $STS(v_2)$ both exist, then there is an $STS(v_1v_2 - v_2 + 1)$. Use this construction to find an $STS(85)$.

(b) Construct an $STS(15)$ on the set $\{0, 1, \ldots, 14\}$ such that it contains a Fano plane on $\{0, 1, \ldots, 6\}$ as a subsystem.

EXAMPLE 19.15. Consider as point set $\mathbb{Z}_{2t} \times \mathbb{Z}_3 \cup \{\infty\}$. Addition of elements is coordinatewise with the extra convention $\infty + (x, i) = \infty$. For notational convenience we sometimes write the second coordinate as an index, i.e. $x_i$ instead of $(x, i)$. We now define four types of "*base blocks*":

(1) $\{0_0, 0_1, 0_2\}$;
(2) $\{\infty, 0_0, t_1\}$, $\quad \{\infty, 0_1, t_2\}$, $\quad \{\infty, 0_2, t_0\}$;
(3) $\{0_0, i_1, (-i)_1\}$, $\{0_1, i_2, (-i)_2\}$, $\{0_2, i_0, (-i)_0\}$, $i = 1, \ldots, t-1$;
(4) $\{t_0, i_1, (1-i)_1\}$, $\{t_1, i_2, (1-i)_2\}$, $\{t_2, i_0, (1-i)_0\}$, $\quad i = 1, \ldots, t$.

We have $6t + 1$ base blocks. For $a = 0, 1, \ldots, t - 1$ we add the element $(a, 0)$ (i.e. $a_0$) to each of the elements of every base block, thus producing $t(6t + 1)$ blocks. We claim that these are the triples of an $STS(6t+1)$. It is trivial that the base blocks of type 2 yield a set of blocks in which every pair of points, one of which is $\infty$, occurs exactly once. The cyclic nature of the definition of the base blocks shows that it is sufficient for us to check that all pairs $\{a_0, b_0\}$ with $a \neq b$ and all pairs $\{a_0, b_1\}$ occur in the triples we have defined. If $a < b$ and $b - a = 2s$, then the pair $\{a_0, b_0\}$ occurs in the triple obtained from $\{0_2, s_0, (-s)_0\}$ "translated" by the element $(b - s, 0)$. Similarly, if $b - a$ is odd, we find the required pair by translating a base block of type 4. Now consider a pair $\{a_0, b_1\}$. If $a = b \leq t - 1$, we find the pair by translating the base block of type 1 by $(a, 0)$. If $a \neq b$ and $a < t$, we have to look for the pair in a translate of a base block of type 2 or of type 3. We must search for a base block in which the difference $b - a$ occurs as $y - x$ for two elements $y_1$, $x_0$. For type 2, this difference is $t$ and in the blocks $\{0_0, i_1, (-i)_1\}$ we find the differences $i$, $1 \leq i \leq t - 1$, and $-i = 2t - i$, $1 \leq i \leq t - 1$, indeed every difference once! Now, the rest of the details can be left as an exercise.

This example shows that if $v = 6t + 1$, then an $STS(v)$ exists. Combined with Example 19.11 we have a construction for every feasible value of $v$.

We end this chapter with an amusing application of the Fano plane. At present the idea is not used in practice but the problem itself has a practical origin, and maybe some day generalizations of the following method will be used. Suppose one wishes to store one of the integers 1 to 7 in a so-called *"write-once memory"*. This is a binary memory, originally filled with zeros, for which it is possible to change certain bits to ones but not back again, i.e. the state 1 is permanent. This happens in practice with paper tape, into which holes are punched, or compact discs, where a laser creates pits in certain positions. In both cases, we cannot erase what was written in the memory. To store the integers 1 to 7, we need a memory of three bits. What if one wishes to use the memory four consecutive times? The simplest solution is to have a 12-bit memory that is partitioned into four 3-bit sections, one for each consecutive usage. We assume that the memory is very expensive and we would like to be able to use a shorter memory for the same purpose. We shall now show that seven bits suffice, a saving of more than 40%.

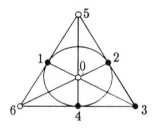

Figure 19.3

Let $\mathcal{P} = \{1, 2, \ldots, 7\}$ be the set of points of $PG_2(2)$ and let $\mathcal{L}$ denote the set of lines. To store one of the integers 1 to 7 in a memory with positions numbered 1 to 7, we use the following rules. As a general rule: if we wish to store $i$ and the memory is in a state corresponding to $i$ (from a previous usage), then we do nothing. Otherwise the rules are:

(1) if the memory is empty, store $i$ by putting a 1 in position $i$;
(2) to store $j$ when the memory is in state $i$, put a 1 in position $k$, where $\{i, j, k\} \in \mathcal{L}$;
(3) to store $i$ when the memory contains two 1's, not correspond-

ing to $i$, put in two more 1's, such that $i$ is one of the four 1's and the other three form a line in $\mathcal{L}$. No matter what the two original 1's were, this is possible (sometimes in two ways);

(4) if the memory contains four 1's, we may assume that we are in the situation of Fig. 19.3. To store 3, we do nothing (by the general rule); to store one of the missing numbers, we put 1's in the other two positions; to store 1, 2, or 4, store a 1 in the empty position on the line through 3 and the number we wish to store.

We leave it as an exercise for the reader to formulate the rules for reading the memory. Note that the memory uniquely reads the integer presently stored in the memory but it cannot see how often an integer has been stored or what was stored on the previous usage.

**Notes.**

The first occurrence of a 2-design may be $AG_2(3)$ in a paper by Plücker (1839). One usually attributes the introduction of Steiner systems to Woolhouse (1844); of course *not* to Steiner! Quite often they are said to originate with a problem of T. P. Kirkman (1847). T. P. Kirkman (1806–1895), a self-educated man, was a minister of the Church of England. He was an amateur mathematician with many contributions to the subject. Probably the best known is his *15 schoolgirls problem.* The problem is to arrange 15 schoolgirls in parties of three for seven days' walks such that every two of them walk together exactly once. This amounts to constructing an $STS(15)$ for which the set of triples can be partitioned into seven "parallel classes".

Jakob Steiner (1796–1863) was an important geometer of his time. He became interested in what we now call Steiner systems in 1853 when he studied the configuration of 28 double tangents of a plane quartic curve.

Sir Ronald A. Fisher (1890–1962) is considered to be one of the most prominent statisticians. Besides important contributions to statistics (multivariate analysis) and genetics, he is known for his work on the application of statistical theory to agriculture and the design of experiments. The applications to the design of agricultural experiments account for our usage of $v$ for the number of points of a design (*varieties*) and $r$ for the number of blocks through a point (*replication number*).

Theorem 19.7 is due to M. P. Schutzenberger (1949).

R. A. Fisher was not the only statistician to contribute to the mathematical theory of designs. In fact, we should probably consider the Indian mathematician R. C. Bose (1901–1987) to be the most important one. Many of the construction methods described in this chapter (such as the method of differences) are due to him.

G. Fano (1871–1952), whose name has become attached to the plane $PG_2(2)$, was important in the Italian school of projective geometry.

Projective planes over finite fields were first studied by K. G. C. von Staudt (1798–1867) in his book *Geometrie der Lage* (1856).

The first example of a nonembeddable quasiresidual design was given by Bhattacharya (1944). It was also a 2-(16,6,3) design. However, Example 19.9 is a much simpler example.

J. L. Lagrange (1736–1813) was born and educated in Italy but he is considered a French mathematician (who studied in Berlin). Besides many important contributions to analysis, he is known for several theorems in number theory.

The Bruck-Ryser-Chowla theorem, Theorem 19.11, is so well known that it is often referred to as just BRC.

The idea of Example 19.15 is due to Skolem (1958). His method was actually slightly different. It has led to the term *Skolem sequences*. These have other applications, e.g. in radioastronomy. Here is the idea. Partition the set $\{1, 2, \ldots, 2n\}$ into pairs $\{a_i, b_i\}$ such that $b_i - a_i = i$, $1 \leq i \leq n$. This is a Skolem sequence. For example, $\{9, 10\}, \{2, 4\}, \{5, 8\}, \{3, 7\}, \{1, 6\}$ is such a partition for $n = 5$. Now form the triples $\{0, a_i + n, b_i + n\}$ and consider these as base blocks (mod $6n + 1$). Since all the differences $1, 2, \ldots, 3n$ and their negatives occur exactly once, the blocks form an $STS(6n + 1)$.

For an interesting application of the Golay code (see Chapter 20) to write-once memories, see Cohen *et al.* (1986). They show the possibility of three successive writings of 11 bits on 23 positions.

**References.**

W. O. Alltop (1972), An infinite class of 5-designs, *J. Combinatorial Theory* (A) **12**, 390–395.

K. N. Bhattacharya (1944), A new balanced incomplete block design, *Science and Culture* **9**, 108.

R. H. Bruck and H. J. Ryser (1949), The non-existence of certain finite projective planes, *Canad. J. Math.* **1**, 88–93.

N. G. de Bruijn and P. Erdős (1948), On a combinatorial problem, *Proc. Kon. Ned. Akad. v. Wetensch.* **51**, 1277–1279.

K. Chandrasekharan (1968), *Introduction to Analytic Number Theory*, Springer-Verlag.

S. Chowla and H. J. Ryser (1950), Combinatorial problems, *Canad. J. Math.* **2**, 93–99.

G. D. Cohen, P. Godlewski, and F. Merkx (1986), Linear binary codes for write-once memories, *IEEE Trans. Information Theory* **32**, 697–700.

W. S. Connor, Jr. (1952), On the structure of balanced incomplete block designs, *Ann. Math. Stat.* bf 23, 57–71; correction *ibid.* **24**, 135.

R. H. F. Denniston (1976), Some new 5-designs, *Bull. London Math. Soc.* **8**, 263–267.

G. Fano (1892), *Giornale di Matimatiche* **30**, 114–124.

T. P. Kirkman (1847), On a problem in combinations, *Cambridge and Dublin Math. J.* **2**, 191–204.

D. L. Kreher and S. P. Radziszowski (1986), The existence of simple 6-(14,7,4) designs, *J. Combinatorial Theory* (A) **41**, 237–243.

D. W. Leavitt and S. S. Magliveras (1982), Simple 6-(33,8,36)-designs from $P\Gamma L_2(32)$, pp. 337–352, in: *Computational Group Theory, Proc. Durham 1982*.

W. H. Mills (1978), A new 5-design, *Ars Combinatoria* **6**, 193–195.

A. Ya. Petrenjuk (1968), On Fisher's inequality for tactical configurations (in Russian), *Mat. Zametki* **4**, 417–425.

D. K. Ray-Chaudhuri and R. M. Wilson (1975), On *t*-designs, *Osaka J. Math.* **12**, 737–744.

H. J. Ryser (1963), *Combinatorial Mathematics*, Carus Math. Monograph **14**.

M. P. Schutzenberger (1949), A non-existence theorem for an infinite family of symmetrical block designs, *Ann. Eugenics* **14**, 286–287.

Th. Skolem (1958), Some remarks on the triple systems of Steiner, *Math. Scand.* **6**, 273–280.

J. Steiner (1853), Combinatorische Aufgabe, *J. f. d. reine u. angew. Mathematik* **45**, 181–182.

L. Teirlinck (1987), Nontrivial *t*-designs without repeated blocks exist for all *t*, *Discrete Math.* **65**, 301–311.

J. Tits (1964), Sur les systèmes de Steiner associés aux trois 'grands' groupes de Mathieu, *Rend. Math. e Appl.* (5) **23**, 166–184.

E. Witt (1938), Die 5-fach transitiven Gruppen von Mathieu, *Abh. Math. Sem. Univ. Hamburg* **12**, 256–264.

W. S. B. Woolhouse (1844), Prize question 1733, *Lady's and Gentleman's Diary*.

# 20
# Codes and designs

We introduce some more terminology from the theory of error-correcting codes. In the most general sense, a *code* of length $n$ is simply a subset $C \subseteq S^n$, where $S$ is a finite set (the *alphabet*). Elements of $C$ are called *codewords*. A *binary* code is one with alphabet $S = \{0,1\}$; a *ternary* code is one with $S = \{0,1,2\}$. The *distance* $d(\mathbf{x}, \mathbf{y})$ between two words (vectors) $\mathbf{x}$ and $\mathbf{y}$ in $S^n$ is defined to be the number of positions in which they differ, i.e.

$$(20.1) \qquad d(\mathbf{x}, \mathbf{y}) := |\{i : 1 \leq i \leq n, \ x_i \neq y_i\}|.$$

This is indeed a distance function in the usual sense; check that it satisfies the triangle inequality.

The concept of distance has led to the usage of geometric terminology, e.g. the set $B_r(\mathbf{x}) := \{\mathbf{y} \in \mathbb{F}_q^n : d(\mathbf{x}, \mathbf{y}) \leq r\}$ is called the *sphere* with radius $r$ and center $\mathbf{x}$, though actually the name "ball" would be better.

The *minimum distance* $d$ of the code $C$ is

$$(20.2) \qquad d := \min\{d(\mathbf{x}, \mathbf{y}) : \mathbf{x} \in C, \mathbf{y} \in C, \mathbf{x} \neq \mathbf{y}\}.$$

Much of coding theory is concerned with *linear* codes. By a $q$-ary $[n, k]$ code, we mean a linear subspace $C$ of dimension $k$ of the vector space $\mathbb{F}_q^n$.

The *weight* $w(\mathbf{x})$ of $\mathbf{x}$ is defined by

$$(20.3) \qquad w(\mathbf{x}) := d(\mathbf{x}, \mathbf{0}).$$

This can be defined whenever 0 is one of the symbols (elements of the alphabet), but is especially meaningful for linear codes. When

$C$ is linear, the distance between codewords $\mathbf{x}$ and $\mathbf{y}$ is equal to the weight of $\mathbf{x} - \mathbf{y}$, which is another codeword, and so the minimum distance of $C$ is equal to the *minimum weight*, i.e. the minimum of the weights of nonzero codewords. We use the notation $[n, k, d]$ code for an $[n, k]$ code with minimum distance at least $d$. If $d = 2e + 1$, then $C$ is called an $e$-error-correcting code.

The *covering radius* $\rho(C)$ of the code $C$ is defined to be the minimal $R$ such that the spheres with radius $R$ and codewords as centers cover $S^n$, i.e.

$$(20.4) \qquad \rho(C) := \max\{\min\{d(\mathbf{x}, \mathbf{c}) : \mathbf{c} \in C\} : \mathbf{x} \in S^n\}.$$

In Chapter 18, we mentioned the repetition code in $\{0, 1\}^n$, i.e. the 1-dimensional subspace of $\mathbb{F}_2^n$ containing only $\mathbf{0}$ and $\mathbf{1}$. If $n = 2e+1$ then every word has distance $\leq e$ to exactly one codeword. So this code has covering radius $e$. The two spheres of radius $e$ around the two codewords are disjoint and they cover the space.

In general, we call a not necessarily linear code $C \subseteq S^n$ an $(e$-error-correcting) *perfect code* when $|C| > 1$ and when every $\mathbf{x} \in S^n$ has distance $\leq e$ to *exactly* one codeword. This is equivalent to $C$ having minimum distance $d = 2e + 1$ and covering radius $e$. Clearly, perfect codes are combinatorially interesting objects. However, they are extremely rare.

THEOREM 20.1. *If $C$ is a code in $S^n$ with distance $d \geq 2e+1$, then*

$$(20.5) \qquad |C| \cdot \sum_{i=0}^{e} \binom{n}{i}(q-1)^i \leq q^n.$$

PROOF: The sum on the lefthand side of (20.5) counts the number of words in a sphere of radius $e$. $\qquad\square$

The bound given in this theorem is known as the *sphere packing bound* or as the *Hamming bound*. If equality holds, then the code is perfect.

PROBLEM 20A. Show that if a $[23, 12, 7]$ binary code exists, then this code is perfect.

PROBLEM 20B. By (20.5), a binary code of length 6 and minimum distance 3 has at most 9 codewords. Show that equality cannot hold. (However, 8 is possible.)

Two codes are called *equivalent* if one is obtained from the other by some permutation of the coordinates of $S^n$.

A $k$ by $n$ matrix $G$ is called a *generator matrix* of the $[n, k]$ code $C$ if $C$ is spanned by the rows of $G$. Elementary linear algebra shows that $C$ is equivalent to a code with a generator matrix $G = (I_k\ P)$, where $P$ is some $k$ by $n - k$ matrix. This is called the *reduced echelon form* for a generator.

The *dual* $C^\perp$ of $C$ is defined by

$$(20.6) \qquad C^\perp := \{\mathbf{x} \in \mathbb{F}_q^n : \forall_{\mathbf{c} \in C} \langle \mathbf{x}, \mathbf{c} \rangle = 0\}.$$

If $H$ is a generator matrix for $C^\perp$, then clearly

$$(20.7) \qquad C = \{\mathbf{x} \in \mathbb{F}_q^n : \mathbf{x} H^\top = \mathbf{0}\}.$$

$H$ is called a *parity check matrix* for the code $C$. If $G = (I_k\ P)$ is a generator matrix, then $H = (-P^\top\ I_{n-k})$ is a parity check matrix. If $C = C^\perp$, then $C$ is called a *selfdual code*. If $C \subseteq C^\perp$, then $C$ is called *selforthogonal*.

If $C$ is a linear code in $\mathbb{F}_q^n$, then the *extended code* $\overline{C}$ is defined by
$$(20.8)$$
$$\overline{C} := \{(c_1, \ldots, c_n, c_{n+1}) : (c_1, \ldots, c_n) \in C,\ c_1 + \cdots + c_{n+1} = 0\}.$$

The symbol $c_{n+1}$ is called the *parity check symbol*.

EXAMPLE 20.1. Let $n = (q^k - 1)/(q - 1)$. Consider a matrix $H$ of size $k$ by $n$, with entries in $\mathbb{F}_q$, for which the columns are pairwise linearly independent. Note that this is the maximal value of $n$ for which this is possible. Then $H$ is clearly the parity check matrix of an $[n, n - k]$ code with minimum distance 3. Such a code is called a $q$-ary *Hamming code*. If $\mathbf{c}$ is a codeword, then $|B_1(\mathbf{c})| = 1 + n(q - 1) = q^k$. Since $|C| = q^{n-k}$, we see that this code is perfect (by (20.5)).

PROBLEM 20C. Let $H$ be the ternary [4,2] Hamming code. Define a (nonlinear) ternary code $C$ of length 9 with codewords

$$(x_0, x_1, \ldots, x_4; y_1, \ldots, y_4)$$

by requiring that $\sum_{i=0}^{4} x_i \neq 0$ and that $(y_1 - x_1, \ldots, y_4 - x_4)$ is a codeword in $H$. Show that $C$ has covering radius 1. (No ternary code of length 9 with covering radius 1 and fewer than 9 codewords is known.)

EXAMPLE 20.2. Consider the binary Hamming code $C$ of length $n = 2^k - 1$. By definition, the dual code $C^\perp$ has as generator matrix the $k$ by $n$ matrix that has all possible nonzero vectors of length $k$ as columns. Therefore, the extension of $C^\perp$ is the code $R(1, k)$ of Chapter 18. $C^\perp$ is usually called the *simplex code* of length $n$. We remark that the [8,4] extended binary Hamming code is selfdual.

PROBLEM 20D. Show that the covering radius of the code $R(1, 2k)$ is $2^{2k-1} - 2^{k-1}$. (Hint: use the $\pm 1$ representation instead of $(0,1)$ and work over $\mathbb{Q}$ to show that the covering radius is at most this large. Consider the word $\mathbf{z}$ that is the characteristic function of

$$\{\mathbf{x} \in \mathbb{F}_2^{2k} : x_1 x_2 + \cdots + x_{2k-1} x_{2k} = 1\}$$

to show equality.)

We mention one easily proved bound for arbitrary codes, known as the *Singleton bound*.

THEOREM 20.2. *Let $C$ be any code of length $n$ and minimum distance $d$ over $\mathbb{F}_q$. Then*

$$|C| \leq q^{n-d+1}.$$

PROOF: From each codeword, we delete the last $d - 1$ symbols. The set of "shortened" words consists of words that are pairwise different! There can be no more than $q^{n-d+1}$ of them. □

EXAMPLE 20.3. We give an example that shows a nice connection with Chapter 19. Let $q = 2^a$. Let $S$ be a hyperoval in $PG_2(q)$. Using the terminology of Example 19.7, the elements of $S$ are vectors in $\mathbb{F}_q^3$ with the property that no three of them are linearly dependent. We take the elements of $S$ as the columns of a 3 by $q + 2$ matrix $H$. If we interpret $H$ as the parity check matrix of a $[q + 2, q - 1]$ code over $\mathbb{F}_q$, then this code has minimum distance at least 4. So by Theorem 20.2, the distance is equal to 4. This is one of the rare examples of equality in Theorem 20.2. Codes for which equality holds in the Singleton bound are called *maximum distance separable* codes (MDS codes), another unfortunate name! There are many open problems concerning these codes.

If $C$ is a $q$-ary $[n, k]$ code, and if $A_i$ denotes the number of codewords in $C$ with weight $i$, then

$$(20.9) \qquad A(z) := \sum_{i=0}^{n} A_i z^i$$

is called the *weight enumerator* of $C$. Of course, $A_0 = 1$ and $A(1) = |C| = q^k$.

PROBLEM 20E. Suppose $C$ is a binary code, not necessarily linear, with length 23, minimum distance 7, and $|C| = 2^{12}$. Assume $\mathbf{0} \in C$. First show that $C$ is a perfect code. Let the weight enumerator of $C$ be given by (20.9). Count pairs $(\mathbf{x}, \mathbf{c})$ with $\mathbf{c} \in C$, $w(\mathbf{x}) = 4$ and $d(\mathbf{x}, \mathbf{c}) = 3$ and show that $A_7 = 253$. Then show that the weight enumerator of $C$ is in fact completely determined by the fact that $C$ is a perfect code and $\mathbf{0} \in C$.

The following theorem is one of the most useful in the theory of error-correcting codes. It is due to F. J. MacWilliams (1963).

THEOREM 20.3. *Let $C$ be an $[n, k]$ code over $\mathbb{F}_q$ with weight enumerator $A(z)$ and let $B(z)$ be the weight enumerator of $C^{\perp}$. Then*

$$(20.10) \qquad B(z) = q^{-k} (1 + (q-1)z)^n\, A\left(\frac{1-z}{1+(q-1)z}\right).$$

PROOF: We only give the proof for the case $q = 2$. For other values of $q$, the proof is essentially the same (instead of $(-1)^{\langle \mathbf{u}, \mathbf{v} \rangle}$ used below, one must use $\chi(\langle \mathbf{u}, \mathbf{v} \rangle)$, where $\chi$ is a character on $\mathbb{F}_q$).
Define
$$g(\mathbf{u}) := \sum_{\mathbf{v} \in \mathbb{F}_2^n} (-1)^{\langle \mathbf{u}, \mathbf{v} \rangle} z^{w(\mathbf{v})}.$$

Then

$$\sum_{\mathbf{u} \in C} g(\mathbf{u}) = \sum_{\mathbf{u} \in C} \sum_{\mathbf{v} \in \mathbb{F}_2^n} (-1)^{\langle \mathbf{u}, \mathbf{v} \rangle} z^{w(\mathbf{v})} = \sum_{\mathbf{v} \in \mathbb{F}_2^n} z^{w(\mathbf{v})} \sum_{\mathbf{u} \in C} (-1)^{\langle \mathbf{u}, \mathbf{v} \rangle}.$$

Here, if $\mathbf{v} \in C^\perp$, the inner sum is $|C|$. If $\mathbf{v} \notin C^\perp$, then half of the terms in the inner sum have value $+1$, the other half $-1$. Therefore

$$(20.11) \qquad \sum_{\mathbf{u} \in C} g(\mathbf{u}) = |C| \cdot B(z).$$

Now

$$g(\mathbf{u}) = \sum_{(v_1,v_2,\dots,v_n) \in \mathbb{F}_2^n} \prod_{i=1}^{n} \left( (-1)^{u_i v_i} z^{v_i} \right)$$

$$= \prod_{i=1}^{n} \left( 1 + (-1)^{u_i} z \right)$$

$$= (1-z)^{w(\mathbf{u})} (1+z)^{n-w(\mathbf{u})}.$$

The result follows by substituting this in (20.11). $\qquad\qquad \square$

COROLLARY. *If we write* $B(z) = \sum_{j=0}^{n} B_j z^j$, *then for* $q = 2$, *we find from (20.10):*

$$(20.12) \qquad B_j = 2^{-k} \sum_{i=0}^{n} A_i \sum_{l=0}^{j} (-1)^l \binom{i}{l} \binom{n-i}{j-l}.$$

These relations, known as the *MacWilliams relations*, are linearly independent equations for the coefficients $A_i$, given the coefficients $B_j$.

Many of the known nontrivial 5-designs were found by the following elegant application of MacWilliams' theorem, usually referred to as the *Assmus-Mattson theorem* (1969). Again we restrict the proof to the binary case. For other $q$, the theorem has an obvious generalization with nearly the same proof. We identify the positions of a code of length $n$ with the set $\mathcal{P} := \{1, 2, \dots, n\}$. This allows us to interpret a codeword in a binary code as a subset of $\mathcal{P}$ (i.e. as the characteristic function of a subset). The *support* of a codeword is the set of coordinate positions where the codeword is not zero.

PROBLEM 20F. Let $C$ be a perfect binary $e$-error-correcting code of length $n$. Assume 0 is a symbol and that $\mathbf{0}$ is a codeword. Show that $\mathcal{P}$ together with the supports of codewords of weight $d = 2e+1$ is a $S(e + 1, 2e + 1, n)$.

THEOREM 20.4. *Let $A$ be a binary $[n, k, d]$ code and let $B := A^{\perp}$ be the dual code, an $[n, n - k]$ code. Let $t < d$. Suppose the number of nonzero weights in $B$, that are less than or equal to $n - t$, is $\leq d - t$. Then for each weight $w$, the supports of the words of weight $w$ in $A$ form a $t$-design and the supports of the words of weight $w$ in $B$ form a $t$-design.*

PROOF: If $C$ is any code and a $t$-subset $T$ of $\mathcal{P}$ has been fixed, then we denote by $C'$ the code of length $n - t$ obtained by deleting the coordinates in $T$ from the codewords in $C$. We denote by $C_0$ the subcode of $C'$ obtained by deleting the coordinates in $T$ from the codewords in $C$ that have zeros at all positions of $T$.

The proof is in four steps.

(i) Let $T$ be a subset of $\mathcal{P}$ of size $t$. Since $t$ is less than the minimum distance of $A$, the code $A'$ has as many codewords as $A$, i.e. has dimension $k$ also. In fact, $A'$ still has minimum distance $\geq d - t$. So the dual $(A')^{\perp}$ has dimension $n - k - t$. Clearly, $B_0$ is a subcode of $(A')^{\perp}$. Since the dimension of $B_0$ is at least $n - k - t$, we must have $B_0 = (A')^{\perp}$.

(ii) Let $\sum \alpha_i z^i$ and $\sum \beta_i z^i$ be the weight enumerators for $A'$ and $B_0$, respectively. We claim that these weight enumerators *do not* depend on the particular $t$-subset $T$, but only on the numbers $t$, $n$, $k$, and the weights of words in $B$.

Let $0 < \ell_1 < \ell_2 < \cdots < \ell_r \leq n - t$, where $r \leq d - t$, be the nonzero weights $\leq n - t$ for the code $B$. These are the only possible weights for $B_0$. Then (20.12) gives

$$|B_0|\alpha_j = \binom{n - t}{j} + \sum_{i=1}^{r} \beta_{\ell_i} \sum_{m=0}^{j} (-1)^m \binom{\ell_i}{m} \binom{n - t - \ell_i}{j - m}.$$

By hypothesis, the minimum distance of $A'$ is $\geq r$, so we know the values of $\alpha_j$ for $j < r$, namely $\alpha_0 = 1$, $\alpha_1 = \cdots = \alpha_{r-1} = 0$. Thus we have $r$ linear equations in $r$ unknowns $\beta_{\ell_i}$. These unknowns are uniquely determined if the $r$ by $r$ coefficient matrix $M$ that has $(i,j)$-entry $p_j(\ell_i)$ where

$$p_j(x) := \sum_{m=0}^{j} (-1)^m \binom{x}{m} \binom{n-t-x}{j-m},$$

$1 \le i \le r$, $0 \le j \le r-1$, is nonsingular. But $p_j(x)$ is a polynomial in $x$ of exact degree $j$ (the coefficient of $x^j$ is $(-1)^j 2^j / j!$), so elementary column operations reduce $M$ to the Vandermonde matrix

$$\begin{pmatrix} 1 & \ell_1 & \ell_1^2 & \cdots & \ell_1^{r-1} \\ 1 & \ell_2 & \ell_2^2 & \cdots & \ell_2^{r-1} \\ \vdots & \vdots & \vdots & & \vdots \\ 1 & \ell_r & \ell_r^2 & \cdots & \ell_r^{r-1} \end{pmatrix},$$

which is nonsingular since the $\ell_i$'s are distinct.

Thus the weight enumerator $\sum \beta_i z^i$ does not depend on the choice of the subset $T$. Since $A'$ is the dual of $B_0$, $\sum \alpha_i z^i$ is also independent of $T$.

(iii) Let $\mathcal{E}$ be the collection of words of weight $w$ in $B$, interpreted as subsets of $\mathcal{P}$. The number of members of $\mathcal{E}$ that *miss* all coordinates in $T$ is the number of words of weight $w$ in $B_0$, and this is independent of $T$. That is, the *complement* of $(\mathcal{P}, \mathcal{E})$ is a $t$-design. By Problem 19C, the sets in $\mathcal{E}$ also form the blocks of a $t$-design. This proves the second assertion of the theorem. (Remark: This might be criticized if $w > n - t$. But in this case our argument when applied to $t' := n - w$ shows that either every $w$-subset is the support of a codeword, or no $w$-subset is the support of a codeword; so the words of weight $w$, if any, form a trivial $t$-design.)

(iv) To prove the first assertion of the theorem, we proceed by induction. We start with $w = d$. Let $\mathcal{D}$ be the collection of words of weight $d$ in $A$. The number of sets in $\mathcal{D}$ that contain a given $t$-subset $T$ of $\mathcal{P}$ is equal to the number of words of weight $d-t$ in $A'$, and as we saw above, this number does not depend on the choice of $T$. So $\mathcal{D}$ is a $t$-design. Let $w > d$ and suppose the assertion is true for all $w'$ with $w \ge w' > d$. Now let $\mathcal{D}$ denote the collection of words of weight $w$ in $A$. In this case the number of subsets in

$\mathcal{D}$ that contain a given $t$-subset $T$, is equal to the number of words of weight $w - t$ in $A'$ corresponding to codewords of weight $w$ in $A$. By (iii), the total number of words of weight $w - t$ in $A'$ does not depend on the choice of $T$. By the induction hypothesis and (19.6), the number of words of weight $w - t$ in $A'$ corresponding to codewords of weight less than $w$ in $A$ does not depend either on $T$. This proves the assertion. $\qquad\qquad\square$

PROBLEM 20G. What is the weight enumerator of the dual of the binary Hamming code of lengths $2^r - 1$? Derive an expression for the weight enumerator of the Hamming code itself.

EXAMPLE 20.4. Let $A$ be the extended $[8, 4]$ binary Hamming code. We know that $A = A^{\perp}$. In the notation of Theorem 20.4, we have $d = e = 4$. Take $t = 3$. The conditions of Theorem 20.4 are satisfied. Hence the words of weight 4 form a 3-design, which is of course the Hadamard 3-(8,4,1) design corresponding to $R(1, 3)$.

EXAMPLE 20.5. After the next example, we shall treat the famous binary Golay code and show that the corresponding extended code $G_{24}$ is a selfdual $[24, 12, 8]$ code with weights $0, 8, 12, 16$ and $24$. So Theorem 20.4 shows that the words of weight 8 in this code form a 5-design. From Problem 20E, we know that there are 759 blocks and then (19.1) shows that $\lambda = 1$. This also follows from the fact that the code has distance 8, so two blocks have at most four points in common. This design is the Witt design $S(5, 8, 24)$ mentioned in Chapter 19. The words of other weights also yield 5-designs.

EXAMPLE 20.6. The following 5-designs, and several others, were found by V. Pless (1972). Consider the 18 by 18 Paley matrix $C$ given by (18.5). We consider a ternary $[36, 18]$ code $Sym_{36}$ with generator $G = (I_{18} \ C)$. Such a code is called a *symmetry code*. Since $C$ is a Paley matrix, we have $GG^{\top} = O$, i.e. $Sym_{36}$ is a selfdual code. This implies that all weights in the code are divisible by 3. We claim that all the words in $Sym_{36}$ have weight at least 12. Observe that since $C$ is symmetric, the matrix $(-C \ I_{18})$ is a parity check matrix for the code, and, because the code is selfdual, this means that that matrix is also a generator matrix for the code. If $(\mathbf{a}, \mathbf{b})$, where $\mathbf{a}$ and $\mathbf{b}$ are vectors in $\mathbb{F}_3^{18}$, is a codeword, then $(-\mathbf{b}, \mathbf{a})$ is also a codeword. This shows that if there is a codeword with weight less than 12, there is such a codeword that is a linear combination of at most four rows of $G$. These are easily checked by hand as follows; the reader should do it as an exercise. The fact that $C$ is a Paley matrix and the argument used in the proof of Theorem 18.1 (and in Problem 18B) show that a linear combination of 1, 2, or 3 rows of $G$ has weight 18, respectively 12, respectively 12 or 15. It remains to check combinations of four rows. Now use the fact that $Q$ in (18.5) is cyclic, which implies that only a few essentially different combinations have to be examined. One can also extend the array used in the proof of Theorem 18.1 by one row. Both methods involve very little work and produce the result that the minimum weight is 12. (Originally this was done by computer.)

We now use the generalization of Theorem 20.4 to ternary codes. If one considers the words of some fixed weight in $Sym_{36}$ (the generalization holds for weights $12, 15, 18$, and $21$), and if we replace each word by the set of positions where the nonzero coordinates occur, we find 5-designs. Since the codewords $\mathbf{c}$ and $2\mathbf{c}$ yield the same set, we only consider this set as one block.

We now come to the most famous of all binary codes: the *binary Golay code* $G_{23}$. There are very many constructions of this code, some of them quite elegant and with short proofs of its properties. We show only one of these constructions related to design theory.

We consider the incidence matrix $N$ of the (unique) 2-(11,6,3) design; see Problem 19D. We have $NN^{\top} = 3I + 3J$. Consider $N$ as a matrix with entries in $\mathbb{F}_2$. Then $NN^{\top} = I + J$. So $N$ has rank 10 and the only vector $\mathbf{x}$ with $\mathbf{x}N = \mathbf{0}$ is $\mathbf{1}$. The design properties imply trivially that any row has weight 6 and that the sum of two rows of $N$ has weight 6. We also know that the sum of three or four rows of $N$ is not $\mathbf{0}$.

Next, let $G$ be the 12 by 24 matrix over $\mathbb{F}_2$ given by $G = (I_{12}\ P)$, where

$$(20.13) \qquad P = \begin{pmatrix} 0 & 1 & \cdots & 1 \\ 1 & & & \\ \vdots & & N & \\ 1 & & & \end{pmatrix}.$$

Every row of $G$ has a weight $\equiv 0 \pmod 4$. Any two rows of $G$ have inner product 0. This implies that the weight of any linear combination of the rows of $G$ is $\equiv 0 \pmod 4$; prove this by induction. The observations made about $N$ then show that a linear combination of any number of rows of $G$ has weight at least 8. Consider the code generated by $G$ and call it $G_{24}$. Delete any coordinate to find a binary $[23, 12]$ code with minimum distance at least 7. By Problem 20A, this code must have minimum distance equal to 7 and, furthermore, it is a *perfect code*! We denote it by $G_{23}$. These two notations will be justified below, where we show that $G_{24}$, the extended binary Golay code, is unique, from which the uniqueness of $G_{23}$ can be shown.

THEOREM 20.5. *If $C$ is a binary code of length 24, with $|C| = 2^{12}$, minimum distance 8, and if $\mathbf{0} \in C$, then $C$ is equivalent to $G_{24}$.*

PROOF: (i) The difficult part of the proof is to show that $C$ must be a linear code. To see this, observe that deleting any coordinate produces a code $C'$ of length 23 and distance 7 with $|C'| = 2^{12}$. By Problem 20E, the weight enumerator of this perfect code is determined: $A_0 = A_{23} = 1$, $A_7 = A_{16} = 253$, $A_8 = A_{15} = 506$, $A_{11} = A_{12} = 1288$. From the fact that this is the case, no matter which of the 24 positions is deleted from $C$, it follows that all codewords in $C$ have weight 0, 8, 12, 16, or 24. Furthermore a change of origin (i.e. adding a fixed codeword to all the codewords) shows that we can also infer that the distance of any two codewords is 0, 8, 12, 16, or 24. Since all weights and all distances are $\equiv 0 \pmod 4$, any two codewords have inner product 0. Therefore the words of $C$ span a code that is selforthogonal. However, such a code can have at most $2^{12}$ words. Therefore, $C$ itself must be a linear and selfdual code.

(ii) We form a generator matrix of $C$ by taking as first row any word of weight 12. So after a permutation of positions, we have:

$$G = \begin{pmatrix} 1 & \cdots & 1 & 0 & \cdots & 0 \\ & A & & & B & \end{pmatrix}.$$

We know that any linear combination of the rows of $B$ must have even weight $\neq 0$, so $B$ has rank 11. Therefore the code generated by $B$ is the $[12, 11, 2]$ even weight code. We conclude that we may assume that $B$ is the matrix $I_{11}$, bordered by a column of 1's. Another permutation of columns yields a generator matrix $G'$ of the form $(I_{12}\ P)$, where $P$ has the same form as in (20.13). What do we know about the matrix $N$? Clearly any row of $N$ must have weight 6. Furthermore, the sum of any two rows of $N$ must have weight at least 6. By Problem 19D, $N$ is the incidence matrix of the unique 2-(11,6,3) design, i.e. $C$ is equivalent to $G_{24}$.  □

As we saw in Example 20.5, the words of weight 8 in $G_{24}$ form the blocks of the Witt design $\mathcal{D} = S(5, 8, 24)$. Denote by $\{0, 1, \ldots, 23\}$ the point set of $\mathcal{D}$ and consider $I = \{21, 22, 23\}$. By the Corollary to Theorem 19.3, $\mathcal{D}_I$ is an $S(2, 5, 21)$, i.e. a projective plane of order 4 (which is also known to be unique). The following problem shows the beautiful combinatorial structure of the design $\mathcal{D}$.

PROBLEM 20H. Let $B$ be a block of $\mathcal{D}$ with $|B \cap I| = \alpha$ and define $B^* := B \backslash I$. We saw that if $\alpha = 3$, then $B^*$ is a line in $PG_2(4)$. Show that

  (i) $\alpha = 2$ implies that $B^*$ is a hyperoval in $PG_2(4)$;
  (ii) $\alpha = 0$ implies that $B^*$ is the symmetric difference of two lines.

(The ambitious reader may wish to show that if $\alpha = 1$, then the seven points of $B^*$ and the lines of $PG_2(4)$ containing at least two such points, form a Fano plane. Such a plane is called a *Baer subplane* of $PG_2(4)$.)

By counting the hyperovals, pairs of lines, etc., one can show that each of the geometric configurations mentioned above is one of the sets $B^*$. In fact, one of the well known constructions of $S(5, 8, 24)$ starts with these objects and produces the design by appending suitable subsets of $I$. If one does not use the automorphism group of $PG_2(4)$ in the argument, then this construction is a nontrivial combinatorial problem. (Again: the ambitious reader should try it.)

PROBLEM 20I. Let $\mathbb{F}_4 = \{0, 1, \omega, \overline{\omega}\}$. Let $C$ be the $[6, 3]$ code over $\mathbb{F}_4$ with codewords $(a, b, c, f(1), f(\omega), f(\overline{\omega}))$, where $f(x) := ax^2 + bx + c$.

  (i) Show that $C$ has minimum weight 4 and no words of weight 5.

Let $G$ be the binary code with as codewords all 4 by 6 (0,1)-

matrices $A$ with rows $\mathbf{a}_0, \mathbf{a}_1, \mathbf{a}_\omega, \mathbf{a}_{\overline{\omega}}$ such that:

(1) every column of $A$ has the same parity as its first row $\mathbf{a}_0$,
(2) $\mathbf{a}_1 + \omega\mathbf{a}_\omega + \overline{\omega}\mathbf{a}_{\overline{\omega}} \in C$.

Show that $G$ is a $[24, 12, 8]$ code, i.e. $G = G_{24}$.

PROBLEM 20J. As in Example 20.6, we construct a ternary code $Sym_{12}$ by using the Paley matrix $C$ of order 6. Show that $Sym_{12}$ is a $[12, 6, 6]$ selfdual code. Puncture the code (i.e. delete some coordinate) to obtain a $[11, 6, 5]$ ternary code $G_{11}$. Show that this code is perfect. It is the *ternary Golay code*.

We have now given several examples of designs constructed by using a suitable code. We reverse the procedure and study codes generated by the (characteristic functions of the) blocks of a design. Let $N$ be the incidence matrix of projective plane of order $n$. We consider the subspace $C$ of $\mathbb{F}_2^v$, where $v = n^2 + n + 1$, generated by the rows of $N$. If $n$ is odd, $C$ is not very interesting. Namely, if we take the sum of the rows of $N$ that have a 1 in a fixed position, the result is a row with a 0 in that position and 1's elsewhere. These vectors generate the $[v, v-1, 2]$ even weight code and this must be $C$, since $C$ obviously has no words of odd weight. If $n$ is even, the problem becomes more interesting. We restrict ourselves to $n \equiv 2$ (mod 4).

THEOREM 20.6. *If $n \equiv 2$ (mod 4) the rows of the incidence matrix $N$ of a projective plane of order $n$ generate a binary code $C$ with dimension $\frac{1}{2}(n^2 + n + 2)$.*

PROOF: (i) Since $n$ is even, the code $\overline{C}$ is selforthogonal because every line has an odd number of points and any two lines meet in one point. Therefore $\dim C \leq \frac{1}{2}(n^2 + n + 2)$.

(ii) Let $\dim C = r$ and let $k := n^2 + n + 1 - r = \dim C^\perp$. Let $H$ be a parity check matrix for $C$. Assume that the coordinate places have been permuted in such a way that $H$ has the form $(I_k \ P)$. Define $A := \begin{pmatrix} I_k & P \\ O & I_r \end{pmatrix}$. Interpret the (0,1)-matrices $N$ and $A$ as matrices over $\mathbb{Q}$. Then

$$\det NA^\top = \det N = (n+1)n^{\frac{1}{2}(n^2+n)}.$$

Since all entries in the first $k$ columns of $NA^\top$ are even integers, $\det N$ is apparently divisible by $2^k$. So $\frac{1}{2}(n^2 + n) \geq k$, i.e. $r \geq \frac{1}{2}(n^2 + n + 2)$.

The result follows from (i) and (ii). $\quad\square$

The theorem shows that the code $C$ generated by the rows of $N$ has the property that $\overline{C}$ is selfdual. We shall now show an even more interesting property of this code, namely that one can recover the plane from the code.

THEOREM 20.7. *The code $C$ of Theorem 20.6 has minimum weight $n+1$ and every codeword of minimum weight corresponds to a line in the plane of order $n$.*

PROOF: As before, we interpret a codeword as a subset of the plane. So it will be clear what we mean by saying that a point is on the codeword $\mathbf{c}$. Let $\mathbf{c}$ be a codeword with $w(\mathbf{c}) = d$. Since $n$ is even, the codewords corresponding to lines in the plane have a 1 as parity check symbol in the extended code $\overline{C}$. This code is selfdual and this implies:

(1) if $d$ is odd then $\mathbf{c}$ meets every line at least once;
(2) if $d$ is even then every line through a fixed point of $\mathbf{c}$ meets $\mathbf{c}$ in a second point.

In case (2) we immediately see that $d > n + 1$. In case (1) we find: $(n+1)d \geq n^2 + n + 1$, i.e. $d \geq n + 1$. If $w(\mathbf{c}) = d + 1$ then there is a line $L$ of the plane that meets $\mathbf{c}$ in at least 3 points. If some point of $L$ is not a point of $\mathbf{c}$, then every line $\neq L$ through that point must meet $\mathbf{c}$ by (1). This would imply that $d \geq n + 3$. So $\mathbf{c}$ must be the line $L$. $\quad\square$

Recall that in a projective plane of even order $n$, a *hyperoval* is a set of $n + 2$ points such that no three are on one line.

THEOREM 20.8. *The codewords of weight $n + 2$ in the code $C$ of Theorem 20.6 are precisely all the hyperovals of the plane.*

PROOF: (i) Let $\mathbf{v} \in C$ and $w(\mathbf{v}) = n + 2$. Every line meets $\mathbf{v}$ in an even number of points. Let $L$ be a line of the plane and suppose that $\mathbf{v}$ and $L$ have $2a$ points in common. Each of the $n$ lines $\neq L$ through one of these $2a$ points meets $\mathbf{v}$ at least once more. Therefore $2a + n \leq n + 2$, i.e. $a = 0$ or $a = 1$.

(ii) Let $V$ be a hyperoval. Let $S$ be the set of $\binom{n+2}{2}$ secants of $V$. Each point not in $V$ is on $\frac{1}{2}(n+2)$ such lines; each point of $V$ is on $n+1$ secants. Since $n \equiv 2 \pmod 4$, the sum of the codewords corresponding to secants is the characteristic function of $V$. So this is a codeword.                                                    □

Theorems like the previous two yielded enough information about the code corresponding to a projective plane of order 10 to make possible the computer search that we mentioned earlier. The first important step was due to F. J. MacWilliams, N. J. A. Sloane, and J. G. Thompson (1973) who showed that this code could have no words of weight 15. Presently, the only known projective plane with order $n \equiv 2 \pmod 4$ is the Fano plane.

**Notes.**

We have not been concerned with the error-correcting properties of codes in this chapter. For a systematic treatment of coding theory we refer to Van Lint (1982). Much more material on the relation between design theory and coding theory can be found in Cameron and Van Lint (1980). The best reference book for coding theory is MacWilliams and Sloane (1977).

For some of the history of the origin of coding theory, we refer to Thompson (1983). There is some controversy about priority, but it is clear that both R. W. Hamming and M. J. E. Golay contributed to the "discovery" of this fascinating topic in 1947 and 1948. The monumental paper by C. E. Shannon (1948) really started things going. It seems that Hamming was irritated by the fact that his computer kept stopping when it detected an error. He correctly decided that if it could detect errors, it should be able to locate them and correct them and then get on with the job! Golay published the two Golay codes by giving their generator matrices but without proof of the properties.

M. J. E. Golay (1902–1989) was a Swiss physicist who worked in many different fields. He is known for his work on infrared spectroscopy and the invention of the capillary column but to mathematicians mainly for his discovery of the two Golay codes.

For more about MDS codes, we refer to MacWilliams and Sloane (1977).

F. J. MacWilliams (1917–1990) made many contributions to coding theory. Of these the theorem known by her name is the most important. Her book with N. J. A. Sloane is the most important reference book on coding theory. As is the case with many important coding theorists, she spent most of her career at Bell Laboratories.

As mentioned earlier, there are many constructions of the Golay code $G_{24}$. Each of them shows that some group of permutations is contained in the automorphism group of the code. For example, one of the most common constructions displays an automorphism of order 23. Since the code is unique, its full automorphism group must contain all these groups as subgroups. In this way, one can prove that this automorphism group is the famous *Mathieu group* $M_{24}$. This group is of order $24 \cdot 23 \cdot 22 \cdot 21 \cdot 20 \cdot 16 \cdot 3$ and acts 5-transitively on the 24 coordinate positions of the codewords.

It has been shown that there are no other perfect $e$-error-correcting codes with $e > 2$ than the ones mentioned in this chapter. For more on this subject, see Van Lint (1982). Also see Chapter 30.

**References.**

E. F. Assmus, Jr. and H. F. Mattson, Jr. (1969), New 5-designs, *J. Combinatorial Theory* **6**, 122–151.

P. J. Cameron and J. H. van Lint (1980), *Graphs, Codes and Designs*, London Math. Soc. Lecture Note Series **23**, Cambridge University Press.

J. H. van Lint (1982), *Introduction to Coding Theory*, Springer-Verlag.

F. J. MacWilliams (1963), A theorem on the distribution of weights in a systematic code, *Bell Syst. Tech. J.* **42**, 79–94.

F. J. MacWilliams and N. J. A. Sloane (1977), *The Theory of Error-Correcting Codes*, North-Holland.

F. J. MacWilliams, N. J. A. Sloane and J. G. Thompson (1973), On the existence of a projective plane of order 10, *J. Combinatorial Theory* (A) **14**, 66–78.

V. Pless (1972), Symmetry codes over $GF(3)$ and new 5-designs, *J. Combinatorial Theory* (A) **12**, 119–142.

C. E. Shannon (1948), A mathematical theory of communication, *Bell Syst. Tech. J.* **27**, 379–423 and 623–656.

T. M. Thompson (1983), *From Error-Correcting Codes through Sphere Packings to Simple Groups*, Carus Math. Monograph **21**.

# 21

# Strongly regular graphs and partial geometries

A *strongly regular graph* $srg(v, k, \lambda, \mu)$ is a graph with $v$ vertices that is regular of degree $k$ and that has the following properties:

  (1) For any two adjacent vertices $x, y$, there are exactly $\lambda$ vertices adjacent to $x$ and to $y$.
  (2) For any two nonadjacent vertices $x, y$, there are exactly $\mu$ vertices adjacent to $x$ and to $y$.

A trivial example is a pentagon, an $srg(5, 2, 0, 1)$. Perhaps the most famous example is the graph of Fig. 1.4, the Petersen graph, an $srg(10, 3, 0, 1)$.

Clearly a graph that is the union of $m$ complete graphs $K_k$ is an $srg(km, k-1, k-2, 0)$. Sometimes we shall exclude trivial examples by requiring that a strongly regular graph and its complement are *connected*, i.e. we assume

$$(21.1) \qquad\qquad 0 < \mu < k < v - 1.$$

(We remark that the fact that $\mu = 0$ implies that the graph is a union of complete graphs is most easily seen from (21.4) below.) It is not difficult to see that the complement $\overline{G}$ of an $srg(v, k, \lambda, \mu)$ $G$ is an

$$(21.2) \qquad srg(v, v - k - 1, v - 2k + \mu - 2, v - 2k + \lambda)$$

and since the parameters are nonnegative, we find a simple condition on the parameters, namely

$$(21.3) \qquad\qquad v - 2k + \mu - 2 \geq 0.$$

Another relation between the parameters is easily found as follows. Consider any vertex $x$ and partition the other vertices into the set $\Gamma(x)$ of vertices joined to $x$ and the set $\Delta(x)$ of vertices not joined to $x$. By the definition of strongly regular graphs, $\Gamma(x)$ consists of $k$ vertices, each of which is joined to $\lambda$ vertices of $\Gamma(x)$. Each vertex in $\Delta(x)$ is joined to $\mu$ vertices in $\Gamma(x)$. Counting edges with one end in $\Gamma(x)$ and one end in $\Delta(x)$ in two ways, we find

$$(21.4) \qquad k(k - \lambda - 1) = \mu(v - k - 1).$$

PROBLEM 21A. Show that a strongly regular graph is extremal in the following sense. Let $G$ be a graph with $v$ vertices, each of degree at most $k$. Suppose that any two adjacent vertices, respectively nonadjacent vertices, have at least $\lambda$, respectively $\mu$, common neighbors. Then

$$k(k - 1 - \lambda) \geq \mu(v - k - 1)$$

and equality implies that $G$ is strongly regular.

Before going into the fascinating theory of these graphs, we mention several classes of examples.

EXAMPLE 21.1. The *triangular graph* $T(m)$, $m \geq 4$, has as vertices the 2-element subsets of a set of cardinality $m$; two distinct vertices are adjacent if and only if they are not disjoint. $T(m)$ is an $srg(\binom{m}{2}, 2(m - 2), m - 2, 4)$. The Petersen graph is $\overline{T(5)}$ (see Problem 1A).

EXAMPLE 21.2. The *lattice graph* $L_2(m)$, $m \geq 2$, has as vertex set $S \times S$, where $S$ is a set of cardinality $m$; two distinct vertices are adjacent if and only if they have a common coordinate. $L_2(m)$ is an $srg(m^2, 2(m - 1), m - 2, 2)$. $L_2(2)$ is a quadrangle which is a trivial example because its complement is not connected.

EXAMPLE 21.3. Let $q$ be a prime power with $q \equiv 1 \pmod 4$. The *Paley graph* $P(q)$ has the elements of $\mathbb{F}_q$ as vertices; two vertices are adjacent if and only if their difference is a nonzero square in $\mathbb{F}_q$. That this is an $srg(q, \frac{1}{2}(q - 1), \frac{1}{4}(q - 5), \frac{1}{4}(q - 1))$ is a direct consequence of (18.4), but it is easier to show this using the matrix $Q$ of (18.5) as we shall see below. Note that $P(5)$ is the pentagon.

EXAMPLE 21.4. The Clebsch graph has as vertices all subsets of even cardinality of the set $\{1, \ldots, 5\}$; two vertices are joined if and only if their symmetric difference has cardinality 4. This is an $srg(16, 5, 0, 2)$. One can also describe the vertices as the words of even weight in $\mathbb{F}_2^5$, with an edge if the distance is 4. For any vertex $x$, the induced subgraph on $\Delta(x)$ is the Petersen graph. See Fig. 21.1 below.

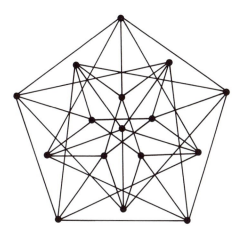

Figure 21.1

Let us define the *adjacency matrix* $A$ of a graph $G$ with $v$ vertices $1, \ldots, v$ to be the $v$ by $v$ (0,1)-matrix with $a_{ij} = a_{ji} = 1$ if and only if the vertices $i$ and $j$ are joined. Clearly $A$ is symmetric with zeros on the diagonal. The statement that $G$ is an $srg(v, k, \lambda, \mu)$ is equivalent to

$$(21.5) \qquad AJ = kJ, \qquad A^2 + (\mu - \lambda)A + (\mu - k)I = \mu J.$$

If $Q$ is the matrix occurring in (18.5), then by (18.4) we have $Q^2 = qI - J$, and the adjacency matrix of the graph $P(q)$ defined in Example 21.3 is $A = \frac{1}{2}(Q + J - I)$. So $A$ satisfies (21.5) with $k = \frac{1}{2}(q - 1)$, $\lambda = \frac{1}{4}(q - 5)$, $\mu = \frac{1}{4}(q - 1)$.

We have already seen one connection between strongly regular graphs and topics treated in earlier chapters. Several more will follow. The theory of strongly regular graphs shows a completely different aspect of graph theory from what we studied in Chapters 1–4 and 8. We now rely heavily on *algebraic* methods. Nevertheless, we also wish to show some nice counting arguments that are used in this theory.

We define the *Bose-Mesner algebra* $\mathfrak{A}$ of an *srg* $G$ to be the 3-dimensional algebra $\mathfrak{A}$ of linear combinations of $I$, $J$, and $A$. That this is indeed an algebra is a consequence of (21.5). This algebra consists of symmetric commuting matrices and therefore there is an orthogonal matrix that simultaneously diagonalizes them. This can also be seen in an elementary way from (21.5). In fact, we shall see in the next theorem that $A$ has three distinct eigenspaces in $\mathbb{R}^v$ and each of them is an eigenspace for any element of $\mathfrak{A}$.

THEOREM 21.1. *If there is an $srg(v, k, \lambda, \mu)$, then the numbers*

$$f := \frac{1}{2}\left\{v - 1 + \frac{(v-1)(\mu - \lambda) - 2k}{\sqrt{(\mu - \lambda)^2 + 4(k - \mu)}}\right\}$$

*and*

$$g := \frac{1}{2}\left\{v - 1 - \frac{(v-1)(\mu - \lambda) - 2k}{\sqrt{(\mu - \lambda)^2 + 4(k - \mu)}}\right\}$$

*are nonnegative integers.*

PROOF: Let $A$ be the adjacency matrix of the graph. By (21.5) the all-one vector $\mathbf{j} := (1, 1, \ldots, 1)^\top$ is an eigenvector of $A$, with eigenvalue $k$, and of course it is also an eigenvector of $I$ and of $J$. Application of (21.5) yields a second proof of (21.4). The multiplicity of this eigenvalue is one because the graph is connected. Any other eigenvector, say with eigenvalue $x$, is orthogonal to $\mathbf{j}$ and therefore we find from (21.5),

$$x^2 + (\mu - \lambda)x + (\mu - k) = 0.$$

This equation has two solutions

(21.6)        $$r, s = \frac{1}{2}\left\{\lambda - \mu \pm \sqrt{(\lambda - \mu)^2 + 4(k - \mu)}\right\}$$

Let $f$ and $g$ be the multiplicities of $r$ and $s$ as eigenvalues of $A$. Then we have

$$1 + f + g = v \quad \text{and} \quad tr(A) = k + fr + gs = 0.$$

If we solve these two linear equations, we find the assertion of the theorem. $\qquad\square$

Note that the multiplicities can also be expressed as

$$(21.7) \qquad f = \frac{-k(s+1)(k-s)}{(k+rs)(r-s)} \quad \text{and} \quad g = \frac{k(r+1)(k-r)}{(k+rs)(r-s)}.$$

From (21.6) we can draw a further (surprising) conclusion. If $f \neq g$, then the square root in the denominator of the expressions for $f$ and for $g$ must be an integer, i.e. $(\mu - \lambda)^2 + 4(k - \mu)$ is a perfect square. It then follows from (21.6) that the eigenvalues $r$ and $s$ are *integers*!

The other case, i.e. when $f = g$, is usually called the *half-case*. We then have an $srg(4\mu + 1, 2\mu, \mu - 1, \mu)$. The Paley graphs are examples of the half-case. In his paper on conference telephony, mentioned in Chapter 18, Belevitch observed that a necessary condition for the existence of a conference matrix of order $n$ is that $n - 1$ is the sum of two squares. See Problem 19M. We note that the parameters $v = 21$, $k = 10$, $\lambda = 4$, $\mu = 5$ satisfy all the necessary conditions for the existence of a strongly regular graph that we stated above, but the graph does not exist because, using (18.5), it would imply the existence of a conference matrix of order 22 and since 21 is not the sum of two squares, this is impossible.

The condition of Theorem 21.1 is known as the *integrality condition*. We shall call a parameter set $(v, k, \lambda, \mu)$ that satisfies these conditions and the earlier necessary conditions, a *feasible* set.

PROBLEM 21B. Show that if an $srg(k^2 + 1, k, 0, 1)$ exists, then $k = 1, 2, 3, 7$ or 57. (See the notes to Chapter 4.)

We have seen that the adjacency matrix of an $srg(v, k, \lambda, \mu)$ has three eigenvalues, one of which is $k$. There is a partial converse: If $G$ is a connected regular graph of degree $k$ with an adjacency matrix $A$ with exactly three distinct eigenvalues, then $G$ is strongly regular. This is Problem 31F of Chapter 31.

To obtain some more examples of strongly regular graphs, we consider another connection with an earlier topic, namely designs. This idea is due to J.-M. Goethals and J. J. Seidel (1970). A 2-design is called *quasisymmetric* if the cardinality of the intersection of two distinct blocks takes only two distinct values, say $x > y$. We introduce a graph, called the *block graph* of the design; the vertices are the blocks of the design and two vertices are adjacent if and only if their intersection has cardinality $y$.

THEOREM 21.2. *The block graph of a quasisymmetric design is strongly regular.*

PROOF: Let $N$ be the $v$ by $b$ incidence matrix of the design and $A$ the adjacency matrix of its block graph $G$. We have (using the parameters $v$, $k$, $b$, $r$, $\lambda$ of the 2-design):

$$NN^\top = (r - \lambda)I + \lambda J,$$
$$N^\top N = kI + yA + x(J - I - A).$$

(The first equation is (19.7), the second is the definition of $A$.) We know that both $NN^\top$ and $N^\top N$ have all-one eigenvectors $\mathbf{j}$ (of different lengths!) with eigenvalue $kr$. Also, we know that $NN^\top$ has only the eigenvalue $r - \lambda$ on $\mathbf{j}^\perp$, with multiplicity $v-1$. Therefore $N^\top N$ has this same eigenvalue, with the same multiplicity, and the eigenvalue 0 with multiplicity $b - v$. Since $x \neq y$, $A$ is a linear combination of $I$, $J$, and $N^\top N$. Therefore $A$ has eigenvector $\mathbf{j}$ and only two eigenvalues on the space $\mathbf{j}^\perp$. By our observation above, $G$ is strongly regular. $\qquad\square$

Now that we know that $G$ is strongly regular, it is easy to calculate the parameters. We do not give the distasteful formulae.

EXAMPLE 21.5. Consider the $S(5, 8, 24)$ of Chapter 20. We fix two points and consider the residual design with respect to these two points. This is a 3-(22,6,1). From Problem 20F we know that the 21 blocks of this design that contain a given point are the lines of $PG_2(4)$ (i.e. after the point is removed) and the 56 blocks that do not contain that point are hyperovals in this plane. They form a 2-(21,6,4) design. From the properties of $S(5, 8, 24)$ it follows that any two of these hyperovals meet in 0 or in 2 points. Therefore, this derived design is a quasiresidual design. From Theorem 21.2 it follows that the block graph of the design is a $(56, k, \lambda, \mu)$. Here, $k = 10$ because there are $45 = (4 - 1) \cdot \binom{6}{2}$ hyperovals that meet a given hyperoval in 2 points. From the design properties or from the conditions on strongly regular graphs, one then easily finds that $\lambda = 0$, $\mu = 2$. The graph constructed in this example is known as the *Gewirtz graph*. Note that if $A$ is the adjacency matrix of the Gewirtz graph, then by (21.5) we have $(I + A)^2 = 9I + 2J$, and that means that $N := I + A$ is the incidence matrix of a 2-(56,11,2) design (a so-called *biplane*). So in this case, we also find a new design from the strongly regular graph.

PROBLEM 21C. Let $\mathcal{D} := (\mathcal{P}, \mathcal{B}, \mathbf{I})$ be the 3-(22,6,1) design mentioned above; it is unique. We form a graph with as vertex set $\mathcal{P} \cup \mathcal{B} \cup \{\infty\}$. The vertex $\infty$ is joined to every element of $\mathcal{P}$. An element of $\mathcal{P}$ is joined to an element of $\mathcal{B}$ if they are incident. Finally, two elements of $\mathcal{B}$ are joined if and only if the blocks are disjoint. Show that this defines an $srg(100, 22, 0, 6)$. This graph is called the *Higman-Sims graph*.

If at this point we were to make a list of feasible parameter sets for strongly regular graphs, then the list would contain many sets that do not correspond to such a graph. We shall prove a few of the theorems that exclude these sets. Again, the methods are algebraic. As preparation, the reader should try to do the following problem using purely combinatorial arguments.

PROBLEM 21D. Show that an $srg(28, 9, 0, 4)$ does not exist.

That this graph does not exist is a consequence of the next theorem, known as the *Krein condition*.

THEOREM 21.3. *Let $G$ be a strongly regular graph with adjacency matrix $A$, having eigenvalues $k$, $r$, and $s$. Then*

$$(r + 1)(k + r + 2rs) \leq (k + r)(s + 1)^2,$$

*and*

$$(s + 1)(k + s + 2sr) \leq (k + s)(r + 1)^2.$$

PROOF: Let $B := J - I - A$. We know that the matrices of $\mathfrak{A}$ have three common eigenspaces of dimension 1, $f$, and $g$, respectively.

Call these spaces $V_0$, $V_1$, and $V_2$. Here $V_0$ is the space spanned by $\mathbf{j}$, and $V_1$ and $V_2$ correspond to the eigenvalues $r$ and $s$ of $A$. For $i = 0, 1, 2$ let $E_i$ be the matrix of the projection onto $V_i$, i.e. $E_i$ has eigenvalue 1 on $V_i$ and 0 on the other two eigenspaces. These matrices are what is called a basis of minimal *idempotents* of $\mathfrak{A}$. We now consider the same set $\mathfrak{A}$, but as multiplication we take the Hadamard product (see Problem 21E). It is obvious that any two of $I$, $A$, and $B$ have product $O$. Any (0,1)-matrix is idempotent for Hadamard multiplication. So we conclude that $\mathfrak{A}$ is closed under Hadamard multiplication and that the matrices $I$, $A$, and $B$ are a basis of minimal idempotents.

Note that the definition of $E_i$ ($i = 0, 1, 2$) implies that

$$I = E_0 + E_1 + E_2, \quad A = kE_0 + rE_1 + sE_2,$$

and

$$B = (v - k - 1)E_0 + (-r - 1)E_1 + (-s - 1)E_2.$$

From this, we can express the $E_i$ in $I$, $A$, and $B$.

Consider the behavior of the matrices $E_i$ under the Hadamard product. Since they form a basis of $\mathfrak{A}$, we have

$$E_i \circ E_j = \sum_{k=0}^{2} q_{ij}^k E_k,$$

where $q_{ij}^k$ is the eigenvalue of $E_i \circ E_j$ on $V_k$. It is a tedious calculation, but the numbers $q_{ij}^k$ can be expressed in terms of the parameters of $G$ using the relations given above.

At this point we need the result of Problem 21E. The Hadamard product $E_i \circ E_j$ is a principal submatrix of the Kronecker product $E_i \otimes E_j$. This matrix is idempotent, hence its eigenvalues are 0 and 1. By the theorem alluded to in Problem 21E, the eigenvalues $q_{ij}^k$ have to be between 0 and 1. It turns out (after one has done all the calculations) that all except two of the inequalities that one finds in this way are satisfied. These two are $q_{11}^1 \geq 0$ and $q_{22}^2 \geq 0$. These are the two equations of the assertion. $\qquad\qquad\square$

PROBLEM 21E. Let $A$ and $B$ be two symmetric $n$ by $n$ matrices with eigenvalues $\lambda_1, \ldots, \lambda_n$, respectively $\mu_1, \ldots, \mu_n$. Determine the eigenvalues of $A \otimes B$. We define the Hadamard product $A \circ B$ of $A$ and $B$ to be the matrix with entries $a_{ij}b_{ij}$. Show that $A \circ B$ is a principal submatrix of $A \otimes B$. What can you conclude about the eigenvalues of $A \circ B$?

To appreciate the next theorem, the reader should first convince himself that the parameter set (50,21,4,12) is feasible and that it satisfies the Krein conditions. What should one count to show that the set nevertheless does not correspond to a strongly regular graph?

THEOREM 21.4. *Let $k, r, s$ be the eigenvalues of the adjacency matrix $A$ of an $srg(v, k, \lambda, \mu)$ and let the multiplicities be 1, $f$ and $g$. Then*

$$v \leq \frac{1}{2}f(f+3) \quad and \quad v \leq \frac{1}{2}g(g+3).$$

PROOF: Let $B$ be $J - I - A$ and let the spaces $E_i$ $(i = 0, 1, 2)$ be as in the proof of the previous theorem. Let

$$E_1 = \alpha I + \beta A + \gamma B.$$

Since $E_1$ is symmetric, there is an orthogonal matrix $(H_1 \ K_1)$ such that

$$E_1 = (H_1 \ K_1) \begin{pmatrix} I & O \\ O & O \end{pmatrix} \begin{pmatrix} H_1^\top \\ K_1^\top \end{pmatrix} = H_1 H_1^\top.$$

Here $H_1$ is an $n$ by $f$ matrix with $H_1^\top H_1 = I$. We consider the rows of $H_1$ as $v$ vectors in $\mathbb{R}^f$. It follows that each of these vectors has length $\alpha^{\frac{1}{2}}$ and any two distinct vectors from this set (call it $S$) have inner product $\beta$ or $\gamma$. Such a set is called a *spherical 2-distance set* because we can interpret $S$ as a set of points on a sphere with only two distinct (angular) distances. We must show that the cardinality of $S$ is at most $\frac{1}{2}f(f+3)$.

We normalize and obtain a set $S'$ of $v$ vectors on the unit sphere $\Omega$ in $\mathbb{R}^f$ with only two inner products, say $b$ and $c$. For every $\mathbf{v} \in S'$ we define a function $f_{\mathbf{v}} : \Omega \to \mathbb{R}$ by

$$f_{\mathbf{v}}(\mathbf{x}) := \frac{(\langle \mathbf{v}, \mathbf{x} \rangle - b)(\langle \mathbf{v}, \mathbf{x} \rangle - c)}{(1-b)(1-c)}.$$

These functions are polynomials of degree 2 in the coordinates of $\mathbf{x}$. If $\mathbf{v} \in S$, $\mathbf{w} \in S$, $\mathbf{v} \neq \mathbf{w}$, then $f_{\mathbf{v}}(\mathbf{v}) = 1$ and $f_{\mathbf{v}}(\mathbf{w}) = 0$. Therefore these functions are linearly independent. The spaces of homogeneous linear and quadratic forms on $\Omega$ have dimensions $f$ and $\frac{1}{2}f(f+1)$ respectively; since $x_1^2 + \cdots + x_f^2 = 1$ on $\Omega$, we can express constants in forms of degree 2 and 1. From the linear independence of the functions $f_{\mathbf{v}}$, it follows that there can be at most $f + \frac{1}{2}f(f+1) = \frac{1}{2}f(f+3)$ of them. $\square$

This theorem is known as the *absolute bound*. It was shown by A. Neumaier (1980) that the absolute bound can be improved to

$$v \le \frac{1}{2}f(f+1) \text{ unless } q_{11}^1 = 0$$

(and similarly for the other inequality).

Before turning to the relation between strongly regular graphs and certain incidence structures, we show that nice counting arguments also play a role in this area.

THEOREM 21.5. *Let $G$ be a strongly regular graph with the same parameters as $T(n)$, i.e. $G = srg(\binom{n}{2}, 2(n-2), n-2, 4)$. If $n > 8$, then $G$ is isomorphic to $T(n)$.*

PROOF: Fix a vertex $x$ and denote by $\Gamma$ the induced subgraph on $\Gamma(x)$. This is a regular graph on $2(n-2)$ vertices, with degree $n-2$. Let $y$ and $z$ be nonadjacent vertices of $\Gamma$ and let there be $m$ vertices in $\Gamma$ adjacent to both. Since $\mu = 4$ and $x$ is adjacent to $y$ and $z$, we have $m \le 3$. In the graph $\Gamma$, there are $n-2-m$ vertices adjacent to $y$ but not to $z$, and the same number adjacent only to $z$. Hence there are $m-2$ vertices adjacent to neither of them. So $m \ge 2$. Suppose $m = 3$. Consider the unique vertex $w$ adjacent to neither $y$ nor $z$. Every vertex adjacent to $w$ in $\Gamma$ is adjacent to $y$ or to $z$, which implies that $n - 2 \le 3 + 3 = 6$, a contradiction. Hence $m = 2$, and we also see that there are no triangles in the complement $\overline{\Gamma}$.

We now show that $\overline{\Gamma}$ is bipartite. On the contrary, assume there is a circuit of odd length in this graph. Choose such a circuit $C = (x_0, x_1, \ldots, x_k = x_0)$ with $k$ minimal. From the argument above we know that $k \neq 3$. In the graph $\Gamma$, the vertices $x_0$ and $x_1$

are nonadjacent and are both adjacent to $x_3, x_4, \ldots, x_{k-2}$. By the argument above, $k \leq 6$ and since $k$ is odd, $k = 5$. The vertex $x_0$ has degree $n - 3$ in $\overline{\Gamma}$, i.e. it is adjacent to $n - 5$ vertices, besides $x_1$ and $x_4$. These must be nonadjacent in $\overline{\Gamma}$ to both $x_1$ and $x_4$. A similar assertion can be made for $x_2$, yielding $n - 5$ vertices outside $C$ nonadjacent in $\overline{\Gamma}$ to both $x_1$ and $x_3$. There are exactly $n - 4$ vertices outside $C$ nonadjacent to $x_1$; hence at least $n - 6$ vertices nonadjacent in $\overline{\Gamma}$ to both $x_3$ and $x_4$. The result of the first paragraph implies that $n - 6 \leq 1$, a contradiction.

The result of the second paragraph means that $\Gamma$ contains two disjoint *cliques* (sets of vertices, any two of which are adjacent) of size $n - 2$. Since $x$ was arbitrary, we have shown that any vertex in $G$ lies in two cliques of size $n - 1$ (cliques of this size will be called *grand cliques*). The same argument shows that any edge is in one of these grand cliques. The number of grand cliques is $2\binom{n}{2}/(n - 1) = n$. Since any two grand cliques have at most one vertex in common, they must have exactly one vertex in common. If we consider grand cliques as "points" and vertices as "blocks", then we have just shown that these points and blocks form a 2-$(n, 2, 1)$ design, i.e. the trivial design of all pairs from an $n$-set. $G$ is the block graph of this design, i.e. $G$ is indeed isomorphic to $T(n)$. $\qquad\square$

We remark that the theorem can also be proved, by case analysis, for $n < 8$ and that for $n = 8$ there are three other graphs with the parameters of $T(8)$, known as the Chang graphs.

PROBLEM 21F. Let $G$ be a strongly regular graph with the same parameters as $L_2(n)$, i.e. $G = srg(n^2, 2(n-1), n-2, 2)$. Prove that if $n > 4$, then $G$ is isomorphic to $L_2(n)$.

R. C. Bose (1963) studied large cliques in more general strongly regular graphs. This led him to the concept of a *partial geometry*. A partial geometry $pg(K, R, T)$ is an incidence structure of points and lines with the following properties:

(1) every line has $K$ points and every point is on $R$ lines;
(2) any two points are incident with at most one line;
(3) if point $p$ is not on line $L$, then there are exactly $T$ lines through $p$ that meet $L$.

If two points $x$ and $y$ are on a line, we say that they are *collinear* and write $x \sim y$.

PROBLEM 21G. Determine the number of points and lines of a partial geometry $pg(K, R, T)$.

By interchanging the roles of points and lines in a $pg(K, R, T)$, we find the so-called *dual* partial geometry, a $pg(R, K, T)$.

We introduce the *point graph* of a partial geometry as the graph with the points of the geometry as vertices, and an edge $\{x, y\}$ if and only if $x \sim y$.

PROBLEM 21H. Show that the point graph of a $pg(K, R, T)$ is a (possibly trivial) $srg(v, k, \lambda, \mu)$, with:

$$v = K \left( 1 + \frac{(K-1)(R-1)}{T} \right), \qquad k = R(K-1),$$

$$\lambda = (K-2) + (R-1)(T-1), \qquad \mu = RT,$$

$$r = K - 1 - T, \qquad s = -R.$$

If a strongly regular graph has parameters such that it could be the point graph of some partial geometry, then the graph is called *pseudo-geometric*, and it is called *geometric* if it is indeed the point graph of a partial geometry. The idea of introducing grand cliques and then showing that the grand cliques and the points form a design, as used above, was used by Bose to prove the following theorem.

THEOREM 21.6. *If a strongly regular graph is pseudo-geometric, corresponding to $pg(K, R, T)$, and if $2K > R(R-1) + T(R+1)(R^2 - 2R+2)$, then the graph is geometric.*

We do not give a proof here. The ideas of this proof were extended by A. Neumaier (1979), and, after an improvement by A. E. Brouwer, his result obtained the following form (known as the *claw bound*).

THEOREM 21.7. *If, for a strongly regular graph with the usual parameters, $\mu \neq s^2$ and $\mu \neq s(s+1)$, then $2(r+1) \leq s(s+1)(\mu+1)$.*

The idea of the proof is to show that if $r$ is large, then the graph is the point graph of a partial geometry; this is done by counting arguments. Then the absolute bound and Krein conditions are applied to the point graph of the dual partial geometry. These show the inequality stating that $r$ cannot be too large.

The parameters $(2058, 242, 91, 20)$ are feasible and satisfy all the other necessary conditions stated in this chapter, except the claw bound. So a graph with these parameters does not exist.

PROBLEM 21I. Consider an $srg(v, k, \lambda, 1)$. Show that the induced subgraph on $\Gamma(x)$ is a union of cliques. Count the number of $(\lambda+2)$-cliques in the graph. Thus show that both $k/(\lambda+1)$ and $vk/\{(\lambda+1)(\lambda+2)\}$ are integers. Apply to the set $(209, 16, 3, 1)$.

The partial geometries can be divided into four classes:

(1) A partial geometry with $T = K$ is a 2-$(v, K, 1)$ design.
(2) A partial geometry with $T = R - 1$ is called a *net*; dually, if $T = K - 1$ we speak of a *transversal design*.
(3) A partial geometry with $T = 1$ is called a *generalized quadrangle*; the notation $GQ(K - 1, R - 1)$ is commonly used for a $pg(K, R, 1)$.
(4) If $1 < T < \min\{K - 1, R - 1\}$, then we call the partial geometry *proper*.

EXAMPLE 21.6. Consider the affine plane $AG_2(n)$. From Problem 19K we know that the lines can be divided into equivalence classes of parallel lines, each containing $n$ lines. Consider all the points of the plane and take the lines of $m$ parallel classes. Then it is clear that these form a $pg(n, m, m - 1)$, i.e. a net.

EXAMPLE 21.7. Consider a Latin square of order $n$. Let the $n^2$ cells be the vertices of a graph; two vertices are joined by an edge if and only if they are in the same row or in the same column, or if they have the same entry. The graph is regular, with degree $3(n - 1)$. If two vertices are joined, we may assume without loss of generality that they are in the same row (think of the Latin square as an $OA(n, 3)$). Then they have $n - 2$ mutual neighbors.

If two vertices are not joined, then they clearly have six common neighbors. So this is an $srg(n^2, 3(n-1), n-2, 6)$. This is called a *Latin square graph* and the notation is $L_3(n)$, in accordance with the notation of Example 21.2. Generalizations will occur in a later chapter. The graph $L_3(n)$ is geometric, corresponding to a partial geometry $pg(n, 3, 2)$, i.e. a net. Of course lines correspond to rows, columns, and symbols of the square. We shall see in Chapter 22 that in fact nets and orthogonal arrays are equivalent concepts.

EXAMPLE 21.8. A quadrangle satisfies the conditions for a partial geometry. It is a $pg(2, 2, 1)$ which accounts for the name generalized quadrangle.

Consider an hyperoval $O$ in $PG_2(4)$. We take as point set the 15 points of the plane that are not on $O$. Lines will be the secants of $O$. Then every line has 3 points, and every point is on 3 lines. In fact this is a $pg(3, 3, 1)$, i.e. $GQ(2, 2)$.

Note that the same construction applied to $PG_2(q)$ for $q$ even yields a $pg(q-1, \frac{1}{2}(q+2), \frac{1}{2}(q-2))$.

EXAMPLE 21.9. Consider $PG_2(q)$ where $q$ is even. As in Example 19.7, this is an incidence structure whose objects are the 1-dimensional subspaces and 2-dimensional subspaces of $\mathbb{F}_q^3$. Again, let $O$ be a hyperoval in the projective plane. Let all 1-dimensional subspaces that are in $O$ and their cosets in $\mathbb{F}_q^3$ be lines; points are the points of the vector space. Every line has $q$ points; every point is on $q+2$ lines. The fact that $O$ is a hyperoval implies that for any point $p$ not on line $L$, there is a unique line through $p$ that meets $L$. So we have defined a $GQ(q-1, q+1)$.

EXAMPLE 21.10. Consider the subgroup $G$ of $\mathbb{Z}_3^6$ generated by the element $(1,1,1,1,1,1)$. For each coset $\mathbf{a} + G$, the sum of the coordinates of the points is a constant $i$. We say that the coset is of type $i$. Let $\mathcal{A}_i$ be the set of cosets of $G$ of type $i$. We define a tripartite graph $\Gamma$ by joining the coset $\mathbf{a}+G$ to the coset $\mathbf{a}+\mathbf{b}+G$ for each $\mathbf{b}$ that has only one nonzero coordinate. Clearly, any element of $\mathcal{A}_i$ has six neighbors in $\mathcal{A}_{i+1}$ and six in $\mathcal{A}_{i+2}$. We construct a partial geometry by taking some $\mathcal{A}_i$ as point set and one of the other two classes $\mathcal{A}_j$ as line set. Incidence corresponds to adjacency. That $K = R = 6$ is clear. It is an easy exercise to show that $T = 2$.

This defines a $pg(6, 6, 2)$.

We now give a nice application of Theorem 21.5. We shall prove the theorem, mentioned in Chapter 19 without proof, which states that a quasiresidual design with $\lambda = 2$ is a residual design. Since the cases $k \le 6$ require separate treatment, we restrict ourselves to blocksize $> 6$.

THEOREM 21.8. *Let $\mathcal{D}$ be a 2-$(v, k, 2)$ design with $k > 6$ and $v = \frac{1}{2}k(k + 1)$ (so $\mathcal{D}$ is quasiresidual). Then $\mathcal{D}$ is a residual of a symmetric 2-design.*

PROOF: Let $B$ be a block and let $a_i$ denote the number of blocks $(\neq B)$ that meet $B$ in $i$ points. Just as in Theorem 19.9 we find

$$\sum a_i = \frac{1}{2}k(k+3), \quad \sum ia_i = k(k+1), \quad \sum i(i-1)a_i = k(k-1).$$

This implies that $\sum(i-1)(i-2)a_i = 0$. Therefore any two distinct blocks meet in 1 or in 2 points, i.e. the design $\mathcal{D}$ is quasisymmetric. From Theorem 21.2 we find that the block graph $G$ of $\mathcal{D}$ is strongly regular. Calculation of the parameters shows that $G$ has the same parameters as $T(k + 2)$. So by Theorem 21.5, $G$ is isomorphic to $T(k + 2)$. This means that we can label the blocks of $\mathcal{D}$ with 2-subsets of $S := \{1, 2, \ldots, k+2\}$ in such a way that two blocks meet in $i$ points whenever their labels meet in $2 - i$ points $(i = 1, 2)$. We adjoin the set $S$ to the point set of $\mathcal{D}$, and we adjoin to each block its label, a 2-subset of $S$. Finally, we consider $S$ as a new block. This produces the required symmetric design with $\lambda = 2$ that has $\mathcal{D}$ as residual with respect to the block $S$. $\square$

We list below a table of feasible parameter sets with $v < 30$ and what we have learned about them in this chapter. The only one left as a challenge for the reader is no. 14, i.e. $GQ(2, 4)$ for which no construction was given. The corresponding graph is known as the *Schlaefli graph*.

| No. | $v$ | $k$ | $\lambda$ | $\mu$ | Description |
|-----|-----|-----|-----------|-------|-------------|
| 1 | 5 | 2 | 0 | 1 | $P(5)$ |
| 2 | 9 | 4 | 1 | 2 | $L_2(3)$ |
| 3 | 10 | 3 | 0 | 1 | Petersen, $\overline{T(5)}$ |
| 4 | 13 | 6 | 2 | 3 | $P(13)$ |
| 5 | 15 | 6 | 1 | 3 | $GQ(2,2)$ |
| 6 | 16 | 5 | 0 | 2 | Clebsch |
| 7 | 16 | 6 | 2 | 2 | $L_2(4)$ |
| 8 | 17 | 8 | 3 | 4 | $P(17)$ |
| 9 | 21 | 10 | 3 | 6 | $T(7)$ |
| 10 | 21 | 10 | 4 | 5 | does not exist, conference |
| 11 | 25 | 8 | 3 | 2 | $L_2(5)$ |
| 12 | 25 | 12 | 5 | 6 | $L_3(5)$ |
| 13 | 26 | 10 | 3 | 4 | $STS(13)$, Theorem 21.2 |
| 14 | 27 | 10 | 1 | 5 | $GQ(2,4)$ |
| 15 | 28 | 9 | 0 | 4 | does not exist, Thms. 21.3, 21.4 |
| 16 | 28 | 12 | 6 | 4 | $T(8)$ |
| 17 | 29 | 14 | 6 | 7 | $P(29)$ |

**Notes.**

Strongly regular graphs and partial geometries were introduced by R. C. Bose in 1963. A much more general concept, namely that of an *association scheme*, had been introduced by Bose and Shimamoto in 1952; see Chapter 30 here and Cameron and Van Lint (1980), Ch. 17.

For a survey of construction methods for strongly regular graphs, see Hubaut (1975) and Brouwer and Van Lint (1982).

The Clebsch graph may be defined by the 16 lines in the Clebsch quartic surface, a pair of lines being adjacent if and only if they are skew. Alfred Clebsch (1833–1872) was a mathematician and physicist who worked in Karlsruhe, Giessen, and Göttingen. He was one of the founders of the famous journal *Mathematische Annalen*.

The algebra $\mathfrak{A}$ was introduced by R. C. Bose and D. M. Mesner in 1959.

The reference Goethals and Seidel (1970) contains, besides Theorem 21.2, many interesting connections between strongly regular graphs and other combinatorial designs. Hadamard matrices, Steiner systems and the Golay code all contribute to the theory. For these connections, again see Cameron and Van Lint (1980).

The Higman-Sims graph is connected to their famous finite simple group (see Higman and Sims (1968)).

Theorem 21.3 became known as the Krein condition because a special case of this theorem was proved by L. L. Scott by applying a result of M. G. Krein concerning topological groups to a problem on finite groups. The simple proof given in this chapter is due to D. G. Higman (1975) and P. Delsarte (1973). The case of equality in the Krein conditions was treated by Delsarte, Goethals and Seidel (1977) and Cameron, Goethals and Seidel (1978).

The idea of the proof of Theorem 21.4 is due to T. H. Koornwinder (1976).

Theorem 21.5 was proved several times: see for example, L. C. Chang (1959), W. S. Connor (1958), A. J. Hoffman (1960). The three exceptional graphs for $n = 8$ are known as the *Chang graphs*.

The most recent survey on partial geometries is Van Lint (1983).

Nets were introduced by R. H. Bruck in 1951. Bose's result on pseudo-geometric graphs was inspired by Bruck's work (e.g. the idea of grand cliques).

For all that one would like to know about generalized quadrangles (and more) we refer to the book *Finite Generalized Quadrangles* by S. E. Payne and J. A. Thas (1984).

Only two proper partial geometries with $T = 2$ are known. The geometry $pg(6, 6, 2)$ of Example 21.9 was first constructed by Van Lint and Schrijver (1981); the construction that we described is due to Cameron and Van Lint (1982). The other known example is $pg(5, 18, 2)$.

A big step in the direction of the proof of the nonexistence of a projective plane of order 10, mentioned in Chapter 19, was the proof of the nonexistence of $pg(6, 9, 4)$. This implied (see Example 21.8) that, if the plane existed, it would not have any hyperovals.

**References.**

R. C. Bose (1963), Strongly regular graphs, partial geometries, and partially balanced designs, *Pacific J. Math.* **13**, 389–419.

R. C. Bose and D. M. Mesner (1959), On linear associative algebras corresponding to association schemes of partially balanced designs, *Ann. Math. Stat.* **30**, 21–38.

A. E. Brouwer and J. H. van Lint (1982), Strongly regular graphs and partial geometries, in: *Enumeration and Design* (D. M. Jackson and S. A. Vanstone, eds.), Academic Press.

R. H. Bruck (1951), Finite nets I, *Canad. J. Math.* **3**, 94–107.

R. H. Bruck (1963), Finite nets II, *Pacific J. Math.* **13**, 421–457.

P. J. Cameron (1978), Strongly regular graphs, in: *Selected Topics in Graph Theory* (L. W. Beineke and R. J. Wilson, eds.), Academic Press.

P. J. Cameron, J.-M. Goethals, and J. J. Seidel (1978), The Krein condition, spherical designs, Norton algebras and permutation groups, *Proc. Kon. Ned. Akad. v. Wetensch.* **81**, 196–206.

P. J. Cameron and J. H. van Lint (1980), *Graphs, Codes and Designs*, London Math. Soc. Lecture Note Series **43**, Cambridge University Press.

P. J. Cameron and J. H. van Lint (1982), On the partial geometry $pg(6,6,2)$, *J. Combinatorial Theory* (A) **32**, 252–255.

L. C. Chang (1959), The uniqueness and nonuniqueness of triangular association schemes, *Sci. Record Peking Math.* **3**, 604–613.

W. S. Connor (1958), The uniqueness of the triangular association scheme, *Ann. Math. Stat.* **29**, 262–266.

P. Delsarte (1973), An algebraic approach to the association schemes of coding theory, *Philips Res. Repts. Suppl.* **10**.

P. Delsarte, J.-M. Goethals and J. J. Seidel (1977), Spherical codes and designs, *Geometriae Dedicata* **6**, 363–388.

A. Gewirtz (1969), The uniqueness of $g(2,2,10,56)$, *Trans. New York Acad. Sci.* **31**, 656–675.

J.-M. Goethals and J. J. Seidel (1970), Strongly regular graphs derived from combinatorial designs, *Canad. J. Math.* **22**, 597–614.

D. G. Higman (1975), Invariant relations, coherent configurations and generalized polygons, in: *Combinatorics* (M. Hall, Jr. and J. H. van Lint, eds.), D. Reidel.

D. G. Higman and C. C. Sims (1968), A simple group of order 44,352,000, *Math. Z.* **105**, 110–113.

A. J. Hoffman (1960), On the uniqueness of the triangular association scheme, *Ann. Math. Stat.* **31**, 492–497.

X. Hubaut (1975), Strongly regular graphs, *Discrete Math.* **13**, 357–381.

T. H. Koornwinder (1976), A note on the absolute bound for systems of lines, *Proc. Kon. Ned. Akad. v. Wetensch.* **79**, 152–153.

J. H. van Lint (1983), Partial geometries, *Proc. Int. Congress of Math.*, Warsaw.

J. H. van Lint and A. Schrijver (1981), Construction of strongly regular graphs, two-weight codes and partial geometries by finite fields, *Combinatorica* **1**, 63–73.

J. H. van Lint and J. J. Seidel (1969), Equilateral point sets in elliptic geometry, *Proc. Kon. Ned. Akad. v. Wetensch.* **69**, 335–348.

A. Neumaier (1979), Strongly regular graphs with smallest eigenvalue $-m$, *Archiv der Mathematik* **33**, 392–400.

A. Neumaier (1980), New inequalities for the parameters of an association scheme, in: *Combinatorics and Graph Theory*, Lecture Notes in Math. **885**, Springer-Verlag.

S. E. Payne and J. A. Thas (1984), *Finite Generalized Quadrangles*, Pitman.

# 22
# Orthogonal Latin squares

Two Latin squares $L_1 : R \times C \to S$ and $L_2 : R \times C \to T$ (with the same row and column sets) are said to be *orthogonal* when for each ordered pair $(s, t) \in S \times T$, there is a unique cell $(x, y) \in R \times C$ so that

$$L_1(x, y) = s \quad \text{and} \quad L_2(x, y) = t.$$

The use of the word "orthogonal" is perhaps unfortunate since it has other meanings in mathematics, but it has become far too commonplace to try to change now.

For example, let

$$A := \begin{bmatrix} A & K & Q & J \\ Q & J & A & K \\ J & Q & K & A \\ K & A & J & Q \end{bmatrix}, \qquad B := \begin{bmatrix} \spadesuit & \heartsuit & \diamondsuit & \clubsuit \\ \clubsuit & \diamondsuit & \heartsuit & \spadesuit \\ \heartsuit & \spadesuit & \clubsuit & \diamondsuit \\ \diamondsuit & \clubsuit & \spadesuit & \heartsuit \end{bmatrix}$$

Figure 22.1

where $R = C = \{1, 2, 3, 4\}$, $S = \{A, K, Q, J\}$, $T = \{\spadesuit, \heartsuit, \diamondsuit, \clubsuit\}$. The orthogonality of $A$ and $B$ is evident when the two squares are superposed and we see that each element of $S \times T$ appears exactly once.

$$\begin{bmatrix} A\spadesuit & K\heartsuit & Q\diamondsuit & J\clubsuit \\ Q\clubsuit & J\diamondsuit & A\heartsuit & K\spadesuit \\ J\heartsuit & Q\spadesuit & K\clubsuit & A\diamondsuit \\ K\diamondsuit & A\clubsuit & J\spadesuit & Q\heartsuit \end{bmatrix}$$

Orthogonality does not depend on the particular symbol set used. We could replace the playing card suits in our example by any

four symbols in any order, and the resulting square would still be
orthogonal to $A$. We have used Latin letters for the symbols of $A$;
if we had used Greek letters for $B$, it would have been clear why
the superposition of two orthogonal Latin squares is often called a
*Græco-Latin square.*

The example we have given with playing cards is similar to what
is known as *Euler's 36 officers problem.* According to folklore,
Euler was asked by Catherine the Great (at whose court he was in
residence) to arrange 36 officers from six different regiments and
of six different ranks (one officer of each rank from each regiment)
in a 6 by 6 array so that each row and each column contained one
officer of each rank and one officer of each regiment. A solution
requires a pair of orthogonal Latin squares of order 6.

Surprisingly, there is no solution. Whether he was asked by
Catherine or not, Euler did consider the problem in 1779 and con-
vinced himself that it was impossible. Euler was capable of monumental calculation, but it is not clear that he had really examined
all cases. This was systematically done in 1900 by G. Tarry. Today
a computer can do this easily.

Euler knew that a pair of orthogonal Latin squares of order $n$
existed for all odd values of $n$, and all $n \equiv 0 \pmod 4$; see Theorem
22.3 below. For examples of orthogonal Latin squares of odd orders
$n$, let $G$ be any group of order $n$ and define squares $L_1$ and $L_2$ with
row, column, and symbol sets $G$ by

$$L_1(x, y) := xy, \qquad L_2(x, y) := x^{-1}y.$$

For completely trivial reasons, there is no pair of orthogonal
Latin squares of order 2. Euler asserted that he was of the opinion that this impossibility for $n = 2, 6$ extended as well to $n =
10, 14, 18, \ldots$, i.e. to orders $n \equiv 2 \pmod 4$. This statement was
known as "Euler's conjecture" for 177 years until it was suddenly
and completely disproved by Bose, Parker, and Shrikhande. We will
prove later their theorem that pairs of orthogonal Latin squares of
orders $n$ exist for *all* $n$ except $n = 2, 6$.

At various times, a more general question has been raised: can
one find sets of many Latin squares, any two of which are orthogonal? We will denote by $N(n)$ the largest integer $k$ for which there

exist $k$ Latin squares of order $n$ that are pairwise orthogonal. For example, we may add

$$
C := \begin{bmatrix} 0 & \alpha & \beta & \gamma \\ \alpha & 0 & \gamma & \beta \\ \beta & \gamma & 0 & \alpha \\ \gamma & \beta & \alpha & 0 \end{bmatrix}
$$

(the addition table of the Klein four-group) to the squares in Fig. 22.1 to obtain three pairwise orthogonal Latin squares of order 4; thus $N(4) \geq 3$. We remark that the definition of orthogonality is trivial or vacuous for $n = 1$ or $0$, and so $N(1) = N(0) = \infty$.

Here is a construction that proves $N(q) \geq q-1$ for a prime power $q$. All row, column, and symbol sets are to be the elements of the field $\mathbb{F}_q$. For each nonzero element $a$ in $\mathbb{F}_q$, define $L_a(x,y) := ax+y$; these $q - 1$ squares are pairwise orthogonal. As we all know, the systems

$$
ax + by = s
$$
$$
cx + dy = t
$$

have unique solutions $(x, y)$ when $ad - bc \neq 0$, so it is easy to tell when two squares defined by linear equations are orthogonal. Here are the squares for $q = 5$:

$$
\begin{bmatrix} 0 & 1 & 2 & 3 & 4 \\ 1 & 2 & 3 & 4 & 0 \\ 2 & 3 & 4 & 0 & 1 \\ 3 & 4 & 0 & 1 & 2 \\ 4 & 0 & 1 & 2 & 3 \end{bmatrix},
\begin{bmatrix} 0 & 1 & 2 & 3 & 4 \\ 2 & 3 & 4 & 0 & 1 \\ 4 & 0 & 1 & 2 & 3 \\ 1 & 2 & 3 & 4 & 0 \\ 3 & 4 & 0 & 1 & 2 \end{bmatrix},
$$

$$
\begin{bmatrix} 0 & 1 & 2 & 3 & 4 \\ 3 & 4 & 0 & 1 & 2 \\ 1 & 2 & 3 & 4 & 0 \\ 4 & 0 & 1 & 2 & 3 \\ 2 & 3 & 4 & 0 & 1 \end{bmatrix},
\begin{bmatrix} 0 & 1 & 2 & 3 & 4 \\ 4 & 0 & 1 & 2 & 3 \\ 3 & 4 & 0 & 1 & 2 \\ 2 & 3 & 4 & 0 & 1 \\ 1 & 2 & 3 & 4 & 0 \end{bmatrix}.
$$

THEOREM 22.1. *For* $n \geq 2$, *we have* $1 \leq N(n) \leq n - 1$.

PROOF: We may change all row, column, and symbol sets in a set of $k$ pairwise orthogonal Latin squares $\{L_i\}_{i=1}^k$ to $\{1, 2, \ldots, n\}$ (for notational convenience), and we might as well rename the symbols if necessary so that the first row of each square is $1, 2, \ldots, n$ in that order. Now we consider the elements

$$L_1(2, 1), \ L_2(2, 1), \ \ldots, \ L_k(2, 1)$$

in the second row, first column. None of these is 1, since 1 occurs already in the first column. They are distinct because if $L_i(2, 1) = L_j(2, 1) = s$, say, then the fact that also $L_i(1, s) = L_j(1, s) = s$ would contradict the orthogonality of $L_i$ and $L_j$. The upper bound follows. $\qquad\square$

In view of our construction from finite fields and the simple theorem above, we have

$$(22.1) \qquad N(q) = q - 1 \quad \text{if } q \text{ is a prime power.}$$

At this point, we want to show that the concept of pairwise orthogonal Latin squares is equivalent to something the reader has already seen in the previous chapter! The assertion of the theorem will be more immediately apparent if we first notice that the property of being a Latin square can itself be described in terms of orthogonality. For example, a square of order 5 is Latin if and only if it is orthogonal to the two squares

$$
\begin{bmatrix}
0 & 0 & 0 & 0 & 0 \\
1 & 1 & 1 & 1 & 1 \\
2 & 2 & 2 & 2 & 2 \\
3 & 3 & 3 & 3 & 3 \\
4 & 4 & 4 & 4 & 4
\end{bmatrix}
\quad \text{and} \quad
\begin{bmatrix}
0 & 1 & 2 & 3 & 4 \\
0 & 1 & 2 & 3 & 4 \\
0 & 1 & 2 & 3 & 4 \\
0 & 1 & 2 & 3 & 4 \\
0 & 1 & 2 & 3 & 4
\end{bmatrix} .
$$

THEOREM 22.2. *A set of k pairwise orthogonal Latin squares of order n exists if and only if an $(n, k + 2)$-net exists.*

PROOF: Suppose $L_i : R \times C \to S_i$, $1 \le i \le k$, are pairwise orthogonal Latin squares. Let $\mathcal{P} = R \times C$, the set of $n^2$ cells. Roughly speaking, we take as lines the rows, the columns, and the "iso-symbol" lines in each square. More formally, let

$$\mathcal{A}_1 = \{\{(x, b) : b \in C\} : x \in R\},$$

$$\mathcal{A}_2 = \{\{(a, y) : a \in R\} : y \in C\},$$

$$\mathcal{A}_{i+2} = \{\{(x, y) : L_i(x, y) = c\} : c \in S_i\}, \quad 1 \leq i \leq k,$$

and let $\mathcal{B} = \cup_{i=1}^{k+2} \mathcal{A}_i$. That $(\mathcal{P}, \mathcal{B})$ is an $(n, k+2)$-net follows from the definitions of Latin and orthogonal.

Given an $(n, k+2)$-net $(\mathcal{P}, \mathcal{B})$, where $\mathcal{B} = \cup_{i=1}^{k+2} \mathcal{A}_i$ is the partition into parallel classes, define $L_i : \mathcal{A}_1 \times \mathcal{A}_2 \to \mathcal{A}_{i+2}$ by declaring that $L_i(A, B)$ is to be the unique line in $\mathcal{A}_{i+2}$ containing the point of intersection of $A$ and $B$. As usual we leave to the reader the details of checking the Latin and orthogonal properties of these squares.

$\square$

COROLLARY. $N(n) = n - 1$ if and only if there exists a projective (or affine) plane of order $n$.

The value of $N(n)$ tells us, in some very imprecise sense, how close to the existence of a projective plane of order $n$ we may get. But we know terribly little about this function. The only value of $n$ other than prime powers for which $N(n)$ is known is $n = 6$.

PROBLEM 22A. (i) Consider the complement of the graph of an $(n, n)$-net and prove that $N(n) \geq n - 2$ implies that $N(n) = n - 1$.

(ii) Use the result of Problem 21F to prove that if $n > 4$, then $N(n) \geq n - 3$ implies that $N(n) = n - 1$.

THEOREM 22.3.

(i) $N(nm) \geq min\{N(n), N(m)\}$.

(ii) If $n = p_1^{e_1} p_2^{e_2} \cdots p_r^{e_r}$ is the factorization of $n$ into powers of primes, then

$$N(n) \geq \min_{1 \leq i \leq r} (p_i^{e_i} - 1).$$

PROOF: Part (ii) follows from part (i) and (22.1), used inductively.

To prove part (i), we first define a Kronecker-like product for squares in the natural way: Given $L_i : R_i \times C_i \to S_i$, $i = 1, 2$, define

$$L_1 \otimes L_2 : (R_1 \times R_2) \times (C_1 \times C_2) \to (S_1 \times S_2)$$

by

$$(L_1 \otimes L_2)((x_1, x_2), (y_1, y_2)) := (L_1(x_1, y_1), L_2(x_2, y_2)).$$

It is straightforward to verify that if $\{A_i\}_{i=1}^k$ are pairwise orthogonal Latin squares of order $n$ and $\{B_i\}_{i=1}^k$ are pairwise orthogonal Latin squares of order $m$, then $\{A_i \otimes B_i\}_{i=1}^k$ are pairwise orthogonal Latin squares of order $nm$. We leave this task to the reader, but advise you just to read on and not bother. □

Theorem 22.3 is known as MacNeish's theorem. MacNeish conjectured in 1922 that equality held in Theorem 22.3(ii). This would imply Euler's conjecture. But combinatorial problems of this type rarely have such simple answers.

The MacNeish conjecture was disproved first for $n = 21$ when three Latin squares of order 21 were constructed with the aid of the 21 point projective plane of order 4; see Example 22.1 below. This was an example of *composition methods*, which also provided the first counterexample to Euler's conjecture ($n = 22$). The first pair of orthogonal Latin squares of order 10 was found by Parker by what might be called *difference methods*. We discuss composition methods first; they have proved more powerful for large values of $n$.

Bose, Parker and Shrikhande gave many constructions for sets of orthogonal Latin squares. We choose to describe two basic and elegant ones in terms of quasigroups. Recall that a *quasigroup* is a Latin square whose row, column, and symbol sets are the same set $X$. A quasigroup $L$ is *idempotent* when $L(x, x) = x$ for all $x \in X$. For example, if $X$ is the finite field $\mathbb{F}_q$ of order $q$, then

$$L_a(x, y) := ax + (1 - a)y$$

defines an idempotent quasigroup $L_a$ if $a \neq 0, 1$. Any two squares of this form are orthogonal.

PROBLEM 22B. For a prime power $q \geq 4$, construct two Latin squares $A, S$ of order $q$ with row and column sets equal to $\mathbb{F}_q$ so that $A$ is orthogonal to its transpose, and to $S$, which is to be symmetric. In addition, ensure that $S$ is idempotent if $q$ is odd and "unipotent", i.e. has constant diagonal, if $q$ is even.

Suppose we have a linear space, cf. Chapter 19, on a point set $X$ with line set $\mathcal{A}$. Further suppose that for each line $A$ in $\mathcal{A}$, we have $k$ pairwise orthogonal idempotent quasigroups $L_1^A, L_2^A, \ldots, L_k^A$ on the set $A$. Then we can construct $k$ pairwise orthogonal idempotent quasigroups $L_1, L_2, \ldots, L_k$ on the entire set $X$ by declaring, for each $i = 1, 2, \ldots, k$, that $L_i(x, x) := x$ for $x \in X$, and for distinct $x, y \in X$, $L_i(x, y) := L_i^A(x, y)$, where $A$ is the unique line in $\mathcal{A}$ which contains both $x$ and $y$. It is a simple matter to check that $L_1, L_2, \ldots, L_k$ are Latin squares and that they are orthogonal; for example, given $i, j, s$ and $t$, $s \neq t$, a cell $(x, y)$ for which

$$L_i(x, y) = s \quad \text{and} \quad L_j(x, y) = t$$

exists, and $x, y$ may be found in that line $B$ which contains $s$ and $t$, because of the orthogonality of $L_i^B$ and $L_j^B$.

THEOREM 22.4. *For every linear space with line set $\mathcal{A}$ on an $n$-set $X$, we have*

$$N(n) \geq \min_{A \in \mathcal{A}} \{ N(|A|) - 1 \}.$$

PROOF: Let $k$ be the indicated minimum. This means we have at least $k + 1$ pairwise orthogonal quasigroups on each $A \in \mathcal{A}$. We will describe how to obtain $k$ pairwise orthogonal *idempotent* quasigroups on each $A$. Then $N(n) \geq k$ follows from the construction above.

   In general, let $H_1, H_2, \ldots, H_{k+1}$ be pairwise orthogonal quasigroups on an $m$-set $B$. Pick any $b \in B$. There will be $m$ cells $(x, y)$ for which $H_{k+1}(x, y) = b$, one in each row and one in each column. Simultaneously permute the columns, say, of all squares so that these cells are on the diagonal of the $(k + 1)$-th square. Orthogonality implies that for each $i \leq k$, all $m$ symbols occur on the diagonal of the $i$-th square. Finally, we permute the symbols independently in each of the first $k$ squares so that the resulting squares are idempotent.                                              □

THEOREM 22.5. *If $\mathcal{A}$ is the set of lines of a linear space on an $n$-set $X$ and $\mathcal{B} \subseteq \mathcal{A}$ is a set of pairwise disjoint lines, then*

$$N(n) \geq \min \left( \{ N(|A|) - 1 : A \in \mathcal{A} \backslash \mathcal{B} \} \cup \{ N(|B|) : B \in \mathcal{B} \} \right).$$

PROOF: Let $k$ be the indicated minimum. As we saw in the proof of the preceding theorem, there exist $k$ pairwise orthogonal idempotent quasigroups $L_i^A$ on each $A$ in $\mathcal{A} \backslash \mathcal{B}$. If necessary, we add singleton sets to $\mathcal{B}$ so that $\mathcal{B}$ becomes a partition of $X$. We have $k$ pairwise orthogonal quasigroups $L_i^B$ (not necessarily idempotent) on each $B \in \mathcal{B}$. We define $L_1, L_2, \ldots, L_k$ on the entire set $X$ by declaring, for each $i = 1, 2, \ldots, k$, that $L_i(x, x) := L_i^B(x, x)$ where $B$ is the unique member of $\mathcal{B}$ that contains $x$, and for distinct $x, y$, $L_i(x, y) := L_i^A(x, y)$, where $A$ is the unique line in $\mathcal{A}$ (whether or not in $\mathcal{B}$) that contains both $x$ and $y$. The quasigroups $L_i$ are orthogonal (the easy details are left to the reader) and thus $N(n) \geq k$. $\qquad\square$

EXAMPLE 22.1. Let $n = 21$ and consider the projective plane of order 4 with 21 lines of size 5. Theorem 22.4 implies that $N(21) \geq 3$ and this disproves MacNeish's conjecture. Delete three noncollinear points. We have a linear space with 18 points, and lines of sizes 5, 4, and 3; the three lines of size 3 are pairwise disjoint. Theorem 22.5 implies that $N(18) \geq 2$ and this disproves Euler's conjecture.

EXAMPLE 22.2. To see how far wrong Euler was, we consider two more values of $n \equiv 2 \pmod 4$. Delete three noncollinear points from a projective plane of order 8 to obtain a linear space on 70 points with lines of sizes 9, 8, and (pairwise disjoint lines of size) 7. Theorem 22.5 shows $N(70) \geq 6$. It is possible to find seven points in the plane constructed from $\mathbb{F}_8$ such that no three are collinear (even 10, see Problem 19I), and deleting these produces a linear space which can be used in Theorem 22.4 to give us $N(66) \geq 5$.

The next construction would be difficult to describe without the use of *transversal designs*. Transversal designs provide a compact and conceptually convenient language with which to manipulate sets of pairwise orthogonal Latin squares. (They also provide a large family of linear spaces useful for the construction of other combinatorial designs or even more orthogonal Latin squares.) A $TD(n, k)$ can be defined as the dual incidence structure of an $(n, k)$-net, i.e. as an $(n, k, k-1)$-partial geometry. So there are $nk$ points and $n^2$ blocks (we will use "blocks" rather than "lines" when discussing transversal designs). Each point is in $n$ blocks; each block contains $k$ points. The points fall into $k$ equivalence classes (called, confusingly, *groups*) of size $n$ so that two points in the same group are not contained in a block while two points in different groups belong to exactly one block. In particular, each block contains exactly one point from each group. We need to refer to the groups so often that we incorporate them into the notation and speak of a transversal design $(X, \mathcal{G}, \mathcal{A})$ where the three coordinates are the sets of points, groups, and blocks, respectively. $(X, \mathcal{G} \cup \mathcal{A})$ is a linear space with blocksizes $k$ and $n$.

EXAMPLE 22.3. Here is a TD(2,3):

$$\text{Groups:} \quad \{a_1, a_2\}, \{b_1, b_2\}, \{c_1, c_2\}$$

$$\text{Blocks:} \quad \{a_1, b_1, c_1\}, \{a_1, b_2, c_2\}, \{a_2, b_1, c_2\}, \{a_2, b_2, c_1\}.$$

In view of Theorem 22.2, the existence of a TD($n, k+2$) is equivalent to the existence of $k$ pairwise orthogonal Latin squares. At the risk of boring some readers, we review the connection by quickly describing how to get the squares from a TD($n, k + 2$): number the groups $\{G_1, G_2, \ldots, G_{k+2}\}$ and define $L_i : G_1 \times G_2 \to G_{i+2}$ by declaring $L_i(x, y)$ to be the point of intersection of $G_{i+2}$ and the block which contains $x$ and $y$.

A TD($n, k$) is said to be *resolvable* when the set $\mathcal{A}$ of blocks can be partitioned into $n$ parallel classes $\mathcal{A}_1, \mathcal{A}_2, \ldots, \mathcal{A}_n$, i.e. each $\mathcal{A}_i$ is a set of $n$ blocks of size $k$ which partition the point set $X$. From a TD($n, k$) $(X, \mathcal{G}, \mathcal{A})$, we can always construct a resolvable TD($n, k - 1$) by deleting the $n$ points of one of the groups $G_0 = \{x_1, x_1, \ldots, x_n\}$, which removes exactly one point from each block; for each $i$, the blocks which have had $x_i$ removed form a parallel class on the point set $X \backslash G_0$ of the TD($n, k - 1$). (Conversely, a resolvable TD($n, k - 1$) can be extended to a TD($n, k$).)

THEOREM 22.6. *If* $0 \leq u \leq t$, *then*

$$N(mt + u) \geq \min\{N(m), N(m + 1), N(t) - 1, N(u)\}.$$

PROOF: Let $k$ be the righthand side above, plus 2. This means that transversal designs TD($m, k$), TD($m + 1, k$), TD($t, k + 1$), and TD($u, k$) exist; and to prove the theorem, we must construct a TD($mt + u, k$).

As the construction is rather technical, we warm up by first describing the construction of a TD($mt, k$) from a TD($t, k$) and various TD($m, k$)'s. This is the degenerate case $u = 0$ in the theorem.

The construction does not require a TD($m+1, k$) or a TD($t, k+1$) and thus it reproves Theorem 22.3(i).

Let $(X, \mathcal{G}, \mathcal{A})$ be a TD($t, k$). To each $x \in X$, associate a set $M_x$ of $m$ new elements so that any two sets $M_x$ are disjoint. For $S \subseteq X$, let $M_S := \cup_{x \in S} M_x$.

We construct a TD($mt, k$) on the point set $M_X$ of size $kmt$ with groups $\{M_G : G \in \mathcal{G}\}$ each of size $mt$; the blocks $\mathcal{B}$ are obtained as follows. For each $A \in \mathcal{A}$, choose blocks $\mathcal{B}_A$ so that

$$(M_A, \{M_x : x \in A\}, \mathcal{B}_A)$$

is a TD($m, k$), and let $\mathcal{B} = \cup_{A \in \mathcal{A}} \mathcal{B}_A$. The verification is straightforward.

To return to the general case, recall that the existence of a TD($t, k+1$) implies the existence of a resolvable TD($t, k$). So we have a TD($t, k$) $(X, \mathcal{G}, \mathcal{A})$ where $\mathcal{A}$ admits a partition into parallel classes $\mathcal{A}_1, \mathcal{A}_2, \dots, \mathcal{A}_t$. We treat the blocks of the first $u$ parallel classes in one way, and the blocks in $\mathcal{B} := \cup_{i=u+1}^{t} \mathcal{A}_i$ in another way.

Let $(U, \mathcal{H}, \mathcal{C})$ be a TD($u, k$). We also require a partition of $U$ into $u$ $k$-subsets $\{K_1, K_2, \dots, K_u\}$; each is to consist of exactly one point from each set $H \in \mathcal{H}$, but these $k$-subsets are not required to be blocks in $\mathcal{C}$.

Let $\mathcal{G} = \{G_1, G_2, \dots, G_k\}$ and $\mathcal{H} = \{H_1, H_2, \dots, H_k\}$ be numberings of the groups. We construct a TD($mt + u, k$) with point set $Y := M_X \cup U$ and groups

$$\mathcal{J} := \{M_{G_1} \cup H_1, M_{G_2} \cup H_2, \dots, M_{G_k} \cup H_k\}.$$

The blocks are obtained as follows. For each block $B \in \mathcal{B}$ as before, let

$$(M_B, \{M_x : x \in B\}, \mathcal{D}_B)$$

be a TD($m, k$). For each block $A \in \mathcal{A}_i$, let

$$(M_A \cup K_i, \{(M_A \cap M_{G_j}) \cup (K_i \cap H_j) : j = 1, 2, \dots, k\}, \mathcal{D}_A)$$

be a TD($m+1, k$) in which $K_i$ occurs as a block, and let $\mathcal{D}'_A$ denote the remaining $(m+1)^2 - 1$ blocks. Then we claim that $(Y, \mathcal{J}, \mathcal{E})$ is the required TD($mt + u, k$), where

$$\mathcal{E} := \mathcal{C} \cup \left( \bigcup_{B \in \mathcal{B}} \mathcal{D}_B \right) \cup \left( \bigcup_{A \in \mathcal{A}_1 \cup \dots \cup \mathcal{A}_u} \mathcal{D}'_A \right).$$

Verification requires consideration of several cases. □

EXAMPLE 22.4. With $m = 3$ in Theorem 22.6, we see that

(22.2)
$$N(3t + u) \geq 2 \text{ whenever } 0 \leq u \leq t, N(t) \geq 3, \text{ and } N(u) \geq 2.$$

We may take $(t, u) = (5, 3), (7, 1), (7, 5),$ and $(9, 3)$ to find that $N(n) \geq 2$ for $n = 18, 22, 26,$ and 30.

THEOREM 22.7. $N(n) \geq 2$ for all $n \neq 2, 6$.

PROOF: We need only consider $n \equiv 2 \pmod 4$. For $n = 10, 14$, see Examples 22.6 and 22.7 below. For $n = 18, 22, 26,$ and 30, see Example 22.4 above. We now assume $n \geq 34$. One of

$$n - 1, n - 3, n - 5, n - 7, n - 9, n - 11$$

is divisible by 3 but not by 9, so we can write $n = 3t + u$ where $u = 1, 3, 5, 7, 9,$ or 11, and $t$ is not divisible by 3. Since $n$ is even, $t$ is also odd, so $N(t) \geq 4$ by Theorem 22.3(ii). A consequence of $n \geq 34$ is $t \geq 11$, so $0 \leq u \leq t$, and then $N(n) \geq 2$ by (22.2). □

THEOREM 22.8. $N(n) \to \infty$ as $n \to \infty$.

PROOF: Let $x$ be a positive integer. We claim that $N(n) \geq x - 1$ whenever

(22.3)
$$n \geq 2 \left( \prod_{p \leq x} p \right)^{2x+1},$$

where the product is extended over all primes $p \leq x$.

Given $n$ satisfying (22.3), let $m$ be chosen (Chinese remainder theorem) so that $0 \leq m \leq (\prod_{p \leq x} p)^x$ and such that for all primes $p \leq x$,

$$m \equiv \begin{cases} -1 \pmod{p^x} & \text{if } p \text{ divides } n, \\ 0 \pmod{p^x} & \text{if } p \text{ does not divide } n. \end{cases}$$

Then choose an integer $t \equiv 1 \pmod{\prod_{p \leq x} p}$ so that

$$0 \leq u := n - mt < \left( \prod_{p \leq x} p \right)^{x+1}.$$

The remainder of the proof is provided by Problem 22C below. □

PROBLEM 22C. With $m$, $t$, and $u$ chosen as above, show that $u \le t$ and that Theorem 22.3(ii) implies that $N(mt + u) \ge x - 1$.

$$* * *$$

We have seen constructions by so-called difference methods in Chapter 19. We use this idea to construct Latin squares here. It is convenient to describe our constructions in terms of *orthogonal arrays*. An $OA(n, k)$ is a $k$ by $n^2$ array (or we prefer here to think of a *set* of $n^2$ column vectors of height $k$ since the order of the columns is immaterial) whose entries are taken from a set $S$ of $n$ symbols, so that for any distinct $i, j$, $1 \le i, j \le k$, and any two symbols $s, t$ from $S$, there is a unique column whose $i$-th coordinate is $s$ and whose $j$-th coordinate is $t$. This generalizes the $OA(n, 3)$'s introduced in Chapter 17.

This is yet another equivalent formulation of orthogonality; the existence of an $OA(n, k + 2)$ is equivalent to the existence of $k$ pairwise orthogonal Latin squares. For example, to get the array from squares $L_1, L_2, \ldots, L_k$, assume without loss of generality that all row, column, and symbol sets are the same set $S$ and take all columns

$$[i, j, L_1(i, j), L_2(i, j), \ldots, L_k(i, j)]^\top, \quad i, j \in S.$$

Readers should immediately be able to write out two orthogonal squares from the $OA(3, 4)$ in Fig. 22.2 (of course, *any* two rows may be used to "coordinatize" the squares).

$$\begin{bmatrix} x & x & x & y & y & y & z & z & z \\ x & y & z & x & y & z & x & y & z \\ x & y & z & z & x & y & y & z & x \\ x & y & z & y & z & x & z & x & y \end{bmatrix}$$

Figure 22.2

EXAMPLE 22.5. The following matrix of elements of $\mathbb{Z}_{15}$ was found with the aid of a computer by Schellenberg, Van Rees, and Vanstone

(1978):

$$
\begin{bmatrix}
0 & 0 & 0 & 0 & 0 & 0 & 0 & 0 & 0 & 0 & 0 & 0 & 0 & 0 & 0 \\
0 & 1 & 2 & 3 & 4 & 5 & 6 & 7 & 8 & 9 & 10 & 11 & 12 & 13 & 14 \\
0 & 2 & 5 & 7 & 9 & 12 & 4 & 1 & 14 & 11 & 3 & 6 & 8 & 10 & 13 \\
0 & 6 & 3 & 14 & 10 & 7 & 13 & 4 & 11 & 2 & 8 & 5 & 1 & 12 & 9 \\
0 & 10 & 6 & 1 & 11 & 2 & 7 & 12 & 3 & 8 & 13 & 4 & 14 & 9 & 5
\end{bmatrix}
$$

It has the property that for any two rows, the differences between the coordinates of those two rows comprise all elements of $\mathbf{Z}_{15}$, each with multiplicity one. It follows that we obtain an $OA(15,5)$ by taking all *translates* of these columns by elements of $\mathbf{Z}_{15}$. Thus three pairwise orthogonal Latin squares of order 15 can be constructed. But the $OA(15,5)$ is resolvable: the 15 translates of any column have the property that in each coordinate, all symbols occur once. As with transversal designs, we can add a new row to obtain an $OA(15,6)$ and hence four pairwise orthogonal Latin squares of order 15. It is not known whether $N(15) \geq 5$.

The last three rows of the above matrix are what we call below pairwise orthogonal orthomorphisms of $\mathbf{Z}_{15}$.

In 1960, Johnson, Dulmage, Mendelsohn and a computer proved by this method that $N(12) \geq 5$, using the group $\mathbf{Z}_2 \oplus \mathbf{Z}_2 \oplus \mathbf{Z}_3$. The cyclic group of order 12 is of no use; see Theorem 22.9 below.

EXAMPLE 22.6. We will describe a construction that proves

$$
N(m) \geq 2 \Rightarrow N(3m+1) \geq 2.
$$

Let $G$ be an abelian group of order $2m+1$ and let $M$ be a system of representatives for the pairs $\{i, -i\}$ ($i \neq 0$) in $G$. That is $|M| = m$ and $G = \{0\} \cup M \cup -M$. For each $i \in M$, introduce a new symbol $\infty_i$ and take $S := G \cup \{\infty_i : i \in M\}$ as the symbols set for an $OA(3m+1,4)$.

Consider the set of all column vectors obtained from

$$
\left\{ \begin{bmatrix} 0 \\ 0 \\ 0 \\ 0 \end{bmatrix} \right\} \cup \left\{ \begin{bmatrix} \infty_i \\ 0 \\ i \\ -i \end{bmatrix}, \begin{bmatrix} 0 \\ \infty_i \\ -i \\ i \end{bmatrix}, \begin{bmatrix} i \\ -i \\ \infty_i \\ 0 \end{bmatrix}, \begin{bmatrix} -i \\ i \\ 0 \\ \infty_i \end{bmatrix} : i \in M \right\}
$$

by translating them by the elements of $G$ with the understanding that the $\infty$'s are fixed, i.e. $\infty_i + g = \infty_i$. This gives us

$(4m + 1)(2m + 1)$ columns. Add to these the $m^2$ columns of an $OA(m, 4)$ to obtain an $OA(3m + 1, 4)$.

To be more explicit in the case $m = 3$, let us use symbols $\mathbb{Z}_7 \cup \{x, y, z\}$ and $M = \{1, 2, 3\}$. We take the 7 translates of the following 13 columns, and adjoin the $OA(3, 4)$ from Fig. 22.2:

$$
\begin{bmatrix}
0 & x & 0 & 1 & 6 & y & 0 & 2 & 5 & z & 0 & 3 & 4 \\
0 & 0 & x & 6 & 1 & 0 & y & 5 & 2 & 0 & z & 4 & 3 \\
0 & 1 & 6 & x & 0 & 2 & 5 & y & 0 & 3 & 4 & z & 0 \\
0 & 6 & 1 & 0 & x & 5 & 2 & 0 & y & 4 & 3 & 0 & z
\end{bmatrix}
$$

Verification of the construction requires special consideration of the symbols $x, y, z$ (which we have used rather than the $\infty$'s), and the observation that the differences between any two of the above rows (when both coordinates are in $\mathbb{Z}_7$) comprise all the elements of $\mathbb{Z}_7$, each with multiplicity one. The resulting squares are shown below.

$$
\begin{bmatrix}
0 & z & 1 & y & 2 & x & 3 & 6 & 5 & 4 \\
4 & 1 & z & 2 & y & 3 & x & 0 & 6 & 5 \\
x & 5 & 2 & z & 3 & y & 4 & 1 & 0 & 6 \\
5 & x & 6 & 3 & z & 4 & y & 2 & 1 & 0 \\
y & 6 & x & 0 & 4 & z & 5 & 3 & 2 & 1 \\
6 & y & 0 & x & 1 & 5 & z & 4 & 3 & 2 \\
z & 0 & y & 1 & x & 2 & 6 & 5 & 4 & 3 \\
1 & 2 & 3 & 4 & 5 & 6 & 0 & x & y & z \\
2 & 3 & 4 & 5 & 6 & 0 & 1 & z & x & y \\
3 & 4 & 5 & 6 & 0 & 1 & 2 & y & z & x
\end{bmatrix}
\quad
\begin{bmatrix}
0 & 4 & x & 5 & y & 6 & z & 1 & 2 & 3 \\
z & 1 & 5 & x & 6 & y & 0 & 2 & 3 & 4 \\
1 & z & 2 & 6 & x & 0 & y & 3 & 4 & 5 \\
y & 2 & z & 3 & 0 & x & 1 & 4 & 5 & 6 \\
2 & y & 3 & z & 4 & 1 & x & 5 & 6 & 0 \\
x & 3 & y & 4 & z & 5 & 2 & 6 & 0 & 1 \\
3 & x & 4 & y & 5 & z & 6 & 0 & 1 & 2 \\
6 & 0 & 1 & 2 & 3 & 4 & 5 & x & y & z \\
5 & 6 & 0 & 1 & 2 & 3 & 4 & y & z & x \\
4 & 5 & 6 & 0 & 1 & 2 & 3 & z & x & y
\end{bmatrix}
$$

EXAMPLE 22.7. We will describe an $OA(14, 4)$ on the symbols $\mathbb{Z}_{11} \cup \{x, y, z\}$. We take the 11 translates by elements of $\mathbb{Z}_{11}$ of the following 17 columns, and adjoin the $OA(3, 4)$ from Fig. 22.2:

$$
\begin{bmatrix}
0 & 0 & 6 & 4 & 1 & x & 1 & 4 & 0 & y & 2 & 6 & 0 & z & 8 & 9 & 0 \\
0 & 1 & 0 & 6 & 4 & 0 & x & 1 & 4 & 0 & y & 2 & 6 & 0 & z & 8 & 9 \\
0 & 4 & 1 & 0 & 6 & 4 & 0 & x & 1 & 6 & 0 & y & 2 & 9 & 0 & z & 8 \\
0 & 6 & 4 & 1 & 0 & 1 & 4 & 0 & x & 2 & 6 & 0 & y & 8 & 9 & 0 & z
\end{bmatrix}
$$

Once again, verification of the construction requires special consideration of the symbols $x, y, z$ (which we have used rather than the $\infty$'s), and the observation that the differences between any two of the above rows (when both coordinates are in $\mathbb{Z}_{11}$) comprise all the elements of $\mathbb{Z}_{11}$, each with multiplicity one.

PROBLEM 22D. Show that the existence of $k$ pairwise orthogonal idempotent quasigroups is equivalent to the existence of a $\mathrm{TD}(n, k+2)$ in which there may be found a parallel class of blocks.

An *orthomorphism* of an abelian group $G$ is a permutation $\sigma$ of the elements of $G$ such that

$$x \mapsto \sigma(x) - x$$

is also a permutation of $G$. The reader can check that the square $L(x, y) := \sigma(x) + y$ is Latin if and only if $\sigma$ is a permutation, and is orthogonal to the addition table $A(x, y) := x + y$ of $G$ if and only if $\sigma$ is an orthomorphism.

We remark that if there is *any* orthogonal mate to $A$, then there is an orthomorphism: the cells in the positions where an orthogonal mate contains a given symbol have the property that there is exactly one in each row and column, and so are of the form $(x, \tau(x))$, $x \in G$, for some permutation $\tau$. Since these cells must contain different symbols in $A$, the mapping $\sigma$ defined by $\sigma(x) := x + \tau(x)$ is a permutation. In view of this remark, the following theorem proves that cyclic squares of even order have no orthogonal mates, i.e. there are no Latin squares orthogonal to it.

THEOREM 22.9. *If an abelian group $G$ admits an orthomorphism, then its order is odd or its Sylow 2-subgroup is not cyclic.*

PROOF: If the Sylow 2-subgroup is cyclic and nontrivial, then there is exactly one element $z$ of $G$ with order 2. When we add all elements of $G$, each element pairs with its additive inverse except $z$ and 0; so the sum of all elements of $G$ is $z$. But if $\sigma$ were an orthomorphism, then

$$z = \sum_{x \in G} (\sigma(x) - x) = \sum_{x \in G} \sigma(x) - \sum_{x \in G} x = z - z = 0,$$

a contradiction to our choice of $z$. $\square$

REMARKS. If $G$ is abelian and either has odd order or has a Sylow 2-subgroup that is not cyclic, then $G$ admits an automorphism fixing only the identity, and such an automorphism is easily seen to be an orthomorphism. Even if $G$ is not abelian, one can define orthomorphisms (also called *complete mappings*) and it remains true that these do not exist if the Sylow 2-subgroup is cyclic and nontrivial—see Hall and Paige (1955), where complete mappings are also shown to exist for solvable groups with trivial or noncyclic Sylow 2-subgroups.

$$* * *$$

H. B. Mann (1950) observed that a Latin square of order $4t + 1$ with a subsquare of order $2t$ has no orthogonal mates. Similarly, a Latin square of order $4t + 2$ with a subsquare of order $2t + 1$ has no orthogonal mates. Both these results are corollaries of part (ii) of the following theorem, which also proves that none of the pairs of orthogonal squares of orders $12t + 10$ we constructed in Example 22.6 can be extended to a set of three or more pairwise orthogonal Latin squares. We leave it to the reader to convince himself that subsquares correspond to subtransversal designs.

THEOREM 22.10. *Let* $(X, \mathcal{G}, \mathcal{A})$ *be a* $TD(n, k)$ *which contains a sub-$TD(m, k)$* $(Y, \mathcal{H}, \mathcal{B})$ *with* $m < n$. *(This means* $Y \subseteq X$, $\mathcal{H} = \{G \cap Y : G \in \mathcal{G}\}$, *and* $\mathcal{B} \subseteq \mathcal{A}$.) *Then*

(i) $m(k - 1) \leq n$;
(ii) *if* $(X, \mathcal{G}, \mathcal{A})$ *is resolvable, then*

$$m^2 \geq n \left\lceil \frac{mk - n}{k - 1} \right\rceil.$$

PROOF: Pick a point $x_0 \in X \backslash Y$. Say $x_0$ belongs to $G_0 \in \mathcal{G}$. For each of the $m(k - 1)$ points $y \in Y \backslash G_0$, there is a unique block $A_y \in \mathcal{A}$ with $\{x_0, y\} \subseteq A_y$. A block containing two points of $Y$ must belong to $\mathcal{B}$ (and could not contain $x_0$), so these $m(k - 1)$ blocks are distinct. This number cannot exceed the total number of blocks on $x_0$, which is $n$.

Suppose that $(X, \mathcal{G}, \mathcal{A})$ is resolvable and that $\{\mathcal{A}_1, \mathcal{A}_2, \ldots, \mathcal{A}_n\}$ is a partition of $\mathcal{A}$ into parallel classes. Let $s_i$ denote the number of blocks of $\mathcal{A}_i$ which are contained in $Y$ and let $t_i$ denote the number of blocks of $\mathcal{A}_i$ which meet $Y$ in exactly one point. Then

$$ks_i + t_i = |Y| = mk \quad \text{and} \quad s_i + t_i \leq n,$$

from which it follows that $(k - 1)s_i \geq mk - n$, or equivalently,

$$s_i \geq \left\lceil \frac{mk - n}{k - 1} \right\rceil.$$

Part (ii) now follows from $m^2 = |\mathcal{B}| = \sum_{i=1}^{n} s_i$. □

**Notes.**

The term "*mutually* orthogonal Latin squares" rather than "pairwise orthogonal Latin squares" and its acronym "MOLS" are in common usage.

We have already mentioned Euler in the notes to Chapter 1. Euler's paper of 1782 was titled "A new type of magic square", because Euler had noticed that special magic squares arise from a pair of orthogonal Latin squares. As an example, change the symbol sets of $A$ and $B$ in Fig. 22.1 to $\{4, 3, 2, 1\}$ and $\{12, 8, 4, 0\}$, respectively, and *add*, rather than superpose, to obtain

$$\begin{pmatrix} 16 & 11 & 6 & 1 \\ 2 & 5 & 12 & 15 \\ 9 & 14 & 3 & 8 \\ 7 & 4 & 13 & 10 \end{pmatrix}.$$

Each line of this matrix sums to 34. (The Latin property does not in general imply that the diagonals sum to the "magic number", but our example is a particularly nice square: the diagonals, the four corners, and many other sets of four entries sum to 34.)

It has been pointed out to the authors that "$A$" is used in Fig. 22.1 both as the name of a matrix and an entry in that same matrix. We are just checking whether readers are on their toes.

Orthogonal Latin squares, often in the form of orthogonal arrays, are of great importance in the statistical theory of design of experiments.

MacNeish's paper of 1922 actually includes a false proof of Euler's conjecture.

The connection between orthogonal Latin squares and finite projective planes was observed by R. C. Bose (1938).

Many results on orthogonal Latin squares were anticipated by E. H. Moore (1896), as has been pointed out to the authors by R. D. Baker. Unfortunately, the term Latin square is nowhere to be found in Moore's paper, so no one noticed his results for a long time. The results in his paper include special cases of Theorem 22.3 and Theorem 22.4 (albeit in entirely different forms) as well as descriptions of projective planes of prime power order, so he could have disproved the MacNeish conjecture had he known of it.

Theorem 22.6 is a special case of a construction of Wilson (1974) which in turn has been generalized in several ways. See T. Beth, D. Jungnickel, and H. Lenz (1986).

Theorem 22.8 is due to Chowla, Erdős, and Straus (1960). They proved more: $N(n) \geq n^{1/91}$ for sufficiently large $n$. Their work is number-theoretical and is based on the Bose-Parker-Shrikhande constructions. Their result has been improved several times, but the best result to date is that of T. Beth (1983). This seems far from the real truth. It is quite possible that $N(n) \geq n/10$, say, for all, or for all but finitely many, values of $n$, but at the present time this appears to be incredibly hard to prove.

## References.

T. Beth (1983), Eine Bemerkung zur Abschätzung der Anzahl orthogonaler lateinischer Quadrate mittels Siebverfahren, *Abh. Math. Sem. Hamburg* **53**, 284–288.

T. Beth, D. Jungnickel, and H. Lenz (1986), *Design Theory*, Bibliographisches Institut.

R. C. Bose (1938), On the application of the properties of Galois fields to the problem of construction of hyper-Graeco-Latin squares, *Sanhkya* **3**, 323–338.

R. C. Bose, S. S. Shrikhande, and E. T. Parker (1960), Further results in the construction of mutually orthogonal Latin squares and the falsity of a conjecture of Euler, *Canad. J. Math.* **12**, 189–203.

S. Chowla, P. Erdős, and E. G. Straus (1960), On the maximal number of pairwise orthogonal Latin squares of a given order, *Canad. J. Math.* **12**, 204–208.

A. L. Dulmage, D. Johnson, and N. S. Mendelsohn (1961), Orthomorphisms of groups and orthogonal Latin squares, *Canadian J. Math.* **13**, 356–372.

M. Hall and L. J. Paige (1955), Complete mappings of finite groups, *Pacific J. Math.* **5**, 541–549.

H. F. MacNeish (1922), Euler squares, *Ann. Math.* **23**, 221–227.

H. B. Mann (1950), On orthogonal Latin squares, *Bull. Amer. Math. Soc.* **50**, 249–257.

E. H. Moore (1896), Tactical memoranda I–III, *Amer. J. Math.* **18**, 264-303.

P. J. Schellenberg, G. M. J. Van Rees, and S. A. Vanstone (1978), Four pairwise orthogonal Latin squares of order 15, *Ars Comb.* **6**, 141–150.

R. M. Wilson (1974), Concerning the number of mutually orthogonal Latin squares, *Discrete Math.* **9**, 181–198.

# 23

# Projective and combinatorial geometries

A *combinatorial geometry* is a pair $(X, \mathcal{F})$ where $X$ is a set of *points* and where $\mathcal{F}$ is a family of subsets of $X$ called *flats* such that

(1)  $\mathcal{F}$ is closed under intersection,
(2)  there are no infinite chains in the poset $\mathcal{F}$,
(3)  $\mathcal{F}$ contains the empty set, all singletons $\{x\}$, $x \in X$, and the set $X$ itself,
(4)  for every flat $E \in \mathcal{F}$, $E \neq X$, the flats that cover $E$ in $\mathcal{F}$ partition the remaining points.

Here, $F$ *covers* $E$ in $\mathcal{F}$ means that $E, F \in \mathcal{F}$, $E \subsetneq F$, but that $E \subsetneq G \subsetneq F$ does not hold for any $G \in \mathcal{F}$. This latter property should be familiar to the reader from geometry: the lines that contain a given point partition the remaining points; the planes that contain a given line partition the remaining points.

A trivial example of a geometry consists of a finite set $X$ and *all* subsets of $X$ as the flats. This is the *Boolean algebra* on $X$.

EXAMPLE 23.1. Every linear space (as introduced in Chapter 19) gives us a combinatorial geometry on its point set $X$ when we take as flats $\emptyset$, all singletons $\{\{x\} : x \in X\}$, all lines, and $X$ itself. The fact that the lines on a given point partition the remaining points is another way of saying that two points determine a unique line.

EXAMPLE 23.2. Every Steiner system $S(t, k, v)$ gives us a combinatorial geometry on its point set $X$ when we take as flats all subsets of cardinality $< t$, all blocks, and $X$. To get a geometry by this construction, it is not necessary that all blocks have the same size, but only that each $t$-subset is contained in a unique block.

EXAMPLE 23.3. Let $V$ be an $n$-dimensional vector space over a field $\mathbb{F}$. By an *affine subspace* of $V$, we mean either the empty set or a coset (or *translate*) of an (ordinary, i.e. linear) subspace of $V$ in the additive group. For example, for $a, b \in \mathbb{F}$, the subset $\{(x, y) : y = ax + b\}$ is an affine subspace of $\mathbb{F}^2$. The set $V$ together with all affine subspaces forms a combinatorial geometry called the *affine geometry* $AG_n(\mathbb{F})$. We write $AG_n(q)$ for $AG_n(\mathbb{F}_q)$; the case $n = 2$ was introduced in Chapter 19.

EXAMPLE 23.4. The *projective geometry* $PG_n(\mathbb{F})$, while more fundamental than $AG_n(\mathbb{F})$, can be a little more awkward to define. Let $V$ be an $(n+1)$-dimensional vector space $V$ over a field $\mathbb{F}$. The point set $X$ of $PG_n(\mathbb{F})$ is to consist of all the 1-dimensional (linear) subspaces of $V$. For example, if $\mathbb{F}$ is the field $\mathbb{F}_q$ of $q$ elements, then the number of projective points of $PG_n(\mathbb{F}_q)$ is

$$\frac{q^{n+1} - 1}{q - 1} = q^n + \cdots + q^2 + q + 1.$$

To each linear subspace $W$ of $V$, we associate a flat $F_W$ consisting of all the 1-dimensional subspaces of $V$ that are contained in $W$, and take $\mathcal{F}$ to be the set of all such flats $F_W$. We write $PG_n(q)$ for $PG_n(\mathbb{F}_q)$; the case $n = 2$ was introduced in Chapter 19.

Projective geometries $PG_n(\mathbb{F})$ can be defined also for $\mathbb{F}$ a *division ring* (lacking commutativity of multiplication); see Crawley and Dilworth (1973).

Let $Y$ be any subset of the elements $X$ of a combinatorial geometry $(X, \mathcal{F})$ and let

$$\mathcal{E} := \{F \cap Y : F \in \mathcal{F}\}.$$

Then $(Y, \mathcal{E})$ is also a combinatorial geometry, the *subgeometry* on $Y$. For example, $AG_n(\mathbb{F})$ is a subgeometry of $PG_n(\mathbb{F})$. There is a standard embedding of the point set of $AG_n(\mathbb{F})$ into the point set of $PG_n(\mathbb{F})$: to each vector $\mathbf{x}$ in an $n$-dimensional vector space $V$, associate the 1-dimensional subspace that is the span of $(\mathbf{x}, 1)$ in the $(n+1)$-dimensional space $V \times \mathbb{F}$. The image consists of all projective points not contained in the hyperplane $V \times \mathbf{0}$.

The set of flats $\mathcal{F}$ of a combinatorial geometry, when ordered by inclusion, has several important properties.

A *lattice* $L$ is a partially ordered set with the property that any finite subset $S \subseteq L$ has a *meet* (or *greatest lower bound*), that is an element $b$ in $L$ so that

$$\forall_{a \in S}[b \leq a] \quad \text{and} \quad (\forall_{a \in S}[c \leq a] \Rightarrow c \leq b),$$

as well as a *join* (or *least upper bound*), that is an element $b$ in $L$ so that

$$\forall_{a \in S}[b \geq a] \quad \text{and} \quad (\forall_{a \in S}[c \geq a] \Rightarrow c \geq b).$$

The meet and join of a two element set $S = \{x, y\}$ are denoted, respectively, by $x \wedge y$ and $x \vee y$. It is easily seen that $\wedge$ and $\vee$ are commutative, associative, idempotent binary operations; moreover, if all two element subsets have meets and joins, then any finite subset has a meet and a join.

The lattices we will consider have the property that there are no infinite chains. Such a lattice has a (unique) least element (that we denote by $0_L$) because the condition that no infinite chains exist allows us to find a minimal element $m$, and any minimal element is minimum since if $m \not\leq a$, then $m \wedge a$ would be less than $m$. Similarly, there is a unique largest element $1_L$.

For elements $a$ and $b$ of a poset, we say that $a$ *covers* $b$ and write $a \gtrdot b$ when $a > b$ but there are no elements $c$ so that $a > c > b$. For example when $U$ and $W$ are linear subspaces of a vector space, $U \gtrdot W$ when $U \supseteq W$ and $\dim(U) = \dim(W) + 1$. Recall that a *chain* in a partially ordered set $P$ is a totally ordered subset of $P$. So a finite chain can be thought of as a sequence $a_0 < a_1 < \cdots < a_n$. A *point* of a lattice $L$ with miminum element $0_L$ is an element that covers $0_L$.

A *geometric lattice* is a lattice $L$ that has no infinite chains, and such that

(1) $L$ is *atomic*, that is, each element of $L$ is the join of points of $L$, and

(2) $L$ is *semimodular*, that is, if $a$ and $b$ are distinct and both cover $c$ in $L$, then $a \vee b$ covers both $a$ and $b$.

THEOREM 23.1. *The set of flats of a combinatorial geometry, ordered by inclusion, is a geometric lattice. Conversely, given a geometric lattice $L$ with points $X$, then $(X, \{F_y : y \in L\})$ is a combinatorial geometry, where*

$$F_y := \{x \in X : x \le y\}.$$

PROOF: Since the set $\mathcal{F}$ of flats of a combinatorial geometry is closed under intersection, the poset of flats is an atomic lattice (the meet of two flats is their intersection; the join of two flats is the intersection of all flats containing both). Suppose that flats $F_1$ and $F_2$ cover a flat $E$ in a combinatorial geometry. Let $x$ be a point in $F_1 \setminus F_2$. There is a flat $G_2$ that covers $F_2$ in $\mathcal{F}$ and contains the point $x$. $G_2$ contains $F_1$ since otherwise $E \subsetneqq F_1 \cap G_2 \subsetneqq F_1$. But by a symmetric argument, there is a flat $G_1$ that covers $F_1$ and also contains $F_1$ and $F_2$. The join $F_1 \vee F_2$ must be contained in both $G_1$ and $G_2$. Since $G_i$ covers $F_i$, $F_1 \vee F_2 = G_1 = G_2$.

Let $L$ be a geometric lattice and define the flats $F_y$ as subsets of the point set $X$ of $L$ as in the statement of the theorem. It is clear that the empty set, any singleton, and $X$ are flats (when $y$ is taken to be $0_L$, a point of $L$, or $1_L$, respectively). Since $x \le y$ and $x \le z$ if and only if $x \le y \wedge z$, we have $F_y \cap F_z = F_{y \wedge z}$; thus $\mathcal{F} := \{F_y : y \in L\}$ is closed under intersection. No point $x$ not in a given flat $F_y$ can be in two flats that cover $F_y$, so it remains only to show that some flat that covers $F_y$ contains $x$. We show that $F_{x \vee y}$ covers $F_y$ to complete the proof.

This is equivalent to showing that $x \vee y > y$ in $L$ whenever $x \not\le y$ in $L$ and $x$ is a point of $L$. Choose a maximal chain

$$0_L \lessdot y_1 \lessdot y_2 \lessdot \cdots \lessdot y_k = y.$$

Since $x$ and $y_1$ cover $0_L$, $x \vee y_1$ covers $y_1$. Since $x \vee y_1$ and $y_2$ cover $y_1$, $x \vee y_2$ covers $y_2$ (clearly $x \vee y_1 \vee y_2 = x \vee y_2$). Inductively, we find that $x \vee y_k$ covers $y_k$. □

In some sense, the difference between geometric lattices and combinatorial geometries is the same as that between incidence structures on the one hand and families of subsets of a set on the other. Incidence structures and lattices are more abstract and must be used to avoid confusion in certain cases (e.g. when discussing duality or intervals), but we prefer the language and notation of sets and subsets for many arguments.

For example, for elements $a$ and $b$ of a partially ordered set, the *interval* $[a,b]$ is defined as $[a,b] := \{x : a \leq x \leq b\}$. Any interval in a lattice is again a lattice. The reader should check that any interval of a geometric lattice is again a geometric lattice. It would be awkward to state the corresponding fact in terms of combinatorial geometries.

In view of Theorem 23.1 (more precisely, the simple correspondences decribed in the proof), it is often convenient (though sometimes confusing) to mix the notation and language of combinatorial geometries and geometric lattices. For example, we can use the same symbol $PG_n(\mathbb{F})$ to denote the combinatorial geometry and the corresponding lattice of subspaces of a vector space. An exercise: Any interval of $PG_n(\mathbb{F})$ is isomorphic as a poset to $PG_m(\mathbb{F})$ for some $m \leq n$. Another exercise: $PG_n(\mathbb{F})$ is isomorphic to its dual poset (where order is reversed).

EXAMPLE 23.5. The partition lattices $\Pi_n$, whose elements are all partitions of an $n$-set $X$, provide another family of geometric lattices, and hence give us combinatorial geometries. The partitions are ordered by *refinement*; $\mathcal{A}$ is a refinement of $\mathcal{B}$ if each block of $\mathcal{B}$ is the union of blocks of $\mathcal{A}$. So the least element (the finest partition) is all singletons $\{\{x\} : x \in X\}$ and the greatest element is $\{\{X\}\}$, consisting of a single block. A moment's thought shows that one partition covers another if the former is obtained by coalescing two blocks of the latter into one. The points are the partitions that consist of a single block of size 2 and $n-2$ singletons.

For example, $\Pi_4$ has 15 elements:

$$\{1234\}$$

$$\{123,4\}, \ \{124,3\}, \ \{134,2\}, \ \{234,1\}, \ \{12,34\}, \ \{13,24\}, \ \{14,23\}$$

$$\{12,3,4\}, \ \{13,2,4\}, \ \{14,2,3\}, \ \{23,1,4\}, \ \{24,1,3\}, \ \{34,1,2\}$$

$$\{1,2,3,4\}$$

To simplify the notation, we have written $\{123,4\}$, for example, for what should be $\{\{1,2,3\},\{4\}\}$.

EXAMPLE 23.6. The points of $\Pi_n$ are in one-to-one correspondence with the edges of $K_n$. Given any simple graph $G$ on $n$ vertices, we obtain a geometric lattice $L(G)$ whose points correspond to the edge set $E(G)$ of $G$ as follows. The elements of $L$ are to be all those partitions $\mathcal{A}$ of $V(G)$ such that the subgraph of $G$ induced by each block of $\mathcal{A}$ is connected. The partitions with this property are exactly those that are the join (in $\Pi_n$) of points (in $\Pi_n$) corresponding to some subset of the edges of $G$. The lattices $L(G)$ are sometimes called the lattices of *contractions* of $G$ (see Chapter 32).

Fig. 23.1 is an attempt to illustrate the combinatorial geometries $L(K_4)$ and $L(K_5)$. $L(K_5)$, for example, has ten points; the dot in the figure labeled 12 represents the point (partition) $\{12, 3, 4, 5\}$. The partitions of the type $\{123, 4, 5\}$, for example, contain three points and are represented by line segments.

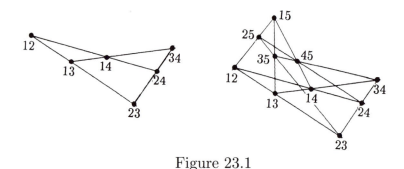

Figure 23.1

For a subset $S$ of the points of a combinatorial geometry, the *closure* $\overline{S}$ is the intersection of all flats containing $S$. For example, in a linear space, the closure of a two point set is the line containing the points. Exercise: show that $S \subseteq \overline{S}$, $A \subseteq B \Rightarrow \overline{A} \subseteq \overline{B}$, and $\overline{\overline{S}} = \overline{S}$. It will be occasionally convenient to use the symbol for join in the lattice of flats:

$$E \vee F := \overline{E \cup F}.$$

A subset $S \subseteq X$ is *independent* when for each $x \in S$, $x \notin \overline{S \setminus \{x\}}$. In Example 23.1, three points in a linear space are independent if and only if they are not contained in a line; no four points are independent. A set of points in $PG_n(\mathbb{F})$ is independent if and only if representative vectors are linearly independent. A set of points in $AG_n(\mathbb{F})$ is independent if and only if the vectors are affinely independent. One viewpoint of combinatorial geometries is that they represent the study of "abstract independence".

By the *rank* of a combinatorial geometry with point set $X$, we mean the maximum size of an independent subset of $X$. $PG_n(\mathbb{F})$ and $AG_n(\mathbb{F})$ both have rank $n + 1$. We will avoid the word *dimension*, but if we forget and use it, it will mean the rank *minus one*. The maximum size of an independent subset of a flat $F$ is the *rank* of that flat. The flats of rank 1 are called *points*, the flats of rank 2 are called *lines*, the flats of rank 3 are called *planes*. The flats of rank 1 less than rank$(X)$ are called *hyperplanes* or *copoints*. Occasionally, we need to use the term *coline* for a flat for which the rank equals rank$(X) - 2$.

From any combinatorial geometry $(X, \mathcal{F})$, we obtain a linear space $(X, \mathcal{L})$ where $\mathcal{L}$ is the set of all lines (flats of rank 2).

In Fig. 23.1, any two points determine a line that may have two or three points; only the latter are drawn in the figure. The planes in $L(K_5)$ have either four or six points; the latter contain four three-point lines.

We collect some simple facts:

LEMMA 23.2.

(1) If $x \notin \overline{A}$ but $x \in \overline{A \cup \{y\}}$, then $y \in \overline{A \cup \{x\}}$ (*the exchange axiom*).
(2) If $S$ is an independent set of points in a geometry and $x \notin \overline{S}$, then $\{x\} \cup S$ is independent.
(3) If $F = \overline{S}$, then $F = \overline{A}$ for any maximal independent subset $A$ of $S$.

PROOF: $F_1 := \overline{A \cup \{x\}}$ is the flat that covers $E := \overline{A}$ and contains $x$; $F_2 := \overline{A \cup \{y\}}$ is the flat that covers $E$ and contains $y$. If $x \in F_2$, it must be that $F_1 = F_2$. This proves (1).

If $\{x\} \cup S$ is not independent, there is some $y \in S$ such that $y \in \overline{(S \cup \{x\}) \backslash \{y\}}$. Let $A := S \backslash \{y\}$. Since $S$ is independent, $y \notin \overline{A}$. But (1) then implies that $x \in \overline{A \cup \{y\}}$, i.e. $x \in \overline{S}$. This proves (2).

Suppose $F = \overline{S}$ and that $A$ is an independent subset of $S$ that is maximal with respect to that property. If $S \not\subseteq \overline{A}$, we could find a larger independent subset of $S$ by (2). So $S \subseteq \overline{A}$ and since $F$ is the smallest flat containing $S$, $F \subseteq \overline{A}$. This proves (3).  □

A *basis* for a flat $F$ is an independent subset $B \subseteq F$ so that $\overline{B} = F$, i.e. any maximal independent subset of $F$. As an exercise, the reader should check that we could also have defined a basis of $F$ as a minimal *spanning* set, i.e. a subset $B$ of $F$ so that $\overline{B} = F$ and that is minimal with respect to this property.

PROBLEM 23A. Let $G$ be a connected simple graph. Show that the bases of the combinatorial geometry $L(G)$ are exactly the edge sets of spanning trees in $G$.

THEOREM 23.3. *All bases of a flat $F$ in a combinatorial geometry have the same finite cardinality (called the* rank *of $F$). For flats $E$ and $F$,*

$$(23.1) \qquad \mathrm{rank}(E) + \mathrm{rank}(F) \geq \mathrm{rank}(E \cap F) + \mathrm{rank}(E \vee F).$$

PROOF: It should be clear that the condition that there are no infinite chains of flats forces all independent sets to be finite.

If it is not true that all bases of $F$ have the same cardinality, choose two, say $B_1$ and $B_2$, of different sizes but so that $|B_1 \cap B_2|$ is as large as possible subject to this constraint. Say $|B_1| > |B_2|$. Pick $x \in B_1 \setminus B_2$. Then $B_1 \setminus \{x\}$ is independent and has a closure which does not contain $F$ and hence does not contain $B_2$. Pick $y \in B_2 \setminus B_1$ such that $y \notin \overline{B_1 \setminus \{x\}}$. Then $(B_1 \setminus \{x\}) \cup \{y\}$ is independent by Lemma 23.2 and is contained in a basis $B_3$ for $F$. Clearly $|B_3| \neq |B_2|$, but $|B_3 \cap B_2| > |B_1 \cap B_2|$, a contradiction.

To prove the second part of the theorem, first note that Lemma 23.2(3) implies that a basis for a flat $E$ can be extended to a basis for any flat $E'$ containing $E$.

Now let $B$ be a basis for $E \cap F$. Extend $B$ to bases $B_1$ and $B_2$ for $E$ and $F$, respectively. Then any flat containing $B_1 \cup B_2$ contains $E$ and $F$ and hence $\overline{E \cup F}$; that is, $\overline{B_1 \cup B_2} = \overline{E \cup F}$ and so $B_1 \cup B_2$

contains a basis for $\overline{E \cup F}$. Then

$$\begin{aligned}
\operatorname{rank}(\overline{E \cup F}) &\leq |B_1 \cup B_2| = |B_1| + |B_2| - |B_1 \cap B_2| \\
&= \operatorname{rank}(E) + \operatorname{rank}(F) - \operatorname{rank}(E \cap F).
\end{aligned}$$

$\square$

The inequality (23.1) is called *the semimodular law*.

We remark that the proof of Theorem 23.3 shows that all maximal independent subsets of any subset $S$, not necessarily a flat, have the same cardinality. This is a corollary of the statement for flats, however, because we can apply Theorem 23.4 to the subgeometry on $S$.

As an exercise, make sure that you understand the following: Let $E$ and $F$ be flats in a geometry with $E \subseteq F$. If $\operatorname{rank}(E) = \operatorname{rank}(F)$, then $E = F$; $F$ covers $E$ if and only if $\operatorname{rank}(F) = \operatorname{rank}(E) + 1$.

For the affine geometries $AG_n(q)$, the number of points on a flat of rank $r$ is $q^r$. For the projective geometries $PG_n(q)$, the number of points on a flat of rank $r$ is

$$(23.2) \qquad \frac{q^{r+1} - 1}{q - 1} = q^r + \cdots + q^2 + q + 1.$$

THEOREM 23.4. *Suppose that every flat of rank $i$ in a combinatorial geometry on $v$ points has exactly $k_i$ points, $i = 0, 1, \ldots, n$. Then the total number of flats of rank $r$ is*

$$(23.3) \qquad \prod_{i=0}^{r-1} \frac{(v - k_i)}{(k_r - k_i)}.$$

*Furthermore, the set of points together with the family of rank $r$ flats is a 2-design.*

PROOF: Since the flats of rank $r + 1$ that contain a flat of rank $r$ partition the remaining $v - k_r$ points into sets of size $k_{r+1} - k_r$, there must be exactly $(v - k_r)/(k_{r+1} - k_r)$ such flats. The formula (23.3) now follows by induction on $r$ when we count the ordered pairs $(E, F)$ of rank $r$ and rank $r + 1$ flats with $E \subseteq F$. Note that $k_0 = 0$ and $k_1 = 1$, so that (23.3) is valid for $r = 1$.

Any two points are contained in a unique rank 2 flat. The above argument implies that any rank 2 flat is contained in the same number of rank $r$ flats. Hence the rank $r$ flats give a 2-design. □

The numbers of rank $r$ flats of $PG_n(q)$ are called *Gaussian numbers*; see Chapter 24. Equations (23.2) and (23.3) imply that the numbers of points and hyperplanes of $PG_n(q)$ are equal (there are other ways of seeing this) and so we have the following corollary.

COROLLARY. *The points and hyperplanes of the projective geometry $PG_n(q)$ form a $(v, k, \lambda)$-symmetric design with*

$$v = (q^{n+1} - 1)/(q - 1),$$
$$k = (q^n - 1)/(q - 1),$$
$$\lambda = (q^{n-1} - 1)/(q - 1).$$

PROBLEM 23B. Show that the incidence structure whose points are the points of $AG_r(2)$ and whose blocks are the planes of $AG_r(2)$ is a Steiner system $S(3, 4, 2^r)$.

The following theorem is due to C. Greene (1970).

THEOREM 23.5. *The number of hyperplanes in a finite combinatorial geometry is at least the number of points.*

PROOF: The proof will be similar to that of Theorem 19.1. First, we shall show that if a point $x$ is not on a hyperplane $H$, then the number $r_x$ of hyperplanes on $x$ is at least the number $k_H$ of points on $H$ by induction on the rank. The assertion is trivial for geometries of rank $\leq 2$. By the induction hypothesis, we know that the number of hyperplanes of the subgeometry on $H$ (i.e. the number of colines $C$ contained in $H$) is at least $k_H$. But for each such coline $C$, we get a hyperplane (the join of $C$ and $x$) on $x$.

Now we repeat from the proof of Theorem 19.1: Let $\mathcal{H}$ denote the set of hyperplanes, $v := |X|$, $b := |\mathcal{H}|$. Suppose $b \leq v$. Then

$$1 = \sum_{x \in X} \sum_{H \not\ni x} \frac{1}{v(b - r_x)} \geq \sum_{H \in \mathcal{H}} \sum_{x \notin H} \frac{1}{b(v - k_H)} = 1,$$

and this implies that in all the inequalities, equality must hold. Therefore $v = b$. □

See the remark following Lemma 23.8 below concerning the case of equality in Theorem 23.5.

In studying $PG_n(\mathbb{F})$, the dimension formula

$$\dim(U \cap W) + \dim(U + W) = \dim(U) + \dim(W)$$

for linear subspaces of a vector space plays a crucial role. Equivalently,

$$(23.4) \qquad \operatorname{rank}(E \cap F) + \operatorname{rank}(E \vee F) = \operatorname{rank}(E) + \operatorname{rank}(F)$$

for flats $E$ and $F$ of $PG_n(\mathbb{F})$. This is a stronger version of the semimodular law. Indeed, a combinatorial geometry in which (23.4) holds is said to be *modular*. In a modular combinatorial geometry, for example, (23.4) implies that any two lines contained in a plane must meet nontrivially (cf. Theorem 19.1).

PROBLEM 23C. Let $(X, \mathcal{B})$ be the linear space consisting of the points and lines of a modular combinatorial geometry of rank 3. Show that $(X, \mathcal{B})$ is either a nearpencil as defined following Theorem 19.1 or else is a projective plane as defined in Chapter 19—that is, show that any two lines have the same cardinality $n+1$ and that there are exactly $n^2 + n + 1$ points for some integer $n \geq 2$.

We have introduced the projective geometries $PG_n(\mathbb{F})$ and have defined projective planes in Chapter 19. Here is the general definition: A *projective geometry* is a modular combinatorial geometry that is *connected* in the sense that the point set cannot be expressed as the union of two proper flats.

We do not have the space or the time to prove here the following fundamental result. See Veblen and Young (1907) or Crawley and Dilworth (1973) for a proof.

THEOREM 23.6. *Every projective geometry of rank $n \geq 4$ is isomorphic to $PG_n(\mathbb{F})$ for some division ring $\mathbb{F}$.*

Combinatorial geometries of rank $\leq 2$ are without great interest, but all are modular and, except for the two point line, are projective geometries. The rank 3 projective geometries are, by Problem 23C, equivalent to projective planes as introduced in Chapter 19 (the nearpencils are not connected).

Every modular combinatorial geometry can be put together from projective geometries. This is the content of the following problem.

PROBLEM 23D.
(i) Let $(X_1, \mathcal{F}_1)$ and $(X_2, \mathcal{F}_2)$ be modular combinatorial geometries. Assuming $X_{-1}$ and $X_{-2}$ are disjoint, show that $(X_1 \cup X_2, \{F_1 \cup F_2 : F_1 \in \mathcal{F}_1, F_2 \in \mathcal{F}_2\})$ is a modular combinatorial geometry.
(ii) Let $(X, \mathcal{F})$ be a modular combinatorial geometry such that $X$ is the union of two flats $X_1$ and $X_2$. Let $\mathcal{F}_i := \{F \cap X_i : F \in \mathcal{F}\}$, $i = 1, 2$. Show that $(X_i, \mathcal{F}_i)$, $i = 1, 2$, is a modular combinatorial geometry and that $\mathcal{F} = \{F_1 \cup F_2 : F_1 \in \mathcal{F}_1, F_2 \in \mathcal{F}_2\}$.

PROBLEM 23E. Prove that every flat $F$ of a combinatorial geometry $(X, \mathcal{F})$ has a *modular complement*, i.e. there exists a flat $E$ such that $E \cap F = \emptyset$, $E \vee F = X$, and $\text{rank}(E) + \text{rank}(F) = \text{rank}(X)$.

The linear spaces consisting of the points and lines of modular combinatorial geometries satisfy the following condition known as the *Pasch axiom*. A line that meets two sides of a triangle also meets the third. More precisely, suppose $A, B, C$ are distinct lines and $a, b, c$ distinct points with incidence as in Fig. 23.2. We require that any line $L$ that contains a point other than $a$ or $b$ from the line $C$ and a point other than $a$ or $c$ from the line $B$ also contains a point other than $b$ or $c$ from the line $A$. See Fig. 23.2. To see that this holds in a modular geometry, notice that $L$ and $A$ must both be contained in the plane $P = \overline{\{a, b, c\}}$, and the modular equation implies that two lines in a plane must meet nontrivially.

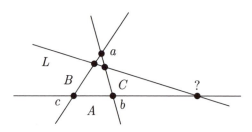

Figure 23.2

PROBLEM 23F. Consider the incidence structure whose points are the $k$-dimensional subspaces of a vector space $V$ and whose lines are the $(k + 1)$-dimensional subspaces of $V$, with incidence being containment. Show that this incidence structure satisfies the Pasch axiom.

THEOREM 23.7. *A finite linear space $(X, \mathcal{A})$ for which the Pasch axiom holds, consists of the points and lines of some modular combinatorial geometry on $X$.*

PROOF: Suppose the Pasch axiom holds. We must construct the flats of a combinatorial geometry, which we do as follows. We say that a subset $S \subseteq X$ is a *flat* when, for any line $L$,

$$|L \cap S| \geq 2 \quad \text{implies} \quad L \subseteq S.$$

Let $\mathcal{F}$ denote the set of all such flats; this includes the empty set, all singletons, and $X$, as well as all lines in $\mathcal{A}$. Check that the definition implies that the intersection of any number of flats is again a flat.

Let $x$ be a point not in a flat $S$. Let $T$ be the union of all lines joining $x$ to the points $s \in S$. We claim that this simple construction produces a flat $T$, and that $T$ covers $S$.

To this end, let $L$ be a line containing two points $t_1, t_2$ of $T$. We want to show $L \subseteq T$. If $t_1, t_2$ both belong to one of the lines on $x$, then $L$ is that line and so $L \subseteq T$. If both $t_1$ and $t_2$ are in $S$, then $L \subseteq T$, so assume $t_1 \notin S$. Otherwise, $t_1$ and $t_2$ belong to distinct lines $M_1$ and $M_2$ on $x$. Say $M_i$ meets $S$ in $s_i$, $i = 1, 2$, and let $N$ be the line joining $s_1$ and $s_2$. We have $N \subseteq S$. The Pasch axiom guarantees that $L$ meets $N$ in some point $z$ of $S$.

Consider any other point $t_3 \in L$. Let $M_3$ be the line joining $x$ and $t_3$. Since $M_3$ meets two sides of the triangle with sides $M_1$, $L$, and $N$ (with vertices $z$, $t_1$, and $s_1$), it must meet the third side $N$ in some point $s_3$ of $S$. Then $M_3$ is one of the lines joining $x$ to points of $S$, so $t_3 \in T$. We have now proved that $T$ is a flat.

That $T$ covers $S$ is easy to see: If $U$ is a flat, $S \subsetneqq U \subseteq T$, pick $y \in U \backslash S$. By construction, the line joining $x$ and $y$ is one of the lines on $x$ that meets $S$; that line contains two points of $U$ and hence is contained in $U$. This means $x \in U$, whence $T \subseteq U$.

Since any point lies in a flat that covers a flat $S$, $(X, \mathcal{F})$ is a combinatorial geometry. (The hypothesis that $X$ is finite is only used here to ensure that there are no infinite chains of flats.)

Let $x$ be a point on a line $L$ and let $H$ be any hyperplane. Since (if $x$ is not already in $H$) the join of $H$ and $x$ must be $X$, $L$ must be one of the lines on $x$ that meet $H$. That is, any line $L$ and any hyperplane $H$ meet nontrivially. This implies modularity by Lemma 23.8 below.        □

LEMMA 23.8.  *A combinatorial geometry is modular if and only if any line and any hyperplane meet nontrivially.*

PROOF: The modular law (23.2) shows that any rank 2 flat and any rank $n-1$ flat in a rank $n$ geometry meet in a flat of rank $\geq 1$.

For the converse, we use induction on the rank. Assume that all lines and hyperplanes in a rank $n$ combinatorial geometry meet nontrivially. Suppose $H$ is a hyperplane and that there exists a line $L \subseteq H$ and a rank $n-2$ flat $C \subseteq H$ that are disjoint. But then for any point $x \notin H$, the hyperplane $C \vee \{x\}$ would be disjoint from the line $L$. So by induction, the subgeometry on any hyperplane $H$ is modular.

Thus if there is a pair of flats $E$ and $F$ that provide a counterexample to the modular law (23.4), then $E \vee F$ is equal to the entire point set $X$. Let $H$ be a hyperplane containing $E$ and let $F' := H \cap F$. Then $E \vee F' \subseteq H$. But rank$(F') \geq$ rank$(F) - 1$ since otherwise a modular complement of $F'$ in $F$ would have rank $\geq 2$ and so would contain a line, one that would be disjoint from $H$. We see that $E$ and $F'$ provide a counterexample to (23.4) contained in $H$, contradicting the modularity of the subgeometry $H$.        □

REMARK. Suppose the number of hyperplanes of a finite combinatorial geometry $(X, \mathcal{F})$ is equal to the number of points. Then, reviewing the proof of Theorem 23.5, $r_x = k_H$ whenever $x \notin H$. If we apply Theorem 23.5 to the interval $[\{x\}, X]$ in any combinatorial geometry, we find that $r_x \geq l_x$, where $l_x$ denotes the number of lines containing $x$ (since these lines are the points of the geometric lattice $[\{x\}, X]$). For $x \notin H$ we obviously have $l_x \geq k_H$ since the lines $\{x\} \vee \{y\}$, $y \in H$, are distinct. In summary, equality in Theorem 23.5 implies that $l_x = k_H$ when $x \notin H$. This means that every line meets every hyperplane and hence $(X, \mathcal{F})$ is modular by Lemma 23.8.

Two triangles $\{a_1, b_1, c_1\}$ and $\{a_2, b_2, c_2\}$ are said to be *perspective from a point* if there exists a point $p$ (the *point of perspectivity*) such that $\{p, a_1, a_2\}$ are collinear, $\{p, b_1, b_2\}$ are collinear, and $\{p, c_1, c_2\}$ are collinear. Two triangles $\{a_1, b_1, c_1\}$ and $\{a_2, b_2, c_2\}$ are said to be *perspective from a line* if there exists a line $L$ (the *line* or *axis of perspectivity*) such that $\{L, A_1, A_2\}$ are concurrent, $\{L, B_1, B_2\}$ are concurrent, and $\{L, C_1, C_2\}$ are concurrent. Here we are using "triangle" to mean three noncollinear points, and $A_i$, $B_i$, and $C_i$ to denote the "sides" of the triangle opposite $a_i$, $b_i$, and $c_i$, respectively. That is, $A_i := \{\overline{b_i, c_i}\}$, $i = 1, 2$, etc.

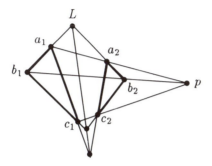

Figure 23.3

In Fig. 23.3, the triangles are perspective from the point $p$ and also perspective from the line $L$. The incidence stucture so illustrated, with ten points and ten lines, each line on three points and each point on three lines, is called the *Desargues configuration*. The triangles need not lie in a plane. The right diagram in Fig. 23.1 is isomorphic (!) to that in Fig. 23.3 as an incidence structure; the former is easier to imagine as 3-dimensional. It can be viewed in many ways, e.g. as showing that the triangles $\{25, 35, 45\}$ and $\{12, 13, 14\}$ are perspective from the point $15$ and also perspective from the line $\{23, 24, 34\}$.

If we choose a basis for the $(n+1)$-dimensional vector space whose subspaces comprise $PG_n(\mathbb{F})$, we can describe projective points by so-called *homogeneous coordinates*. By $\langle x_0, x_1, \ldots, x_n \rangle$, we mean the 1-dimensional subspace spanned by the vector $(x_0, x_1, \ldots, x_n)$. The projective point $\langle x_0, x_1, \ldots, x_n \rangle$ is the same as $\langle y_0, y_1, \ldots, y_n \rangle$ if and only if the vector $(y_0, y_1, \ldots, y_n)$ is a nonzero scalar multiple of $(x_0, x_1, \ldots, x_n)$. The hyperplanes may also be described by homogeneous $(n+1)$-tuples: $[c_0, c_1, \ldots, c_n]$ will denote the hyperplane consisting of points with homogeneous coordinates $\langle x_0, x_1, \ldots, x_n \rangle$ so that $c_0 x_0 + x_1 c_1 + \cdots + x_n c_n = 0$.

THEOREM 23.9 (DESARGUES' THEOREM). *In $PG_n(\mathbb{F})$, if two triangles are perspective from a point, then they are perspective from a line.*

PROOF: Let triangles $\{a_1, b_1, c_1\}$ and $\{a_2, b_2, c_2\}$ be perspective from a point $p$. We are going to assume that no three of $p$, $a_1$, $b_1$ and $c_1$ are collinear so that we do not spend too much time on details. We will also assume that all points lie in a plane, i.e. we prove the theorem for $PG_2(\mathbb{F})$. The other cases are left for the reader.

We choose a basis for a 3-dimensional vector space so that the basis vectors $\mathbf{x}, \mathbf{y}, \mathbf{z}$ span the 1-dimensional subspaces $a_1$, $b_1$, and $c_1$, respectively. Then the homogeneous coordinates for $a_1$, $b_1$, and $c_1$ are $\langle 1, 0, 0 \rangle$, $\langle 0, 1, 0 \rangle$, and $\langle 0, 0, 1 \rangle$, respectively. Then $p$ has homogeneous coordinates $\langle \alpha, \beta, \gamma \rangle$ where all coordinates are nonzero. We replace the original basis vectors by $\alpha \mathbf{x}$, $\beta \mathbf{y}$ and $\gamma \mathbf{z}$; then $a_1$, $b_1$, and $c_1$ are still represented by $\langle 1, 0, 0 \rangle$, $\langle 0, 1, 0 \rangle$, and $\langle 0, 0, 1 \rangle$, respectively, while $p$ now has homogeneous coordinates $\langle 1, 1, 1 \rangle$. This will simplify our computations.

It follows that $a_2 = \langle \alpha, 1, 1 \rangle$, $b_2 = \langle 1, \beta, 1 \rangle$, $c_2 = \langle 1, 1, \gamma \rangle$, for some $\alpha, \beta, \gamma \in \mathbb{F}$. The line joining $a_1$ and $b_1$ is $[0, 0, 1]$; the line joining $a_2$ and $b_2$ is $[1 - \alpha, 1 - \beta, \alpha \beta - 1]$ (check that both $a_2$ and $b_2$ are on this line). The point on both lines is $\langle 1 - \beta, \alpha - 1, 0 \rangle$. Similarly, the point of intersection of the lines joining $b_1$ and $c_1$ with the line joining $b_2$ and $c_2$ is found to be $\langle 0, 1 - \alpha, \gamma - 1 \rangle$, and the point of intersection of the lines joining $c_1$ and $a_1$ with the line joining $c_2$ and $a_2$ is found to be $\langle \beta - 1, 0, 1 - \gamma \rangle$. The three points of intersection are collinear since their coordinates are linearly dependent.

(If we are careful, the proof should go through also for division rings $\mathbb{F}$.) $\square$

PROBLEM 23G. Prove that the linear space of $PG_n(\mathbb{F})$ also satisfies *Pappus' theorem* when $\mathbb{F}$ is a (commutative) field, which is as follows. Let $a_i$, $b_i$, and $c_i$ be collinear points, say on a line $L_i$, for $i = 1, 2$. Suppose $L_1$ and $L_2$ meet in a point (not one of the six named points). Then the lines $\overline{\{a_1, b_2\}}$ and $\overline{\{a_2, b_1\}}$ meet in a point $c$, the lines $\overline{\{b_1, c_2\}}$ and $\overline{\{b_2, c_1\}}$ meet in a point $a$, and the lines

$\overline{\{a_1, c_2\}}$ and $\overline{\{a_2, c_1\}}$ meet in a point $b$; and the three points $a, b, c$ are collinear.

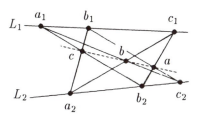

Figure 23.4

Projective planes in which the statement of Theorem 23.9 is valid (if two triangles are perspective from a point, then they are perspective from a line) are called *Desarguesian planes.* Projective planes in which the conclusion of Problem 23G holds are called *Pappian planes.* We do not have the space or time to prove the further following fundamental result. See Crawley and Dilworth (1973).

THEOREM 23.10.

(1) *A projective geometry of rank 3 is isomorphic to $PG_2(\mathbb{E})$ for some division ring $\mathbb{E}$ if and only if it is a Desarguesian plane.*

(2) *A projective geometry of rank 3 is isomorphic to $PG_2(\mathbb{F})$ for some field $\mathbb{F}$ if and only if it is a Pappian plane.*

A corollary of Theorem 23.10 is that every Pappian plane is Desarguesian. For a synthetic (i.e. coordinate-free) proof that Pappus' theorem implies Desargues' theorem, see D. Pedoe (1963). We remark that every finite division ring is a field (Wedderburn's theorem), so that every finite Desarguesian plane is Pappian; see M. Hall, Jr. (1972).

We describe one method of obtaining finite projective planes that are not Desarguesian.

Let $G$ be an abelian group of order $n^2$, and suppose that there exist subgroups $H_0, H_1, \ldots, H_n$ of order $n$ that partition the nonzero elements of $G$, that is such that $H_i \cap H_j = \{0\}$ for $i \neq j$ and, therefore,

$$\bigcup_{i=0}^{n} H_i \setminus \{0\} = G \setminus \{0\}.$$

Then, we claim, the incidence structure whose points are the elements of $G$ and whose lines are all cosets (translates) of all the subgroups $H_i$ is an affine plane. In general, the intersection of cosets of two subgroups $H$ and $K$ is either empty or a coset of $H \cap K$, so lines meet in at most one point. We could now count covered pairs to show that every two distinct points are contained in a line, but it is just as simple to note that given points $x, y \in G$, the coset $H_\ell + y$ contains both points where $H_\ell$ is the subgroup containing the difference $x - y$. An affine plane obtained from a group $G$ and $n + 1$ subgroups in this way is called an *affine translation plane*. It is known that $G$ must be elementary abelian for such subgroups to exist; see J. André (1954). The affine plane may be completed to a projective plane as in Problem 19K; any projective plane obtained in this way is a *translation plane*.

EXAMPLE 23.7. Here is a partition of the nonzero elements of $\mathbb{F}_3^4$ into ten subgroups $H_0, H_1, \ldots, H_9$. (We have denoted the elements as strings, dropping commas and parentheses.)

$$\{0000, 1000, 2000, 0100, 0200, 1100, 2200, 2100, 1200\}$$
$$\{0000, 0010, 0020, 0001, 0002, 0011, 0022, 0021, 0012\}$$
$$\{0000, 1010, 2020, 0101, 0202, 1111, 2222, 2121, 1212\}$$
$$\{0000, 2010, 1020, 0201, 0102, 2211, 1122, 1221, 2112\}$$
$$\{0000, 0110, 0220, 2001, 1002, 2111, 1222, 2221, 1112\}$$
$$\{0000, 0210, 0120, 1001, 2002, 1211, 2122, 1121, 2212\}$$
$$\{0000, 1110, 2220, 2101, 1202, 0211, 0122, 1021, 2012\}$$
$$\{0000, 2210, 1120, 1201, 2102, 0111, 0222, 2021, 1012\}$$
$$\{0000, 2110, 1220, 2201, 1102, 1011, 2022, 0121, 0212\}$$
$$\{0000, 1210, 2120, 1101, 2202, 2011, 1022, 0221, 0112\}$$

The affine plane corresponding to this "spread" of subgroups is the Desarguesian plane.

But note that the elements of the first four subgroups, for example, can be partitioned into subgroups in *another* way, as indicated by the columns below.

$$
\begin{array}{llll}
\{0000, & \{0000, & \{0000, & \{0000, \\
1000, 2000, & 0100, 0200, & 1100, 2200, & 2100, 1200, \\
0010, 0020, & 0001, 0002, & 0011, 0022, & 0021, 0012, \\
1010, 2020, & 0101, 0202, & 1111, 2222, & 2121, 1212, \\
2010, 1020\} & 0201, 0102\} & 2211, 1122\} & 1221, 2112\}
\end{array}
$$

The above example is the special case $q = 3$ of a construction we now describe. Let $V$ be a 2-dimensional vector space over $\mathbb{F}_{q^2}$ and let $H_0, H_1, \ldots, H_{q^2}$ be the 1-dimensional subspaces over $\mathbb{F}_{q^2}$. If we consider $V$ as a 4-dimensional vector space over $\mathbb{F}_q$, the subspaces $H_i$ are 2-dimensional over $\mathbb{F}_q$ (they form a *spread* of lines in $PG_3(q)$; see Theorem 24.3). Let $U$ be any 2-dimensional subspace over $\mathbb{F}_q$ other than one of the $H_i$'s. Then $U$ meets any $H_i$ in $\{0\}$ or in a 1-dimensional subspace over $\mathbb{F}_q$. Say $U$ meets $H_0, H_1, \ldots, H_q$ in $q$ points each and $H_{q+1}, H_{q+2}, \ldots, H_{q^2}$ in the zero vector only. Consider the $\mathbb{F}_q$-subspaces $\alpha U$ as $\alpha$ ranges over the nonzero elements of $\mathbb{F}_{q^2}$; since multiplying by the nonzero elements of $\mathbb{F}_q$ fixes $U$, there are only $(q^2 - 1)/(q - 1)$ distinct such subspaces, say $U_0, U_1, \ldots, U_q$. These $q + 1$ subspaces $U_i$ meet pairwise only in the zero vector and their union is $H_0 \cup H_1 \cup \cdots \cup H_q$. Thus

$$
U_0, U_1, \ldots, U_q, H_{q+1}, H_{q+2}, \ldots, H_{q^2}
$$

is another partition of $V$ into subgroups of order $q^2$.

We shall not prove here that the projective planes resulting from this construction are not Desarguesian (for $q > 2$). See Theorem 10.9 in D. R. Hughes and F. C. Piper (1973). The reader can find counterexamples to Desargues' theorem in Example 23.7 by hand.

We conclude the chapter with proof of an important step in the difficult proof of Theorem 23.6. We need a preliminary observation.

PROPOSITION 23.11. *The subgeometry on any flat of a projective geometry is again a projective geometry.*

PROOF: That the modular law holds in the subgeometry on a flat of a modular projective geometry is immediate. We must show that these subgeometries are connected. To this end, it will suffice to show that if $H$ is a hyperplane in a projective geometry with point set $X$, then the subgeometry on $H$ is connected.

If not, $H = E \cup F$ where $E$ and $F$ are flats. By (23.4), rank$(E)+$ rank$(F) = n-1$ where $n := \text{rank}(X)$. Let $E_1, \ldots, E_s$ and $F_1, \ldots, F_t$ be the flats that cover $E$ and $F$, respectively, that contain points outside of $H$; each family partitions the points outside of $H$. Suppose that both $s, t \geq 2$. By (23.4), there exists $x_i \in E_i \cap F_i$, $i = 1, 2$. The line $\overline{\{x_1, x_2\}}$, again by (23.4), meets $H$ in some point $e \in E$. But then $x_2$, being on the line joining $e$ and $x_1$, must be contained in $E_1$, a contradiction. Thus one of $s$ or $t$ must be 1, say $s = 1$, and then $X$ is the union of the disjoint flats $F$ and $E_1$, contradicting the connectivity of the original projective geometry. $\quad\square$

A corollary of Proposition 23.9 is that every line of a projective geometry has at least three points (since a two-point line is not connected). This is used at a critical point in the proof below. We remark that with the aid of Problem 23C, it is easy to see that all lines in a finite projective geometry have the same size $n + 1$.

THEOREM 23.12. *Desargues' theorem holds in any projective geometry of rank $\geq 4$.*

PROOF: Consider two triangles $\{a_1, b_1, c_1\}$ and $\{a_2, b_2, c_2\}$ that are perspective from a point $x$.

We first assume that the planes $P_1$ and $P_2$ spanned by $\{a_1, b_1, c_1\}$ and $\{a_2, b_2, c_2\}$, respectively, are distinct. Let $T := \overline{\{x, a_1, b_1, c_1\}}$. The lines $\overline{\{x, a_1\}}$, $\overline{\{x, b_1\}}$, and $\overline{\{x, c_1\}}$ contain, respectively, the points $a_2$, $b_2$, and $c_2$, so $P_1$ and $P_2$ are contained in $T$, that evidently has rank 4. So by the modular law, $P_1$ and $P_2$ must meet in a line $L$. We claim that the two original triangles are perspective from the line $L$.

The plane $Q := \overline{\{p, a_1, b_1\}}$ contains $a_2$ and $b_2$ and hence contains both the lines $\overline{\{a_1, b_1\}}$ and $\overline{\{a_2, b_2\}}$ that therefore intersect in a point $q$, say. The point $q$ belongs to both $P_1$ and $P_2$, and hence $q \in L$. Similarly, the lines $\overline{\{b_i, c_i\}}$, $i = 1, 2$, meet in a point on $L$; and the lines $\overline{\{a_i, c_i\}}$, $i = 1, 2$, meet in a point on $L$. The two triangles are thus perspective from $L$.

Now assume that the two original triangles lie in a plane $P$. Let $x_1$ be any point not in $P$ and $x_2$ any other point on the line containing $x_1$ and $p$. The lines $\overline{\{x_1, a_1\}}$ and $\overline{\{x_2, a_2\}}$ are contained in the plane $\overline{\{p, x_1, a_1\}}$ and so meet in a point $a^*$. Similarly, the lines $\overline{\{x_1, b_1\}}$ and $\overline{\{x_2, b_2\}}$ meet in a point $b^*$ and the lines $\overline{\{x_1, c_1\}}$ and $\overline{\{x_2, c_2\}}$ meet in a point $c^*$. The plane $P^* := \overline{\{a^*, b^*, c^*\}}$ and $P$ are contained in a rank 4 flat and so meet in a line $L := P \cap P^*$.

The triangles $\{a_1, b_1, c_1\}$ and $\{a^*, b^*, c^*\}$ are perspective from the point $x_1$ and lie in different planes, and so are perspective from the line $L$. Similarly, the triangles $\{a_2, b_2, c_2\}$ and $\{a^*, b^*, c^*\}$ are perspective from $x_2$, and so are perspective from $L$. The line $\overline{\{a_1, b_1\}}$ meets $L$ in the same point as the line $\overline{\{a^*, b^*\}}$; the line $\overline{\{a_2, b_2\}}$ meets $L$ in the same point as the line $\overline{\{a^*, b^*\}}$; hence the lines $\overline{\{a_1, b_1\}}$ and $\overline{\{a_2, b_2\}}$ meet in a point of $L$. Similarly, the other corresponding "sides" of the two original triangles meet in points of $L$. In summary, the original triangles are perspective from the line $L$.     $\square$

## Notes.

Girard Desargues (1593–1662) was an architect and military engineer who introduced the term *involution* in his work on perspective. He worked in Lyons and Paris. The work of Pappus of Alexandria (fl. 320), the last significant Greek geometer, was considerably earlier. The mathematical discipline of projective geometry did not develop until the 19th century.

Moritz Pasch was originally an algebraist but then became interested in non-euclidean geometry. He came to Giessen at the age of 27 as Privatdozent and stayed there until his death (60 years later). He was Rektor of the university 1893–1894.

An interesting family of combinatorial geometries called transversal geometries arise from bipartite graphs; see Crapo and Rota (1970). The independent sets of these geometries are those subsets of one part $X$ of a bipartition $(X, Y)$ with the property that they can be matched into $Y$.

We have seen that the concept of combinatorial geometries and geometric lattices are "cryptomorphic", i.e. they are essentially different axiom systems for the same structures. Crapo and Rota (1970) give further cryptomorphic versions. For example, "matroids" are more-or-less the same thing as combinatorial geometries, though the emphasis of the theory of matroids is different.

Some characterizations of the symmetric design consisting of the points and hyperplanes of a finite projective space may be found in Section 2.1 of P. Dembowski (1968).

### References.

J. André (1954), Über nicht-Desarguessche Ebenen mit transitiver Translationgruppe, *Math. Zeitschr.* **60**, 156–186.

L. M. Batten (1986), *Combinatorics of Finite Geometries*, Cambridge University Press.

H. Crapo and G.-C. Rota (1970), *Combinatorial Geometries*, MIT Press.

P. Crawley and R. P. Dilworth (1973), *Algebraic Theory of Lattices*, Prentice-Hall.

P. Dembowski (1968), *Finite Geometries*, Springer-Verlag.

C. Greene (1970), A rank inequality for finite geometric lattices, *J. Combinatorial Theory* **9**, 357–364.

M. Hall, Jr. (1972), *The Theory of Groups*, 2nd edn., Chelsea.

D. R. Hughes and F. C. Piper (1973), *Projective Planes*, Springer-Verlag.

D. Pedoe (1963), *An Introduction to Projective Geometry*, Macmillan.

O. Veblen and J. W. Young (1907), *Projective Geometries* (2 vols.), Ginn Co.

# 24

# Gaussian numbers and $q$-analogues

There exist many analogies between the partially ordered set of all subsets of a finite set and the partially ordered set of all subspaces of a finite vector space. This is primarily because they are both examples of "matroid designs" as defined in the previous chapter. Let $V_n(q)$ denote an $n$-dimensional vector space over the field $\mathbb{F}_q$ of $q$ elements. We use the term $k$-subspace as a short form of $k$-dimensional subspace.

We begin with some counting. To obtain a maximal chain (i.e. a chain of size $n+1$ containing one subspace of each possible dimension) in the partially ordered set of all subspaces of $V_n(q)$, we start with the 0-subspace. After we have chosen an $i$-subspace $U_i$, $0 \le i < n$, we can choose an $(i+1)$-subspace $U_{i+1}$ that contains $U_i$ in $(q^n - q^i)/(q^{i+1} - q^i)$ ways since we can take the span of $U_i$ and any of the $(q^n - q^i)$ vectors *not* in $U_i$—but any $(i+1)$-subspace will arise exactly $(q^{i+1} - q^i)$ times in this manner. In summary, the number of maximal chains of subspaces in $V_n(q)$ is

$$M(n,q) = \frac{(q^n - 1)(q^{n-1} - 1)(q^{n-2} - 1)\cdots(q^2 - 1)(q - 1)}{(q-1)^n}.$$

We may consider $M(n,q)$ as a polynomial in $q$ for each integer $n$. When the indeterminate $q$ is replaced by a prime power, we have the number of maximal chains in the poset $PG_n(q)$. When $q$ is replaced by 1, we have $M(n,1) = n!$, which is the number of maximal chains in the poset of subsets of an $n$-set.

The *Gaussian number* $\begin{bmatrix} n \\ k \end{bmatrix}_q$ can be defined as the number of $k$-subspaces of $V_n(q)$. These are called *Gaussian coefficients* by some authors to emphasize the analogy with binomial coefficients. To find an expression for $\begin{bmatrix} n \\ k \end{bmatrix}_q$, we count the number $N$ of pairs $(U, \mathcal{C})$ where $U$ is a $k$-subspace and $\mathcal{C}$ is a maximal chain that contains $U$. Of course, every maximal chain contains exactly one subspace of dimension $k$, so $N = M(n, q)$. On the other hand, we get each such maximal chain uniquely by appending to a maximal chain in the partially ordered set of all subspaces of $U$, of which there are $M(k, q)$, a maximal chain in the partially ordered set of all subspaces of $V_n(q)$ that contain $U$; there are $M(n - k, q)$ of these, since the partially ordered set $\{W : U \subseteq W \subseteq V\}$ is isomorphic to the partially ordered set of subspaces of the factor space $V/U$ that has dimension $n - k$. Thus

$$\begin{bmatrix} n \\ k \end{bmatrix}_q = \frac{M(n, q)}{M(k, q)M(n - k, q)} = \frac{(q^n - 1)(q^{n-1} - 1) \cdots (q^{n-k+1} - 1)}{(q^k - 1)(q^{k-1} - 1) \cdots (q - 1)}.$$

For some purposes, it is better to think of $\begin{bmatrix} n \\ k \end{bmatrix}_q$ as a *polynomial* in an indeterminate $q$ rather than as a function of a prime power $q$. That the rational function above is in fact a polynomial can be seen in several ways. For example, it is an easy exercise to see that a rational function in $x$ which is integral for infinitely many integral values of $x$ must be a polynomial in $x$. Perhaps *Gaussian polynomial* is a better term than *Gaussian number* or *coefficient*. As an example,

$$\begin{bmatrix} 6 \\ 3 \end{bmatrix}_q = q^9 + q^8 + 2q^7 + 3q^6 + 3q^5 + 3q^4 + 3q^3 + 2q^2 + q + 1.$$

When the indeterminate $q$ is replaced by 1 in $\begin{bmatrix} n \\ k \end{bmatrix}_q$, we obtain $\binom{n}{k}$. This explains a small part of a tendency for results concerning finite vector spaces to reduce to the corresponding results for sets when $q$ is replaced by 1. It is also possible to have so-called *q-analogues* of results on sets where we try to replace "$k$-subset" by "$k$-subspace". Sometimes these statements are true and have proofs that are similar to the results on sets.

The following is the $q$-analogue of Sperner's theorem, Theorem 6.3.

THEOREM 24.1. *If $\mathcal{A}$ is an antichain in the partially ordered set of all subspaces of $V_n(q)$, then*

$$|\mathcal{A}| \leq \begin{bmatrix} n \\ \lfloor n/2 \rfloor \end{bmatrix}_q.$$

PROOF: Let $\mathcal{A}$ be an antichain and count the number $N$ of pairs $(U, \mathcal{C})$ where $U \in \mathcal{A}$ and $\mathcal{C}$ is a maximal chain that contains $U$. Every maximal chain contains at most one subspace in $\mathcal{A}$, so $N \leq M(n, q)$. On the other hand, each $k$-subspace in $\mathcal{A}$ lies in exactly $M(k, q)M(n - k, q)$ maximal chains $\mathcal{C}$. Thus

$$M(n, q) \geq N = \sum_{k=0}^{n} c_k M(k, q) M(n - k, q),$$

where $c_k$ is the number of $k$-dimensional subspaces belonging to $\mathcal{A}$. The proof is completed in a manner analogous to the proof of Theorem 6.3 if we believe that $\begin{bmatrix} n \\ k \end{bmatrix}_q \leq \begin{bmatrix} n \\ \lfloor n/2 \rfloor \end{bmatrix}_q$ for all $k$, and this is left to the reader to verify. $\qquad\qquad\square$

The following theorem gives a combinatorial interpretation of the coefficients of the $\begin{bmatrix} n \\ k \end{bmatrix}_q$ as a polynomial in $q$, and thus proves they are all positive integers.

THEOREM 24.2. *Let*

$$\begin{bmatrix} n \\ k \end{bmatrix}_q = \sum_{\ell=0}^{k(n-k)} a_\ell q^\ell.$$

*Then the coefficient $a_\ell$ is the number of partitions of $\ell$ whose Ferrers diagrams fit in a box of size $k$ by $n - k$.*

PROOF: We can work with the vector space $\mathbb{F}_q^n$ of $n$-tuples over $\mathbb{F}_q$. It is well known that every $k$-subspace of $\mathbb{F}_q^n$ arises uniquely as the rowspace of a $k$ by $n$ matrix over $\mathbb{F}_q$ that (i) has rank $k$ and (ii) is a so-called *row-reduced echelon form*. This means that the leading entry in each row is a 1, the entries above a leading 1 are 0's, and the leading 1 in row $i$ is further to the right than the leading 1 in row $i - 1$, $i = 2, 3, \ldots, k$.

Suppose the leading 1 in row $i$ occurs in column $c_i$, $i = 1, 2, \ldots, k$. Then $(n - k + 1 - c_1, n - k + 2 - c_2, \ldots, n - 1 - c_{k-1}, n - c_k)$ is a nonincreasing sequence of nonnegative numbers and so, when terminal zeros are dropped, corresponds to a partition of some number into at most $k$ parts of size at most $n - k$. Conversely, such a partition gives us the positions of leading ones in a *class* of echelon forms.

For example, there are 20 classes of echelon forms in the case $n = 6$, $k = 3$; see Fig. 24.1.

$$
\begin{bmatrix} 1 & 0 & 0 & \bullet & \bullet & \bullet \\ 0 & 1 & 0 & \bullet & \bullet & \bullet \\ 0 & 0 & 1 & \bullet & \bullet & \bullet \end{bmatrix}
\begin{bmatrix} 1 & 0 & \bullet & 0 & \bullet & \bullet \\ 0 & 1 & \bullet & 0 & \bullet & \bullet \\ 0 & 0 & 0 & 1 & \bullet & \bullet \end{bmatrix}
\begin{bmatrix} 1 & 0 & \bullet & \bullet & 0 & \bullet \\ 0 & 1 & \bullet & \bullet & 0 & \bullet \\ 0 & 0 & 0 & 0 & 1 & \bullet \end{bmatrix}
\begin{bmatrix} 1 & 0 & \bullet & \bullet & \bullet & 0 \\ 0 & 1 & \bullet & \bullet & \bullet & 0 \\ 0 & 0 & 0 & 0 & 0 & 1 \end{bmatrix}
\begin{bmatrix} 1 & \bullet & 0 & 0 & \bullet & \bullet \\ 0 & 0 & 1 & 0 & \bullet & \bullet \\ 0 & 0 & 0 & 1 & \bullet & \bullet \end{bmatrix}
$$

$$
\begin{bmatrix} 1 & \bullet & 0 & \bullet & 0 & \bullet \\ 0 & 0 & 1 & \bullet & 0 & \bullet \\ 0 & 0 & 0 & 0 & 1 & \bullet \end{bmatrix}
\begin{bmatrix} 1 & \bullet & 0 & \bullet & \bullet & 0 \\ 0 & 0 & 1 & \bullet & \bullet & 0 \\ 0 & 0 & 0 & 0 & 0 & 1 \end{bmatrix}
\begin{bmatrix} 1 & \bullet & \bullet & 0 & 0 & \bullet \\ 0 & 0 & 0 & 1 & 0 & \bullet \\ 0 & 0 & 0 & 0 & 1 & \bullet \end{bmatrix}
\begin{bmatrix} 1 & \bullet & \bullet & 0 & \bullet & 0 \\ 0 & 0 & 0 & 1 & \bullet & 0 \\ 0 & 0 & 0 & 0 & 0 & 1 \end{bmatrix}
\begin{bmatrix} 1 & \bullet & \bullet & \bullet & 0 & 0 \\ 0 & 0 & 0 & 0 & 1 & 0 \\ 0 & 0 & 0 & 0 & 0 & 1 \end{bmatrix}
$$

$$
\begin{bmatrix} 0 & 1 & 0 & 0 & \bullet & \bullet \\ 0 & 0 & 1 & 0 & \bullet & \bullet \\ 0 & 0 & 0 & 1 & \bullet & \bullet \end{bmatrix}
\begin{bmatrix} 0 & 1 & 0 & \bullet & 0 & \bullet \\ 0 & 0 & 1 & \bullet & 0 & \bullet \\ 0 & 0 & 0 & 0 & 1 & \bullet \end{bmatrix}
\begin{bmatrix} 0 & 1 & 0 & \bullet & \bullet & 0 \\ 0 & 0 & 1 & \bullet & \bullet & 0 \\ 0 & 0 & 0 & 0 & 0 & 1 \end{bmatrix}
\begin{bmatrix} 0 & 1 & \bullet & 0 & 0 & \bullet \\ 0 & 0 & 0 & 1 & 0 & \bullet \\ 0 & 0 & 0 & 0 & 1 & \bullet \end{bmatrix}
\begin{bmatrix} 0 & 1 & \bullet & 0 & \bullet & 0 \\ 0 & 0 & 0 & 1 & \bullet & 0 \\ 0 & 0 & 0 & 0 & 0 & 1 \end{bmatrix}
$$

$$
\begin{bmatrix} 0 & 1 & \bullet & \bullet & 0 & 0 \\ 0 & 0 & 0 & 0 & 1 & 0 \\ 0 & 0 & 0 & 0 & 0 & 1 \end{bmatrix}
\begin{bmatrix} 0 & 0 & 1 & 0 & 0 & \bullet \\ 0 & 0 & 0 & 1 & 0 & \bullet \\ 0 & 0 & 0 & 0 & 1 & \bullet \end{bmatrix}
\begin{bmatrix} 0 & 0 & 1 & 0 & \bullet & 0 \\ 0 & 0 & 0 & 1 & \bullet & 0 \\ 0 & 0 & 0 & 0 & 0 & 1 \end{bmatrix}
\begin{bmatrix} 0 & 0 & 1 & \bullet & 0 & 0 \\ 0 & 0 & 0 & 0 & 1 & 0 \\ 0 & 0 & 0 & 0 & 0 & 1 \end{bmatrix}
\begin{bmatrix} 0 & 0 & 0 & 1 & 0 & 0 \\ 0 & 0 & 0 & 0 & 1 & 0 \\ 0 & 0 & 0 & 0 & 0 & 1 \end{bmatrix}
$$

Figure 24.1

In each matrix of Fig. 24.1, the positions of the dots, when the intervening columns are deleted and we reflect through a vertical axis, describe the Ferrers diagram of a partition (of some number $\leq 9$) that fits in a 3 by 3 box. The class represented by the last matrix in the first row, for example, contains $q^7$ echelon forms.

In general, the class of echelon forms where the leading 1 in row $i$ occurs in column $c_i$, $i = 1, 2, \ldots, k$, contains $q^\ell$ matrices for some $\ell$ since the positions not containing 1's or required to be 0's may be filled arbitrarily with elements of $\mathbb{F}_q$. To be precise,

$$\ell = (n - k + 1 - c_1) + \cdots + (n - 1 - c_{k-1}) + (n - c_k)$$

because for each $i = 1, 2, \ldots, k$, there are $n - (k - i) - c_i$ positions in the $i$-th row that may be arbitrarily filled. That is, the class consists of $q^\ell$ matrices where the partition into at most $k$ parts of size at most $n-k$ that corresponds to the class is in fact a partition of the number $\ell$.

Thus when $a_\ell$ is *defined* as the number of partitions of $\ell$ whose Ferrers diagrams fit in a box of size $k$ by $n-k$, we have a polynomial $\sum_{\ell=0}^{k(n-k)} a_\ell q^\ell$ that agrees with the polynomial $\begin{bmatrix} n \\ k \end{bmatrix}_q$ when evaluated at any prime power $q$; hence the two polynomials are equal.    □

We remark that when we set $q=1$ in the statement of Theorem 24.2, we obtain as a corollary the result that the total number of partitions whose Ferrers diagram fits in a box of size $k$ by $n-k$ is $\binom{n}{k}$.

Next, we derive a recursion (24.1) for the Gaussian numbers. This provides another way to see that they are polynomials in $q$ (by induction on $n$, say). Pick a hyperplane $H$, i.e. an $(n-1)$-subspace, of $V_n(q)$. Some $k$-subspaces are contained in $H$ (their number is $\begin{bmatrix} n-1 \\ k \end{bmatrix}_q$) and the rest meet $H$ in a $(k-1)$-subspace. Each of the $\begin{bmatrix} n-1 \\ k-1 \end{bmatrix}_q$ $(k-1)$-subspaces in $H$ is contained in

$$\begin{bmatrix} n-k+1 \\ 1 \end{bmatrix}_q = \frac{q^{n-k+1}-1}{q-1}$$

$k$-subspaces of $V$, of which

$$\begin{bmatrix} n-k \\ 1 \end{bmatrix}_q = \frac{q^{n-k}-1}{q-1}$$

are contained in $H$; this leaves $q^{n-k}$ that are not contained in $H$. Thus

(24.1)
$$\begin{bmatrix} n \\ k \end{bmatrix}_q = \begin{bmatrix} n-1 \\ k \end{bmatrix}_q + q^{n-k} \begin{bmatrix} n-1 \\ k-1 \end{bmatrix}_q .$$

PROBLEM 24A. Determine the exponents $e_i$ (they may be functions of $m$, $n$, and $k$ as well as $i$) so that the following identity is valid:

$$\begin{bmatrix} n+m \\ k \end{bmatrix}_q = \sum_{i=0}^{k} q^{e_i} \begin{bmatrix} n \\ i \end{bmatrix}_q \begin{bmatrix} m \\ k-i \end{bmatrix}_q .$$

(One way to solve this involves echelon forms; another uses (24.1).)

The equation of Problem 24A is the $q$-analoque of equation (10.5) for binomial coefficients. The $q$-analogue of inclusion-exclusion appears in the next chapter; see Theorem 25.2 for an application.

What is the $q$-analogue of (simple) $t$-designs? This would be a family $\mathcal{B}$ of $k$-subspaces of $V_v(q)$ such that each $t$-subspace is contained in exactly $\lambda$ members of $\mathcal{B}$. No nontrivial examples of such "$q$-$S_\lambda(t, k, v)$'s" were known with $t \geq 2$ until 1986 when S. Thomas described a family with

$$q = 2, \quad \lambda = 7, \quad t = 2, \quad k = 3, \quad v \equiv \pm 1 \pmod 6.$$

The case $t = \lambda = 1$ is already nontrivial for vector spaces. For sets, an $S(1, k, v)$ is a partition of a $v$-set into $k$-subsets, and this exists if and only if $k$ divides $v$.

THEOREM 24.3.  *There exists a family of $k$-subspaces of $V_v(q)$ such that every 1-subspace is contained in exactly one member of the family (a so-called* spread *of $k$-subspaces) if and only if $k$ divides $v$.*

PROOF: In other words, we are interested in families of $k$-subspaces such that any two of them intersect in the 0-subspace but such that their union is the entire $v$-dimensional vector space.

The number of $k$-subspaces that we require is the total number of nonzero vectors divided by the number of nonzero vectors in a $k$-subspace, and this is $(q^v - 1)/(q^k - 1)$. This is an integer if and only if $k$ divides $v$.

Suppose $v = km$ where $m$ is an integer. As a $v$-dimensional vector space over $\mathbb{F}_q$, we take an $m$-dimensional vector space $V$ over $\mathbb{F}_{q^k}$. Let $\mathcal{B}$ be the family of 1-subspaces of $V$ as a vector space over $\mathbb{F}_{q^k}$. (So there are $q^{k(m-1)} + q^{k(m-2)} + \cdots + q^k + 1$ of these and each contains $q^k$ vectors.) Now think of $V$ as a vector space over $\mathbb{F}_q$, a subfield of $\mathbb{F}_{q^k}$; the members of $\mathcal{B}$ are $k$-dimensional subspaces over $\mathbb{F}_q$ and provide the required partition. $\qquad\square$

PROBLEM 24B.  Let $\mathcal{B}$ be a spread of $k$-subspaces in $V_v(q)$. Let $\mathcal{A}$ consist of all cosets of all members of $\mathcal{B}$. Show that $\mathcal{A}$ is the block set of an $S(2, q^k, q^v)$ on the set of points $V_v(q)$. (The case $v = 2k$

is particularly interesting because we obtain affine planes of order $q^k$. Planes obtained from spreads are called *translation planes*.)

**Notes.**

It has been suggested that one should use the notation $n^{\underline{q}}$ for $M(n,q)$ (which is to be pronounced "*n q-torial*"). Then we have

$$\begin{bmatrix} n \\ k \end{bmatrix}_q = \frac{n^{\underline{q}}}{k^{\underline{q}}\,(n-k)^{\underline{q}}}.$$

Also note

$$\lim_{q\to 1} n^{\underline{q}} = n^{\underline{1}} \quad \text{(that is, } n!\text{)}.$$

We do not know who first proposed this.

W. N. Hsieh (1975) proved a $q$-analogue of the Erdős-Ko-Rado theorem. See P. Frankl and R. M. Wilson (1986) for a stronger version.

The $q$-analogue of Ramsey's theorem was a long outstanding problem until it was solved by R. L. Graham, K. Leeb, and B. L. Rothschild in 1972.

Carl Friedrich Gauss (1777–1855) was possibly the greatest mathematician (scientist) of all time. He made extremely important contributions to number theory. He was the first to study properties of the expressions that are now called Gaussian polynomials or Gaussian numbers.

**References.**

P. Frankl and R. M. Wilson (1986), The Erdős-Ko-Rado theorem for vector spaces, *J. Combinatorial Theory* (A) **43**, 228–236.

R. L. Graham, K. Leeb, B. L. Rothschild (1972), Ramsey's theorem for a class of categories, *Adv. Math.* **8**, 417–433.

R. L. Graham, B. L. Rothschild, and J. Spencer (1980), *Ramsey Theory*, Wiley.

W. N. Hsieh (1975), Intersection theorems for systems of finite vector spaces, *Discrete Math.* **12**, 1–16.

S. Thomas (1986), Designs over finite fields, *Geometriae Dedicata* **24**, 237–242.

# 25

# Lattices and Möbius inversion

One of the techniques belonging to the foundations of combinatorics is the principle of Möbius inversion over partially ordered sets. This can be thought of as a generalization of inclusion-exclusion as well as an extension of the inversion with the classical Möbius function of number theory, which was discussed in Chapter 10.

Let $P$ be a finite partially ordered set. We will consider matrices $\alpha$ whose rows and columns are indexed by the elements of $P$, that is mappings from $P \times P$ to the rationals or complex numbers. The *incidence algebra* $\mathfrak{A}(P)$ consists of all matrices $\alpha$ such that $\alpha(x, y) = 0$ unless $x \leq y$ in $P$. By the definition of matrix multiplication,

$$(\alpha\beta)(x, y) = \sum_{z \in P} \alpha(x, z)\beta(z, y).$$

If $\alpha, \beta \in \mathfrak{A}(P)$, then the above sum need be extended over only those $z$ in the interval $[x, y] := \{x \leq z \leq y\}$. It is easily seen that $\mathfrak{A}(P)$ is closed under multiplication, as well as addition and scalar multiplication.

An element of $\mathfrak{A}(P)$ that will play an important role in the following theory is $\zeta$ (the *zeta function* of $P$) which is defined by

$$\zeta(x, y) = \begin{cases} 1 & \text{if } x \leq y \text{ in } P, \\ 0 & \text{otherwise.} \end{cases}$$

We claim that $\zeta$ is invertible and that its inverse, which is called the *Möbius function* of $P$ and is denoted by $\mu$, is integral and lies in $\mathfrak{A}(P)$.

This is simple to verify. The equation $\mu\zeta = I$ (the identity) requires that

$$(25.1) \qquad \sum_{x \leq z \leq y} \mu(x, z) = \begin{cases} 1 & \text{if } x = y, \\ 0 & \text{otherwise,} \end{cases}$$

and this can be ensured by simply *defining* $\mu$ inductively, by declaring $\mu(x, x) := 1$, $\mu(x, y) := 0$ if $x \not\leq y$, and

$$(25.2) \qquad \mu(x, y) := - \sum_{x \leq z < y} \mu(x, z) \quad \text{for } x < y \text{ in } P.$$

For example, when $P = \{1, 2, 3, 4, 6, 12\}$, the lattice of all (positive) divisors of the integer 12,

$$\zeta = \begin{pmatrix} 1 & 1 & 1 & 1 & 1 & 1 \\ 0 & 1 & 0 & 1 & 1 & 1 \\ 0 & 0 & 1 & 0 & 1 & 1 \\ 0 & 0 & 0 & 1 & 0 & 1 \\ 0 & 0 & 0 & 0 & 1 & 1 \\ 0 & 0 & 0 & 0 & 0 & 1 \end{pmatrix}, \quad \mu = \begin{pmatrix} 1 & -1 & -1 & 0 & 1 & 0 \\ 0 & 1 & 0 & -1 & -1 & 1 \\ 0 & 0 & 1 & 0 & -1 & 0 \\ 0 & 0 & 0 & 1 & 0 & -1 \\ 0 & 0 & 0 & 0 & 1 & -1 \\ 0 & 0 & 0 & 0 & 0 & 1 \end{pmatrix}.$$

The top row of $\mu$ was computed as follows:

$$\begin{aligned} \mu(1, 1) &= +1, \\ \mu(1, 2) &= -\mu(1, 1) = -1, \\ \mu(1, 3) &= -\mu(1, 1) = -1, \\ \mu(1, 4) &= -\mu(1, 1) - \mu(1, 2) = 0, \\ \mu(1, 6) &= -\mu(1, 1) - \mu(1, 2) - \mu(1, 3) = +1, \\ \mu(1, 12) &= -\mu(1, 1) - \mu(1, 2) - \mu(1, 3) - \mu(1, 4) - \mu(1, 6) = 0. \end{aligned}$$

We remark that a somewhat different equation concerning $\mu$ results if we consider instead the relation $\zeta\mu = I$:

$$(25.3) \qquad \sum_{x \leq z \leq y} \mu(z, y) = \begin{cases} 1 & \text{if } x = y, \\ 0 & \text{otherwise.} \end{cases}$$

Another much more complicated way of looking at the matter of the invertibility of $\zeta$ is as follows. We first remark that a finite partial order can always be "dominated" by a total order; that is, there exists an indexing $P = \{x_1, x_2, \dots, x_n\}$ such that $x_i \le x_j$ in $P$ implies that $i \le j$. (Proof: Let $x$ be any maximal element of $P$, index $P\backslash\{x\}$ by induction, and put $x$ at the end.) With respect to such an indexing of $P$, the matrices in $\mathfrak{A}(P)$ are upper triangular. The zeta function has 1's on the diagonal and so has determinant 1 and hence an integral inverse by Cramer's formula. Then $\zeta^{-1} = \mu$ lies in $\mathfrak{A}(P)$ because the inverse of any matrix (or element of a finite–dimensional commutative algebra) is a polynomial in that matrix (or element).

The equation (25.2) implies that $\mu(x, y) = -1$ if $x \lessdot y$. Also note that $\mu(x, y) = k - 1$ if the interval $[x, y]$ is a $k$-point line, i.e. the interval consists of $x$, $y$, and $k$ pairwise noncomparable elements $z_1, \dots, z_k$, each with $x \lessdot z_i \lessdot y$.

In the following theorem, we list the values of the Möbius function of some common partially ordered sets. The proof of Theorem 25.1 will not be given until the end of the chapter because we wish to talk about the implications of Möbius inversion first. Clever and/or industrious readers will be able to prove some parts of Theorem 25.1 with the recursion (25.2) and simple induction arguments, but we prefer to wait until we have Weisner's theorem, Theorem 25.3, available. We do not prove part (v) here—see the notes.

THEOREM 25.1. (i) *For the lattice of all subsets of an $n$-set $X$,*

$$\mu(A, B) = \begin{cases} (-1)^{|B|-|A|} & \text{if } A \subseteq B, \\ 0 & \text{otherwise.} \end{cases}$$

(ii) *For the lattice of all (positive) divisors of an integer $n$,*

$$\mu(a, b) = \begin{cases} (-1)^r & \text{if } \frac{b}{a} \text{ is the product of } r \text{ distinct primes,} \\ 0 & \text{otherwise, i.e. if } a \nmid b \text{ or } \frac{b}{a} \text{ is not squarefree.} \end{cases}$$

(*That is, $\mu(a, b) = \mu(\frac{b}{a})$ where the latter $\mu$ is the classical function of a single variable in (10.8).*)

(iii) *For the lattice of all subspaces of a finite-dimensional vector space $V$ over the field $\mathbb{F}_q$ of $q$ elements,*

$$\mu(U, W) = \begin{cases} (-1)^k q^{\binom{k}{2}} & \text{if } U \subseteq W \text{ and } \dim(U) - \dim(W) = k, \\ 0 & \text{if } U \not\subseteq W. \end{cases}$$

(iv) *For the lattice $\Pi_n$ of all partitions of an $n$-set $X$,*

$$\mu(\mathcal{A}, \mathcal{B}) = (-1)^{|\mathcal{A}|-|\mathcal{B}|} \prod_{B \in \mathcal{B}} (n_B - 1)!$$

*for elements $\mathcal{A}$ and $\mathcal{B}$ of $\Pi_n$ with $\mathcal{A} \preceq \mathcal{B}$, where $n_B$ denotes the number of blocks of $\mathcal{A}$ that are contained in a block $B$ of $\mathcal{B}$.*

(v) *For the lattice of faces of a convex polytope,*

$$\mu(A, B) = \begin{cases} (-1)^{\dim(B)-\dim(A)} & \text{if } A \subseteq B, \\ 0 & \text{otherwise.} \end{cases}$$

We now state the principle of *Möbius inversion*. Let P be a finite partially ordered set and $\mu$ its Möbius function. Let $f, g, h : P \to \mathbb{R}$ (or into any additive group) be functions such that the relations

$$g(x) = \sum_{a:a \leq x} f(a) \quad \text{and} \quad h(x) = \sum_{b:b \geq x} f(b)$$

hold for all $x \in P$. Then, for all $x \in P$,

(25.4) $$f(x) = \sum_{a:a \leq x} \mu(a, x) g(a),$$

and

(25.5) $$f(x) = \sum_{b:b \geq x} \mu(x, b) h(b).$$

These are readily verified by direct substitution. For example, the righthand side of (25.4) is

$$\sum_{a:a \leq x} \mu(a, x) \left( \sum_{b:b \leq a} f(b) \right) = \sum_{b:b \leq x} f(b) \left( \sum_{a:b \leq a \leq x} \mu(a, x) \right).$$

But the inner sum on the right is zero by (25.3) unless $b = x$, so only the term $f(x)$ survives. Or, in matrix notation (when we regard $f$, $g$, and $h$ as row or column vectors—whichever is appropriate—whose coordinates are indexed by $P$), the relations assumed for $g$ and $h$ are equivalent to $g = \zeta f$, so $f = \mu g$, and $h = f\zeta$, so $f = h\mu$.

When the principle of Möbius inversion is applied to the lattice of subsets of a set $I$, we recover inclusion-exclusion. Let $(A_i : i \in I)$ be a finite family of subsets of a finite set $X$. For $J \subseteq I$, let $f(J)$ equal the number of elements of $X$ that belong to *exactly* the sets $A_j$ with $j \in J$ and to no others. (Think of the size of one of the regions in a Venn diagram of the sets.) Let $g(J)$ equal the number of elements in $\cap_{j \in J} A_j$. Then $g(J) = \sum_{K:J \subseteq K} f(K)$, so Möbius inversion gives

$$f(J) = \sum_{K:J \subseteq K} (-1)^{|K \setminus J|} g(K).$$

In the case $J = \emptyset$, the above can be written

$$\left| X - \bigcup_{i \in I} A_i \right| = \sum_{K \subseteq I} (-1)^{|K|} \left| \bigcap_{j \in K} A_j \right|,$$

which is a cryptic way of stating inclusion-exclusion.

When the principle of Möbius inversion is applied to the lattice of (positive) divisors of an integer $n$, we recover the classical number-theoretic Möbius inversion. If $f$ and $g$ satisfy

$$g(m) = \sum_{k|m} f(k) \quad \text{for all } m \text{ dividing } n,$$

then Möbius inversion gives

$$f(m) = \sum_{k|m} \mu(k, m) g(k) \quad \text{for all } m \text{ dividing } n,$$

and this, in view of Theorem 25.1(ii), is equivalent to Theorem 10.4.

We give a $q$-analogue of Example 10.2, where we used inclusion-exclusion to find the expression $\sum_{k=0}^{m} (-1)^{m-k} \binom{m}{k} k^n$ for the number of surjections from an $n$-set to an $m$-set.

THEOREM 25.2. *The number of surjective linear transformations from an n-dimensional vector space to an m-dimensional vector space $V$ over $\mathbb{F}_q$ is*

$$\sum_{k=0}^{m}(-1)^{m-k}\begin{bmatrix}m\\k\end{bmatrix}_q q^{nk+\binom{m-k}{2}}.$$

PROOF: For a subspace $U \subseteq V$, let $f(U)$ denote the number of linear transformations whose image is $U$. Let $g(U)$ denote the number of linear transformations whose image is contained in $U$. Clearly,

$$g(U) = \sum_{W:W\subseteq U} f(W)$$

and $g(U)$ is $q^{nr}$ if $\dim(U) = r$. By Möbius inversion on the lattice of subspaces of $V$,

$$f(U) = \sum_{W:W\subseteq U} \mu(W,U)q^{n\dim(W)}.$$

Take $U = V$ and use Theorem 25.1(iii) to produce the stated result.
□

COROLLARY. *The number of $n$ by $m$ matrices over the field $\mathbb{F}_q$ that have rank $r$ is*

$$\begin{bmatrix}m\\r\end{bmatrix}_q \sum_{k=0}^{r}(-1)^{r-k}\begin{bmatrix}r\\k\end{bmatrix}_q q^{nk+\binom{r-k}{2}}.$$

We note that the number of injective linear transformations has a relatively simple form. If we fix a basis for an $n$-dimensional vector space and consider injections into an $m$-dimensional vector space, the image of the $i$-th basis vector must be chosen as one of the $(q^m - q^{i-1})$ vectors not in the span of the images of the previous basis vectors. In summary, there are $(q^m - 1)(q^m - q) \cdots (q^m - q^{n-1})$ injective linear transformations. Since Theorem 25.2 with $m = n$ also gives an expression for this number, we have proved an identity.

PROBLEM 25A. Use Möbius inversion to derive an expression for the number of $k$-subspaces that meet trivially a given $r$-subspace of an $n$-dimensional vector space over $\mathbb{F}_q$. For the special case when $r + k = n$, show that there are exactly $q^{rk}$ such subspaces by consideration of $r$ by $n$ matrices over $\mathbb{F}_q$ of the form $(I \ M)$, where $I$ is an identity of order $r$.

$$* * *$$

The partially ordered sets mentioned in the statement of Theorem 25.1 are all lattices. The following useful theorem was found by L. Weisner (1935).

THEOREM 25.3. *Let $\mu$ be the Möbius function of a finite lattice $L$ and let $a \in L$ with $a > 0_L$. Then*

$$\sum_{x:x \vee a = 1_L} \mu(0_L, x) = 0.$$

PROOF: Fix $a$ and consider

$$S := \sum_{x,y \in L} \mu(0, x)\zeta(x, y)\zeta(a, y)\mu(y, 1) = \sum_{x \in L} \sum_{\substack{y \geq x, \\ y \geq a}} \mu(0, x)\mu(y, 1).$$

Now on the one hand,

$$S = \sum_x \mu(0, x) \sum_{\substack{y \geq x \\ y \geq a}} \mu(y, 1);$$

but $y \geq a$ and $y \geq x$ if and only if $y \geq x \vee a$, and the inner sum is

$$\sum_{y \geq x \vee a} \mu(y, 1) = \begin{cases} 1 & \text{if } x \vee a = 1, \\ 0 & \text{if } x \vee a < 1. \end{cases}$$

Thus $S$ is the sum in the statement of the theorem. On the other hand,

$$S = \sum_{y \geq a} \mu(y, 1) \sum_{0 \leq x \leq y} \mu(0, x),$$

and the inner sum is always 0 since $y > 0$. □

COROLLARY. *For elements $x, y$ of a geometric lattice $L$ with $x \leq y$, $\mu(x, y)$ has sign $(-1)^{\text{rank}(y) - \text{rank}(x)}$ and, in particular, is never 0.*

PROOF: We show that $\mu(0_L, 1_L)$ has sign $(-1)^{\text{rank}(L)}$ by induction on the rank of $L$. Pick a point $p \in L$. By semimodularity, $a \vee p = 1_L$ only if $a$ is a copoint not on $a$; so Weisner's theorem, Theorem 25.3, gives

$$(25.6) \qquad \mu(0_L, 1_L) = - \sum_{h : h \lessdot 1_L, h \not\geq p} \mu(0_L, h).$$

Since all terms on the righthand side have sign $(-1)^{\text{rank}(L) - 1}$ by the induction hypothesis, the proof is complete. $\qquad\qquad\square$

The number of proper colorings $\chi_G(x)$ of a graph $G$ with $x$ colors can be found by Möbius inversion on the lattice $L(G)$ introduced in Chapter 23, although this is not necessarily a very practical method. Recall that the elements of $L(G)$ are the partitions $\mathcal{A}$ of the vertex set of $G$, all of whose blocks induce connected subgraphs of $G$. For $\mathcal{A}$ in $L(G)$, let $g(\mathcal{A})$ denote the number of mappings from the vertex set of $G$ into a set of $x$ colors (i.e. colorings) so that all vertices in each block of $\mathcal{A}$ receive the same color. Clearly, $g(\mathcal{A}) = x^{|\mathcal{A}|}$. Let $f(\mathcal{A})$ denote the number of mappings that are constant on each block of $\mathcal{A}$, but such that the endpoints of edges joining distinct blocks of $\mathcal{A}$ receive different colors. Given a coloring "counted by" $g(\mathcal{A})$, a moment's thought shows that there is a unique coarser partition $\mathcal{B}$ so that the coloring is "counted by" $f(\mathcal{B})$ (one must coalesce two blocks of $\mathcal{A}$ of the same color if an edge joins them). Thus, $g(\mathcal{A}) = \sum_{\mathcal{B} \succeq \mathcal{A}} f(\mathcal{B})$, so Möbius inversion gives

$$f(\mathcal{A}) = \sum_{\mathcal{B} \succeq \mathcal{A}} \mu(\mathcal{A}, \mathcal{B}) g(\mathcal{B}).$$

The number of proper colorings is $f$ evaluated on $0_{L(G)}$ (the partition into singletons), and this is

$$\chi_G(x) = \sum_{\mathcal{B}} \mu(0_{L(G)}, \mathcal{B}) x^{|\mathcal{B}|} = \sum_{k=1}^{n} \left( \sum_{|\mathcal{B}| = k} \mu(0_{L(G)}, \mathcal{B}) \right) x^k.$$

The polynomial $\chi_G(x)$ is called the *chromatic polynomial* of the graph. The corollary above leads directly to the following statement.

THEOREM 25.4. *The number of proper colorings of a graph $G$ on $n$ vertices in $x$ colors is given by a monic polynomial $\chi_G(x)$ of degree $n$ whose coefficients alternate in sign.*

PROBLEM 25B. Let $G$ be a simple graph with $n$ vertices and $m$ edges. Show that the coefficient of $x^{n-1}$ in $\chi_G(x)$ is $-m$, and that the coefficient of $x^{n-2}$ is $m(m-1)/2$ minus the number of triangles in $G$.

T. Dowling and R. M. Wilson (1975) proved the following theorem and its corollary, the latter being a generalization of the inequality of Theorem 19.1 on linear spaces.

THEOREM 25.5. *If $L$ is a finite lattice so that $\mu(x, 1_L) \neq 0$ for all $x \in L$, then there exists a permutation $\pi$ of the elements of $L$ so that $x \vee \pi(x) = 1_L$ for all $x \in L$.*

PROOF: Before we begin the proof, we remark that an example of a lattice that does not satisfy our hypothesis is a chain with more than two elements. And, of course, there is no permutation $\pi$ with the property stated above. On the other hand, the lattice of all subsets of an $n$-set admits a unique permutation with the property above, namely the permutation that takes each subset to its complement.

All matrices here will have rows and columns indexed by $L$. Let

$$\eta(x, y) := \begin{cases} 1 & \text{if } x \vee y = 1_L, \\ 0 & \text{otherwise.} \end{cases}$$

Let $\delta_1$ be the diagonal matrix with $\delta_1(x, x) := \mu(x, 1_L)$. Then $\zeta \delta_1 \zeta^\top = \eta$ because

$$\zeta \delta_1 \zeta^\top (x, y) = \sum_{a,b} \zeta(x, a) \delta_1(a, b) \zeta(y, b)$$

$$= \sum_{a: a \geq x \text{ and } a \geq y} \delta_1(a, a)$$

$$= \sum_{a: a \geq x \vee y} \mu(a, 1_L)$$

and the last sum is, by (25.1), 1 if $x \vee y = 1_L$, and 0 otherwise.

Under our hypothesis that $\mu(x, 1_L) \neq 0$, the matrix $\delta_1$ is nonsingular. Since $\zeta$ is nonsingular too, we conclude that $\eta$ is also. Hence some term in its determinant expansion does not vanish, and this implies the conclusion of the theorem. $\square$

COROLLARY. *In a finite geometric lattice of rank $n$, the number of elements of rank $\geq n - k$ is at least the number of elements of rank $\leq k$, $0 \leq k \leq n$.*

PROOF: Consider a permutation $\pi$ as in Theorem 25.5 (which applies because of the Corollary to Theorem 25.3). The semimodular law

$$\text{rank}(x) + \text{rank}(\pi(x)) \geq \text{rank}(x \vee \pi(x)) + \text{rank}(x \wedge \pi(x)) \geq n$$

implies that the image of an element of rank $\leq k$ is one of rank $\geq n - k$. $\square$

We give a similar matrix proof of a theorem of T. Dowling (1977) on *complementing permutations* below.

THEOREM 25.6. *If $L$ is a finite lattice such that $\mu(x, 1_L) \neq 0$ and $\mu(0_L, x) \neq 0$ for all $x \in L$, then there exists a permutation $\pi$ of the elements of $L$ so that*

$$x \vee \pi(x) = 1_L \quad \text{and} \quad x \wedge \pi(x) = 0_L$$

*for all $x \in L$.*

PROOF: Let $\delta_1$ be as in the proof of the previous theorem and let $\delta_0$ be the diagonal matrix with $\delta_0(x, x) := \mu(0_L, x)$. Now consider $\kappa := \zeta \delta_1 \zeta^\top \delta_0 \zeta$. Our hypotheses imply that $\kappa$ is nonsingular. We claim that $\kappa(x, y) = 0$ unless $x$ and $y$ are complements. Then any permutation corresponding to a nonvanishing term in the determinant expansion of $\kappa$ will be a complementing permutation.

To establish our claim, first note that $\kappa = \eta \delta_0 \zeta$, so

$$\kappa(x, y) = \sum_{z : z \vee x = 1_L, z \leq y} \mu(0_L, z).$$

If this sum is not zero, then there exists some $z$ with $z \vee x = 1_L$ and $z \leq y$, which implies that $y \vee x = 1_L$. By duality, $\eta' = \zeta^\top \delta_0 \zeta$, where

$$\eta'(x,y) := \begin{cases} 1 & \text{if } x \wedge y = 0_L, \\ 0 & \text{otherwise.} \end{cases}$$

Now note that $\kappa = \zeta \delta_1 \eta'$ and, similarly, $\kappa(x,y) \neq 0$ will imply that $x \wedge y = 0_L$.     $\square$

Finally, we give the proof we promised at the beginning of the chapter.

PROOF OF THEOREM 25.1: (i) Since the interval $[A,B]$ is isomorphic to the lattice of subsets of $B \backslash A$, it suffices to show $\mu(\emptyset, C) = (-1)^{|C|}$. We use the equation (25.6) and proceed by induction on $|C|$. Let $p$ be a point of $C$. There is only one copoint not on $p$, namely the complement of $\{p\}$. So (25.6) implies that

$$\mu(\emptyset, C) = -\mu(\emptyset, C \backslash \{p\}) = -(-1)^{(|C|-1)} = (-1)^{|C|}.$$

(ii) Again, it will suffice to calculate $\mu(1,m)$, and we use induction on $m$. Let $p$ be a prime divisor of $m$. Weisner's theorem asserts

$$\mu(1,m) = - \sum_{\text{lcm}(a,p)=m, a<m} \mu(0,a).$$

If $p^2$ divides $m$, then the sum on the right is empty, and thus $\mu(1,m) = 0$. If $p^2$ does not divide $m$, then there is one term, namely when $a = m/p$.

(iii) It will suffice to show that $\mu(0,V) = (-1)^k q^{\binom{k}{2}}$ for $V$ of dimension $k$. We proceed by induction on $k$. Let $P$ be a 1-dimensional subspace of $V$. By Weisner's theorem,

$$\mu(0,V) = - \sum_{U:U\vee P=V, U\neq V} \mu(0,U).$$

By the induction hypothesis, each term $\mu(0,U)$ in the above sum is equal to $(-1)^{k-1} q^{\binom{k-1}{2}}$. The only subspaces $U$ other than $V$ such that $U \vee P = V$ are those $U$ of dimension $k-1$ that do not contain $P$; their number is

$$\begin{bmatrix} k \\ 1 \end{bmatrix}_q - \begin{bmatrix} k-1 \\ 1 \end{bmatrix}_q = q^{k-1}.$$

This completes the proof of part (iii).

(iv) Consider two partitions $\mathcal{A}$ and $\mathcal{B}$ with $\mathcal{A} \preceq \mathcal{B}$. Say $\mathcal{B}$ has $k$ blocks that, when numbered, are the unions of $n_1, n_2, \dots, n_k$ blocks of $\mathcal{A}$, respectively. Then the interval $[\mathcal{A}, \mathcal{B}]$ is isomorphic to the direct product of partition lattices

$$\Pi_{n_1} \times \Pi_{n_2} \times \cdots \times \Pi_{n_k},$$

since a partition $\mathcal{C}$ with $\mathcal{A} \preceq \mathcal{C} \preceq \mathcal{B}$ is specified by a partition of the $n_i$ blocks of $\mathcal{A}$ that lie in the $i$-th block of $\mathcal{B}$, for each $i = 1, 2, \dots, k$.

We now mention the fact that the Möbius function $\mu_{P \times Q}$ of the direct product $P \times Q$ of two partially ordered sets is the (Kronecker or tensor) product of the Möbius functions $\mu_P$ and $\mu_Q$ of $P$ and $Q$; that is,

$$\mu_{P \times Q}((a_1, b_1), (a_2, b_2)) = \mu_P(a_1, a_2)\mu_Q(b_1, b_2).$$

(We leave the reader to think about this; just check that defining $\mu_{P \times Q}$ as above produces an inverse of $\zeta_{P \times Q}$.) Thus $\mu(\mathcal{A}, \mathcal{B})$ is the product of $\mu(0_{\Pi_{n_i}}, 1_{\Pi_{n_i}})$ for $i = 1, 2, \dots, k$.

We now show that $\mu(0_{\Pi_n}, 1_{\Pi_n}) = (-1)^{n-1}(n-1)!$ by induction on $n$. Let $P$ be a point of $\Pi_n$, i.e. a partition of an $n$-set into a 2-set $\{x, y\}$ and $n-2$ singletons. The partitions $\mathcal{A}$ so that $P \vee \mathcal{A} = 1_{\Pi_n}$ are the $2^{n-2}$ partitions with two blocks that separate $x$ and $y$; there are $\binom{n-2}{i}$ such partitions where the block containing $x$ has size $i+1$ and the block containing $y$ has size $n-1-i$. From (25.6) and the induction hypothesis,

$$\mu(0_{\Pi_n}, 1_{\Pi_n}) = -\sum_{i=0}^{n-2} \binom{n-2}{i}(-1)^i(i)!(-1)^{n-2-i}(n-2-i)!$$
$$= (-1)^{n-1}(n-1)!.$$

$\square$

PROBLEM 25C. The *dual* of Weisner's theorem (derived by applying Theorem 25.3 to the dual lattice) asserts

$$\sum_{x:x\wedge a=0_L} \mu(x, 1_L) = 0 \qquad \text{for all } a \in L \text{ with } a < 1_L.$$

Take $a$ to be a partition with one block of size $n-1$ and one of size 1 (in the partition lattice) and use the above to give a somewhat different proof of $\mu(0_{\Pi_n}, 1_{\Pi_n}) = (-1)^{n-1}(n-1)!$.

We use Möbius inversion over the lattice of partitions to get an expression for the number of connected labeled simple graphs on $n$ vertices. Let $X$ be an $n$-set. To each graph $G$ with vertex set $X$, there corresponds the partition $\mathcal{C}_G$ of $X$ whose blocks are the vertex sets of the connected components of $G$. We let $g(\mathcal{B})$ denote the number of simple graphs $G$ with $V(G) = X$ and $\mathcal{C}_G = \mathcal{B}$. We let $f(\mathcal{B})$ denote the number of simple graphs $G$ with $V(G) = X$ and where $\mathcal{C}_G$ is a refinement of $B$. Clearly,

$$f(\mathcal{B}) = \sum_{\mathcal{A} \preceq \mathcal{B}} g(\mathcal{A}).$$

We are interested in $g(1_{\Pi_n})$, where $1_{\Pi_n}$ is the partition with one block. But what is easy to calculate is $f(\mathcal{B})$; we get a graph counted by $f(\mathcal{B})$ by arbitrarily choosing a simple graph on each block of $\mathcal{B}$, so if $\mathcal{B}$ has $k_i$ blocks of size $i$, then

$$f(\mathcal{B}) = 2^{k_2\binom{2}{2}} 2^{k_3\binom{3}{2}} \cdots 2^{k_n\binom{n}{2}}.$$

By Möbius inversion,

$$g(\mathcal{B}) = \sum_{\mathcal{A} \preceq \mathcal{B}} \mu(\mathcal{A}, \mathcal{B}) f(\mathcal{A}).$$

Specializing to $\mathcal{B} = 1_{\Pi_n}$, we have

(25.7) $$g(1_{\Pi_n}) = \sum_{\mathcal{A}} \mu(\mathcal{A}, 1_{\Pi_n}) f(\mathcal{A}).$$

Recall that the number of partitions of $X$ of type $(k_1, k_2, \ldots, k_n)$ is

$$\frac{n!}{(1!)^{k_1} k_1! (2!)^{k_2} k_2! \cdots (n!)^{k_n} k_n!};$$

cf. (13.3). And using Theorem 25.1(iv), we find the number of connected labeled graphs on $n$ vertices to be

$$\sum (-1)^{k_1+k_2+\cdots+k_n-1} \frac{n! \, (k_1 + \cdots + k_n - 1)!}{(1!)^{k_1} k_1! \cdots (n!)^{k_n} k_n!} 2^{k_2 \binom{2}{2} + k_3 \binom{3}{2} + \cdots + k_n \binom{n}{2}}$$

where the summation is extended over all partitions $(k_1, k_2, \ldots, k_n)$ of the integer $n$, that is all nonnegative $n$-tuples with $1k_1 + 2k_2 + \cdots + nk_n = n$.

We do not mean to imply that this is a particularly nice answer to our original question. Indeed, it is a summation over a large collection of objects, whose number exceeds $e^{\sqrt{n}}$ for large $n$. But it is a natural illustration of Möbius inversion and does greatly reduce the problem. For $n = 5$, we have

| partition of 5 | number of $\mathcal{A}$ | $f(\mathcal{A})$ | $\mu(\mathcal{A}, 1_{\Pi_5})$ |
|:---:|:---:|:---:|:---:|
| 5 | 1 | 1024 | 1 |
| 41 | 5 | 64 | $-1$ |
| 32 | 10 | 16 | $-1$ |
| 311 | 10 | 8 | 2 |
| 221 | 15 | 4 | 2 |
| 2111 | 10 | 2 | $-6$ |
| 11111 | 1 | 1 | 24 |

from which table and (25.7) we see that 728 out of the 1024 simple labeled graphs on 5 vertices are connected.

**Notes.**

The expression of the chromatic polynomial of a graph in terms of the Möbius function of $L(G)$ is due to G.-C. Rota (1964).

Theorem 25.1 (v) is essentially Euler's formula for polytopes:

$$f_0 - f_1 + f_2 - f_3 + \cdots + (-1)^n f_n = 0,$$

where $f_i$ is the number of faces of rank $i$, or dimension $i - 1$ ($f_0 = f_n = 1$). See B. Grünbaum (1967). Also see R. Stanley (1986), where posets with the property that $\mu(x, y) = (-1)^d$ whenever the ranks of $x$ and $y$ differ by $d$ are called *Eulerian* posets.

**References.**
T. Dowling (1977), A note on complementing permutations, *J. Combinatorial Theory* (B) **23**, 223–226.

T. Dowling and R. M. Wilson (1975), Whitney number inequalities for geometric lattices, *Proc. Amer. Math. Soc.* **47**, 504–512.

B. Grünbaum (1967), *Convex Polytopes*, J. Wiley (Interscience).

G.-C. Rota (1964), On the foundations of combinatorial theory I. Theory of Möbius functions, *Z. Wahrscheinlichkeitstheorie* **2**, 340–368.

R. P. Stanley (1986), *Enumerative Combinatorics*, Vol. 1, Wadsworth.

L. Weisner (1935), Abstract theory of inversion of finite series, *Trans. Amer. Math. Soc.* **38**, 474–484.

# 26

## Combinatorial designs and projective geometries

Geometries over finite fields are a rich source of combinatorial designs and related combinatorial configurations. We begin with two topics (arcs and subplanes) that we choose to mention only in projective planes before going on to discuss quadrics and other configurations in general projective spaces.

An $(m, k)$-*arc* in a projective plane is a set of $m$ points, no $k+1$ of which are collinear. We were concerned with $(m, 2)$-arcs in Problem 19I.

Let $A$ be an $(m, k)$-arc in a projective plane of order $n$ and let $x$ be a point in $A$. The $n + 1$ lines on $x$ each contain at most $k - 1$ other points of $A$, so

$$m \leq 1 + (n + 1)(k - 1).$$

An $(m, k)$-arc $A$ is called *perfect* when equality holds above. Any line that contains a point of a perfect $(m, k)$-arc evidently contains exactly $k$ points of the arc; that is,

$$|L \cap A| = 0 \text{ or } k$$

for any line $L$. Clearly, the nonempty intersections of lines with a perfect $(m, k)$-arc provide the blocks of a Steiner system $S(2, k, m)$.

A single point is a perfect $(1,1)$-arc. The set of $n^2$ points not on a fixed line of a projective plane of order $n$ is a perfect $(n^2, n)$-arc and the corresponding Steiner system is an affine plane of order $n$. The hyperovals in Problem 19I are perfect $(q + 2, 2)$-arcs. The corresponding designs are trivial. But these have "dual" arcs where the corresponding Steiner systems are interesting—see Problem 26A.

The construction of perfect $(m, k)$-arcs in Desarguesian planes of even order given in part (2) of the following theorem is due to R. H. F. Denniston (1969). No examples of perfect $(m, k)$-arcs, $1 < k < n$, in projective planes of odd orders $n$ are known at this time.

THEOREM 26.1.

  (1) *If there exists a perfect $(m, k)$-arc in a projective plane of order $n$, then $k$ divides $n$.*
  (2) *If $q$ is a power of 2 and $k$ divides $q$, then there exists a perfect $(m, k)$-arc in $PG_2(q)$.*

PROOF: Let $x$ be a point not in a perfect $(m, k)$-arc $A$ in a projective plane of order $n$. The lines on $x$ partition the remaining points, but each line on $x$ contains exactly 0 or $k$ points of $A$. So $k$ must divide $m = 1 + (n + 1)(k - 1)$. It follows $k$ divides $n$.

Now let $q$ be a power of 2 and $k$ a divisor of $q$. Let $f(x, y) = \alpha x^2 + \beta xy + \gamma y^2$ be any irreducible quadratic over $\mathbb{F}_q$, and let $H$ be any subgroup of the additive group of $\mathbb{F}_q$ with order $k$. In the affine plane $AG(2, q)$, let

$$A := \{(x, y) : f(x, y) \in H\}.$$

We claim that any (affine) line meets $A$ in 0 or $k$ points. When the affine plane is embedded in a $PG_2(q)$, $A$ will be a perfect $(m, k)$-arc.

Consider a line $L = \{(x, y) : y = mx + b\}$ with $b$ and $m$ both nonzero. (We leave consideration of lines of the forms $\{(x, y) : y = mx\}$ and $\{(x, y) : x = c\}$ to the reader.)

The intersection $L \cap A$ is the set of points $(x, mx + b)$, where

$$\alpha x^2 + \beta x(mx + b) + \gamma(mx + b)^2 \in H, \quad \text{or}$$

$$F(x) \in H, \quad \text{where} \quad F(x) := (\alpha + \beta m + \gamma m^2)x^2 + \beta bx + \gamma b^2.$$

We are working over a field of characteristic 2, so

$$x \mapsto (\alpha + \beta m + \gamma m^2)x^2 + \beta bx$$

is a linear mapping; it has kernel of order 2 since the irreducibility of $f(x, y)$ ensures that both $\beta$ and the coefficient of $x^2$ are nonzero.

The image $K_F$ of this mapping is thus a subgroup of order $q/2$ of the additive group of $\mathbb{F}_q$. The image of $F$ is a coset of $K_F$, but since $F(x)$ is never 0 (again by the irreducibility of $f(x,y)$), the image of $F$ is the complement $\mathbb{F}_q \setminus K_F$. In summary,

$$|\{x : F(x) = a\}| = \begin{cases} 2 & \text{if } x \in K_F, \\ 0 & \text{if } x \notin K_F. \end{cases}$$

Thus $|L \cap A| = 2|H \cap (\mathbb{F}_q \setminus K_F)|$. The subgroup $H$ is either contained in $K_F$, in which case $L \cap A = \emptyset$, or intersects $K_F$ in a subgroup of order $k/2$, in which case $|L \cap A| = k$.    □

As a corollary of this theorem (with $n = 2^{m+1}$, $k = 2^m$), we obtain Steiner systems

$$S(2, 2^m, 2^{2m+1} - 2^m).$$

Steiner systems $S(2, k, v)$ are in many ways most interesting when $v$ is close to $k^2$, as is the case for affine and projective planes (Fisher's inequality, Theorem 19.6, shows $v \geq k^2 - k + 1$). These examples have $v < 2k^2$, which is still impressive.

PROBLEM 26A. Let $A$ be a perfect $(m, k)$-arc in a projective plane $P$ of order $n$, $1 \leq k \leq n$. Let $A^*$ be the set of lines that do not meet $A$. Show that $A^*$ is a $[n/k]$-arc in the dual plane $P^*$.

PROBLEM 26B. A *parallel class* in a Steiner system $S(2, k, m)$ is a set $\mathcal{A}$ of $m/k$ blocks so that each point is incident with exactly one block in $\mathcal{A}$. An $S(2, k, m)$ is *resolvable* when the blocks can be partitioned into parallel classes (cf. Problem 19K).

Let $A$ be a perfect $(m, k)$-arc in a projective plane $P$ of order $n$. Explain why the Steiner system $S(2, k, m)$ whose blocks are the nontrivial intersections of lines with $A$ is resolvable.

A *subplane* **S** of a projective plane **P** is a substructure of **P** that is a projective plane in its own right. Recall that a *substructure* of an incidence structure $(\mathcal{P}, \mathcal{B}, \mathbf{I})$ is an incidence structure $(\mathcal{P}_0, \mathcal{B}_0, \mathbf{I}_0)$ where $\mathcal{P}_0 \subseteq \mathcal{P}$, $\mathcal{B}_0 \subseteq \mathcal{B}$, and $\mathbf{I}_0 = \mathbf{I} \cap (\mathcal{P}_0 \times \mathcal{B}_0)$. Note that given an automorphism (or *collineation*) $\alpha$ of a projective plane **P**, the substructure **S** consisting of the points of **P** fixed by $\alpha$ and the lines of **P** fixed by $\alpha$ has the property that two points of **S** are incident with a unique line of **S**, and two lines of **S** are incident with a unique point of **S**. If **S** contains four points, no three collinear, then **S** is a subplane; but it may be a nearpencil or have one or no lines.

EXAMPLE 26.1. Let $V := \mathbb{F}_{q^n}^3$. The points and lines of $PG_2(q^n)$ are the 1-dimensional and 2-dimensional $\mathbb{F}_{q^n}$-subspaces of $V$. Define a substructure **S** of $PG_2(q)$ whose points and lines are those 1- and 2-dimensional $\mathbb{F}_{q^n}$-subspaces of $V$ that admit bases consisting of vectors whose entries are in the subfield $\mathbb{F}_q$. Then **S** is a subplane. A line of $S$ is incident with $q^n + 1$ points in $PG_2(q^n)$, but only $q + 1$ of those are points of $S$.

THEOREM 26.2. *If a projective plane* **P** *of order* $n$ *contains a subplane* **S** *of order* $m < n$, *then either*
   (i) $n = m^2$, *or*
   (ii) $n \geq m^2 + m$.

PROOF: Let $L$ be a line of the subplane **S** and $x$ a point on $L$ in **P** but not in **S**. The $n$ other lines on $x$ can contain at most one point of **S** (since a line $M$ in **P** containing at least two points of **S** would necessarily belong to the subplane, and then the point $x$ in common to $M$ and $L$ would also belong to the subplane). The line $L$ contains $m + 1$ points of **S**, which has a total of $m^2 + m + 1$ points, each of which is on some line through $x$, so $m^2 \leq n$.

Equality will imply that every line of **P** will meet **S** (evidently in one or $m + 1$ points) since if there were a line $N$ disjoint from **S**, and $x$ were taken as the point of intersection of $L$ and $N$, each of the $m^2$ points of **S** not on $L$ would belong to one of $n - 1$ lines.

Now assume $m^2 < n$, so that there does exist a line $N$ incident with no points of $S$. Each of the $m^2 + m + 1$ lines of **S** contains one point of $N$ and no two such lines can contain the same point, so $m^2 + m + 1 \leq n + 1$. $\qquad\square$

PROBLEM 26C. Show that if $PG_2(\mathbb{F})$ contains the Fano configuration $PG_2(2)$, then $\mathbb{F}$ has characteristic 2. Suggestion: Without loss of generality, four points of the Fano configuration are $\langle 1, 0, 0 \rangle$, $\langle 0, 1, 0 \rangle$, $\langle 0, 0, 1 \rangle$, and $\langle 1, 1, 1 \rangle$. Calculate the homogeneous coordinates of the other three points (that must lie on a line).

PROBLEM 26D. A *blocking set* in a projective plane is a set $S$ of points that contains no line but such that every line meets $S$ in at least one point. Show that a blocking set in a projective plane of order $n$ contains at least $n + \sqrt{n} + 1$ points, and that equality holds if and only if $S$ is the set of points of a Baer subplane.

A *quadratic form* in indeterminates $x_1, x_2, \ldots, x_n$ over a field $\mathbb{F}$ is a homogeneous polynomial of degree 2 in those indeterminates, i.e.

$$(26.1) \qquad f(\mathbf{x}) = f(x_1, x_2, \ldots, x_n) = \sum_{i,j=1}^{n} c_{ij} x_i x_j$$

where the coefficients $c_{ij}$ are in $\mathbb{F}$. The same quadratic form is defined by more than one matrix $C = ((c_{ij}))$ of coefficients, since it is only the diagonal entries and the sums $c_{ij} + c_{ji}$ that are important. We could require that $c_{ij} = 0$ for $i > j$ and then quadratic forms in $x_1, x_2, \ldots, x_n$ are in one-to-one correspondence with upper triangular $n$ by $n$ matrices $C$ over $\mathbb{F}$.

A *quadric* in $PG_n(\mathbb{F})$ is a set of projective points

$$Q = Q(f) := \{\langle \mathbf{x} \rangle : f(\mathbf{x}) = 0\}$$

where $f$ is a quadratic form in $n + 1$ indeterminates. We must choose a basis and identify the vectors with $(n + 1)$-tuples over $\mathbb{F}$ in order to make this definition.

Two quadratic forms $f$ and $g$ are *projectively equivalent* when $g$ can be obtained from $f$ by an "invertible linear substitution", or in matrix notation, $g(\mathbf{x}) = f(\mathbf{x}A)$ for some nonsingular $n$ by $n$ matrix $A$ over $\mathbb{F}$. So if $f$ is given by a matrix $C$ as in (26.1), i.e. $f(\mathbf{x}) = \mathbf{x} C \mathbf{x}^\top$, then $g$ is given by the matrix $ACA^\top$. For example, for any positive integer $n$, $nx_1^2 + nx_2^2 + nx_3^2 + nx_4^2$ is projectively equivalent to $x_1^2 + x_2^2 + x_3^2 + x_4^2$ over the rationals; see equation (19.12).

We remark that the matrix notation does not lend itself well to changes of the names or number of the indeterminates, which we wish to allow. For example, replacing $x_1$ in a quadratic form $f$ by $y_1 + y_2$, where $y_1$ and $y_2$ are new indeterminates, produces a projectively equivalent form.

The *rank* of a quadratic form is the least number of indeterminates that occur (with nonzero coefficients) in any projectively equivalent quadratic form. For example, $(x_1 + \cdots + x_n)^2$ has rank 1. Projectively equivalent quadratic forms have the same rank. A quadratic form in $r$ indeterminates is said to be *nondegenerate* when it has rank $r$.

EXAMPLE 26.2. Consider quadratic forms in two indeterminates

$$f(x, y) = ax^2 + bxy + cy^2.$$

The form has rank 0 if and only if $a = b = c = 0$. The corresponding quadric contains all points of $PG_1(\mathbb{F})$ (the projective line). If not zero, it will have rank 1 if and only if it is a scalar multiple of the square of a linear form $dx + ey$, which is the case if and only if the discriminant $b^2 - 4ac$ is 0. The corresponding quadric in $PG_1(\mathbb{F})$ will then consist of a single point. A rank 2 quadratic form in two indeterminates is either irreducible or can be factored into two distinct linear forms. In the first case, the corresponding quadric in $PG_1(\mathbb{F})$ will be empty, and in the second case, the corresponding quadric in $PG_1(\mathbb{F})$ will consist of two points.

It is clear that an irreducible quadratic form of rank 2 is not projectively equivalent to a reducible quadratic form, since the latter has zeros in $\mathbb{F}$ and the former does not. A reducible quadratic form is projectively equivalent to $x_1 x_2$.

PROBLEM 26E. Show that over a field $\mathbb{F}$ of odd characteristic, a quadratic form $f$ as in (26.1) is degenerate if and only the symmetric matrix $C + C^\top$ is singular. For a field $\mathbb{F}$ of characteristic 2, show that $f$ is degenerate if and only if $C + C^\top$ is singular *and* $\mathbf{x}(C + C^\top) = 0$ for some $\mathbf{x}$ with $f(\mathbf{x}) = 0$.

It is important to understand that the intersection of a quadric with a flat $U$ of $PG_n(\mathbb{F})$ is a quadric in that flat. For example, suppose that $U$ is a projective line $PG_1(\mathbb{F})$ and that $f$ is a quadratic form in $x_0, x_1, \ldots, x_n$. The homogeneous coordinates of points of $U$ are $\langle y\mathbf{a} + z\mathbf{b} \rangle$ where $\mathbf{a} = (a_0, a_1, \ldots, a_n)$ and $\mathbf{b} = (b_0, b_1, \ldots, b_n)$ are linearly independent vectors. These points are in one-to-one correspondence with the homogeneous coordinates $\langle (y, z) \rangle$ of points of $PG_1(\mathbb{F})$. Say $f = \sum_{0 \le i \le j \le n} c_{ij} x_i x_j$. Then

$$g(y, z) := \sum_{1 \le i \le j \le n} c_{ij}(ya_i + zb_i)(ya_j + zb_j)$$

is a quadratic form in $y$ and $z$ and defines a quadric in $PG_1(\mathbb{F})$. The quadratic form $g$ may be degenerate even if $f$ is not. In view of Example 26.2, we see that a line is either contained fully in a quadric $Q$, or meets $Q$ in 0, 1, or 2 points.

LEMMA 26.3. *Any quadratic form $f$ of rank $n \ge 3$ is projectively equivalent to*

$$(26.2) \qquad x_1 x_2 + g(x_3, \ldots, x_n)$$

*for some quadratic form $g(x_3, \ldots, x_n)$.*

PROOF: First assume $q$ is odd. It is easy to see that a quadratic form $f$ is projectively equivalent to some $\sum_{1 \le i \le j \le n} c_{ij} x_i x_j$ where $c_{11} \ne 0$. Let $y := x_1 + \frac{1}{2c_{11}}(c_{12}x_2 + \cdots + c_{1n}x_n)$. Then $f = c_{11}y^2 + g(x_2, \ldots, x_n)$. Inductively, if the rank of $f$ is at least 3, we find that $f$ is projectively equivalent to

$$h(\mathbf{x}) = ax_1^2 + bx_2^2 + cx_3^2 + g'(x_4, \ldots, x_n)$$

with $a, b, c$ all nonzero. The three scalars can be changed by any nonzero square factor and a projectively equivalent quadratic form results; some pair must differ by a square factor, so we can assume that $b = c$, say.

We claim that there exist $s, t \in \mathbb{F}_q$ so that $s^2 + t^2 = -b^{-1}a$. One way to see this is to consider the addition table of $\mathbb{F}_q$, which is a Latin square. The number of squares in $\mathbb{F}_q$, including 0, is $(q + 1)/2$. The element $-b^{-1}a$ (or any other) occurs $(q - 1)/2$ times in the $(q - 1)/2$ columns of the Latin square indexed by the nonsquares, and $(q - 1)/2$ times in the $(q - 1)/2$ rows indexed by the nonsquares; so $-b^{-1}a$ must occur at least once in the submatrix with rows and columns indexed by the squares. With $s$ and $t$ so chosen, $h$ is projectively equivalent to

$$ax_1^2 + b(sx_2 + tx_3)^2 + b(tx_2 - sx_3)^2 + g'(x_4, \ldots, x_n)$$
$$= ax_1^2 - ax_2^2 - ax_3^2 + g'(x_4, \ldots, x_n)$$
$$= (ax_1 + ax_2)(x_1 - x_2) - ax_3^2 + g'(x_4, \ldots, x_n),$$

and the latter is clearly projectively equivalent to (26.2).

The case $q$ even is a little more tedious; see the proof of Theorem 5.1.7 in J. W. P. Hirschfeld (1979). $\square$

THEOREM 26.4.

(i) *Any quadratic form $f$ of odd rank $n$ is projectively equivalent to*

$$(26.3) \qquad f_0(\mathbf{x}) := x_1 x_2 + \cdots + x_{n-2} x_{n-1} + c x_n^2$$

*for some scalar $c$.*

(ii) *Any quadratic form $f$ of even rank $n$ is projectively equivalent to either*

$$(26.4) \qquad f_1(\mathbf{x}) := x_1 x_2 + \cdots + x_{n-3} x_{n-2} + x_{n-1} x_n$$

*or*

$$(26.5) \qquad f_2(\mathbf{x}) := x_1 x_2 + \cdots + x_{n-3} x_{n-2} + p(x_{n-1}, x_n)$$

*where $p(x_{n-1}, x_n)$ is an irreducible quadratic form in two indeterminates.*

PROOF: This follows from Lemma 26.3 and induction. $\square$

Quadratic forms (and the corresponding quadrics) of odd rank are called *parabolic*. Quadratic forms of even rank projectively equivalent to (26.4) are called *hyperbolic*, and those equivalent to (26.5) are called *elliptic*. Any two hyperbolic quadratic forms of a given rank are projectively equivalent to (26.4) and hence to each other. It is also true that all parabolic quadratic forms of a given even rank are projectively equivalent, i.e. we may take $c = 1$ in (26.3), and that all elliptic quadratic forms of a given rank are projectively equivalent. See J. W. P. Hirschfeld (1979) for this and more on canonical forms. That hyperbolic and elliptic quadratic forms are not projectively equivalent is a consequence of the following theorem.

THEOREM 26.5. *A nondegenerate quadric $Q$ in $PG_n(q)$ has cardinality*

$$
\begin{cases}
\frac{q^n-1}{q-1} & \text{if } n \text{ is even, i.e. } Q \text{ is parabolic,} \\
\frac{(q^{(n+1)/2}-1)(q^{(n-1)/2}+1)}{q-1} & \text{if } n \text{ is odd and } Q \text{ is hyperbolic,} \\
\frac{(q^{(n+1)/2}+1)(q^{(n-1)/2}-1)}{q-1} & \text{if } n \text{ is odd and } Q \text{ is elliptic.}
\end{cases}
$$

PROOF: In general, if

$$
f(x_1, \ldots, x_n) = x_1 x_2 + g(x_3, \ldots, x_n)
$$

and there are $N$ vectors $(x_3, \ldots, x_n)$ such that $g(x_3, \ldots, x_n) = 0$, then there are $(2q-1)N + (q-1)^2(q^{n-2} - N)$ vectors $(x_1, \ldots, x_n)$ such that $f(x_1, \ldots, x_n) = 0$. This allows us to verify by induction the following formulae for the number of zeros of a rank $r$ quadratic form $f$ in $r$ indeterminates as the following:

$$
\begin{cases}
q^{r-1} & \text{if } r \text{ is odd, i.e. } f \text{ is parabolic,} \\
q^{r-1} + q^{r/2} - q^{r/2-1} & \text{if } r \text{ is even and } f \text{ is hyperbolic,} \\
q^{r-1} - q^{r/2} + q^{r/2-1} & \text{if } r \text{ is even and } f \text{ is elliptic.}
\end{cases}
$$

Of course, the number of projective points on the corresponding quadrics is found from the above numbers by subtracting 1 (the zero vector) and then dividing by $q - 1$.    □

THEOREM 26.6. *Let $Q$ be a nondegenerate quadric in $PG_r(q)$. The maximum projective dimension of a flat $F$ with $F \subseteq Q$ is*

$$
\begin{cases}
r/2 - 1 & \text{if } r \text{ is even, i.e. } Q \text{ is parabolic,} \\
(r-1)/2 & \text{if } r \text{ is odd and } Q \text{ is hyperbolic,} \\
(r-3)/2 & \text{if } r \text{ is odd and } Q \text{ is elliptic.}
\end{cases}
$$

PROOF: Let $f$ be a nondegenerate quadratic form in $n$ indeterminates. The statement of the theorem is equivalent to the statement that the maximum dimension of a subspace $U$ of $\mathbb{F}_q^n$ such that $f$ vanishes on $U$ is

$$
\begin{cases}
(n-1)/2 & \text{if } n \text{ is odd, i.e. } f \text{ is parabolic,} \\
n/2 & \text{if } n \text{ is even and } f \text{ is hyperbolic,} \\
n/2 - 1 & \text{if } n \text{ is even and } f \text{ is elliptic.}
\end{cases}
$$

First note that if $f = f_0$ in (26.3), then $f(\mathbf{x}) = 0$ for any $\mathbf{x} \in \operatorname{span}(\mathbf{e}_2, \mathbf{e}_4, \ldots, \mathbf{e}_{n-1})$, where $\mathbf{e}_1, \mathbf{e}_2, \ldots, \mathbf{e}_n$ is the standard basis for $\mathbb{F}_q^n$. If $f = f_1$ in (26.4), then $f(\mathbf{x}) = 0$ for any $\mathbf{x} \in \operatorname{span}(\mathbf{e}_2, \mathbf{e}_4, \ldots, \mathbf{e}_n)$. If $f = f_2$ in (26.5), then $f(\mathbf{x}) = 0$ for any $\mathbf{x} \in \operatorname{span}(\mathbf{e}_2, \mathbf{e}_4, \ldots, \mathbf{e}_{n-2})$. These subspaces have dimensions $(n-1)/2$, $n/2$, and $n/2 - 1$, respectively.

It remains to show that $f$ cannot vanish on any subspace of larger dimension in any of these cases. We will use Theorem 26.4 and induction. The cases $n = 1$ and $n = 2$ are trivial. Suppose $f(x_1, \ldots, x_n) = x_1 x_2 + g(x_3, \ldots, x_n)$ and that $f(\mathbf{x}) = 0$ for all $\mathbf{x}$ in some subspace $U \subseteq \mathbb{F}_q^n$ of dimension $k$. To complete the proof, we will show that there exists a subspace $U' \subseteq \mathbb{F}_q^{n-2}$ of dimension $\geq k - 1$ so that then $g(\mathbf{y}) = 0$ for all $\mathbf{y} \in U'$.

Clearly, $g(\mathbf{y}) = 0$ for all $\mathbf{y}$ in

$$U_0 := \{(x_3, \ldots, x_n) : (0, 0, x_3, \ldots, x_n) \in U\}.$$

We are done if $\dim(U_0) \geq k - 1$, so assume $\dim(U_0) = k - 2$. Then there exist vectors

$$(1, 0, a_3, \ldots, a_n) \quad \text{and} \quad (0, 1, b_3, \ldots, b_n)$$

in $U$. Then $(1, 1, a_3 + b_3, \ldots, a_n + b_n) \in U$ so $g(a_3 + b_3, \ldots, a_n + b_n) = -1$. The form $g$ evidently vanishes on all vectors in

$$\operatorname{span}((a_3, \ldots, a_n)) + U_0 \quad \text{and} \quad \operatorname{span}((b_3, \ldots, b_n)) + U_0.$$

One of these subspaces must have dimension $> k - 2$ because otherwise $(a_3 + b_3, \ldots, a_n + b_n) \in U_0$, which contradicts $g(a_3 + b_3, \ldots, a_n + b_n) = -1$. $\qquad\square$

EXAMPLE 26.3. Let $f$ be a quadratic form in three indeterminates and consider the quadric $Q(f)$ in the projective plane $PG_2(q)$. If $f$ is nondegenerate, there are $q + 1$ projective points on $Q(f)$. As we remarked above, any line $L$ of $PG_2(q)$ meets $Q(f)$ in a quadric in $L$ ($Q(f)$ cannot contain $L$ by Theorem 26.6) and so $Q(f)$ is a set of $q + 1$ points, no three of which are collinear, i.e. $Q(f)$ is an oval. See Problem 19I.

If $f$ has rank 0, $Q(f)$ is all of $PG_2(q)$. If $f$ has rank 1, $Q(f)$ consists of the points of a line in $PG_2(q)$. If $f$ has rank 2, there are two cases. If $f$ is reducible, say $f$ is projectively equivalent to $xy$, then $Q(f)$ consists of the points on the union of two lines of $PG_2(q)$; if $f$ is irreducible, then $Q(f)$ is a single point in $PG_2(q)$.

EXAMPLE 26.4. The first order Reed-Muller codes were introduced in Chapter 18. Here is one way to introduce the entire family: Let $V = \mathbb{F}_2^m$. We consider vectors of length $2^m$ whose coordinates are indexed by elements of $V$; for concreteness, write $V = \{\mathbf{v}_0, \mathbf{v}_1, \ldots, \mathbf{v}_{2^m-1}\}$. The *k-th order Reed-Muller code* $RM(k,m)$ is defined to consist of all vectors (of length $2^m$)

$$(f(\mathbf{v}_0), f(\mathbf{v}_1), \ldots, f(\mathbf{v}_{2^m-1}))$$

where $f$ ranges over all polynomials in $x_1, \ldots, x_m$ of degree at most $k$.

Since a linear form $x_{i_1} + \cdots + x_{i_k}$ gives the same function as the quadratic form $x_{i_1}^2 + \cdots + x_{i_k}^2$ over $\mathbb{F}_2$, the words in the second order code $RM(2,m)$ are given by binary quadratic forms $f(x_1, \ldots, x_n)$ and their "complements" $f(x_1, \ldots, x_n)+1$. Theorems 26.4 and 26.5 are useful for determining the weights that occur for codewords in $RM(2,m)$.

We must consider degenerate forms also; e.g. $x_1x_2 + x_3x_4$ corresponds to a codeword of weight 24 in $RM(2,6)$. It will be found that the weights of codewords in $RM(2,6)$ are 0, 16, 24, 28, 32, 36, 40, 48, and 64.

EXAMPLE 26.5. Let $Q_3$ be a nondegenerate elliptic quadric in projective 3-space $PG_3(q)$. By Theorem 26.5, $|Q_3| = q^2 + 1$. By Theorem 26.6, $Q_3$ contains no lines. Any plane $P$ meets $Q_3$ in a quadric in that plane; in view of Example 26.3, every plane must meet $Q_3$ in either an oval in that plane or a single point. Any three points of $Q_3$ are contained in a unique plane, so it follows that the nontrivial intersections of planes with $Q_3$ provide the blocks of a Steiner system

$$S(3, q + 1, q^2 + 1).$$

In general, a set of $q^2 + 1$ points, no three collinear, in $PG_3(q)$ is called an *ovoid* and an $S(3, n + 1, n^2 + 1)$ is called a *Möbius plane* or an *inversive plane*.

EXAMPLE 26.6. Let $Q_4$ be a nondegenerate quadric in projective 4-space $PG_4(q)$. By Theorem 26.5, $|Q_4| = q^3 + q^2 + q + 1$. Let $\mathcal{Q}$ be the incidence structure whose points are the elements of $Q_4$ and whose blocks are the lines of $PG_4(q)$ fully contained in $Q_4$. Each point of $\mathcal{Q}$ is on exactly $q+1$ blocks of $\mathcal{Q}$ (see Problem 26F). Given a point $x$ and a block $L$ of $\mathcal{Q}$, the intersection of the plane $P := \{x\} \vee L$ with $Q_4$ is a quadric $Q_2$ in $P$ that contains, obviously, a line and a point off that line. From Example 26.3, $Q_2$ must consist of the points on two (intersecting) lines. This implies that $\mathcal{Q}$ is a partial geometry $pg(r, k, t)$, as defined in Chapter 21, where

$$r = q+1, \ k = q+1, \ t = 1.$$

EXAMPLE 26.7. Let $Q_5$ be a nondegenerate elliptic quadric in $PG_5(q)$. By Theorem 26.5, $|Q_5| = (q + 1)(q^3 + 1)$. By Theorem 26.6, $Q_5$ contains no planes. Let $\mathcal{Q}$ be the incidence structure whose points are the elements of $Q_5$ and whose blocks are the lines of $PG_5(q)$ fully contained in $Q_5$. By Problem 26F, each point of $\mathcal{Q}$ is on $q^2 + 1$ lines of $\mathcal{Q}$. By an argument similar to that of Example 26.6, $\mathcal{Q}$ is a partial geometry $pg(r, k, t)$ where

$$r = q^2 + 1, \ k = q+1, \ t = 1.$$

Partial geometries with $t = 1$ are called generalized quadrangles. See L. M. Batten (1986) for further results and references.

PROBLEM 26F. Let $f$ be a nondegenerate quadratic form in $n$ indeterminates given by a matrix $C = ((c_{ij}))$ over $\mathbb{F}_q$ as in (26.1). Let $Q = Q(f)$ be the corresponding quadric in $PG_{n-1}(\mathbb{F}_q)$. Let $p = \langle \mathbf{x} \rangle$ be one of the points on $Q$. Let

$$T_p := \{\langle \mathbf{y} \rangle : \mathbf{x}(C + C^\top)\mathbf{y}^\top = 0\}.$$

Then $T_p$ is a hyperplane in $PG_{n-1}(\mathbb{F}_q)$ by Problem 26E. Show that $T_p \cap Q$ consists exactly of $p$ and the union of any lines on $p$ that are contained in $Q$. Further show that if $W$ is any hyperplane of $T_p$ not containing $p$, then $Q' := W \cap Q$ is a nondegenerate quadric in $W (= PG_{n-3}(\mathbb{F}_q))$ and that $Q'$ is parabolic, hyperbolic, or elliptic according to whether $Q$ is parabolic, hyperbolic, or elliptic. We see, in particular, that the number of lines on $p$ that lie entirely in $Q$ is equal to $|Q'|$.

A *hermitian form* in indeterminates $x_1, \ldots, x_n$ over $\mathbb{F}_{q^2}$ is an expression of the form

$$(26.6) \qquad h(\mathbf{x}) = h(x_1, \ldots, x_n) = \sum_{i,j=1}^{n} c_{ij} x_i x_j^q,$$

where the coefficients $c_{ij}$ are from $\mathbb{F}_{q^2}$ and where $c_{ji} = c_{ij}^q$. In particular, the diagonal coefficients $c_{ii}$, being fixed by the Frobenius automorphism $x \mapsto x^q$, lie in $\mathbb{F}_q$.

Two hermitian forms $f$ and $g$ over $\mathbb{F}_{q^2}$ are *projectively equivalent* when $g$ can be obtained from $f$ by an invertible linear substitution, or, in matrix notation, $g(\mathbf{x}) = f(\mathbf{x}A)$ for some nonsingular $n$ by $n$ matrix $A$ over $\mathbb{F}_{q^2}$. So if $f$ is defined by a matrix $C$ as in (26.6), i.e. $f(\mathbf{x}) = \mathbf{x}C\mathbf{x}^\top$, then $g$ is defined by the matrix $ACA^*$, where $A^*$ is the *conjugate transpose* of $A$; if $A = ((a_{ij}))$, then $A^* := ((a_{ji}^q))$.

The *rank* of a hermitian form is the least number of indeterminates that occur (with nonzero coefficients) in any projectively equivalent hermitian form.

A *hermitian variety* in $PG_n(q^2)$ is a set of projective points

$$H = H(f) := \{\langle \mathbf{x} \rangle : f(\mathbf{x}) = 0\}$$

where $f$ is a hermitian form in $n+1$ indeterminates. We must choose a basis and identify the vectors with $(n+1)$-tuples over $\mathbb{F}_{q^2}$ in order to make this definition. It can be seen that the intersection of a hermitian variety in $PG_n(q^2)$ with a flat is a hermitian variety in that flat.

THEOREM 26.6. *A hermitian form of rank $n$ is projectively equivalent to*

$$(26.7) \qquad x_1^{q+1} + x_2^{q+1} + \cdots + x_n^{q+1}.$$

PROOF: It is not hard to see that any nonzero hermitian form is projectively equivalent to $h$ as in (26.6) where $c_{11} \neq 0$; we leave this to the reader. Let $y := c_{11}x_1 + c_{12}x_2 + \cdots + c_{1n}x_n$. Then

$h = c_{11}^{-1} y y^q + g(x_2, \ldots, x_n)$ where $g$ is a hermitian form in $x_2, \ldots, x_n$. Since $c_{11} \in \mathbb{F}_q$, there exists $a \in \mathbb{F}_{q^2}$ such that $a^{q+1} = c_{11}^{-1}$ and then $h = z^{q+1} + g(x_2, \ldots, x_n)$, where $z = ay$.

The theorem follows from this step and induction. $\qquad\square$

THEOREM 26.8. *The number of points on a nondegenerate hermitian variety $H$ in $PG_n(q^2)$ is*

$$\frac{(q^{n+1} + (-1)^n)(q^n - (-1)^n)}{q^2 - 1}.$$

PROOF: For each nonzero $a \in \mathbb{F}_q$, there are $q + 1$ values of $x \in \mathbb{F}_{q^2}$ such that $x^{q+1} = a$. If

$$f(x_1, \ldots, x_n) = x_1^{q+1} + g(x_2, \ldots, x_n)$$

and there are $N$ vectors $(x_2, \ldots, x_n)$ such that $g(x_2, \ldots, x_n) = 0$, then there are $N + (q+1)(q^{2(n-1)} - N)$ vectors $(x_1, \ldots, x_n)$ such that $f(x_1, \ldots, x_n) = 0$. Inductively, there are $q^{2n-1} + (-1)^n(q^n - q^{n-1})$ vectors in $\mathbb{F}_{q^2}^n$ that are zeros of (26.7) and the theorem follows. $\quad\square$

EXAMPLE 26.8. Consider hermitian varieties in the projective line $PG_1(q^2)$. If a hermitian form $f$ in two indeterminates has rank 0, then the variety $H(f)$ contains all $q^2 + 1$ points; if $f$ has rank 1, $H(f)$ consists of one point; if $f$ has rank 2, $H(f)$ contains $q + 1$ points. It follows that if $H$ is a hermitian variety in $PG_n(q^2)$ for any $n$, then lines of $PG_n(q^2)$ meet $H$ in 1, $q + 1$, or $q^2 + 1$ points.

Now consider a nondegenerate hermitian variety $H_2$ in the projective plane $PG_2(q^2)$. It has $q^3 + 1$ points by Theorem 26.8. Any line $L$ meets $H_2$ in a hermitian variety $L \cap H_2$ in that line. As an exercise, the reader should check that no lines are contained in $H_2$, so $|L \cap H_2| = 1$ or $q + 1$. It follows that the nontrivial intersections of lines of $PG_2(q^2)$ with $H_2$ provide the blocks of a Steiner system

$$S(2, q + 1, q^3 + 1)$$

on the point set $H_2$.

Designs with these parameters are called *unitals*. Further analysis shows that the designs constructed above are resolvable; see R. C. Bose (1959).

We conclude this chapter with a construction of higher dimensional analogues of the Steiner systems $S(3, q+1, q^2+1)$ of Example 26.5; those constructed below have been called *circle geometries*.

THEOREM 26.9. *If $q$ is a power of a prime and $n$ a positive integer, then there exists a Steiner system $S(3, q+1, q^n+1)$.*

PROOF: Let $V$ be a 2-dimensional vector space over $\mathbb{F}_{q^n}$ and let $\mathcal{X}$ be the set of 1-dimensional subspaces of $V$ over $\mathbb{F}_{q^n}$ (i.e. the $q^n+1$ points of $PG_1(q^n)$, the projective line of order $q^n$).

Now think of $V$ as a $2n$-dimensional vector space over $\mathbb{F}_q$. The blocks of our Steiner system will come from the 2-dimensional subspaces $U$ over $\mathbb{F}_q$ of $V$ that are not contained in any member of $\mathcal{X}$. For every such subspace $U$, let

$$B_U := \{P \in \mathcal{X} : P \cap U \neq \{\mathbf{0}\}\}.$$

Note that each $P \in B_U$ meets $U$ in a 1-dimensional subspace over $\mathbb{F}_q$ (that contains $q-1$ of the $q^2-1$ nonzero vectors of $U$), so $|B_U| = q + 1$. If $W = \lambda U$ for some nonzero scalar $\lambda \in \mathbb{F}_{q^n}$, then $B_U = B_W$; we take only the distinct sets $B_U$ as blocks.

Consider three distinct points $P_i \in \mathcal{X}$, $i = 1, 2, 3$. Say $P_i$ is the $\mathbb{F}_{q^n}$-span of a vector $\mathbf{x}_i$, $i = 1, 2, 3$. These three vectors are linearly dependent over $\mathbb{F}_{q^n}$, say

$$\mathbf{x}_3 = \alpha \mathbf{x}_1 + \beta \mathbf{x}_2$$

with $\alpha, \beta \in \mathbb{F}_{q^n}$. But then it is clear that $U := \text{span}_{\mathbb{F}_q}\{\alpha \mathbf{x}_1, \beta \mathbf{x}_2\}$ meets $P_1$, $P_2$, and $P_3$ all nontrivially. Suppose some 2-dimensional subspace $W$ over $\mathbb{F}_q$ meets each of the $P_i$ nontrivially, say $W$ contains $\gamma_i \mathbf{x}_i$ with $0 \neq \gamma_i \in \mathbb{F}_{q^n}$, $i = 1, 2, 3$. Then these vectors are linearly dependent over $\mathbb{F}_q$, say

$$\gamma_3 \mathbf{x}_3 = a\gamma_1 \mathbf{x}_1 + b\gamma_2 \mathbf{x}_2$$

where $a, b \in \mathbb{F}_q$. Since $\mathbf{x}_1$ and $\mathbf{x}_2$ are linearly independent over $\mathbb{F}_{q^n}$, we have $\gamma_3 \alpha = a\gamma_1$ and $\gamma_3 \beta = b\gamma_2$. It follows that $\gamma_3 U = W$, and we see that three distinct points are contained in a unique block. $\square$

**Notes.**

The term "maximal $(m, k)$-arc" is often used in the literature for what we have called "perfect $(m, k)$-arcs", but this is too much of an abuse of the word "maximal" for us to tolerate.

The combinatorial properties of quadrics and hermitian varieties are discussed in more detail in D. K. Ray-Chaudhuri (1962) and R. C. Bose and I. M. Chakravarti (1966).

**References.**

L. M. Batten (1986), *Combinatorics of Finite Geometries*, Cambridge University Press.

R. C. Bose (1959), On the application of finite projective geometry for deriving a certain series of balanced Kirkman arrangements, *Golden Jubilee Commemoration Volume (1958–59)*, Calcutta Math. Soc., pp. 341–354.

R. C. Bose and I. M. Chakravarti (1966), Hermitian varieties in a finite projective space $PG(N, q^2)$, *Canad. J. Math* **18**, 1161–1182.

P. Dembowski (1968), *Finite Geometries*, Springer-Verlag.

R. H. F. Denniston (1969), Some maximal arcs in finite projective planes, *J. Combinatorial Theory* **6**, 317–319.

J. W. P. Hirschfeld (1979), *Projective Geometries over Finite Fields*, Clerendon Press.

D. K. Ray-Chaudhuri (1962), Some results on quadrics in finite projective geometry based on Galois fields, *Canad. J. Math.* **14**, 129-138.

# 27

## Difference sets and automorphisms

A class of symmetric designs arises from *difference sets* (defined below) in abelian groups. One such design appeared in Example 19.6. The group reappears in the automorphism group of the design. We begin with a theorem on automorphisms of symmetric designs in general.

THEOREM 27.1. *Let* $\mathcal{S} = (X, \mathcal{A})$ *be a symmetric* $(v, k, \lambda)$-*design and* $\alpha$ *an automorphism of* $\mathcal{S}$. *Then the number of points fixed by* $\alpha$ *is equal to the number of blocks fixed by* $\alpha$.

PROOF: Let $N$ be the incidence matrix of $\mathcal{S}$. Define a permutation matrix $P$ whose rows and columns are indexed by the points and where

$$P(x, y) := \begin{cases} 1 & \text{if } \alpha(x) = y, \\ 0 & \text{otherwise.} \end{cases}$$

Define a permutation matrix $Q$ whose rows and columns are indexed by the blocks and where

$$Q(A, B) := \begin{cases} 1 & \text{if } \alpha(A) = B, \\ 0 & \text{otherwise.} \end{cases}$$

Note that the trace of $P$ is equal to the number of fixed points, and the trace of $Q$ is equal to the number of fixed blocks of $\alpha$.

Now we have

$$PNQ^\top(x, A) = \sum_{y \in X, B \in \mathcal{A}} P(x, y) N(y, B) Q(A, B)$$

$$= N(\alpha(x), \alpha(A)) = N(x, A).$$

That is, $PNQ^\top = N$. Equivalently, $P = NQN^{-1}$. Thus $P$ and $Q$, being similar matrices, have the same trace and the theorem is proved. $\square$

COROLLARY. *The type of the cycle decomposition of $\alpha$ on the point set $X$ is the same as the type of the cycle decomposition of $\alpha$ on the block set $\mathcal{A}$.*

PROOF: By Theorem 27.1, $\alpha^i$ has the same number of fixed points as fixed blocks, for each $i = 1, 2, \ldots$.

Suppose a permutation $\beta$ has $c_i$ cycles of length $i$ on some set $S$, $i = 1, 2, \ldots, |S|$. Let $f_j$ denote the number of fixed points of $\beta^j$. Then

$$f_j = \sum_{i \mid j} i c_i,$$

and by Möbius inversion, Theorem 10.4,

$$c_j = \sum_{i \mid j} \mu\left(\frac{j}{i}\right) f_i.$$

The point is that the numbers of cycles of each length (i.e. the type of $\beta$) are determined completely by the numbers of fixed points of the powers of $\beta$. $\square$

COROLLARY. *If $\Gamma$ is a group of automorphisms of a symmetric design, then the number of orbits of $\Gamma$ on the point set $X$ is the same as the number of orbits of $\Gamma$ on the block set $\mathcal{A}$. In particular, $\Gamma$ is transitive on the points if and only if $\Gamma$ is transitive on the blocks.*

PROOF: By Burnside's lemma, Theorem 10.5, the number of orbits of a group $\Gamma$ of permutations of a set $S$ is determined exactly by the multiset $(f(\alpha) : \alpha \in \Gamma)$ where $f(\alpha)$ is the number of elements of $S$ fixed by $\alpha$. $\square$

Let $G$ be an abelian group of order $v$. A $(v, k, \lambda)$-*difference set* in $G$ is a $k$-subset $D \subseteq G$ such that each nonzero $g \in G$ occurs exactly $\lambda$ times in the multiset $(x - y : x, y \in D)$ of differences from $D$. More formally, we are requiring that the number of ordered pairs $(x, y)$ with $x, y \in D$ and $x - y = g$ is $\lambda$ when $g \neq 0$ and this number is $k$ for $g = 0$. Evidently, $\lambda(v - 1) = k(k - 1)$.

EXAMPLE 27.1. Examples of difference sets include:

$(7,3,1)$   $\{1,2,4\}$ in $\mathbb{Z}_7$

$(13,4,1)$ $\{0,1,3,9\}$ in $\mathbb{Z}_{13}$

$(11,5,2)$ $\{1,3,9,5,4\}$ in $\mathbb{Z}_{11}$

$(16,6,2)$ $\{10,20,30,01,02,03\}$ in $\mathbb{Z}_4 \times \mathbb{Z}_4$

$(16,6,2)$ $\{0000,0001,0010,0100,1000,1111\}$ in $\mathbb{Z}_2 \times \mathbb{Z}_2 \times \mathbb{Z}_2 \times \mathbb{Z}_2$

A difference set is *nontrivial* when $1 < k < v - 1$. A difference set with $\lambda = 1$ is sometimes called *planar* or *simple*.

Let $G$ be an abelian group of order $v$. For $S \subseteq G$, $g \in G$, we denote by $S + g$ the *translate*, or *shift*,

$$S + g := \{x + g : x \in S\}$$

of $S$ by $g$. Let $D$ be a $k$-subset of $G$, and $x, y \in G$. In general, we claim that the number of shifts $D + g$ that contain both $x$ and $y$ is equal to the number of times $d := x - y$ occurs as a difference within $D$. This is because $g \mapsto (x - g, y - g)$ is a one-to-one correspondence between the set $\{g \in G : \{x, y\} \subseteq D + g\}$ and the set of ordered pairs $(a, b)$ of elements of $D$ such that $a - b = x - y$. (The reader may check that this common number is also equal to the cardinality of the intersection $(D + x) \cap (D + y)$.)

In particular, $(G, \{D + g : g \in G\})$ is a symmetric $(v, k, \lambda)$-design if and only if $D$ is a $(v, k, \lambda)$-difference set.

PROBLEM 27A. A $(v, k, \lambda)$-*quotient set* in an arbitrary group $G$ of order $v$ (written multiplicatively) is a $k$-subset $D \subseteq G$ such that any one of the following conditions holds:

(1) Each nonidentity element $g \in G$ occurs exactly $\lambda$ times in the list $(x^{-1}y : x, y \in D)$ of "left" quotients from $D$.

(2) Each nonidentity element $g \in G$ occurs exactly $\lambda$ times in the list $(xy^{-1} : x, y \in G)$ of "right" quotients from $D$.

(3) $|D \cap (Dg)| = \lambda$ for each nonidentity $g \in G$.

(4) $|D \cap (gD)| = \lambda$ for each nonidentity $g \in G$.

(5) $(G, \{Dg : g \in G\})$ is a symmetric $(v, k, \lambda)$-design.

(6) $(G, \{gD : g \in G\})$ is a symmetric $(v, k, \lambda)$-design.

Show that the above six conditions are equivalent.

THEOREM 27.2. *Let $G$ be a group of order $v$. The existence of a $(v, k, \lambda)$-quotient set in $G$ is equivalent to the existence of a sym-*

*metric* $(v, k, \lambda)$-*design that admits a group* $\hat{G}$ *of automorphisms that is isomorphic to* $G$ *and regular, i.e. sharply transitive, on the points of the design.*

PROOF: Let $D$ be a $(v, k, \lambda)$-quotient set in $G$. Then $(G, \{gD : g \in G\})$ is a symmetric $(v, k, \lambda)$-design. For $g \in G$, define a permutation $\hat{g}$ of $G$ by $\hat{g}(x) = gx$. Then each $\hat{g}$ is in fact an automorphism of $(G, \{gD : g \in G\})$ and the group $\hat{G} = \{\hat{g} : g \in G\}$ of automorphisms is clearly isomorphic to $G$ and regular on the points.

Conversely, let $G$ be given and let $(X, \mathcal{A})$ be a symmetric $(v, k, \lambda)$-design with regular group $\hat{G}$ of automorphisms of $(X, \mathcal{A})$ that is isomorphic to $G$. It will be sufficient to exhibit a $(v, k, \lambda)$-quotient set in $\hat{G}$.

Fix a point $x_0 \in X$ and a block $A_0 \in \mathcal{A}$. Let

$$D := \{\sigma \in \hat{G} : \sigma(x_0) \in A_0\}.$$

We claim that $D$ is a $(v, k, \lambda)$-quotient set in $\hat{G}$. Since $\hat{G}$ is regular and $|A_0| = k$, we have $|D| = k$. Let $\alpha$ be a nonidentity element of $\hat{G}$. Then $\alpha D = \{\alpha\sigma : \sigma(x_0) \in A_0\} = \{\tau : \tau(x_0) \in \alpha(A_0)\}$, so

$$D \cap (\alpha D) = \{\tau : \tau(x_0) \in A_0 \cap \alpha(A_0)\}.$$

Now since $\hat{G}$ is regular, $\alpha$ has no fixed points and hence, by Theorem 27.1, fixes no blocks. Thus the block $\alpha(A_0)$ is distinct from $A_0$, so $|A_0 \cap \alpha(A_0)| = \lambda$, and by regularity, $|D \cap (\alpha D)| = \lambda$. This holds for all nonidentity elements $\alpha$ and establishes our claim. $\qquad\square$

In particular, the existence of a *cyclic* $(v, k, \lambda)$-*difference set*, i.e. a difference set in $\mathbb{Z}_v$, is equivalent to the existence of a symmetric $(v, k, \lambda)$-design that admits a cyclic automorphism, i.e. an automorphism with cycle decomposition on the points—or blocks—consisting of one cycle of length $v$.

We now restrict our attention to difference sets in abelian groups. Ideally, we would like to describe and classify all difference sets—to find out which groups have difference sets, how many there are, etc. This we shall do for various small parameter triples, but in general the existence problem alone is already extremely difficult.

Observe that $D \subseteq G$ is a $(v, k, \lambda)$-difference set if and only if $G \setminus D$ is a $(v, v-k, v-2k+\lambda)$-difference set. Thus we may confine our attention to the case $k < \frac{1}{2}v$.

Also note that $D$ is a difference set if and only if every translate of $D$ is a difference set.

In the case that $(v, k) = 1$, it happens that we can choose a natural representative from the class of all translates; this can be a help in classification and will be useful in the next chapter. Call a subset of an abelian group $G$ *normalized* in the case that the sum of its elements is zero.

PROPOSITION 27.3. *Let $D$ be a $k$-subset of an abelian group $G$ of order $v$. If $(v, k) = 1$, then $D$ has a unique normalized translate.*

PROOF: Let $h$ be the sum of the elements of $D$. Then the sum of the elements of a translate $D + g$ is $h + kg$. Since $(v, k) = 1$, there is a unique group element $g$ with $h + kg = 0$. □

The reader is invited to verify to his or her satisfaction that the normalized difference sets with parameters $(v, k, \lambda) = (7, 3, 1)$, $(13, 4, 1)$, $(11, 5, 2)$, $(21, 5, 1)$ are, respectively:

$$\{1, 2, 4\}, \ \{3, 5, 6\} \text{ in } \mathbb{Z}_7;$$
$$\{0, 1, 3, 9\}, \ \{0, 2, 5, 6\}, \ \{0, 4, 10, 12\}, \ \{0, 7, 8, 11\} \text{ in } \mathbb{Z}_{13};$$
$$\{1, 3, 4, 5, 9\}, \ \{2, 6, 7, 8, 10\} \text{ in } \mathbb{Z}_{11};$$
$$\{7, 14, 3, 6, 12\}, \ \{7, 14, 9, 15, 18\} \text{ in } \mathbb{Z}_{21}.$$

(This will be easy after Theorem 28.3—see Example 28.2.) Of course, for $(v, k) = 1$ the total number of difference sets is $v$ times the number of normalized difference sets, e.g. there are 52 difference sets with parameters $(13, 4, 1)$ in $\mathbb{Z}_{13}$.

A final preliminary observation is that if $\alpha$ is any automorphism of the group $G$, then a subset $D \subseteq G$ is a difference set if and only if $\alpha(D)$ is a difference set. Thus from a given difference set we can obtain others by taking translates and by means of the symmetries of $G$. We say that difference sets $D_1, D_2$ in $G$ are *equivalent* when there exists $\alpha \in \text{Aut}(G)$ and $g \in G$ such that $D_2 = \alpha(D_1) + g$. (Check that this is an equivalence relation.) The normalized difference sets shown above are, for each parameter triple, equivalent;

indeed, each can be obtained from the others by multiplication by some integer relatively prime to the order of the respective group.

PROBLEM 27B. Show that all $(16, 6, 2)$-difference sets in $\mathbb{Z}_2 \times \mathbb{Z}_2 \times \mathbb{Z}_2 \times \mathbb{Z}_2$ are equivalent.

PROBLEM 27C. Recall from Example 19.4 that symmetric designs with parameters $v = 4t^2$, $k = 2t^2 - t$, $\lambda = t^2 - t$ are related to regular Hadamard matrices. Let $A$ and $B$ be, respectively, $(4x^2, 2x^2 - x, x^2 - x)$- and $(4y^2, 2y^2 - 2, y^2 - y)$-difference sets in groups $G$ and $H$ (admit $x$ or $y = 1$). Show that

$$D := (A \times (H \setminus B)) \cup ((G \setminus A) \times B)$$

is a $(4z^2, 2z^2 - z, z^2 - z)$-difference set in $G \times H$ where $z = x \times y$. (Thus if $G$ is the direct product of $m$ groups of order 4, then there is a $(4^m, 2 \cdot 4^{m-1} - 2^{m-1}, 4^{m-1} - 2^{m-1})$-difference set in $G$.)

We describe several known families of difference sets. The known constructions all seem to involve finite fields and/or vector spaces. The first of these examples is essentially contained in Chapter 18 on Hadamard matrices in the discussion of Paley matrices, but it is worth stating explicitly. Difference sets with parameters

$$(v, k, \lambda) = (4n - 1, 2n - 1, n - 1)$$

are often called *Hadamard* difference sets.

THEOREM 27.4 (PALEY, TODD). *Let $q = 4n - 1$ be a prime power. Then the set $D$ of nonzero squares in $\mathbb{F}_q$ is a $(4n - 1, 2n - 1, n - 1)$-difference set in the additive group of $\mathbb{F}_q$.*

PROOF: Clearly, $|D| = 2n - 1$.

Since $D$ is invariant under multiplication by elements of the set $S$ of nonzero squares, the multiset $M$ of differences from $D$ also has this property. Also, $M$ is obviously invariant under multiplication by $-1$. Since $q \equiv 3 \pmod 4$, $-1 \notin S$ and every nonzero element of $\mathbb{F}_q$ is either in $S$ or of the form $-s$ for some $s \in S$. In summary, $M$ is invariant under multiplication by all nonzero elements of $\mathbb{F}_q$, and so $D$ is a $(4n - 1, 2n - 1, \lambda)$-difference set for some $\lambda$. The relation $\lambda(v - 1) = k(k - 1)$ forces $\lambda = n - 1$. $\square$

We obtain difference sets $\{1, 2, 4\}$ in $\mathbb{Z}_7$, $\{1, 3, 4, 5, 9\}$ in $\mathbb{Z}_{11}$, and $\{1, 4, 5, 6, 7, 9, 11, 16, 17\}$ in $\mathbb{Z}_{19}$ from Theorem 27.4. The $(27, 13, 6)$-difference set later in the series will be in the elementary abelian group of order 27, *not* $\mathbb{Z}_{27}$.

PROBLEM 27D. Show that the Paley-Todd difference sets are normalized if $q > 3$.

Stanton and Sprott (1958) found another family of Hadamard difference sets.

THEOREM 27.5. *If $q$ and $q + 2$ are both odd prime powers, then with $4n - 1 := q(q+2)$, there exists a $(4n-1, 2n-1, n-1)$-difference set in the additive group of the ring $R := \mathbb{F}_q \times \mathbb{F}_{q+2}$.*

PROOF: Let $U := \{(a, b) \in R : a \neq 0, b \neq 0\}$ be the group of invertible elements of $R$. Let $V$ be the subgroup of $U$ consisting of those pairs $(a, b)$ such that both $a$ and $b$ are squares in the respective fields $\mathbb{F}_q$ and $\mathbb{F}_{q+2}$ or both $a$ and $b$ are nonsquares. Check that $V$ is an index 2 subgroup of $U$, and also that $(-1, -1) \notin V$. Put $T := \mathbb{F}_q \times \{0\}$. We claim that $D := T \cup V$ is a difference set as required. We do have $|D| = q + \frac{1}{2}(q - 1)(q + 1) = 2n - 1$.

Since $D$ is invariant under multiplication by elements of $V$, the multiset of differences from $D$ also has this property as well as the property that it is invariant under multiplication by $(-1, -1)$. So the multiset of differences from $D$ is invariant under multiplication by elements of all of $U$. Thus every element of $U$ occurs as a difference the same number of times, say $\lambda_1$. Every element $(x, 0)$ of $R$ with $x \neq 0$ will occur, say, $\lambda_2$ times; and every element $(0, y)$ of $R$ with $y \neq 0$ will occur, say, $\lambda_3$ times as a difference from $D$. Of course, we have

(27.1)     $k(k - 1) = (q - 1)(q + 1)\lambda_1 + (q - 1)\lambda_2 + (q + 1)\lambda_3.$

It is easy to evaluate $\lambda_2$ and $\lambda_3$. Differences of the form $(x, 0)$, $x \neq 0$ (there are $q - 1$ elements of this form) arise $q(q - 1)$ times from $T$, never as a difference between an element of $T$ and $V$, and $(q + 1) \cdot (\frac{1}{2}(q - 1))(\frac{1}{2}(q - 1) - 1)$ times as differences of two elements of $V$; thus

$$(q - 1)\lambda_2 = q(q - 1) + (q + 1) \cdot (\frac{1}{2}(q - 1))(\frac{1}{2}(q - 1) - 1)$$

from which we conclude $\lambda_2 = \frac{1}{4}(q+3)(q-1)$. In a similar manner (Problem 27E), the reader will find that $\lambda_3 = \frac{1}{4}(q+3)(q-1)$. Then (27.1) implies that $\lambda_1 = \frac{1}{4}(q+3)(q-1)$ also. $\qquad\square$

PROBLEM 27E. With the notation as in the proof of Theorem 27.5 prove that $\lambda_3 = \frac{1}{4}(q+3)(q-1)$.

In the case that $q$ and $q+2$ are both primes (twin primes), Theorem 27.5 yields cyclic difference sets. For if $s, t$ are relatively prime integers, $\mathbb{Z}_s \times \mathbb{Z}_t$ and $\mathbb{Z}_{st}$ are isomorphic as additive groups— and also as rings. An isomorphism $\mathbb{Z}_{st} \to \mathbb{Z}_s \times \mathbb{Z}_t$ is provided by $x \ (\mathrm{mod}\ st) \mapsto (x \ (\mathrm{mod}\ s), x \ (\mathrm{mod}\ t))$ . We obtain the $(15, 7, 3)$-difference set $\{0, 5, 10, 1, 2, 4, 8\}$ in $\mathbb{Z}_{15}$.

PROBLEM 27F. List the elements of a cyclic $(35, 17, 8)$-difference set.

As a special case of Singer's theorem, Theorem 27.6, we will obtain another family of Hadamard difference sets: cyclic difference sets with

$$v = 2^t - 1, \quad k = 2^{t-1} - 1, \quad \lambda = 2^{t-2} - 1.$$

Recall from Chapter 23 that the points and the hyperplanes of $PG(n, q)$ form a symmetric design with

$$(27.2) \qquad v = \frac{q^{n+1} - 1}{q - 1}, \quad k = \frac{q^n - 1}{q - 1}, \quad \lambda = \frac{q^{n-1} - 1}{q - 1}.$$

THEOREM 27.6. *For any prime power $q$ and positive integer $n$, there is a difference set $D$ with parameters as in (27.2) in the cyclic group of order $v$ so that the resulting symmetric design is isomorphic to the points and hyperplanes of $PG(n, q)$.*

PROOF: In view of Theorem 27.2, we need only show that there exists an automorphism of $PG(n, q)$ that permutes the points in a single cycle of length $v$, or equivalently so that the powers of the automorphism act transitively on the projective points. The points of $PG(n, q)$ are the 1-dimensional subspaces of an $(n+1)$-dimensional vector space $V$ over $\mathbb{F}_q$. Any nonsingular linear transformation $T$

from $V$ to itself will take subspaces to subspaces of the same dimension and thus gives us an automorphism of $PG(n,q)$.

As an $(n+1)$-dimensional vector space over $\mathbb{F}_q$, we choose $V := \mathbb{F}_{q^{n+1}}$ as a vector space over its subfield $\mathbb{F}_q$. Let $\omega$ be a primitive element of $\mathbb{F}_{q^{n+1}}$ and consider the linear transformation $T : x \mapsto \omega x$ of $V$ over $\mathbb{F}_q$. It is clear that $T$ is nonsingular and that its powers are transitive on the projective points (even on the nonzero vectors!).

$\square$

The difference sets constructed in the proof of Theorem 27.6 above are called *Singer difference sets*—see Singer (1938). We give a more concrete discussion and an example. Let $\omega$ be a primitive element of $\mathbb{F}_{q^{n+1}}$ and define $v := (q^{n+1} - 1)/(q - 1)$. The cyclic multiplicative group $\langle \omega \rangle$ of $\mathbb{F}_{q^{n+1}}$ has a unique subgroup of order $q - 1$, namely,

$$\langle \omega^v \rangle = \{\omega^0 = 1, \omega^v, \omega^{2v}, \ldots, \omega^{(q-2)v}\}.$$

But the multiplicative group of the subfield $\mathbb{F}_q$ has order $q - 1$, so we conclude $\mathbb{F}_q = \{0, \omega^0, \omega^v, \omega^{2v}, \ldots, \omega^{(q-2)v}\}$.

Now two "vectors" $\omega^i$ and $\omega^j$ in $\mathbb{F}_{q^{n+1}}$, considered as a vector space over $\mathbb{F}_q$, represent the same 1-dimensional subspace of $\mathbb{F}_{q^{n+1}}$ if and only if $\omega^i = \alpha \omega^j$ for some $0 \neq \alpha \in \mathbb{F}_q$, that is, if and only if $i \equiv j \pmod{v}$. Thus we have a one-to-one correspondence between the set $X$ of 1-dimensional subspaces (projective points) and the group $\mathbb{Z}_v$ of residue classes modulo $v$:

$$0 \leftrightarrow x_0 = \{0, \omega^0, \omega^v, \omega^{2v}, \ldots, \omega^{(q-2)v}\}$$
$$1 \leftrightarrow x_1 = \{0, \omega^1, \omega^{v+1}, \omega^{2v+1}, \ldots, \omega^{(q-2)v+1}\}$$
$$\vdots$$
$$i \leftrightarrow x_i = \{0, \omega^i, \omega^{v+i}, \omega^{2v+i}, \ldots, \omega^{(q-2)v+i}\}$$
$$\vdots$$
$$v - 1 \leftrightarrow x_{v-1} = \{0, \omega^{v-1}, \omega^{2v-1}, \omega^{3v-1}, \ldots, \omega^{(q-1)v-1}\}.$$

The map $x_i \mapsto x_{i+1}$ (subscripts modulo $v$) is an automorphism of the projective space. To obtain a difference set, let $U$ be any $n$-dimensional subspace of $\mathbb{F}_{q^{n+1}}$ and let $D := \{i \in \mathbb{Z}_v : x_i \in U\}$. A normalized difference set is obtained if $U$ is taken to be the subspace of elements with trace zero (the trace from $\mathbb{F}_{q^{n+1}}$ to $\mathbb{F}_q$).

EXAMPLE 27.2. Consider $n = 2$, $q = 5$. We construct a $(31, 6, 1)$-difference set. A zero $\omega$ of the polynomial $y^3 + y^2 + 2$ (coefficients in $\mathbb{F}_5$) is a primitive element of $\mathbb{F}_{5^3}$ and $1, \omega, \omega^2$ furnish a basis for $\mathbb{F}_{5^3}$ as a vector space over $\mathbb{F}_5$. The 124 nonzero elements of $\mathbb{F}_{5^3}$ fall into 31 cosets modulo the subgroup $\langle \omega^{31} \rangle = \{3, 4, 2, 1\} = \mathbb{F}_5 \setminus \{0\}$, each coset being the nonzero elements of a 1-dimensional subspace. Let us take $U := \mathrm{span}\{1, \omega\}$ as our 2-dimensional subspace. Representatives of the projective points on $U$ are $1, \omega, \omega + 1, \omega + 2, \omega + 3, \omega + 4$ and, after some computation,

$$1 = \omega^0$$
$$\omega = \omega^1$$
$$\omega + 1 = \omega^{29}$$
$$\omega + 2 = \omega^{99}$$
$$\omega + 3 = \omega^{80}$$
$$\omega + 4 = \omega^{84}.$$

The resulting Singer difference set is

$$\{0, 1, 29, 6, 18, 22\} \quad \text{in } \mathbb{Z}_{31}.$$

PROBLEM 27G. Find a $(57, 8, 1)$-difference set in $\mathbb{Z}_{57}$. (Some will probably not want to do this by hand.)

Consider $n = 3$, $q = 2$. Here the points of $PG(3, 2)$ are in one-to-one correspondence with the nonzero elements of $\mathbb{F}_{2^4}$ that in turn are in one-to-one correspondence with the residues modulo 15. It may be instructive to write out all lines and planes of this smallest projective 3-space so that the cyclic automorphism is clear.

A zero $\omega$ of $y^4 + y + 1$ (coefficients in $\mathbb{F}_2$) is a primitive element of $\mathbb{F}_{2^4}$ and $\omega^3, \omega^2, \omega, 1$ from a basis for $\mathbb{F}_{2^4}$ over $\mathbb{F}_2$. Any element of $\mathbb{F}_{2^4}$ can be written uniquely as $a_3\omega^3 + a_2\omega^2 + a_1\omega + a_0$ that we abbreviate as $a_3 a_2 a_1 a_0$ below. We have $\omega^4 + \omega + 1 = 0$, or $\omega^4 = 0011$. We first construct a table of vector representations of powers of $\omega$.

$$\mathbb{F}_{2^4}$$

$$
\begin{array}{lll}
\omega^0 = 0001 & \omega^5 = 0110 & \omega^{10} = 0111 \\
\omega^1 = 0010 & \omega^6 = 1100 & \omega^{11} = 1110 \\
\omega^2 = 0100 & \omega^7 = 1011 & \omega^{12} = 1111 \\
\omega^3 = 1000 & \omega^8 = 0101 & \omega^{13} = 1101 \\
\omega^4 = 0011 & \omega^9 = 1010 & \omega^{14} = 1001
\end{array}
$$

$$PG(3,2)$$

| Lines | | | Planes |
|---|---|---|---|
| {0,5,10} | {0,1,4} | {0,2,8} | {1,2,4,8,0,5,10} |
| {1,6,11} | {1,2,5} | {1,3,9} | {2,3,5,9,1,6,11} |
| {2,7,12} | {2,3,6} | {2,4,10} | {3,4,6,10,2,7,12} |
| {3,8,13} | {3,4,7} | {3,5,11} | {4,5,7,11,3,8,13} |
| {4,9,14} | {4,5,8} | {4,6,12} | {5,6,8,12,4,9,14} |
| | {5,6,9} | {5,7,13} | {6,7,9,13,5,10,0} |
| | {6,7,10} | {6,8,14} | {7,8,10,14,6,11,1} |
| | {7,8,11} | {7,9,0} | {8,9,11,0,7,12,2} |
| | {8,9,12} | {8,10,1} | {9,10,12,1,8,13,3} |
| | {9,10,13} | {9,11,2} | {10,11,13,2,9,14,4} |
| | {10,11,14} | {10,12,3} | {11,12,14,3,10,0,5} |
| | {11,12,0} | {11,13,4} | {12,13,0,4,11,1,6} |
| | {12,13,1} | {12,14,5} | {13,14,1,5,12,2,7} |
| | {13,14,2} | {13,0,6} | {14,0,2,6,13,3,8} |
| | {14,0,3} | {14,1,7} | {0,1,3,7,14,4,9} |

Note that for $q = 2$, the Singer difference sets have Hadamard parameters. We obtain $(31, 15, 7)$-difference sets both from Theorem 27.4 and Theorem 27.6. These two difference sets are not equivalent—and not even isomorphic. (Two difference sets are *isomorphic* when the corresponding symmetric designs are isomorphic. Equivalent difference sets are surely isomorphic, but the converse is not true.) The difference set $D$ of quadratic residues modulo 31 has the property that

$$D \cap (D + 1) \cap (D + 3) = \{5, 8, 10, 19\}.$$

The design $(\mathbb{Z}_{31}, \{D + g : g \in \mathbb{Z}_{31}\})$ cannot be isomorphic to the points and hyperplanes of $PG(4,2)$ since the intersection of flats is again a flat and $PG(4,2)$ has no flats with precisely four points.

Gordon, Mills, and Welch (1962) have shown that the construction of Singer's theorem can be modified in some cases to produce many nonequivalent difference sets with the same parameters.

PROBLEM 27H. Let $D$ be an $(n^2 + n + 1, n + 1, 1)$-difference set in an abelian group $G$. Show that $-D$ is an *oval* in the associated projective plane.

PROBLEM 27I. Let $G = \{0, a_0, a_1, \ldots, a_q\}$ be any group of order $q+2$, where $q$ is a power of a prime. Let $V$ be a 2-dimensional vector space over $\mathbb{F}_q$ and let $U_0, U_1, \ldots U_q$ be its 1-dimensional subspaces. Show that

$$D := \bigcup_{i=0}^{q} \{a_i\} \times U_i$$

is a difference set in $G \times V$. For example, we obtain a (45,12,3)-difference set.

**Notes.**

The idea of group difference sets, as a generalization of cyclic difference sets, is due to R. H. Bruck (1955).

Problem 27I is a result of McFarland (1973).

**References.**

R. H. Bruck (1955), Difference sets in a finite group, *Trans. Amer. Math. Soc.* **78**, 464–481.

B. Gordon, W. H. Mills, and L. R. Welch (1962), Some new difference sets, *Canad. J. Math.* **14**, 614–625.

R. L. McFarland (1973), A family of difference sets in non-cyclic groups, *J. Combinatorial Theory* (A) **15**, 1–10.

J. Singer (1938), A theorem in finite projective geometry and some applications to number theorey, *Trans. Amer. Math. Soc.* **43**, 377–385.

R. G. Stanton and D. A. Sprott (1958), A family of difference sets, *Canad. J. Math.* **10**, 73–77.

# 28

# Difference sets and the group ring

The group ring provides a natural and convenient setting for the study of difference sets. The existence of a difference set will be seen to be equivalent to the existence of a solution to a certain algebraic equation in a group ring. We shall use the group ring to derive the celebrated Multiplier Theorem of M. Hall, Jr. as well as number-theoretic criteria on the parameters of a difference set which are stronger than the Bruck-Ryser-Chowla theorem.

Let $R$ be a ring (commutative with one) and $G$ a finite abelian group (written additively). The elements of the *group ring* $R[G]$ are all formal sums

$$A = \sum_{g \in G} a_g x^g,$$

where $a_g \in R$ for each $g \in G$. (Here the symbol $x$ is just a place holder. The important thing is that we have an element $a_g$ of $R$ for each element $g$ of $G$, i.e. the elements of the group ring are in one-to-one correspondence with mappings $G \to R$.)

We define addition and scalar multiplication in the obvious way:

$$\sum_{g \in G} a_g x^g + \sum_{g \in G} b_g x^g := \sum_{g \in G} (a_g + b_g) x^g,$$

$$c \sum_{g \in G} a_g x^g := \sum_{g \in G} (c a_g) x^g.$$

Multiplication in $R[G]$ is defined by

$$\sum_{g \in G} a_g x^g \sum_{g \in G} b_g x^g := \sum_{g \in G} \left( \sum_{h+h'=g} a_h b_{h'} \right) x^g.$$

With these definitions, $R[G]$ is a commutative, associative $R$-algebra. The notation we have chosen is appropriate for abelian groups and emphasizes an analogy with polynomials. Note that when $G$ is the additive group of the residues modulo $v$, then the group ring $R[\mathbf{Z}_v]$ consists of all sums $\sum_{i=0}^{v-1} r_i x^i$ (exponents modulo $v$) and is isomorphic to the factor ring $R[x]/(x^v - 1)$ of the polynomial ring $R[x]$. In general, we shall even take to denoting elements of arbitrary group rings $R[G]$ as $A(x)$, $B(x), \dots$ when convenient. The element $x^0 \in R[G]$ is the multiplicative identity in $R[G]$ and we denote $x^0$ by 1.

We shall be concerned almost exclusively with the group rings $\mathbf{Z}[G]$ over the integers. For a subset $A \subseteq G$, we define $A(x) \in \mathbf{Z}[G]$ by

$$A(x) = \sum_{g \in A} x^g.$$

In particular, $G(x) = \sum_{g \in G} x^g$. For $A$, $B \subseteq G$,

$$A(x)B(x) = \sum_{g \in G} c_g x^g,$$

where $c_g$ is the number of times $g$ occurs in the multiset

$$(h + h' : h \in A, h' \in B)$$

of sums of elements of $A$ and $B$.

For $A(x) = \sum_{g \in G} a_g x^g \in \mathbf{Z}[G]$, we write

$$A(x^{-1}) := \sum_{g \in G} a_g x^{-g}.$$

So for a $k$-subset $D$ of a group $G$ of order $v$, $D$ is a $(v, k, \lambda)$-difference set in $G$ if and only if the equation

$$D(x)D(x^{-1}) = n + \lambda G(x)$$

holds in the group ring $\mathbf{Z}[G]$, where $n := k - \lambda$.

An important homomorphism from $\mathbf{Z}[G]$ to $\mathbf{Z}$ is given by

$$A(x) \mapsto A(1) := \sum_{g \in G} a_g \in \mathbf{Z}.$$

PROPOSITION 28.1. *Let $v, k, \lambda$ be positive integers such that*

$$\lambda(v - 1) = k(k - 1)$$

*and let $G$ be an abelian group of order $v$. The existence of a $(v, k, \lambda)$-difference set in $G$ is equivalent to the existence of an element $A(x) \in \mathbf{Z}[G]$ satisfying the equation*

(28.1)
$$A(x)A(x^{-1}) = n + \lambda G(x),$$

*where $n := k - \lambda$.*

PROOF: We have already pointed out that for a subset $D$ of $G$, $D(x)$ satisfies (28.1) if and only if $D$ is a $(v, k, \lambda)$-difference set. It remains to show that if there is a solution $A(x)$ to (28.1), then we can find a solution $B(x) = \sum_{g \in G} b_g x^g$ where the coefficients $b_g$ are 0's and 1's.

Assume that $A(x)$ satisfies (28.1) and apply the homomorphism "$x \mapsto 1$" from $\mathbf{Z}[G] \to \mathbf{Z}$. We find

$$(A(1))^2 = n + \lambda v = k^2,$$

so $A(1) = k$ or $-k$. Now if $A(x)$ satisfies (28.1), then so does $B(x) = -A(x)$, so we may assume $A(1) = \sum a_g = k$.

The coefficient of $1 = x^0$ in $A(x)A(x^{-1})$ is $k = \sum_{g \in G} a_g^2$. Thus $\sum_{g \in G} a_g(a_g - 1) = 0$. But $a(a - 1)$ is strictly positive unless the integer $a$ is 0 or 1. □

For any integer $t$, $g \mapsto tg$ is a homomorphism of the group $G$ into itself and induces a ring homomorphism $\mathbf{Z}[G] \to \mathbf{Z}[G]$, namely

$$A(x) = \sum_{g \in G} a_g x^g \quad \mapsto \quad A(x^t) := \sum_{g \in G} a_g x^{tg}.$$

For $A, B \in \mathbf{Z}[G]$ and $n \in \mathbf{Z}$, we say $A \equiv B \pmod{n}$ when $A - B = nC$ for some $C \in \mathbf{Z}[G]$. Here is an easy lemma we will use often.

LEMMA 28.2. *Let $p$ be a prime and $A \in \mathbb{Z}[G]$. Then*

$$(A(x))^p \equiv A(x^p) \pmod{p}.$$

PROOF: We proceed by induction on the number of nonzero coefficients of $A(x)$. The lemma holds for $A(x) = 0$. Now if $A(x) = cx^g + B(x)$ where $B^p(x) \equiv B(x^p) \pmod{p}$, then

$$A^p(x) = (cx^g + B(x))^p \equiv (cx^g)^p + B^p(x)$$
$$= c^p x^{pg} + B^p(x) \equiv cx^{pg} + B(x^p) = A(x^p),$$

where the congruences are modulo $p$. $\qquad\square$

Let $G$ be an abelian group and $D$ a difference set in $G$. An automorphism $\alpha$ of $G$ is said to be a *multiplier* of $D$ if and only if the difference set $\alpha(D)$ is in fact a translate of $D$, i.e. if and only if $\alpha(D) = D + g$ for some $g \in G$. For example, the automorphism $x \mapsto 3x$ of $\mathbb{Z}_{13}$ is a multiplier of the $(13, 4, 1)$-difference set $\{0,2,3,7\}$ since $\{0,6,9,8\} = \{0,2,3,7\} + 6$.

If $t$ is an integer relatively prime to the order of $G$, then $x \mapsto tx$ is an automorphism of $G$ since the maps $x \mapsto t_1 x$ and $x \mapsto t_2 x$ coincide if and only if $t_1 \equiv t_2 \pmod{v^*}$ where $v^*$ is the exponent of $G$, i.e. the least common multiple of the orders of the elements of $G$. If $x \mapsto tx$ is a multiplier of a difference set $D$ in $G$, we say that $t$ is a *numerical multiplier* or *Hall multiplier* of $D$. It is at first surprising that many difference sets (e.g. all known cyclic difference sets) must necessarily have a nontrivial numerical multiplier.

PROBLEM 28A. Let $q = p^t$, $p$ prime. Show that $p$ is a multiplier of the Singer difference sets described in the previous chapter.

Observe that an automorphism $\alpha$ of $G$ is a multiplier of a difference set $D$ in $G$ if and only if in the group ring $\mathbb{Z}[G]$, $D(x^\alpha) = x^g \cdot D(x)$ for some $g \in G$.

THEOREM 28.3 (MULTIPLIER THEOREM, FIRST VERSION). *Let $D$ be a $(v, k, \lambda)$-difference set in an abelian group $G$ of order $v$. Let $p$ be a prime, $p|n$, $(p, v) = 1$, $p > \lambda$. Then $p$ is a numerical multiplier of $D$.*

It will be convenient to isolate part of the proof as a lemma.

LEMMA 28.4. *Let $\alpha$ be an automorphism of $G$ and let $D$ be a $(v, k, \lambda)$-difference set in $G$. Consider*

$$S(x) := D(x^\alpha)D(x^{-1}) - \lambda G(x).$$

*Then $\alpha$ is a multiplier of $D$ if and only if $S(x)$ has nonnegative coefficients.*

PROOF: We begin by remarking that if $\alpha$ is a multiplier, then we have $D(x^\alpha) = x^g \cdot D(x)$ for some $g \in G$, and then

$$D(x^\alpha)D(x^{-1}) = x^g \cdot D(x)D(x^{-1}) = x^g(n + \lambda G(x)) = nx^g + \lambda G(x).$$

So in this case, $S(x)$, as defined above, is equal to $nx^g$ for some $g \in G$. In particular, it has nonnegative coefficients. Note that, conversely, if $D(x^\alpha)D(x^{-1}) = nx^g + \lambda G(x)$, we can multiply this by $D(x)$ to find

$$D(x^\alpha) \cdot (n + \lambda G(x)) = nx^g \cdot D(x) + \lambda D(x)G(x),$$

$$nD(x^\alpha) + \lambda k G(x) = nx^g \cdot D(x) + \lambda k G(x).$$

So $D(x^\alpha) = x^g \cdot D(x)$ and $\alpha$ is a multiplier.

Now "$x \mapsto x^\alpha$" is an automorphism of $\mathbb{Z}[G]$, so $D(x^\alpha)D(x^{-\alpha}) = n + \lambda G(x)$, i.e. $\alpha(D)$ is also a difference set, and

$$\begin{aligned}
S(x)S(x^{-1}) &= \{D(x^\alpha)D(x^{-1}) - \lambda G(x)\}\{D(x^{-\alpha})D(x) - \lambda G(x)\} \\
&= \{n + \lambda G(x)\}^2 - 2\lambda k^2 G(x) + \lambda^2 v G(x) \\
&= n^2 + 2\lambda(n + \lambda v - k^2)G(x) = n^2.
\end{aligned}$$

Suppose that $S(x) = \sum_{g \in G} s_g x^g$ with nonnegative coefficients $s_g$. If $s_g > 0$ and $s_h > 0$ for $g, h \in G$, then the coefficient of $x^{g-h}$ in $S(x)S(x^{-1})$, $= n^2$, is at least $s_g s_h$, i.e. strictly positive, and so $x^{g-h} = x^0$, i.e. $g = h$. So $S(x)$ can have only one positive coefficient, say $S(x) = s_g x^g$. The equation $S(x)S(x^{-1}) = n^2$ forces $s_g = n$ and we have shown $S(x) = nx^g$. As noted above, we may conclude that $\alpha$ is a multiplier.                                        $\square$

PROOF OF THEOREM 28.2: Let $S(x) := D(x^p)D(x^{-1}) - \lambda G(x)$. By Lemma 28.4, it will suffice to show that $S(x)$ has nonnegative coefficients. By Lemma 28.2,

$$D(x^p)D(x^{-1}) \equiv D^p(x)D(x^{-1}) \equiv D^{p-1}(x)D(x)D(x^{-1})$$
$$\equiv D^{p-1}(x) \cdot (n + \lambda G(x)) \equiv nD^{p-1}(x) + \lambda k^{p-1}G(x)$$
$$\equiv \lambda G(x) \pmod{p},$$

since $p$ divides $n$ and $\lambda k^{p-1} \equiv \lambda^p \equiv \lambda \pmod{p}$. Thus the coefficients of $D(x^p)D(x^{-1})$, which are clearly nonnegative, are all congruent to $\lambda$ modulo $p$. Since $p > \lambda$, it must be that the coefficients of $D(x^p)D(x^{-1})$ are greater than or equal to $\lambda$, i.e. $S(x)$ has nonnegative coefficients. □

PROBLEM 28B. Find an element $S(x)$ in $\mathbf{Z}[\mathbf{Z}_7]$ with the property that $S(x)S(x^{-1}) = 4$, but with $S(x) \neq \pm 2x^9$.

Note that the hypothesis $p > \lambda$ was essential for our proof. Yet in every *known* difference set, *every* prime divisor of $n$ (not dividing $v$) is a multiplier. It may be, then, that the hypothesis $p > \lambda$ is unnecessary and by now the question of whether this is so has acquired the status of a classical unsolved problem. There have also been several generalizations of the original Multiplier Theorem, all of which would be trivial if we could eliminate the condition $p > \lambda$, and we state such a generalization below after first giving some applications of the current version.

COROLLARY. *For $\lambda = 1$, every prime divisor of $n$, and hence every divisor, is a multiplier of every $(n^2 + n + 1, n + 1, 1)$-difference set.*

EXAMPLE 28.1. We claim that there are no $(n^2 + n + 1, n + 1, 1)$-difference sets with $n \equiv 0 \pmod 6$. Let $D$ be such a hypothetical difference set, and without loss of generality, assume $D$ is normalized so that it is fixed by all multipliers. Both 2 and 3 would be multipliers. Then for $x \in D$, $2x$ and $3x$ also are in $D$. Then the difference $x$ occurs twice, once as $2x - x$, and once as $3x - 2x$; these are different occurrences as long as $3x \neq 2x$, i.e. $x \neq 0$. This contradicts $\lambda = 1$.

PROBLEM 28C. Show that there are no $(n^2 + n + 1, n + 1, 1)$-difference sets with $n$ divisible by any of 10, 14, 15, 21, 22, 26,

34, 35.

It has been possible to prove, with the Multiplier Theorem and other techniques, that no planar difference sets exist for $n \leq 3600$ unless $n$ is a prime power. But the conjecture that $n$ must be a prime power remains open.

EXAMPLE 28.2. Consider a normalized $(21, 5, 1)$-difference set $D$ in $\mathbb{Z}_{21}$. By the Multiplier Theorem, 2 is a multiplier and hence $2D = D$. Thus $D$ must be the union of the cycles of $x \mapsto 2x$ on $\mathbb{Z}_{21}$. These are

$$\{0\}, \quad \{1, 2, 4, 8, 16, 11\}, \quad \{3, 6, 12\}, \quad \{5, 10, 20, 19, 17, 13\},$$

$$\{7, 14\}, \quad \text{and} \quad \{9, 18, 15\}.$$

But $D$ has 5 elements, so *if* there is such a difference set, $D$ must be $\{7, 14, 3, 6, 12\}$ or $\{7, 14, 9, 18, 15\}$. It turns out that both of these *are* difference sets. One consists of the negatives of the other, so the 42 difference sets with these parameters are all equivalent.

PROBLEM 28D. Find all normalized difference sets with parameters $(7, 3, 1)$, $(11, 5, 2)$, $(13, 4, 1)$, $(19, 9, 4)$, $(31, 10, 3)$, and $(37, 9, 2)$. (Note: no difference sets exist for one of the parameter triples.)

We give two more easy lemmas on the group ring at this point.

LEMMA 28.5. *Let $G$ be an abelian group of order $v$ and $p$ a prime, $p \nmid v$. Let $A \in \mathbb{Z}[G]$ and suppose $A^m \equiv 0 \pmod{p}$ for some positive integer $m$. Then $A \equiv 0 \pmod{p}$.*

PROOF: Choose a power $q = p^e$ of $p$ such that $q \geq m$ and $q \equiv 1 \pmod{v}$. Then surely $A^q(x) \equiv 0 \pmod{p}$. But by Lemma 28.2, $A^q(x) \equiv A(x^q) \pmod{p}$, so $A(x^q) \equiv 0 \pmod{p}$. Since $q \equiv 1 \pmod{v}$, $qg = g$ for every $g \in G$ and $A(x) = A(x^q)$. $\square$

Note that $x^g \cdot G(x) = G(x)$ in $\mathbb{Z}[G]$. It follows that

$$A(x)G(x) = A(1)G(x).$$

For $n \in \mathbb{Z}$ and $A, B \in \mathbb{Z}[G]$, we say $A \equiv B \pmod{n, G}$ when $A - B$ is an element of the ideal in $\mathbb{Z}[G]$ generated by $n$ and $G = G(x)$, or equivalently, when

$$A - B = nC + mG$$

for some $C \in \mathbb{Z}[G]$.

LEMMA 28.6. *Let $G$ be an abelian group of order $v$ and $p$ a prime, $p \nmid v$. If $A \in \mathbb{Z}[G]$ and*

$$A^m \equiv 0 \pmod{p, G}$$

*for some positive integer $m$, then $A \equiv 0 \pmod{p, G}$.*

PROOF: Choose $q = p^e$ with $q \equiv 1 \pmod{v}$ and $q \geq m$. Then $A^q(x) \equiv 0 \pmod{p, G}$ and $A^q(x) \equiv A(x^q) = A(x) \pmod{p}$. □

THEOREM 28.7 (MULTIPLIER THEOREM, SECOND VERSION). *Let $D$ be a $(v, k, \lambda)$-difference set in an abelian group $G$ of exponent $v^*$. Let $t$ be an integer, $(t, v) = 1$, and suppose we can find a divisor $m$ of $n := k - \lambda$ such that $m > \lambda$ and for every prime divisor $p$ of $m$, there is an integer $f$ for which $p^f \equiv t \pmod{v^*}$. Then $t$ is a numerical multiplier of $D$.*

PROOF: The proof will use Lemmas 28.2, 28.4, 28.5 and the following observation:

Let $D$ be a $(v, k, \lambda)$-difference set in $G$, $\alpha$ an automorphism of $G$, and put $S(x) := D(x^\alpha)D(x^{-1}) - \lambda G(x)$. Assume that $\alpha$ has order $e$, so that $\alpha^e = $ identity. Then, we assert, in the group ring $\mathbb{Z}[G]$,

$$S(x)S(x^\alpha)S(x^{\alpha^2}) \cdots S(x^{\alpha^{e-1}}) = n^e.$$

To see this, note that for any integer $i$, we have

$$D(x^{\alpha^i})D(x^{-\alpha^i}) = n + \lambda G(x) \equiv n \pmod{G}$$

and

$$S(x^{\alpha^i}) = D(x^{\alpha^{i+1}})D(x^{-\alpha^i}) - \lambda G(x)$$
$$\equiv D(x^{\alpha^{i+1}})D(x^{-\alpha^i}) \pmod{G}.$$

Then

$$S(x)S(x^\alpha)s(x^{\alpha^2}) \cdots S(x^{\alpha^{e-1}})$$
$$\equiv \{D(x^\alpha)D(x^{-1})\}\{D(x^{\alpha^2})D(x^{-\alpha})\} \cdots \{D(x)D(x^{-\alpha^{e-1}})\}$$
$$\equiv \{D(x)D(x^{-1})\}\{D(x^\alpha)D(x^{-\alpha})\} \cdots \{D(x^{\alpha^{e-1}})D(x^{-\alpha^{e-1}})\}$$
$$\equiv n^e \pmod{G}.$$

Thus $S(x)S(x^\alpha) \cdots S(x^{\alpha^{e-1}}) = n^e + \ell G(x)$ for some integer $\ell$. But $S(1) = (D(1))^2 - \lambda G(1) = k^2 - \lambda v = n$; so applying the homomorphism $x \mapsto 1$, we find $n^e = n^e + \ell v$, and hence $\ell = 0$.

To continue with the proof, let $S(x) := D(x^t)D(x^{-1}) - \lambda G(x)$. By Lemma 28.4, to show $t$ is a multiplier, it will be sufficient to prove that $S(x)$ has nonnegative coefficients. Each coefficient of $S(x)$ is at least $-\lambda$ in value and since $m > \lambda$, the nonnegativity of the coefficients will follow if we can establish that $S(x) \equiv 0 \pmod{m}$. To establish this, it will suffice to show that $S(x) \equiv 0 \pmod{p^i}$ whenever $p$ is prime and $p^i$ divides $m$. This we do below.

Let $e$ be the order of $t$ modulo $v^*$, so that $t^e \equiv 1 \pmod{v^*}$. As shown above,

$$S(x)S(x^t)S(x^{t^2}) \cdots S(x^{t^{e-1}}) = n^e.$$

Let $p$ be a prime divisor of $m$ and let $f$ be such that $p^f \equiv t \pmod{v^*}$. Then

$$S(x)S(x^{p^f})S(x^{p^{2f}}) \cdots S(x^{p^{f(e-1)}}) = n^e.$$

Let $p^i$ be the highest power of $p$ dividing $n$ and let $p^j$ be the highest power of $p$ dividing (all coefficients of) $S(x)$, so $S(x) = p^j T(x)$ where $T(x) \not\equiv 0 \pmod{p}$. Then

$$p^j T(x)\, p^j T(x^{p^f}) \cdots p^j T(x^{p^{f(e-1)}}) = n^e,$$

from which it follows that $p^j$ divides $n$ (so $j \le i$) and

$$T(x)T(x^{p^f}) \cdots T(x^{p^{f(e-1)}}) = \left(\frac{n}{p^j}\right)^e.$$

Suppose that $j < i$, so that $\left(\frac{n}{p^j}\right)^e$ is divisible by $p$. Then

$$\begin{aligned}
0 &\equiv T(x)T(x^{p^f}) \cdots T(x^{p^{f(e-1)}}) \\
&\equiv T(x)T^{p^f}(x) \cdots T(x)T^{p^{f(e-1)}}(x) \\
&\equiv (T(x))^{1+p^f+\cdots+p^{f(e-1)}} \pmod{p}.
\end{aligned}$$

But then, by Lemma 28.5, $T(x) \equiv 0 \pmod{p}$, contradicting the choice of $j$. Thus $i = j$.                    $\square$

COROLLARY. If $n = p^e$, $p$ prime, $(p, v) = 1$, then $p$ is a numerical multiplier of every $(v, k, \lambda)$-difference set.

PROOF: A difference set $D$ and its complement $G \setminus D$ have the same multipliers. Thus we may assume that $k < \frac{1}{2}v$ and hence $n > \lambda$. In Theorem 28.7, take $m = n$, $t = p$.     □

EXAMPLE 28.3. Consider a hypothetical (25,9,3)-difference set. Take $t = 2$ and $m = 6$ in the above theorem. Since $3^3 \equiv 2$ (mod 25), we may conclude that 2 is a multiplier.

PROBLEM 28E. Find all normalized difference sets with parameters $(15, 7, 3)$, $(25, 9, 3)$, $(39, 19, 9)$, $(43, 15, 5)$, and $(61, 16, 4)$. Determine, for each parameter set, whether these normalized difference sets are equivalent. (Note: difference sets do not exist for most parameter sets.)

PROBLEM 28F. Let $D$ be a nontrivial $(v, k, \lambda)$-difference set. Prove that if $-1$ is a multiplier of $D$, then $v$ is even.

PROBLEM 28G. Find a (36,15,6)-difference set in $\mathbb{Z}_6 \times \mathbb{Z}_6$ with $-1$ as a multiplier.

$$* \; * \; *$$

Suppose $D$ is a $(v, k, \lambda)$-difference set in an abelian group $G$ of even order $v$. By Theorem 19.11(i), we know that $n$ is a square. But in the case of difference sets, we can say *what* $n$ is the square *of!* Let $A$ be any subgroup of $G$ of index 2. Say $D$ contains $a$ elements of $A$ and $b$ elements of $B := G \setminus A$. Since every element of $B$ occurs $\lambda$ times as a difference from $D$, and only differences of one element in $A$ and one in $B$ lie in $B$, $2ab = \frac{1}{2}\lambda v$. From this and $a + b = k$, we find that $(a - b)^2 = n$.

Now suppose $v$ is divisible by 3, let $A$ be a subgroup of $G$ of index 3, and let $B$ and $C$ be the cosets of $A$ in $G$. Say $D$ contains $a$ elements of $A$, $b$ elements of $B$, and $c$ elements of $C$. The number of differences from $D$ which lie in $B$, say, is $ba + cb + ac$, which on the other hand must be $\frac{1}{3}\lambda v$. From this and $a + b + c = k$, we can check that $4n = (b + c - 2a)^2 + 3(b - c)^2$. Now it is not true that every integer can be written as the sum of a square and three times another square, so this condition rules out the existence of certain difference sets. For example, there are no $(39, 19, 9)$-difference sets, even though there are symmetric designs with these parameters, because $4n = 40$ cannot be written as above.

We generalize the above necessary conditions on the parameters of difference sets by considering homomorphisms of the group ring. If $\alpha$ is a homomorphism $G \to H$, then $\alpha$ induces a ring homomorphism $\mathbb{Z}[G] \to \mathbb{Z}[H]$:

$$A(x) = \sum a_g x^g \quad \mapsto \quad A(x^\alpha) := \sum a_g x^{\alpha(g)} \in \mathbb{Z}[H].$$

THEOREM 28.8. *Let $D$ be a $(v, k, \lambda)$-difference set in an abelian group $G$. Let $u > 1$ be a divisor of $v$. If $p$ is a prime divisor of $n$ and*

$$p^f \equiv -1 \pmod{u}$$

*for some integer $f$, then $p$ does not divide the squarefree part of $n$.*

PROOF: We prove something stronger. Let $\alpha : G \to H$ be a homomorphism onto a group $H$ of order $u$ and exponent $u^*$, say. Suppose $p$ is prime and $p^f \equiv -1 \pmod{u^*}$ for some $f$. We will show that $n$ is exactly divisible by an even power $p^{2j}$ of $p$ and that for this $j$,

$$D(x^\alpha) \equiv 0 \pmod{p^j, H} \quad \text{in } \mathbb{Z}[H].$$

In other words, all the coefficients of $D(x^\alpha)$, which are the numbers of elements of $D$ belonging to the various cosets of the kernel of $\alpha$, are congruent to each other modulo $p^j$.

Let $p^i$ be the highest power of $p$ which divides $n$. Our hypothesis $p^f \equiv -1 \pmod{u^*}$ means that $D(x^{-\alpha}) = D(x^{p^f \alpha})$ in $\mathbb{Z}[H]$ (where we will do our calculations), so we have

(28.2) $D(x^\alpha)D(x^{p^f \alpha}) =$
$$D(x^\alpha)D(x^{-\alpha}) = n + \lambda \frac{v}{u} H(x) \equiv 0 \pmod{p^i, H}.$$

Let $p^j$ be the highest power of $p$ so that $D(x^\alpha) \equiv 0 \pmod{p^j, H}$; say $D(x^\alpha) \equiv p^j A(x) \pmod{H}$. The reader may check that (28.2) implies that $2j \le i$ and that if $2j < i$, then $A(x)A(x^{p^f}) \equiv 0$

(mod $p, H$). But then Lemmas 28.2 and 28.6 imply in turn that $A(x)^{1+p^f} \equiv 0$ (mod $p, H$) and then $A(x) \equiv 0$ (mod $p, H$). This contradicts the choice of $j$, so $2j = i$. $\qquad\square$

Consequences of Theorem 28.8 are: if $v$ is divisible by 3, then all prime divisors of the squarefree part of $n$ are congruent to 0 or 1 (modulo 3); if $v$ is divisible by 5, then all prime divisors of the squarefree part of $n$ are congruent to 0, 1, or 4 (modulo 5); if $v$ is divisible by 7, then all prime divisors of the squarefree part of $n$ are congruent to 0, 1, 2, or 4 (modulo 7).

EXAMPLE 28.4. We give an application of the stronger claim given in the proof of Theorem 28.8. Consider a hypothetical $(154, 18, 2)$-difference set $D$ in $G$. Let $\alpha : G \to H$ be a homomorphism onto a group $H$ of order $u := 11$. Take $p := 2$. Since $2^5 \equiv -1$ (mod 11), we conclude that $D(x^\alpha) \equiv 0$ (mod $4, H$). This means the 11 coefficients of $D(x^\alpha)$ are all congruent to some integer $h$ modulo 4; since the 11 coefficients sum to 18, $h$ is 2. But then the coefficients would sum to at least 22 and this contradiction shows that no such difference sets exist.

PROBLEM 28H. Let $D$ be a $(q^4+q^2+1, q^2+1, 1)$-difference set in an abelian group $G$. Let $\alpha : G \to H$ be a homomorphism onto a group $H$ of order $u := q^2 - q + 1$. Show that any coset of the kernel of $\alpha$ is the point set of a Baer subplane of the projective plane of order $q^2$ arising from $D$. For example, the 21 points of the projective plane of order 4 are partitioned into three Fano configurations.

PROBLEM 28I. Let $D$ be a $(q^3+q^2+q+1, q^2+q+1, q+1)$-difference set in an abelian group $G$. Let $\alpha : G \to H$ be a homomorphism onto a group $H$ of order $u := q + 1$. Show that the translates of $D$ meet the cosets of the kernel of $\alpha$ in $q + 1$ points or a single point. Show further that if the symmetric design arising from $D$ consists of the points and planes of $PG_3(q)$, then the cosets are ovoids (see Example 26.5).

**Notes.**

The celebrated Multiplier Theorem was first proved in the case of cyclic planar difference sets in Hall (1947). It was generalized to

$\lambda > 1$ by Hall and Ryser (1951) and since then has been extended in many ways—see e.g. Mann (1965), Baumert (1971), and Lander (1983). Algebraic number theory and characters of abelian groups often play a role in the proofs of these results, but we have chosen here to give proofs of Theorem 28.7 and 28.8 using only the group ring.

Marshall Hall (1910–1990) did fundamental work in group theory and coding theory as well as on combinatorial designs. He has been a tremendous influence on many mathematicians, including the authors.

Theorem 28.8 is due to K. Yamamoto (1963).

### References.

L. D. Baumert (1971), *Cyclic Difference Sets*, Lecture Notes in Math. **182**, Springer-Verlag.

M. Hall (1947), Cyclic projective planes, *Duke J. Math.* **14**, 1079–1090.

M. Hall and H. J. Ryser (1951), Cyclic incidence matrices, *Canad. J. Math.* **3**, 495–502.

E. S. Lander (1983), *Symmetric Designs: An Algebraic Approach*, London Math. Soc. Lecture Note Series **74**, Cambridge University Press.

H. B. Mann (1965), *Addition Theorems*, Wiley.

K. Yamamoto (1963), Decomposition fields of difference sets, *Pacific J. Math.* **13**, 337–352.

# 29
# Codes and symmetric designs

In this chapter, we elaborate on some of the material introduced in Chapter 20. We saw in that chapter that the rows of the incidence matrix of a projective plane of order $n \equiv 2 \pmod 4$ span a binary code which, when extended, is selfdual. It was recently observed by H. A. Wilbrink (1985) and others that that result can be used to show that planar difference sets do not exist for values of $n > 2$ with $n \equiv 2 \pmod 4$; see Theorem 29.7.

One can consider the code spanned by the rows of the incidence matrix over other prime fields $\mathbb{F}_p$ as well. With essentially the same proof as that of Theorem 20.6, we have the following theorem.

THEOREM 29.1. *If $p$ divides $n := k - \lambda$, then the $\mathbb{F}_p$-span $C$ of the rows of the incidence matrix $N$ of a symmetric $(v, k, \lambda)$-design has dimension at most $(v + 1)/2$ over $\mathbb{F}_p$. If $(p, k) = 1$ and $p^2$ does not divide $n$, then the dimension of this p-ary code is exactly $(v + 1)/2$.*

PROBLEM 29A. Prove Theorem 29.1.

It is not possible, in general, to extend the code $C$ to get a code of length $v + 1$ that is selforthogonal with respect to the *standard* dot product on $\mathbb{F}_p^v$. But let us consider other "scalar products" for odd primes $p$.

To a nonsingular $m$ by $m$ matrix $B$ over a field $\mathbb{F}$, we may associate the scalar product (or bilinear form)

$$\langle \mathbf{x}, \mathbf{y} \rangle := \mathbf{x} B \mathbf{y}^\top$$

for $\mathbf{x}, \mathbf{y} \in \mathbb{F}^m$. For a subspace $C$ of $\mathbb{F}^m$, let

$$C^B := \{\mathbf{x} : \langle \mathbf{x}, \mathbf{y} \rangle = 0 \text{ for all } \mathbf{y} \in C\}.$$

Then $C$ and $C^B$ have complementary dimensions and $(C^B)^B = C$. We say that $C$ is *totally isotropic* when $C \subseteq C^B$; this is the appropriate terminology for this generalization of selforthogonal. We note without proof the following theorem of Witt.

THEOREM 29.2. *Given a symmetric nonsingular matrix $B$ over a field $\mathbb{F}$ of odd characteristic, there exists a totally isotropic subspace of dimension $m/2$ in $\mathbb{F}^m$ if and only if $(-1)^{m/2} \det(B)$ is a square in $\mathbb{F}$.*

It will be convenient to retain the term *selfdual* for a totally isotropic subspace of dimension equal to half its length. E. S. Lander (1983) has shown how to associate a *family* of $p$-ary codes of length $v + 1$ and a scalar product to a symmetric design in such a way that one of the codes is selfdual when $n$ is exactly divisible by an odd power of $p$. Theorem 29.2 then gives us a condition on the parameters of the symmetric design. It turns out that these conditions are already consequences of the Bruck-Ryser-Chowla theorem, Theorem 19.11. In some sense, then, these selfdual codes provide a "combinatorial interpretation" of part of the BRC theorem. In any case, these codes carry information about the design and have further applications in the theory of symmetric designs—see Lander (1983).

THEOREM 29.3. *Suppose there exists a symmetric $(v, k, \lambda)$-design where $n$ is exactly divisible by an odd power of a prime $p$. Write $n = p^f n_0$ ($f$ odd) and $\lambda = p^b \lambda_0$ with $(n_0, p) = (\lambda_0, p) = 1$. Then there exists a selfdual $p$-ary code of length $v + 1$ with respect to the scalar product corresponding to*

$$B = \begin{cases} \operatorname{diag}(1, 1, \ldots, 1, -\lambda_0) & \text{if } b \text{ is even,} \\ \operatorname{diag}(1, 1, \ldots, 1, n_0\lambda_0) & \text{if } b \text{ is odd.} \end{cases}$$

*Hence from Theorem 29.2,*

$$\begin{cases} -(-1)^{(v+1)/2}\lambda_0 \text{ is a square} \pmod{p} & \text{if } b \text{ is even,} \\ (-1)^{(v+1)/2}n_0\lambda_0 \text{ is a square} \pmod{p} & \text{if } b \text{ is odd.} \end{cases}$$

Towards the proof of Theorem 29.2, we first prove two propositions.

Given any integral $m$ by $m$ matrix $A$, we may consider the $\mathbb{Z}$-module $M(A)$ consisting of all *integral* linear combinations of its rows; that is,

$$M(A) := \{\mathbf{y}A \colon \mathbf{y} \in \mathbb{Z}^m\}.$$

Fix a prime $p$ and for any positive integer $i$ define modules

$$M_i := \{\mathbf{x} \in \mathbb{Z}^m \colon p^i\mathbf{x} \in M(A)\},$$

$$N_i := \{\mathbf{y} \in \mathbb{Z}^m \colon A\mathbf{y}^\top \equiv 0 \pmod{p^{i+1}}\}.$$

We have $M_0 = M(A)$, $M_i \subseteq M_{i+1}$ and $N_i \supseteq N_{i+1}$ for all $i$. Let

(29.1) $$C_i := M_i \pmod{p}, \qquad D_i := N_i \pmod{p}.$$

That is, read all the integer vectors in $M_i$ or $N_i$ modulo $p$ to obtain $C_i$ or $D_i$. Then each $C_i$ and $D_i$ is a subspace of the vector space $\mathbb{F}_p^m$, i.e. a *p-ary linear code*. Clearly,

$$C_0 \subseteq C_1 \subseteq C_2 \subseteq \ldots \quad \text{and} \quad D_0 \supseteq D_1 \supseteq D_2 \ldots.$$

PROPOSITION 29.4. *We have*

$$C_i^\perp = D_i$$

*for all nonnegative integers $i$.*

PROOF: Let $\mathbf{x}$ and $\mathbf{y}$ be integral vectors such that $\mathbf{x} \pmod{p} \in C_i$ and $\mathbf{y} \pmod{p} \in D_i$. This means

$$p^i(\mathbf{x} + p\mathbf{a}) = \mathbf{z}A \quad \text{and} \quad A(\mathbf{y} + p\mathbf{b})^\top \equiv 0 \pmod{p^{i+1}}$$

for some integral vectors $\mathbf{a}$, $\mathbf{b}$, and $\mathbf{z}$. Then

$$p^i(\mathbf{x} + p\mathbf{a}) \cdot (\mathbf{y} + p\mathbf{b})^\top = \mathbf{z}A(\mathbf{y} + p\mathbf{b})^\top \equiv 0 \pmod{p^{i+1}}$$

which implies that $\mathbf{x} \cdot \mathbf{y}^\top = 0$ over $\mathbb{F}_p$.

We complete the proof by showing that $C_i$ and $D_i$ have dimensions which add to $m$. There exist unimodular matrices $E$ and $F$ (integral matrices with integral inverses) such that $S := EAF$

is diagonal with integral diagonal entries $d_1, d_2, \ldots, d_m$ which successively divide one another: $d_1|d_2|\ldots|d_m$. ($S$ is the *Smith normal form* of $A$ and the $d_i$'s are the *invariant factors*.) The reader should verify that the modules $M_i$ and $N_i$ are *equivalent* to

$$M_i' := \{\mathbf{x} : p^i\mathbf{x} \in M(S)\} \quad \text{and}$$

$$N_i' := \{\mathbf{y} : S\mathbf{y}^\top \equiv 0 \pmod{p^{i+1}}\},$$

respectively, in the sense that either can be obtained from the other by application of a unimodular transformation. Hence the dimensions over $\mathbb{F}_p$ of $M_i$ and $M_i'$ are equal, as well as the dimensions of $N_i$ and $N_i'$.

Suppose $p^{i+1}$ does not divide $d_1, \ldots, d_t$ but that $p^{i+1}$ does divide $d_{t+1}, \ldots, d_m$. Then $\mathbf{y} \in N_i'$ implies that the first $t$ coordinates of $\mathbf{y}$ are divisible by $p$; any vector $\mathbf{y}$ with 0's in the first $t$ coordinates is in $N_i'$. Thus $N_i' \pmod{p}$ is the span of the last $m - t$ standard basis vectors and has dimension $m - t$. Also, $\mathbf{x} \in M_i'$ implies that the last $d - t$ coordinates of $\mathbf{x}$ are divisible by $p$ and a little more thought shows that $M_i' \pmod{p}$ is the span of the first $t$ standard basis vectors and has dimension $t$. $\qquad\square$

PROPOSITION 29.5. *Let $A$, $B$, and $U$ be $m$ by $m$ integral matrices with*

(29.2) $$ABA^\top = nU,$$

*and where $U$ and $B$ are nonsingular modulo a prime $p$. Write $n = p^e n_0$ where $(p, n_0) = 1$. Define the sequence $C_i$ of p-ary codes from $A$ as in (29.1). Then $C_e = \mathbb{F}_p^m$ and*

$$C_i^B = C_{e-i-1} \quad \text{for } i = 0, 1, \ldots, e - 1.$$

*In particular, if $e$ is odd, then $C_{\frac{1}{2}(e-1)}$ is a selfdual p-ary code with respect to the scalar product given by $B$ on $\mathbb{F}_p^m$.*

PROOF: Let $\mathbf{x}$ and $\mathbf{y}$ be integral vectors such that $\mathbf{x} \pmod{p} \in C_i$ and $\mathbf{y} \pmod{p} \in C_{e-i-1}$. This means

$$p^i(\mathbf{x} + p\mathbf{a}_1) = \mathbf{z}_1 A \quad \text{and} \quad p^{e-i-1}(\mathbf{y} + p\mathbf{a}_2) = \mathbf{z}_2 A$$

for some integral vectors $\mathbf{z}_1$, $\mathbf{z}_2$, $\mathbf{a}_1$, and $\mathbf{a}_2$. Then

$$p^{e-1}\langle \mathbf{x}, \mathbf{y} \rangle = p^{e-1}\mathbf{x}B\mathbf{y}^\top \equiv \mathbf{z}_1 A B A^\top \mathbf{z}_2{}^\top \equiv 0 \pmod{p^e}$$

in view of (29.2). Thus $\langle \mathbf{x}, \mathbf{y} \rangle = 0$ in $\mathbb{F}_p$ and we see $C_{e-i-1} \subseteq C_i^B$.

Now let $\mathbf{x} \in C_i^B$. This means $\mathbf{x}B \in C_i^\perp$, which is $D_i$ by Proposition 29.4, and so for some integral vector $\mathbf{x}'$ which reduces to $\mathbf{x}$ when read modulo $p$,

$$\mathbf{x}'BA^\top \equiv 0 \pmod{p^{i+1}}.$$

From (29.2), $A \cdot BA^\top U^{-1} = nI$, and since a matrix commutes with its inverse,

$$(29.3) \qquad\qquad BA^\top U^{-1} \cdot A = nI.$$

Since $U$ is nonsingular modulo $p$, $dU^{-1}$ is integral for some $d$ prime to $p$, e.g. $d := \det(U)$. We multiply (29.3) on the left by $d\mathbf{x}'$ to get

$$\mathbf{x}'BA^\top(dU^{-1})A = p^e d n_0 \mathbf{x}',$$

and then

$$\mathbf{z}A = p^{e-i-1} d n_0 \mathbf{x}'$$

where $\mathbf{z} := \frac{1}{p^{i+1}}\mathbf{x}'BA^\top(dU^{-1})$ is integral. This means $p^{e-i-1}dn_0\mathbf{x}'$ is in $M_{e-i-1}$ and hence $\mathbf{x} \in C_{e-i-1}$.

The assertion that $C_e = \mathbb{F}_p^m$ is left as an easy problem. $\qquad\square$

PROBLEM 29B. Prove that $C_e = \mathbb{F}_p^m$.

PROOF OF THEOREM 29.3: Let $N$ be the incidence matrix of a symmetric $(v, k, \lambda)$-design and let $p$ be a prime. Assume $\lambda = p^{2a}\lambda_0$ where $(\lambda_0, p) = 1$ and $a \geq 0$; we will explain later what to do when $\lambda$ is exactly divisible by an odd power of $p$. Let
(29.4)

$$A := \begin{pmatrix} & & & p^a \\ & N & & \vdots \\ & & & p^a \\ p^a\lambda_0 & \cdots & p^a\lambda_0 & k \end{pmatrix}, \quad B := \begin{pmatrix} 1 & & & 0 \\ & \ddots & & \\ & & 1 & \\ 0 & & & -\lambda_0 \end{pmatrix}.$$

The reader should verify, using the properties of $N$ and the relation $\lambda(v-1) = k(k-1)$, that $ABA^\top = nB$.

In the case $\lambda$ is exactly divisible by an even power of $p$, we apply Proposition 29.5 with the matrices $A$ and $B$ as in (29.4), and where $U := B$.

If $\lambda$ is exactly divisible by an odd power of $p$, we apply the above case to the *complement* of the given symmetric design, which is a symmetric $(v, v - k, \lambda')$-design where $\lambda' = v - 2k + \lambda$. Say $\lambda' = p^c \lambda_0'$ where $(\lambda_0', p) = 1$. From $\lambda\lambda' = n(n-1)$, it follows that $c$ is odd and that

$$\lambda_0 \lambda_0' = n_0(n-1) \equiv -n_0 \pmod{p}.$$

We have replaced what would be $-\lambda_0'$ in the conclusion by $\lambda_0 n_0$, which is allowed since they differ by a square factor modulo $p$, in order to express the result in terms of the original parameters.  □

The following theorem is a consequence of Problem 19M, but we give a proof similar to that of Theorem 29.3.

THEOREM 29.6.  *If there exists a conference matrix of order $n \equiv 2$ (mod 4), then no prime $p \equiv 3$ (mod 4) can divide the squarefree part of $n - 1$.*

PROOF:  A conference matrix of order $n$ is, in particular, an integral matrix $A$ such that $AA^\top = (n-1)I_n$. By Proposition 29.5 with $A = U = I$, there exists a selfdual $p$-ary code (selfdual with respect to the standard inner product) of length $n$ for every prime divisor $p$ of the squarefree part of $n - 1$. Theorem 29.2 then implies that every such prime is $\equiv 1$ (mod 4).                    □

$$* * *$$

When $p$ is a prime not dividing the order $v$ of an abelian group $G$, the group ring $\mathbb{F}_p[G]$ is a *semi-simple* algebra over $\mathbb{F}_p = \mathbb{Z}_p$. This means that there exist no nonzero nilpotents, i.e. nonzero elements $a$ such $a^m = 0$ for some positive integer $m$. The nonexistence of nonzero nilpotents is proved in Lemma 28.5. By Wedderburn's theorem, every finite-dimensional commutative semi-simple algebra with identity $\mathcal{A}$ over a field $F$ is isomorphic to the direct product of fields, each an extension of $F$. It follows that each ideal $\mathcal{I}$ of $\mathcal{A}$

is principal and is generated by an *idempotent e*, i.e. an element $e$ with $e^2 = e$. See any advanced text on algebra for proofs.

We do not need all of this information, but here are some facts about principal ideals which are generated by idempotents that we will need. These make good exercises.

First, if $\mathcal{I} = \langle e_1 \rangle$ and also $\mathcal{I} = \langle e_2 \rangle$, where both $e_1$ and $e_2$ are idempotent, then $e_1 = e_2$. Suppose $\mathcal{I}_1 = \langle e_1 \rangle$ and $\mathcal{I}_2 = \langle e_2 \rangle$, where $e_1$ and $e_2$ are idempotent. Then

$$\mathcal{I}_1 \cap \mathcal{I}_2 = \langle e_1 e_2 \rangle \quad \text{and} \quad \mathcal{I}_1 + \mathcal{I}_2 = \langle e_1 + e_2 - e_1 e_2 \rangle.$$

Note that $e_1 e_2$ and $e_1 + e_2 - e_1 e_2$ are again idempotents.

THEOREM 29.7. *Let $D$ be an $(n^2 + n + 1, n + 1, 1)$-difference set in an abelian group $G$ of order $v := n^2 + n + 1$. If $n \equiv 0 \pmod 2$ but $n \not\equiv 0 \pmod 4$, then $n = 2$. If $n \equiv 0 \pmod 3$ but $n \not\equiv 0 \pmod 9$, then $n = 3$.*

PROOF: Let $D$ be a $(n^2 + n + 1, n + 1, 1)$-difference set in an abelian group $G$ and let $p$ be a prime divisor of $n$. By Theorem 28.3, $p$ is a multiplier of $D$ and we will assume from now on that $D$ is fixed by $p$. We work in the $\mathbb{F}_p$-algebra $\mathbb{F}_p[G]$.

Let $\mathcal{I}_1$ be the ideal in $\mathbb{F}_p[G]$ generated by $D(x)$ and $\mathcal{I}_2$ the ideal generated by $D(x^{-1})$.

We have $D(x^p) = D(x)$, but $D^p(x) = D(x^p)$ by Lemma 28.2, so $D^p(x) = D(x)$ in $\mathbb{F}_p[G]$. Then $D^{p-1}(x)$ is idempotent and will also generate $\mathcal{I}_1$. Similarly, $D^{p-1}(x^{-1})$ is an idempotent generator for $\mathcal{I}_2$. So the idempotent generator of $\mathcal{I}_1 \cap \mathcal{I}_2$ is

$$D^{p-1}(x) D^{p-1}(x^{-1}) = (n + G(x))^{p-1} = G(x),$$

and the idempotent generator of $\mathcal{I}_1 + \mathcal{I}_2$ is

$$D^{p-1}(x) + D^{p-1}(x^{-1}) - G(x).$$

We now wish to consider the dimensions of $\mathcal{I}_1$ and $\mathcal{I}_2$ over $\mathbb{F}_p$. In general, the rank of a principal ideal generated by $A(x)$ is the rank of the $v$ by $v$ matrix whose rows are the coefficients of $x^g A(x)$, $g \in G$. This matrix is the incidence matrix of a symmetric $(n^2 +$

$n + 1, n + 1, 1)$-design when $A(x) = D(x)$ or $D(x^{-1})$. If we now assume that $p^2$ does not divide $n$, then by Theorem 29.1, $\mathcal{I}_1$ and $\mathcal{I}_2$ have dimension $(v + 1)/2$. The intersection $\mathcal{I}_1 \cap \mathcal{I}_2$ has dimension 1, so their sum must have dimension $v$. The idempotent generator of the whole group ring, as an ideal in itself, is 1; so we conclude that

$$(29.5) \qquad D^{p-1}(x) + D^{p-1}(x^{-1}) - G(x) = 1 \quad \text{in } \mathbb{F}_p[G].$$

We are only able to exploit the above equation when $p = 2$ or 3. When $p = 2$, (29.5) asserts $D(x) + D(x^{-1}) \equiv 1 + G(x) \pmod{2}$ in $\mathbb{Z}[G]$. The number of odd coefficients of $1 + G(x)$ is $v - 1 = n^2 + n$, but the number of odd coefficients of $D(x) + D(x^{-1})$ cannot exceed $2(n + 1)$; it follows that $n \leq 2$.

When $p = 3$, (29.5) asserts $D^2(x) + D^2(x^{-1}) \equiv 1 + G(x) \pmod{3}$ in $\mathbb{Z}[G]$. We claim that the nonzero coefficients of $D^2(x)$ and $D^2(x^{-1})$ consist of $n + 1$ coefficients that are 1's and $\binom{n+1}{2}$ coefficients that are 2's. If $C$ is any planar difference set, there will be a term $x^{2g}$ in $C^2(x)$ for each $g \in C$ and a term $2x^{g+h}$ for each unordered pair $\{g, h\} \subseteq C$; note that $\lambda = 1$ implies, e.g., that $g_1 + h_1 \neq g_2 + h_2$ unless $\{g_1, h_1\} = \{g_2, h_2\}$. The sum of two such group ring elements cannot have more than $\binom{n+1}{2} + 2(n + 1)$ coefficients that are $\equiv 1 \pmod{3}$, but $1 + G(x)$ has $n^2 + n$ coefficients that are $\equiv 1$; it follows that $n \leq 4$. $\qquad \square$

PROBLEM 29C. Suppose $D$ is a difference set with $n \equiv 2 \pmod{4}$ and that 2 is a multiplier of $D$. What are the parameters of $D$, as functions of $n$?

## Notes.

Some of the material preceding the proof of Theorem 29.3 is more elegant when $p$-adic numbers are introduced as in Lander (1983), but we have chosen to present the material without them.

## References.

D. Jungnickel and K. Vedder (1984), On the geometry of planar difference sets, *European J. Combinatorics* **5**, 143–148.

E. S. Lander (1983), *Symmetric Designs: An Algebraic Approach,* London Math. Soc. Lecture Note Series **74**, Cambridge University Press.

V. Pless (1986), Cyclic projective planes and binary extended cyclic self-dual codes, *J. Combinatorial Theory* (A) **43**, 331–333.

H. A. Wilbrink (1985), A note on planar difference sets, *J. Combinatorial Theory* (A) **38**, 94–95.

# 30

# Association schemes

Given two $k$-subsets $A, B$ of an $n$-set, $n \geq 2k$, there are $k + 1$ possible relations between them: they may be equal, they may intersect in $k - 1$ elements, they may intersect in $k - 2$ elements, ..., or they may be disjoint.

Given two words ($k$-tuples) $\mathbf{a}, \mathbf{b} \in A^k$, where $A$ is an "alphabet" of size at least 2, there are $k + 1$ possible relations between them: they may be equal, they may agree in $k - 1$ coordinates, they may agree in $k - 2$ coordinates, ..., or they may disagree in all coordinates.

These instances of a set together with a list of mutually exclusive and exhaustive binary relations are examples of *association schemes*, which we define shortly. Association schemes provide one of the foundations of combinatorics and so we include this chapter even though it will be difficult reading. They have been implicit in many of the previous chapters; we have explicitly discussed 2-class association schemes, as they are equivalent to the strongly regular graphs discussed in Chapter 21. This chapter elaborates on some the material of Chapter 21 but has different goals.

Association schemes arose first in the statistical theory of design of experiments, but the work of Ph. Delsarte (1973) has shown how they serve to unify many aspects of our subject. In particular, certain results of coding theory and the theory of $t$-designs—which were originally discovered independently—are now seen to be "formally dual" aspects of the same ideas in association schemes. For example, Fisher's inequality and its generalization, Theorem 19.8, is formally dual to the sphere packing bound, Theorem 21.1. We use the machinery of association schemes in this chapter to give proofs of Lloyd's theorem on perfect codes and its formal dual theorem for tight designs and orthogonal arrays. Delsarte's inequalities, Theorem 30.3, on the distribution vector of a subset of an association scheme provide a "linear programming bound" on the size of codes and are also of interest in extremal set theory.

By a *binary relation* on a set $\mathfrak{X}$, we mean a subset of $\mathfrak{X} \times \mathfrak{X}$. A *k-class association scheme*, sometimes we say just *scheme*, on a set $\mathfrak{X}$ of *points* consists of $k+1$ nonempty symmetric binary relations $R_0, R_1, \ldots, R_k$ on $\mathfrak{X}$ which partition $\mathfrak{X} \times \mathfrak{X}$, where $R_0 = \{(x,x) : x \in \mathfrak{X}\}$ is the identity relation, and such that for some nonnegative integers $p_{ij}^{\ell}$, $0 \le \ell, i, j \le k$, the following system of axioms holds: given any $(x,y) \in R_{\ell}$, there are exactly $p_{ij}^{\ell}$ elements $z \in \mathfrak{X}$ such that $(x,z) \in R_i$ and $(z,y) \in R_j$. We say $x, y \in \mathfrak{X}$ are *i-th associates* when $(x,y) \in R_i$.

The numbers $p_{ij}^{\ell}$, $0 \le \ell, i, j \le k$, are the *parameters* of the scheme. That $p_{ii}^0$ exists means that there is a constant number of $i$-th associates of any element of $\mathfrak{X}$, which is usually denoted by $n_i$. We have

(30.1)     $$p_{ii}^0 = n_i \quad \text{and} \quad p_{ij}^0 = 0 \text{ for } i \ne j$$

and

$$n_0 = 1, \quad n_0 + n_1 + \cdots + n_k = N.$$

where $N := |\mathfrak{X}|$. The numbers $n_0, n_1, \ldots, n_k$ are called the *degrees* of the scheme.

EXAMPLE 30.1. The Johnson schemes $J(k,v)$. The points of the scheme $J(k,v)$ are the $\binom{v}{k}$ $k$-subsets of a $v$-set $S$. Two $k$-subsets $A, B$ are declared to be $i$-th associates when $|A \cap B| = k - i$. Thus 0-th associates are equal. The parameters $p_{ij}^{\ell}$ exist "by symmetry" and may be expressed as sums of products of binomial coefficients, but we will not bother to write these out in general.

The scheme $J(3,6)$, for example, is a 3-class scheme with 20 points. The reader should check that $n_1 = n_2 = 9$, $n_3 = 1$. A few of the other parameters are $p_{11}^1 = 4$, $p_{11}^2 = 4$, $p_{11}^3 = 0$.

EXAMPLE 30.2. The Hamming schemes $H(n,q)$. The points of $H(n,q)$ are the $q^n$ words of length $n$ over an alphabet of size $q$.

Two $n$-tuples $x, y$ are declared to be $i$-th associates when they disagree in exactly $i$ coordinates. Thus 0-th associates are equal. The parameters $p_{ij}^\ell$ exist "by symmetry" and may be expressed as sums of products of binomial coefficients and powers of $q - 1$, but we will not bother to write these out in general.

The scheme $H(5, 3)$, for example, is a 5-class scheme with 125 points. The reader should check that $n_1 = 5 \cdot 2$, $n_2 = 10 \cdot 4$, etc.

Each of the relations $R_i$ may be thought of as the adjacency relation of a graph $G_i$ on the vertex set $\mathfrak{X}$. (A scheme is a special kind of partition of the edges—or coloring of the edges—of a complete graph.) It should be clear that if we start with a 2-class scheme, then $G_1$ is a strongly regular graph with degree $n_1$, $\lambda = p_{11}^1$, and $\mu = p_{11}^2$. In fact, any strongly regular graph gives rise to a 2-class association scheme when we declare that two distinct vertices are 1st associates when they are adjacent in $G$ and 2nd associates if not adjacent; the other parameters exist, i.e. are constant, and can be computed from the parameters of the graph. For example, $p_{12}^1 = k - \lambda - 1$.

A *distance regular graph* is a graph $G$ such that the number of vertices at distance $i$ to $x$ and distance $j$ to $y$ depends only on the distance $\ell$ between the vertices $x$ and $y$, not on the particular vertices. That is, defining two vertices to be $i$-th associates if and only if their distance in $G$ is $i$ produces an association scheme (where the number of classes will be the diameter of the graph). See Brouwer, Cohen, and Neumaier (1989). Schemes which arise in this way are called *metric*. The schemes mentioned in Examples 30.1 and 30.2 above are metric, as are those in Examples 30.3 and 30.4 below.

It is important to know that the parameters $p_{ij}^\ell$ exist; it is not so important to know their exact values. This is lucky because there is often no convenient expression for these parameters. In the scheme $J(k, v)$, for example, a triple sum of binomial coefficients seems to be required for the general $p_{ij}^\ell$. In the examples we have just given, we know that the $p_{ij}^\ell$'s exist because of "symmetry". More precisely, in each case there exists a group $G$ of permutations of $\mathfrak{X}$ so that two ordered pairs of points $(x_1, y_1)$ and $(x_2, y_2)$ are in the same relation $R_i$ if and only if there exists $\sigma \in G$ such that $\sigma(x_1) = x_2$ and $\sigma(y_1) = y_2$. That is, the relations $R_0, R_1, \ldots, R_k$ are the *orbits* of $G$

on $\mathfrak{x} \times \mathfrak{x}$ (with $R_0$ the trivial orbit of all $(x,x), x \in \mathfrak{x}$). In Example 30.1, $G$ is the symmetric group $S_v$ acting on all $k$-subsets of a $v$-set; we can find a permutation which takes an ordered pair of $k$-subsets to another if and only if the size of the intersection of each pair is the same. In Example 30.2, $G$ is the wreath product of $S_q$ with $S_n$ (that is, we allow any permutation of the $n$ coordinates followed by independent permutations of the $q$ symbols in each coordinate); we can find such a transformation which takes an ordered pair of $n$-tuples to another if and only if the pairs agree in the same number of coordinates.

In general, if $G$ is a transitive group of permutations on a set $\mathfrak{x}$ such that the orbits of $G$ on $\mathfrak{x} \times \mathfrak{x}$ are symmetric, they may be taken as the relations of an association scheme on $\mathfrak{x}$. Our next three examples also arise in this way.

EXAMPLE 30.3. This is the $q$-analogue of the Johnson scheme: Take the $k$-subspaces of a $v$-space $V$ over $\mathbb{F}_q$ as points. Two $k$-subspaces $A, B$ are declared to be $i$-th associates when $\dim(A \cap B) = k - i$.

EXAMPLE 30.4. Take the $k$ by $m$ matrices over $\mathbb{F}_q$ as the points, where $k \le m$, say. Two matrices $A, B$ are declared to be $i$-th associates when the rank of $A - B$ is $k - i$.

EXAMPLE 30.5. The *cyclotomic schemes* are obtained as follows. Let $q$ be a prime power and $k$ a divisor of $q - 1$. Let $C_1$ be the subgroup of the multiplicative group of $\mathbb{F}_q$ of index $k$, and let $C_1, C_2, \ldots, C_k$ be the cosets of $C_1$. The points of the scheme are to be the elements of $\mathbb{F}_q$, and two points $x, y$ are declared to be $i$-th associates when $x - y \in C_i$ (and 0-th associates when $x - y = 0$). In order to have a scheme with the above definition, we require $-1 \in C_1$ so that the relations will be symmetric, i.e. $2k$ must divide $q - 1$ if $q$ is odd. Cf. Example 21.3 which is the case $k = 2$.

We introduce the *association matrices* $A_0, A_1, \ldots, A_k$ (also called the *adjacency matrices*) of an association scheme. These matrices are square with both rows and columns indexed by the elements of the point set $\mathfrak{x}$ of a scheme. For $i = 0, 1, \ldots, k$, we define

$$A_i(x,y) = \begin{cases} 1 & \text{if } (x,y) \in R_i, \\ 0 & \text{otherwise.} \end{cases}$$

The matrices $A_i$ are symmetric $(0, 1)$-matrices and

$$A_0 = I, \quad A_0 + A_1 + \cdots + A_k = J$$

where $J$ is the all-one matrix of size $N$ by $N$. We denote by $\mathfrak{A}$ the linear span over the reals of $A_0, A_1, \ldots, A_k$. These matrices are linearly independent since each contains at least one 1; and a position in which $A_i$ has a 1 contains a 0 in every other association matrix. The axioms of an association scheme are exactly what is required to ensure that $\mathfrak{A}$ is closed under matrix multiplication. To see this, it suffices to show that the product of any two of the basis matrices is in $\mathfrak{A}$ and, in fact, we have

$$(30.2) \qquad\qquad A_i A_j = \sum_{\ell=0}^{k} p_{ij}^{\ell} A_\ell$$

because $A_i A_j(x, y)$ is the number of $z$ such that $A_i(x, z) = 1$ and $A_j(z, y) = 1$, and this number is $p_{ij}^{\ell}$ where $\ell$ is such that $A_\ell(x, y) = 1$.

The algebra $\mathfrak{A}$ is called the *Bose-Mesner* algebra of the scheme; this algebra was introduced for strongly regular graphs in Chapter 21. We note at this point that not only is $\mathfrak{A}$ closed under normal matrix multiplication, but it is also closed under *Hadamard multiplication*. The Hadamard product $A \circ B$ of two matrices is the matrix obtained by coordinate-wise multiplication:

$$(A \circ B)(x, y) := A(x, y) B(x, y).$$

As an algebra with respect to Hadamard multiplication, $\mathfrak{A}$ is almost trivial. We have

$$A_i \circ A_j = \begin{cases} A_i & \text{if } i = j, \\ O & \text{if } i \neq j \end{cases}$$

(that is, $A_0, A_1, \ldots, A_k$ are *orthogonal idempotents*), and the sum of the $A_i$'s is $J$, the identity with respect to Hadamard multiplication. So Hadamard multiplication is extremely simple when matrices in $\mathfrak{A}$ are expressed with respect to the basis $A_0, A_1, \ldots, A_k$ of $\mathfrak{A}$.

However, a well known result of matrix theory (an extension of the spectral theorem which says that a symmetric real matrix has an orthogonal basis of eigenvectors) asserts that a *commutative* algebra of real symmetric matrices has a basis of orthogonal idempotents *with respect to ordinary matrix multiplication* which sum to the identity. More geometrically, there exists an orthogonal decomposition

$$\mathbb{R}^{\mathfrak{x}} = V_0 \oplus V_1 \oplus \cdots \oplus V_k$$

of the Euclidean space $\mathbb{R}^{\mathfrak{x}}$, the space of all vectors whose coordinates are indexed by the elements of $\mathfrak{x}$ with the standard inner product, such that the orthogonal projections $E_0, E_1, \ldots, E_k$ from $\mathbb{R}^{\mathfrak{x}}$ onto the subspaces $V_0, V_1, \ldots, V_k$, respectively, are a basis for $\mathfrak{A}$. We have

$$E_i E_j = \begin{cases} E_i & \text{if } i = j, \\ O & \text{if } i \neq j, \end{cases}$$

and

$$E_0 + E_1 + \cdots + E_k = I.$$

Of course, when matrices are expressed with respect to the basis $E_0, E_1, \ldots, E_k$, ordinary multiplication is also extremely simple.

The subspaces $V_0, V_1, \ldots, V_k$ are called the *eigenspaces* of the scheme: in a linear combination $M = \sum_{i=0}^{k} \lambda_i E_i$, each vector in $V_i$ is an eigenvector of value $\lambda_i$ for $M$. There is no natural numbering of the eigenspaces *in general* with one exception: since $J \in \mathfrak{A}$, and $J$ has the vector $\mathbf{j}$ of all 1's as an eigenvector of value $N$ and all vectors orthogonal to $\mathbf{j}$ as eigenvectors of value 0, it must be that one of the eigenspaces consists of scalar multiples of $\mathbf{j}$ alone—we shall always assume that this is $V_0$. Then the orthogonal projection onto $V_0$ (which has $\mathbf{j}$ as an eigenvector of value 1 and all vectors orthogonal to $\mathbf{j}$ as eigenvectors of value 0) is

$$E_0 = \frac{1}{N} J.$$

We let $m_i$ denote the dimension of $V_i$. Then

$$m_0 = 1, \quad m_0 + m_1 + \cdots + m_k = N.$$

Note that $m_i$ is the trace of $E_i$ since the eigenvalues of $E_i$ are 1 (with multiplicity equal to the dimension of $V_i$) and 0. The numbers $m_0, m_1, \ldots, m_k$ are called the *multiplicities* of the scheme.

EXAMPLE 30.6. We can explicitly describe the eigenspaces of the Hamming scheme $H(n,2)$. The points of this scheme are the binary $n$-tuples (or words) $a$ of $\mathbb{F}_2^n$.

We claim that $V_i$ may be taken to be the span of all vectors $\mathbf{v}_a$ as $a$ ranges over the words of weight $i$, $i = 0, 1, \ldots, n$, and where the entry in coordinate $b$ of $\mathbf{v}_a$ is

$$\mathbf{v}_a(b) := (-1)^{\langle a,b \rangle}.$$

We note that these vectors $v_a$, $a \in \mathbb{F}_2^n$, are orthogonal and are thus the rows of a Hadamard matrix; see Fig. 30.1 and Chapter 18.

We check that each $\mathbf{v}_a$ is an eigenvector of all association matrices $A_j$. Let $a$ have weight $\ell$. Then

$$(\mathbf{v}_a A_j)(b) = \sum_c \mathbf{v}_a(c) A_j(c,b)$$

$$= \sum_{c:d(b,c)=j} (-1)^{\langle a,c \rangle} = (-1)^{\langle a,b \rangle} \sum_{c:d(b,c)=j} (-1)^{\langle a,b+c \rangle}$$

$$= \mathbf{v}_a(b) \sum_{u:\mathrm{wt}(u)=j} (-1)^{\langle a,u \rangle} = \mathbf{v}_a(b) \sum_{i=0}^n (-1)^i \binom{\ell}{i} \binom{n-\ell}{j-i}.$$

This calculation shows, with $E_i$ the matrix of the orthogonal projection onto $V_i$, that

$$A_j = \sum_{\ell=0}^n \left( \sum_{i=0}^n (-1)^i \binom{\ell}{i} \binom{n-\ell}{j-i} \right) E_\ell$$

since the two sides give the same result when premultiplied by any $\mathbf{v}_a$. So $\mathfrak{A} \subseteq \mathrm{span}\{E_0, \ldots, E_n\}$. Equality holds because both spaces have dimension $n+1$.

The eigenvalues of the schemes $H(n,q)$ for general $q$ are given in Theorem 30.1 below.

$$\begin{pmatrix} 1 & 1 & 1 & 1 & 1 & 1 & 1 & 1 \\ 1 & -1 & 1 & 1 & 1 & -1 & -1 & -1 \\ 1 & 1 & -1 & 1 & -1 & 1 & -1 & -1 \\ 1 & 1 & 1 & -1 & -1 & -1 & 1 & -1 \\ 1 & 1 & -1 & -1 & 1 & -1 & -1 & 1 \\ 1 & -1 & 1 & -1 & -1 & 1 & -1 & 1 \\ 1 & -1 & -1 & 1 & -1 & -1 & 1 & 1 \\ 1 & -1 & -1 & -1 & 1 & 1 & 1 & -1 \end{pmatrix}$$

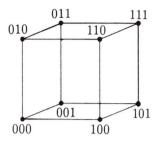

Figure 30.1

For example, if we order the words of length $n = 3$ as

$$000, \ 100, \ 010, \ 001, \ 011, \ 101, \ 110, \ 111,$$

then the eigenspaces $V_0, V_1, V_2, V_3$ of the cube $(= A_1$ of $H(3,2))$ shown at the right of Fig. 30.1 are spanned respectively by the first row, the next three rows, the next three rows, and the last row of the Hadamard matrix at the left of Fig. 30.1.

EXAMPLE 30.7. We now describe the eigenspaces of the Johnson scheme $J(k,v)$. We omit the proofs; see Delsarte (1973) or Wilson (1984b).

The points of this scheme are the $k$-subsets $S$ of a $v$-set $\mathfrak{X}$. For each subset $T$ of size $\leq k$, let $\mathbf{e}_T$ be the vector of length $\binom{v}{k}$ where

$$\mathbf{e}_T(S) := \begin{cases} 1 & \text{if } T \subseteq S, \\ 0 & \text{otherwise.} \end{cases}$$

Let $U_i$ be the span of $\{\mathbf{e}_T : T \subseteq \mathfrak{X}, |T| = i\}$ for $i = 0, 1, \ldots, k$. We claim that $U_0 \subseteq U_1 \subseteq \cdots \subseteq U_k = \mathbb{R}^{\mathfrak{X}}$ and that $U_i$ has dimension $\binom{v}{i}$. Let $V_0 := U_0$ (the constant vectors) and for $i > 0$, let $V_i$ be the orthogonal complement of $U_{i-1}$ in $U_i$, i.e.

$$V_i := U_i \cap U_{i-1}^{\perp}.$$

It is clear that $V_0, V_1, \ldots, V_k$ are orthogonal and sum to $\mathbb{R}^{\mathfrak{X}}$. It can be shown that each vector in $V_i$ is an eigenvector of $A_j$ of value $P_j(i)$ as displayed in Theorem 30.1(i) below.

Since we have two bases of the vector space $\mathfrak{A}$, we may consider the transformation matrices between them, which are called the *eigenmatrices* of the scheme. Define $P$ (the *first* eigenmatrix) and $Q$ (the *second* eigenmatrix) as the $k+1$ by $k+1$ matrices with rows and columns indexed by $0, 1, \ldots, k$ such that

$$(A_0, A_1, \ldots, A_k) = (E_0, E_1, \ldots, E_k)\, P$$

and

$$N\,(E_0, E_1, \ldots, E_k) = (A_0, A_1, \ldots, A_k)\, Q.$$

We write $P_\ell(i)$ for the $(i, \ell)$ entry of $P$, and similarly, we let $Q_\ell(i)$ be the $(i, \ell)$ entry of $Q$, so that

(30.3)          $A_\ell = P_\ell(0)E_0 + P_\ell(1)E_1 + \cdots + P_\ell(k)E_k,$

and

(30.4)          $N\, E_\ell = Q_\ell(0)A_0 + Q_\ell(1)A_1 + \cdots + Q_\ell(k)A_k.$

Of course, we have

$$Q = N\, P^{-1}, \quad P = N\, Q^{-1}.$$

The $\ell$-th column of $P$ consists of the eigenvalues of $A_\ell$.

While it is not so important to know the parameters $p_{ij}^\ell$ of a scheme, it is important for applications to know the eigenmatrices $P$ and $Q$. See Bannai and Ito (1984) or Delsarte (1973) for proofs of the following theorem and the determination of the eigenmatrices of other schemes.

THEOREM 30.1. (i) *For the Johnson scheme* $J(k, v)$, *the degrees are* $n_\ell = \binom{k}{\ell}\binom{v-k}{\ell}$ *and the multiplicities are* $m_\ell = \binom{v}{\ell} - \binom{v}{\ell-1}$, $\ell = 0, 1, \ldots, k$. *The entries of the first eigenmatrix are* $P_\ell(i)$ *where*

$$P_\ell(x) = \sum_{\alpha=0}^{\ell} (-1)^{\ell-\alpha} \binom{k-\alpha}{\ell-\alpha}\binom{k-x}{\alpha}\binom{v-k+\alpha-x}{\alpha}.$$

(ii) *For the Hamming scheme* $H(n, q)$, *the degrees and multiplicities are* $n_\ell = m_\ell = \binom{n}{\ell}(q-1)^\ell$, $\ell = 0, 1, \ldots, n$. *The entries of the first eigenmatrix are* $P_\ell(i)$ *where*

$$P_\ell(x) = \sum_{\alpha=0}^{\ell} (-q)^\alpha (q-1)^{\ell-\alpha} \binom{n-\alpha}{\ell-\alpha}\binom{x}{\alpha}.$$

PROBLEM 30A. Calculate the eigenmatrix $P$ of $J(3,8)$ in the following way. First calculate $A_1 A_2$ as a linear combination of $A_0$, $A_1$, $A_2$, and $A_3$, and then fill in the missing line in the following table:

$$A_1^0 = A_0,$$
$$A_1^1 = A_1,$$
$$A_1^2 = 15A_0 + 6A_1 + 4A_2,$$
$$A_1^3 = \qquad\qquad,$$
$$A_1^4 = 1245A_0 + 1036A_1 + 888A_2 + 720A_3.$$

From this table, it is only mildly tedious to derive the minimal polynomial of $A_1$. Find the eigenvalues of $A_1$. Now express $A_2$ and $A_3$ as polynomials in $A_1$, and so find their eigenvalues. Find the multiplicities. Check your work by verifying the orthogonality relations of Theorem 30.2 or just calculating the values from Theorem 30.1(i).

PROBLEM 30B. Show how to calculate all the parameters $p_{ij}^\ell$ of a scheme given the eigenmatrix $P$. That is, prove that they are uniquely determined by $P$.

PROBLEM 30C. The *Latin square graphs* are $srg(v, k, \lambda, \mu)$'s where for some integers $n$ and $r$,

$$v = n^2, \quad k = r(n-1), \quad \lambda = (n-2)+(r-1)(r-2), \quad \mu = r(r-1).$$

Find the eigenmatrices $P$ and $Q$ for the 2-class schemes corresponding to these graphs.

The so-called *negative Latin square graphs* are strongly regular graphs $srg(v, k, \lambda, \mu)$ whose parameters are obtained from the above by replacing $n$ by $-n$ and $r$ by $-r$. So $v = (-n)^2$, $k = (-r)(-n-1)$, etc. (It is strange that this should yield parameters satisfying (21.4), but it does.) Find the eigenmatrices $P$ and $Q$ for the corresponding 2-class schemes.

THEOREM 30.2. *The eigenmatrices of a scheme satisfy the orthogonality relations*

$$P^\top \begin{pmatrix} 1 & 0 & \cdots & 0 \\ 0 & m_1 & & 0 \\ \vdots & & \ddots & \vdots \\ 0 & 0 & \cdots & m_k \end{pmatrix} P = N \begin{pmatrix} 1 & 0 & \cdots & 0 \\ 0 & n_1 & & 0 \\ \vdots & & \ddots & \vdots \\ 0 & 0 & \cdots & n_k \end{pmatrix}$$

and

$$
Q^\top \begin{pmatrix} 1 & 0 & \cdots & 0 \\ 0 & n_1 & & 0 \\ \vdots & & \ddots & \vdots \\ 0 & 0 & \cdots & n_k \end{pmatrix} Q = N \begin{pmatrix} 1 & 0 & \cdots & 0 \\ 0 & m_1 & & 0 \\ \vdots & & \ddots & \vdots \\ 0 & 0 & \cdots & m_k \end{pmatrix}.
$$

PROOF: The vector space $\mathfrak{A}$ can be equipped with an inner product in a more-or-less natural way: We define $\langle A, B \rangle$ to be the sum of the entries of the Hadamard product $A \circ B$. Check that this is the same as the trace of the matrix product $AB^\top$ (which is the same as $\mathrm{tr}(AB)$ when $B$ is symmetric, of course).

With respect to this inner product, the basis $A_0, A_1, \ldots, A_k$ is an orthogonal basis (but not orthonormal); we have $\langle A_i, A_i \rangle = N\, n_i$. But $E_0, E_1, \ldots, E_k$ is also an orthogonal basis since $E_i E_j = 0$ for $i \neq j$; we have $\langle E_i, E_i \rangle = \mathrm{tr}(E_i) = m_i$. The derivation of the theorem from this is now elementary linear algebra.

Consider the first relation. The entry in the $(\alpha, \beta)$ position on the right is

$$
\langle A_\alpha, A_\beta \rangle = \Big\langle \sum_i P_\alpha(i) E_i, \sum_j P_\beta(j) E_j \Big\rangle
$$

$$
= \sum_{i,j} P_\alpha(i) P_\beta(i) \langle E_i, E_j \rangle = \sum_i m_i P_\alpha(i) P_\beta(i),
$$

and this last expression is the entry in the $(\alpha, \beta)$ position on the left. The second relation is derived similarly. □

Another way to express the content of the above theorem is

$$
Q = \begin{pmatrix} 1 & 0 & \cdots & 0 \\ 0 & n_1 & & 0 \\ \vdots & & \ddots & \vdots \\ 0 & 0 & \cdots & n_k \end{pmatrix}^{-1} P^\top \begin{pmatrix} 1 & 0 & \cdots & 0 \\ 0 & m_1 & & 0 \\ \vdots & & \ddots & \vdots \\ 0 & 0 & \cdots & m_k \end{pmatrix}.
$$

Equivalently,

(30.5) $$m_j P_i(j) = n_i Q_j(i)$$

for all $i, j = 0, 1, \ldots, k$.

With Theorems 30.1 and 30.2, we can find the second eigenmatrix $Q$ for the schemes $J(k, v)$ and $H(n, q)$. It is somewhat surprising that $P = Q$ for $H(n, q)$; the reader should check this.

PROBLEM 30D. Explain why the zeroth row and column of $P$ and $Q$ are as indicated below.

$$(30.6) \qquad P = \begin{pmatrix} 1 & n_1 & \cdots & n_k \\ 1 & & & \\ \vdots & & & \\ 1 & & & \end{pmatrix}, \qquad Q = \begin{pmatrix} 1 & m_1 & \cdots & m_k \\ 1 & & & \\ \vdots & & & \\ 1 & & & \end{pmatrix}.$$

Delsarte (1973) observed that the columns of the second eigen-matrix $Q$ provide a system of linear constraints, which we will call *Delsarte's inequalities*, on, what he calls, the "inner distri-bution vector" of a nonempty subset $Y$ of the point set $\mathfrak{X}$ of an association scheme. We define the *distribution vector* of $Y$ to be $\mathbf{a} = (a_0, a_1, \ldots, a_k)$ where

$$a_i := \frac{1}{|Y|} |(Y \times Y) \cap R_i|;$$

that is, $a_i$ is the average number of $i$-th associates $y \in Y$ of an element $x \in Y$. We have

$$a_0 = 1, \quad a_0 + a_1 + \cdots + a_k = |Y|.$$

For many interesting subsets $Y$, the number of $i$-th associates $y \in Y$ of an element $x \in Y$ is constant, i.e. it does not depend on which $x$ in $Y$ is chosen. For example, this is true if $C$ is a *linear* code in $H(n, q)$ in which case the distribution vector coincides with what we called the weight enumerator of $C$ in Chapter 21, and also for other beautiful configurations which have been discussed in previous chapters and whose distribution vectors we list below. Here Hamming(7) is the Hamming code of length 7, "Golay" is short for "Golay code", "X" means "extended", and a "*" denotes "dual code".

| Object | Scheme | Distribution vector |
|--------|--------|---------------------|
| $S(2,3,7)$ | $J(3,7)$ | (1,0,6,0) |
| $S_2(2,5,11)$ | $J(5,11)$ | (1,0,0,10,0,0) |
| $S_2(3,6,12)$ | $J(6,12)$ | (1,0,0,20,0,0,1) |
| $S(5,6,12)$ | $J(6,12)$ | (1,0,45,40,45,0,1) |
| $S(4,7,23)$ | $J(7,23)$ | (1,0,0,0,140,0,112,0) |
| $S(5,8,24)$ | $J(8,24)$ | (1,0,0,0,280,0,448,0,30) |
| Hamming(7) | $H(7,2)$ | (1,0,0,7,7,0,0,1) |
| X-Hamming(7) | $H(8,2)$ | (1,0,0,0,14,0,0,0,1) |
| Binary Golay | $H(23,2)$ | (1,0,0,0,0,0,0,0,253,506,0,0,1288, ... ) |
| X-Binary Golay | $H(24,2)$ | (1,0,0,0,0,0,0,0,0,759,0,0,0,2576,0, ... ) |
| Ternary Golay | $H(11,3)$ | (1,0,0,0,0,132,132,0,330,110,0,24) |
| *-Ternary Golay | $H(11,3)$ | (1,0,0,0,0,0,132,0,0,110,0,0) |

Note the occurrence of many zeros in these distribution vectors.

THEOREM 30.3. *The distribution vector* **a** *of a nonempty subset of an association scheme satisfies*

$$\mathbf{a}Q \geq \mathbf{0}$$

*where* **0** *is the row vector of* $n+1$ *zeros.*

PROOF: Let $\phi \in \mathbb{R}^{\mathfrak{X}}$ be the characteristic vector of $Y$. That is,

$$\phi(x) = \begin{cases} 1 & \text{if } x \in Y, \\ 0 & \text{if } x \notin Y. \end{cases}$$

Then

$$a_i = \frac{1}{|Y|}\phi A_i \phi^\top.$$

Since $E_\ell$ is idempotent and symmetric,

$$0 \leq \|\phi E_\ell\|^2 = (\phi E_\ell)(\phi E_\ell)^\top = \phi E_\ell \phi^\top$$

$$= \frac{1}{N}\phi \left( \sum_{i=0}^{k} Q_\ell(i)A_i \right) \phi^\top = \frac{|Y|}{N} \sum_{i=0}^{k} Q_\ell(i)a_i.$$

$\square$

We note for further reference that the $\ell$-th inequality holds with equality if and only if the projection $\phi E_\ell$ is the zero vector. The zeroth inequality is trivial since the zeroth coordinate of $\mathbf{a}Q$ is $a_0 + a_1 + \cdots + a_k$, which of course is nonnegative.

EXAMPLE 30.8. For $J(3,8)$, Delsarte's inequalities when fractions are cleared are

$$15 + 7a_1 - a_2 - 9a_3 \geq 0,$$
$$30 + 2a_1 - 5a_2 + 9a_3 \geq 0,$$
$$10 - 2a_1 + a_2 - a_3 \geq 0.$$

These can be viewed as necessary conditions for the existence of a family $\mathcal{F}$ of 3-subsets of an 8-set with a given distribution vector $(1, a_1, a_2, a_3)$.

We can ask what they imply about $|\mathcal{F}|$ when we assume some of the $a_i$'s are zero. For example, assume $a_3 = 0$ (no two members of $\mathcal{F}$ are disjoint). The maximum value of $1 + a_1 + a_2 \ (= |\mathcal{F}|)$ subject to these inequalities is 21; this is a linear programming problem. We already knew $|\mathcal{F}| \leq 21$ because of the Erdős-Ko-Rado theorem, Theorem 6.4. Theorem 30.3 implies Theorem 6.4; see the notes.

Theorem 30.3 leads to a *linear programming bound* on the cardinalities of codes. For the Hamming scheme $H(n, q)$, given an integer $d$, we may consider the linear programming problem

"maximize $1 + a_d + a_{d+1} + \cdots + a_n$ subject to $a_i \geq 0$ and $(1, 0, \ldots, 0, a_d, a_{d+1}, \ldots, a_n) \, Q \geq \mathbf{0}$"

where $Q$ is the $n+1$ by $n+1$ second eigenmatrix of the scheme. If LPB denotes the maximum value of $1 + a_d + a_{d+1} + \cdots + a_n$ under these conditions and $C$ is any $q$-ary code of length $n$ with minimum distance at least $d$, then $|C| \leq$ LPB since the distribution vector $\mathbf{a}$ of $C$ satisfies the conditions and the coordinates sum to $|C|$.

We mention that for the Hamming schemes, if $\mathbf{a}$ is the distribution vector of a *linear* code $C$, then $\frac{1}{|C|}\mathbf{a}Q$ is the distribution vector of the dual code $C^{\perp}$. To see this, just compare the formulae in MacWilliams' theorem, Theorem 20.3, with those in Theorem 30.1(ii). This certainly explains why $\mathbf{a}Q \geq \mathbf{0}$ for the distribution vector $\mathbf{a}$ of a linear code, but we have proved above that it holds even if $C$ is not linear.

| $n$ | $d$ | SPB | LPB | $a_0$ | $a_1$ | $a_2$ | $a_3$ | $a_4$ | $a_5$ | $a_6$ | $a_7$ | $a_8$ | $a_9$ | $a_{10}$ | $a_{11}$ | $a_{12}$ | $a_{13}$ | $a_{14}$ | $a_{15}$ |
|---|---|---|---|---|---|---|---|---|---|---|---|---|---|---|---|---|---|---|---|
| 11 | 3 | 170.7 | 170.7 | 1 | 0 | 0 | 18.3 | 36.7 | 29.3 | 29.3 | 36.7 | 18.3 | 0 | 0 | 1 | | | | |
| 11 | 5 | 30.6 | 24 | 1 | 0 | 0 | 0 | 0 | 11 | 11 | 0 | 0 | 0 | 0 | 1 | | | | |
| 11 | 7 | 8.8 | 4 | 1 | 0 | 0 | 0 | 0 | 0 | 0 | 2 | 1 | 0 | 0 | 0 | | | | |
| 12 | 3 | 315.1 | 292.6 | 1 | 0 | 0 | 20 | 45 | 48 | 56 | 65.1 | 40.7 | 11.4 | 3.4 | 1.7 | 0.1 | | | |
| 12 | 5 | 51.9 | 40 | 1 | 0 | 0 | 0 | 0 | 15 | 17.5 | 0 | 0 | 5 | 1.5 | 0 | 0 | | | |
| 12 | 7 | 13.7 | 5.3 | 1 | 0 | 0 | 0 | 0 | 0 | 0 | 2.7 | 1.7 | 0 | 0 | 0 | 0 | | | |
| 13 | 3 | 585.1 | 512 | 1 | 0 | 0 | 22 | 55 | 72 | 96 | 116 | 87 | 40 | 16 | 6 | 1 | 0 | | |
| 13 | 5 | 89.0 | 64 | 1 | 0 | 0 | 0 | 0 | 18 | 24 | 4 | 3 | 10 | 4 | 0 | 0 | 0 | | |
| 13 | 7 | 21.7 | 8 | 1 | 0 | 0 | 0 | 0 | 0 | 0 | 4 | 3 | 0 | 0 | 0 | 0 | 0 | | |
| 14 | 3 | 1092.3 | 1024 | 1 | 0 | 0 | 28 | 77 | 112 | 168 | 232 | 203 | 112 | 56 | 28 | 7 | 0 | 0 | |
| 14 | 5 | 154.6 | 128 | 1 | 0 | 0 | 0 | 0 | 28 | 42 | 8 | 7 | 28 | 14 | 0 | 0 | 0 | 0 | |
| 14 | 7 | 34.9 | 16 | 1 | 0 | 0 | 0 | 0 | 0 | 0 | 8 | 7 | 0 | 0 | 0 | 0 | 0 | 0 | |
| 15 | 3 | 2048 | 2048 | 1 | 0 | 0 | 35 | 105 | 168 | 280 | 435 | 435 | 280 | 168 | 105 | 35 | 0 | 0 | 1 |
| 15 | 5 | 270.8 | 256 | 1 | 0 | 0 | 0 | 0 | 42 | 70 | 15 | 15 | 70 | 42 | 0 | 0 | 0 | 0 | 1 |
| 15 | 7 | 56.9 | 32 | 1 | 0 | 0 | 0 | 0 | 0 | 0 | 15 | 15 | 0 | 0 | 0 | 0 | 0 | 0 | 1 |

Figure 30.2

It is difficult to extract useful information in general from the linear programming bound (LPB), but it can be shown that it is always at least as good as the sphere packing bound (SPB), Theorem 21.1; see Delsarte (1973). For small values of the parameters, one can use the simplex algorithm and expicitly evaluate this bound. We have done so for some values of $n$ and $d$ in the case of binary codes and display the results in the table of Fig. 30.2. The table gives a nonnegative vector $(1, 0, \ldots, 0, a_d, a_{d+1}, \ldots, a_n)$ whose sum is maximum subject to $\mathbf{a}Q \geq \mathbf{0}$. It is a good exercise for the reader to try to decide whether codes meeting the LPB as in the table actually exist. We mention that several can be obtained from an interesting binary code of length 16 called the Nordstrom-Robinson code, which has 256 codewords and minimum distance 6. It is not linear. See Van Lint (1982).

There appears to be a kind of "duality" between the two bases of, and between ordinary and Hadamard multiplication on, the Bose-Mesner algebra. It sometimes happens that there exists a second association scheme so that the $E_i$'s of the latter under Hadamard multiplication act like the $A_i$'s of the former under ordinary multiplication (apart from a scalar factor), and vice versa. Two schemes are said to be *formally dual* when the first eigenmatrix $P$ of one is equal to the second eigenmatrix $Q$ of the other (cf. Problem 30E below). Many examples of pairs of formally dual strongly regular graphs are known. The Hamming scheme, as mentioned above, is formally dual to itself. The Johnson scheme usually has no formal dual, because in general the Krein parameters defined below are not integers.

The *Krein parameters* of an association scheme are the $(k+1)^3$ numbers $q_{ij}^\ell$ defined by

$$(30.7) \qquad N\, E_i \circ E_j = \sum_{\ell=0}^{k} q_{ij}^\ell E_\ell.$$

If the scheme has a formal dual, these are the parameters $p_{ij}^\ell$ for that formally dual scheme and hence are nonnegative integers. We always have

$$q_{ij}^\ell \geq 0, \quad \text{for all } 0 \leq i, j, \ell \leq k,$$

because the $q_{ij}^\ell$'s are eigenvalues of the Hadamard product of two positive semidefinite matrices and are hence nonnegative; cf. Problem 21E. In principle, the Krein parameters are functions of the original parameters $p_{ij}^\ell$ (see Problem 30E) and their nonnegativity can be viewed as a necessary condition for the existence of a scheme with given parameters; we did this for strongly regular graphs in Theorem 21.3.

PROBLEM 30E. Show how to calculate all the parameters $q_{ij}^\ell$ of a scheme given the eigenmatrix $Q$. That is, prove that they are uniquely determined by $Q$.

We need to know later that

$$(30.8) \qquad q_{ii}^0 = m_i \quad \text{and} \quad q_{ij}^0 = 0 \text{ for } i \neq j.$$

This follows from (30.7) when we consider the sum of all entries of the matrices on both sides; the sum of the entries of the lefthand side is $N$ times

$$\langle E_i, E_j \rangle = \begin{cases} m_i & \text{if } i = j, \\ 0 & \text{otherwise.} \end{cases}$$

For some applications to codes and designs, we need to know more about a scheme. However, here is one interesting result we can prove for general schemes before we specialize.

Given a $k$-class scheme and a subset $K$ of $\{1, 2, \ldots, k\}$, a subset $Y$ of the point set is called a $K$-*clique* when for every pair of distinct elements $x, y \in Y$, $x$ and $y$ are $j$-th associates for some $j \in K$. The subset $Y$ is a $K$-*coclique* when no pair of elements $x, y \in Y$ are $j$-th associates for any $j \in K$. If $G$ is the graph whose adjacency matrix is $\sum_{j \in K} A_j$, then cliques in $G$ are the same as $K$-cliques in the scheme, and cocliques (independent sets of vertices) are the same as $K$-cocliques in the scheme.

THEOREM 30.4. *Let $A \subseteq \mathfrak{x}$ be a $K$-coclique and $B \subseteq \mathfrak{x}$ a $K$-clique in a $k$-class association scheme on a set $\mathfrak{x}$ where $K \subseteq \{1, 2, \ldots, k\}$. Then*

$$|A|\,|B| \le N.$$

PROOF: Let $\mathbf{a} = (a_0, a_1, \ldots, a_k)$ be the distribution vector of $A$ and let $\mathbf{b} = (b_0, b_1, \ldots, b_k)$ be the distribution vector of $B$. Take the inverse of the first equation of Theorem 30.2, premultiply by $\mathbf{a}$ and postmultiply by $\mathbf{b}^\top$ to get

$$(30.9) \quad \mathbf{a}Q \begin{pmatrix} 1 & 0 & \cdots & 0 \\ 0 & m_1 & & 0 \\ \vdots & & \ddots & \vdots \\ 0 & 0 & \cdots & m_k \end{pmatrix}^{-1} (\mathbf{b}Q)^\top = N\,\mathbf{a} \begin{pmatrix} 1 & 0 & \cdots & 0 \\ 0 & n_1 & & 0 \\ \vdots & & \ddots & \vdots \\ 0 & 0 & \cdots & n_k \end{pmatrix}^{-1} \mathbf{b}^\top.$$

By Theorem 30.3, both $\mathbf{a}Q$ and $\mathbf{b}Q$ are nonnegative vectors. Their zeroth coordinates are $|A|$ and $|B|$, respectively, so the scalar on the left of (30.9) is at least $|A|\,|B|$. Our hypothesis implies that $a_i b_i = 0$ for $i > 0$, so the scalar on the right of (30.9) is $N$. $\square$

An example of equality in Theorem 30.4 in $J(k, v)$ occurs when there exists an $S(t, k, v)$. Take $K = \{1, 2, \ldots, k - t\}$. Then the block set of the $S(t, k, v)$ is a $K$-coclique. The set of all $k$-subsets containing a fixed $t$-subset is a $K$-clique. An example of equality in Theorem 30.4 in $H(n,q)$ occurs when there exists a perfect $e$-error-correcting code $C$. Take $K = \{1, 2, \ldots, 2e\}$. Then $C$ is a $K$-coclique. A sphere of radius $e$ about a fixed word is a $K$-clique.

PROBLEM 30F. Find an example of a regular graph $G$, a clique $A$, and a coclique $B$ such that $|A|\,|B| > |G|$.

PROBLEM 30G. Prove that if $A$ is a clique and $B$ a coclique in a graph $G$ which admits a transitive group of automorphisms, then $|A|\,|B| \leq |G|$.

We now define *polynomial schemes*, of which there are two types. An association scheme (with a particular numbering $A_1, A_2, \ldots, A_k$ of its association matrices) is said to be *P-polynomial* when $A_i$ is a polynomial of degree $i$ in $A_1$ for $i = 0, 1, 2, \ldots, k$. An association scheme (with a particular numbering $E_1, E_2, \ldots, E_k$ of its idempotents) is said to be *Q-polynomial* when $E_i$ is a Hadamard polynomial of degree $i$ in $E_1$ for $i = 0, 1, 2, \ldots, k$ (we mean that there is a polynomial of degree $i$ so that $E_i$ results by applying the polynomial element-wise to, i.e. to *each entry* of, $E_1$).

PROBLEM 30H. Prove that an association scheme is $P$-polynomial if and only if it is a metric scheme.

The $Q$-polynomial schemes are also called *cometric*, but there seems to no simple geometric interpretation of cometric. We use the terms metric and cometric rather than $P$- and $Q$-polynomial.

It can be seen that the Hamming and Johnson schemes are cometric; see Delsarte (1973).

We define a *d-code* in a metric scheme to be a subset $S \subseteq \mathfrak{x}$ whose characteristic vector $\phi$ satisfies

$$\phi A_i \phi^\top = 0 \quad \text{for } i = 1, 2, \ldots, d-1.$$

We define a *t-design* in a cometric scheme to be a subset $S \subseteq \mathfrak{x}$ whose characteristic vector $\phi$ satisfies

$$\phi E_i \phi^\top = 0 \quad \text{for } i = 1, 2, \ldots, t.$$

The combinatorial significance of $d$-codes is straightforward: $S$ is a $d$-code if and only if no two distinct elements of $S$ are $i$-th associates for $i < d$. So in the Hamming scheme, $S$ is a $(2e + 1)$-code if and only if it is $e$-error correcting, i.e. has minimum distance at least $2e + 1$. The combinatorial significance of $t$-designs is not so clear and can only be understood with more knowledge of the particular scheme. Note that $S$ is a $t$-design if and only if the first $t$ nontrivial inequalities on the distribution vector of $S$ in Theorem 30.3 hold with equality, i.e. that $\phi E_i = \mathbf{0}$ for $i = 1, \ldots, t$.

The following theorem explains that designs in the Johnson and Hamming schemes correspond to the classical concepts of $t$-designs and orthogonal arrays, respectively. An *orthogonal array* of index $\lambda$ and strength $t$ on a set $A$ of $q$ symbols is a subset $C \subseteq A^n$ of $\lambda q^t$ words so that for every choice of $t$ of the $n$ coordinates, each possible $t$-tuple of elements of $A$ occurs exactly $\lambda$ times in those $t$ coordinates among the members of $C$.

THEOREM 30.5.

(i) A family $S$ of $k$-subsets of a $v$-set is a $t$-design when considered as a subset of $J(k, v)$ if and only if it is the set of blocks of a $t$-design in the classical sense.

(ii) A family $S$ of $n$-tuples from an alphabet $A$ of $q$ elements is a $t$-design when considered as a subset of $H(n, q)$ if and only if it is the set of columns of an orthogonal array of strength $t$.

PARTIAL PROOF: We will show that the definition of $t$-design in the schemes implies that the family $S$ is a classical $t$-design or orthogonal array and leave the converses of both parts (i) and (ii) for the reader.

Let $S$ be a $t$-design in $J(k, v)$ in the sense of association schemes and $\phi$ its characteristic vector. In Example 30.7, we saw that $E_0 + E_1 + \cdots + E_t$ is the orthogonal projection onto the space spanned by $\mathbf{e}_T$ as $T$ ranges over the $t$-subsets of the $v$-set. The number of members of $S$ containing a $t$-subset $T$ is

$$\phi \mathbf{e}_T^\top = \phi(E_0 + E_1 + \cdots + E_t)\mathbf{e}_T^\top = \phi E_0 \mathbf{e}_T^\top.$$

Since $E_0$ is a scalar multiple of the all-one matrix $J$, this number is independent of the particular $t$-subset $T$.

Let $S$ be a $t$-design in $H(n, q)$ in the sense of association schemes and $\phi$ its characteristic vector. In Example 30.6, we described the eigenspaces for $q = 2$, and we prove the theorem only in this case. A complete proof may be found in Delsarte (1973).

With the notation of Example 30.6,

$$\langle \phi, \mathbf{v}_a \rangle = \sum_{b \in S} (-1)^{\langle a,b \rangle},$$

so $S$ is a $t$-design if and only if

$$\sum_{b \in S} (-1)^{\langle a,b \rangle} = 0 \qquad \text{for all nonzero } a \text{ of weight } \le t.$$

For example, when $a$ is taken of weight 1, the above equation implies that in any given coordinate position, half the members of $S$ have entry 0 and half have entry 1. When $a$ is taken of weight 2, e.g. as $(1, 1, 0, 0, \ldots, 0)$, the equation implies that the number of members of $S$ that begin with 00 or 11 is equal to the number that begin with 10 or 01; it follows from this and the previous sentence that exactly $1/4$ of the $n$-tuples in $S$ begin with each of 00, 10, 01, 11.

In general, consider a $t$-subset $T \subseteq \{1, 2, \ldots, n\}$ of coordinate positions. For each subset $I$ of $T$, let $\lambda_I$ denote the number of $a \in S$ with entry 1 in coordinates of $I$ and entry 0 in coordinates of $T \setminus I$. For each subset $I$ of $T$, let $a_J$ denote the binary $n$-tuple with 1's in coordinates of $J$ and 0's in the remaining $n - |J|$ coordinates. We have $2^t$ linear equations, one for each $J \subseteq T$:

$$\sum_{|I \cap J| \equiv 0 \; (\mathrm{mod} \; 2)} \lambda_I \; - \sum_{|I \cap J| \equiv 1 \; (\mathrm{mod} \; 2)} \lambda_I \; = \begin{cases} |S| & \text{if } J = \emptyset \\ 0 & \text{otherwise.} \end{cases}$$

Clearly $\lambda_I = |S|/2^t$ for all $I \subseteq T$ is a solution of this system. But the coefficient matrix of this system is a Hadamard matrix of order $2^t$ (see Chapter 18) and in particular is nonsingular, so the solution is unique. Thus $S$ is an orthogonal array of strength $t$. $\qquad \square$

The next two theorems are "formal duals". That is, their proofs are similar but the roles of the bases of $A_i$'s and $E_i$'s are interchanged, as is the role of ordinary and Hadamard multiplication of matrices. For the Hamming schemes, Theorem 30.6(i) reduces to the sphere packing bound, Theorem 21.1, and part (iii) gives a very strong condition, due to S. P. Lloyd (1957), for equality to hold (perfect codes–see Chapter 20). For the Johnson schemes, Theorem 30.7(i) reduces to Theorem 19.8, and part (iii) gives a very strong condition, due to Ray-Chaudhuri and Wilson (1975), for equality to hold (tight designs–see Chapter 19).

THEOREM 30.6. *Let $C$ be a $(2e+1)$-code in a $k$-class metric association scheme on a set $\mathfrak{X}$. Let $\phi \in \mathbb{R}^{\mathfrak{X}}$ be the characteristic vector of $C$.*

(i) *We have*

$$|C| \le N/(1 + n_1 + n_2 + \cdots + n_e).$$

(ii) *There are at least $e$ indices $i \in \{1, 2, \ldots, k\}$ such that*

$$\phi E_i \phi^\top \ne 0.$$

(iii) *Equality holds in (i) if and only if equality holds in (ii), in which case the $e$ indices $i$ such that $\phi E_i \phi^\top \ne 0$ are exactly those integers $i$ for which*

$$\sum_{\ell=0}^{e} P_\ell(i) = 0.$$

PROOF: Let $\phi$ be the characteristic vector of a $(2e+1)$-code $C$ and consider the expression

$$\alpha := \phi(c_0 A_0 + c_1 A_1 + \cdots + c_e A_e)^2 \phi^\top$$

where $c_0, c_1, \ldots, c_e$ are scalars. We shall evaluate $\alpha$ in two ways.
    Introduce

$$f(i) := c_0 P_0(i) + c_1 P_1(i) + \cdots + c_e P_e(i)$$

as a function of the $c_i$'s. By (30.3), $c_0 A_0 + c_1 A_1 + \cdots + c_e A_e = \sum_{i=0}^{k} f(i) E_i$ so

$$\alpha = \left( \sum_{i=0}^{k} f(i)\phi E_i \right) \left( \sum_{i=0}^{k} f(i)\phi E_i \right)^\top = \sum_{i=0}^{k} f(i)^2 \phi E_i \phi^\top.$$

On the other hand, $A_i$ is a polynomial of degree $i$ in $A_1$, so $(A_0 + A_1 + \cdots + A_e)^2$ is a polynomial of degree $2e$ in $A_1$ and hence a linear combination of $A_0, A_1, \ldots, A_{2e}$. Our hypothesis implies that $\phi A_i \phi^\top = 0$ for $i = 1, 2, \ldots, 2e$. Thus to evaluate $\alpha$, we need only the coefficient of $A_0$ when $(A_0 + A_1 + \cdots + A_e)^2$ is written as a linear combination of $A_0, A_1, \ldots, A_{2e}$; by (30.2) and (30.1),

$$\alpha = \left( \sum_{i,j=0}^{e} c_i c_j p_{ij}^0 \right) \phi A_0 \phi^\top = \left( \sum_{i=0}^{e} c_i^2 n_i \right) |C|.$$

Note that by (30.6), $f(0) = c_0 n_0 + c_1 n_1 + \cdots + c_e n_e$. We now combine the two values for $\alpha$, remembering that $E_0 = \frac{1}{N} J$ so that $\phi E_0 \phi^\top = \frac{1}{N} |C|^2$, to obtain

$$(30.10) \quad (c_0^2 + c_1^2 n_1 + \cdots + c_e^2 n_e)|C| = \sum_{i=0}^{k} f(i)^2 \phi E_i \phi^\top$$

$$\geq \frac{1}{N} (c_0 + c_1 n_1 + \cdots + c_e n_e)^2 |C|^2.$$

Everything will follow from (30.10). Part (i) follows when we take all $c_i := 1$.

To prove part (ii), suppose for contradiction that there are fewer than $e$ indices $i \geq 1$ such that $\phi E_i \phi^\top \neq 0$. By elementary linear algebra, there exist scalars $c_0, \ldots, c_e$, not all zero, such that $f(i) = 0$ for $i = 0$ and all $i$ such that $\phi E_i \phi^\top \neq 0$. But then (30.10) gives $|C| \sum_{i=0}^{e} n_i c_i^2 = 0$, a contradiction.

Assume equality holds in (i). Then (30.10), with all $c_i$'s equal to 1, shows that $f(i) = \sum_{\ell=0}^{e} P_\ell(i) = 0$ for the $e$ or more values of $i$ for which $\phi E_i \phi^\top \neq 0$. We claim that $f(i) = 0$ cannot hold for more than $e$ values of $i$. This is because $f(0), f(1), \ldots, f(k)$ are the eigenvalues of $A_0 + A_1 + \cdots + A_e$, which is a polynomial of degree $e$ in $A_1$; that is, $f(0), \ldots, f(k)$ arise from evaluating a polynomial of degree $e$ on the eigenvalues $P_1(0), \ldots, P_1(k)$ of $A_1$. (The matrix $A_1$ has $k + 1$ distinct eigenvalues since it generates an algebra of dimension $k+1$.) Our claim follows from the fact that a polynomial of degree $e$ has at most $e$ roots. So (ii) holds with equality and (iii) is proved in this case.

Finally, assume equality holds in (ii). Choose scalars $c_0, \ldots, c_e$ so that $f(i) = 0$ for all $i$ such that $\phi E_i \phi^\top \neq 0$ and such that $f(0) = 1$. Then (30.10) yields $|C| = N(c_0^2 + c_1^2 n_1 + \cdots + c_e^2 n_e)$. The Cauchy-Schwartz inequality shows that

$$1 = \left( \sum_{i=0}^{e} c_i n_i \right)^2 \le \left( \sum_{i=0}^{e} c_i^2 n_i \right) \left( \sum_{i=0}^{e} n_i \right)$$

so $|C| \ge N/(1 + n_1 + n_2 + \cdots + n_e)$. Thus equality holds in (i). $\square$

COROLLARY (LLOYD'S THEOREM). *If a perfect e-error-correcting code of length n over an alphabet of size q exits, then*

$$L_e(x) := \sum_{i=0}^{e} (-1)^i \binom{n-x}{e-i} \binom{x-1}{i} (q-1)^{e-i}$$

*has e distinct integral zeros.*

PROOF: This is just Theorem 30.6(iii) stated for the Hamming scheme. The sum $\sum_{\ell=0}^{e} P_\ell(x)$ where $P_\ell(x)$ is as in Theorem 30.1(ii) simplifies to $L_e(x)$ as stated above. $\square$

EXAMPLE 30.9. We show that there are no nontrivial perfect 2-error-correcting binary codes. If such a code has length $n > 2$, then its cardinality is $2^n/(1+n+\binom{n}{2})$, so $1+n+\binom{n}{2} = 2^r$ for some integer $r$. (That $n$ has this property is unusual—but it may happen, e.g. for $n = 90$.) From Theorem 30.1(ii), for $H(n,2)$,

$$P_0(x) + P_1(x) + P_2(x) = 2x^2 - 2(n+1)x + 1 + n + \binom{n}{2}$$

and Theorem 30.6(iii) then asserts that

$$x^2 - (n+1)x + 2^{r-1} = 0$$

has two integral roots $x_1$ and $x_2$. We must have $x_1 = 2^a$ and $x_2 = 2^b$ for positive integers $a$ and $b$ where $a+b = r-1$ and $2^a + 2^b = n+1$. Then it may be easily verified that

$$(2^{a+1} + 2^{b+1} - 1)^2 = 2^{a+b+4} - 7.$$

If $a$ and $b$ are both $\geq 2$, then the lefthand side is $\equiv 1 \pmod{16}$ while the righthand side is $\equiv 9 \pmod{16}$, a contradiction. The reader may consider other cases and will find that the only possibility is $\{a, b\} = \{1, 2\}$ in which case $n = 5$.

THEOREM 30.7. *Let $D$ be a 2s-design in a k-class cometric association scheme on a set $\mathfrak{X}$. Let $\phi \in \mathbb{R}^{\mathfrak{X}}$ be the characteristic vector of $D$.*

(i) *We have*

$$|D| \geq 1 + m_1 + m_2 + \cdots + m_s.$$

(ii) *There are at least $s$ indices $i \in \{1, 2, \ldots, k\}$ such that*

$$\phi A_i \phi^\top \neq 0.$$

(iii) *Equality holds in (i) if and only if equality holds in (ii), in which case the $s$ indices $i$ such that $\phi A_i \phi^\top \neq 0$ are exactly those integers $i$ for which*

$$\sum_{\ell=0}^{s} Q_\ell(i) = 0.$$

PROOF: Let $\phi$ be the characteristic vector of a 2s-design $D$ and consider the expression

$$\beta := \phi(c_0 E_0 + c_1 E_1 + \cdots + c_s E_s) \circ (c_0 E_0 + c_1 E_1 + \cdots + c_s E_s)\phi^\top$$

where $c_0, c_1, \ldots, c_s$ are scalars. We shall evaluate $\beta$ in two ways.
    Introduce

$$g(i) := \frac{1}{N}(c_0 Q_0(i) + c_1 Q_1(i) + \cdots + c_s Q_s(i))$$

as a function of the $c_i$'s. By (30.4), $c_0 E_0 + c_1 E_1 + \cdots + c_s E_s = \sum_{i=0}^{k} g(i)A_i$ so

$$\beta = \left(\sum_{i=0}^{k} g(i)\phi A_i\right) \circ \left(\sum_{i=0}^{k} g(i)\phi A_i\right)^\top = \sum_{i=0}^{k} g(i)^2 \phi A_i \phi^\top.$$

On the other hand, $E_i$ is a Hadamard polynomial of degree $i$ in $E_1$, so the Hadamard square of $E_0 + E_1 + \cdots + E_s$ is a Hadamard polynomial of degree $2s$ in $E_1$ and hence a linear combination of $E_0, E_1, \ldots, E_{2s}$. Our hypothesis implies that $\phi E_i \phi^\top = 0$ for $i = 1, 2, \ldots, 2s$. Thus to evaluate $\beta$, we need only the coefficient of $E_0$ when the Hadamard square of $E_0 + E_1 + \cdots + E_s$ is written as a linear combination of $E_0, E_1, \ldots, E_{2s}$; by (30.7) and (30.8),

$$\beta = \left( \frac{1}{N} \sum_{i,j=0}^{s} c_i c_j q_{ij}^0 \right) \phi E_0 \phi^\top = \frac{1}{N^2} \left( \sum_{i=0}^{s} c_i^2 m_i \right) |C|^2.$$

Note that by (30.6), $g(0) = \frac{1}{N}(c_0 m_0 + c_1 m_1 + \cdots + c_s m_s)$. We now combine the two values for $\beta$ to obtain

$$(30.11) \quad \frac{1}{N^2}(c_0^2 + c_1^2 m_1 + \cdots + c_s^2 m_s)|C|^2 = \sum_{i=0}^{k} g(i)^2 \phi A_i \phi^\top$$

$$\geq \frac{1}{N}(c_0 + c_1 m_1 + \cdots + c_s m_s)^2 |C|.$$

Everything will follow from (30.11) in a manner similar to that of the proof of Theorem 30.6.                                        □

**Notes.**

Association schemes were introduced by statisticians at the same time as *partially balanced designs*. An incidence structure is *partially balanced* with respect to an association scheme on its points when each pair of $i$-th associates occurs in a constant number $\lambda_i$ of blocks. Examples include the partial geometries of Chapter 21. Partially balanced designs were introduced to get around Fisher's inequality $b \geq v$, or equivalently $r \geq k$, which need not hold for them. Replications cost money ($r$ stands for "replication number") or take time, so for practical reasons, one wants $r$ small.

To do the analysis of an experiment for which a certain design (incidence structure) has been used, it is necessary to do calculations with $NN^\top$ where $N$ is the incidence matrix; these calculations are vastly simplified for partially balanced designs because in an $m$-class scheme

$$NN^\top = \lambda_0 A_0 + \lambda_1 A_1 + \cdots + \lambda_m A_m$$

lies in the small-dimensional Bose-Mesner algebra even though the size of the matrices may be very much more than the dimension $m + 1$.

Delsarte's inequalities imply the general Erdős-Ko-Rado theorem: If $n \geq (t + 1)(k - t + 1)$ and $\mathcal{F}$ is a family of $k$-subsets of an $n$-set so that any two of members of $\mathcal{F}$ meet in at least $t$ points, then $|\mathcal{F}| \leq \binom{n-t}{k-t}$. See Wilson (1984a).

See Van Lint (1982) for more on the nonexistence of perfect codes.

## References.

E. Bannai and T. Ito (1984), *Association Schemes*, Benjamin/ Cummings.

A. E. Brouwer, A. M. Cohen, and A. Neumaier (1989), *Distance Regular Graphs*, Springer-Verlag.

Ph. Delsarte (1973), *An Algebraic Approach to the Association Schemes of Coding Theory*, Philips Res. Rep. Suppl. **10**.

Ph. Delsarte (1975), The association schemes of coding theory, in: *Combinatorics* (Proc. Nijenrode Conf.), M. Hall, Jr. and J. H. van Lint, eds., D. Reidel.

J. H. van Lint (1982), *Introduction to Coding Theory*, Springer-Verlag.

S. P. Lloyd (1957), Binary block coding, *Bell System Tech. J.* **36**, 517–535.

D. K. Ray-Chaudhuri and R. M. Wilson (1975), On $t$-designs, *Osaka J. Math.* **12**, 737–744.

R. M. Wilson (1984a), The exact bound in the Erdős-Ko-Rado theorem, *Combinatorica* **4**, 247–257.

R. M. Wilson (1984b), On the theory of $t$-designs, pp. 19–50 in: *Enumeration and Design* (Proceedings of the Waterloo Silver Jubilee Conference), D. M. Jackson and S. A. Vanstone, eds., Academic Press.

# 31

## Algebraic graph theory: eigenvalue techniques

We have used linear algebraic techniques on the adjacency matrices of graphs in Chapter 9, and extensively in Chapter 21. We collect here several more elegant applications; (see also Chapter 34.)

A *tournament* is an orientation of a complete graph; that is, a directed graph such that for any two distinct vertices $x$ and $y$, there is either an edge from $x$ to $y$, or an edge from $y$ to $x$, but not both. Tournaments were introduced briefly in Problem 3D. The *adjacency matrix* of a digraph has a 1 in position $(x, y)$ when there is an edge from $x$ to $y$, and 0 otherwise.

LEMMA 31.1. *The rank of the adjacency matrix $A$ of a tournament on $n$ vertices is either $n$ or $n - 1$.*

PROOF: The definition of tournament ensures that $A + A^\top = J - I$, where all matrices are $n$ by $n$. Suppose the rank of $A$ is at most $n - 2$. Then there exists a nonzero row vector $\mathbf{x}$ such that both $\mathbf{x}A = 0$ and $\mathbf{x}J = 0$. We then compute

$$0 = \mathbf{x}(A + A^\top)\mathbf{x}^\top = \mathbf{x}(J - I)\mathbf{x}^\top = -\mathbf{x}\mathbf{x}^\top < 0,$$

which contradicts the existence of $\mathbf{x}$. □

The following theorem is due to R. L. Graham and H. O. Pollak. Their original proof applied Sylvester's law (see Chapter 9).

THEOREM 31.2. *Suppose that the complete graph $K_n$ can be expressed as the union of $k$ edge-disjoint subgraphs $H_1, H_2, \ldots, H_k$ where each $H_i$ is a complete bipartite graph. Then $k \geq n - 1$.*

PROOF: Orient each complete bipartite subgraph $H_i$ by directing all edges from one of the two color classes to the other. This produces a tournament on $n$ vertices whose adjacency matrix $A$ is the sum of the $k$ adjacency matrices $A_i$ of the digraphs $H_i$ (augmented to $n$ by $n$ matrices by including all vertices). The adjacency matrix of such a complete directed bipartite subgraph, after a suitable renumbering of the vertices, has the block form

$$\begin{pmatrix} 0 & J & 0 \\ 0 & 0 & 0 \\ 0 & 0 & 0 \end{pmatrix}$$

and in particular has rank 1. This implies that the rank of $A$ is at most $k$, and Lemma 31.1 completes the proof.     □

This brings to mind the De Bruijn-Erdős theorem, Theorem 19.1, which asserts, in a quite different terminology, and which was given a quite different proof, that if $K_n$ is the union of $k$ edge-disjoint complete subgraphs, then $k \geq n$. But no proof of Theorem 31.2 is known which does not use linear algebra.

As a real symmetric matrix, the adjacency matrix $A = A(G)$ of a finite graph $G$ has an orthogonal basis of eigenvectors. (Eigenvectors corresponding to different eigenvalues are necessarily orthogonal.) By the "eigenvalues of $G$" we mean the eigenvalues of $A(G)$. Here is a short table of the spectra, i.e. the complete lists of eigenvalues, of several graphs:

| graph | spectrum |
|---|---|
| $K_5$ | $4, -1, -1, -1, -1$ |
| $K_{3,3}$ | $3, 0, 0, 0, 0, -3$ |
| Cube | $3, 1, 1, 1, -1, -1, -1, -3$ |
| Pentagon | $2, \frac{1}{2}(-1 + \sqrt{5}), \frac{1}{2}(-1 + \sqrt{5}),$ $\frac{1}{2}(-1 - \sqrt{5}), \frac{1}{2}(-1 - \sqrt{5})$ |
| Petersen graph | $3, 1, 1, 1, 1, 1, -2, -2, -2, -2$ |
| $L_2(3)$ | $4, 1, 1, 1, 1, -2, -2, -2, -2$ |
| Heawood graph | $3, \sqrt{2}, \sqrt{2}, \sqrt{2}, \sqrt{2}, \sqrt{2}, \sqrt{2},$ $-\sqrt{2}, -\sqrt{2}, -\sqrt{2}, -\sqrt{2}, -\sqrt{2}, -\sqrt{2}, -3$ |

Nonisomorphic graphs may have the same spectra. For example, we mentioned in Chapter 21 that there exist four nonisomorphic strongly regular graphs with the parameters of $T(8)$; cf. Theorem 21.5.

PROBLEM 31A. Show that it is not possible to find three edge-disjoint copies of the Petersen graph in $K_{10}$. Hint: if $A_1$, $A_2$, and $A_3$ are the adjacency matrices of three edge-disjoint Petersen subgraphs of $K_{10}$, then all three matrices have the same spectrum, all have the all-one vector $\mathbf{j}$ as an eigenvector, and $A_1 + A_2 + A_3 = J - I$.

Let $S$ be a symmetric matrix. The expression $\mathbf{x}S\mathbf{x}^\top / \mathbf{x}\mathbf{x}^\top$, for $\mathbf{x} \neq \mathbf{0}$, is called a *Raleigh quotient*. Let $\mathbf{e}_1, \mathbf{e}_2, \ldots, \mathbf{e}_n$ be an orthonormal basis of eigenvectors with corresponding eigenvalues

$$\lambda_1 \geq \lambda_2 \geq \cdots \geq \lambda_n.$$

If we write $\mathbf{x}$ in this basis, say $\mathbf{x} = a_1\mathbf{e}_1 + \cdots + a_n\mathbf{e}_n$, then

$$(31.1) \qquad \frac{\mathbf{x}S\mathbf{x}^\top}{\mathbf{x}\mathbf{x}^\top} = \frac{\lambda_1 a_1^2 + \lambda_2 a_2^2 + \cdots + \lambda_n a_n^2}{a_1^2 + a_2^2 + \cdots + a_n^2}.$$

In particular, for any nonzero $\mathbf{x}$,

$$\lambda_1 \geq \frac{\mathbf{x}S\mathbf{x}^\top}{\mathbf{x}\mathbf{x}^\top} \geq \lambda_n.$$

While most graphs we will consider are simple, we remark that there is nothing to prevent the consideration of adjacency matrices of multigraphs below (where the entry in row $x$ and column $y$ is the number of edges joining $x$ and $y$), or the adjacency matrices of "weighted" graphs (where the entry in row $x$ and column $y$ is a "weight" associated to the edge joining $x$ and $y$).

It is at first surprising that the spectrum of a graph has anything at all to do with the more geometric or combinatorial properties of the graph. One of the first connections observed was the following theorem of A. J. Hoffman. A *coclique*, or an *independent* set of vertices, in a graph $G$ is a set of vertices no two of which are adjacent.

THEOREM 31.3. *Let $G$ be a graph on $n$ vertices which is regular of degree $d$ and let $\lambda_{\min}$ be the least eigenvalue of $G$, so $\lambda_{\min}$ is negative. Then for any coclique $S$ in $G$,*

$$|S| \leq \frac{-n\lambda_{\min}}{d - \lambda_{\min}}.$$

PROOF: Let $A$ be the adjacency matrix of $G$ and $\lambda := \lambda_{\min}$. Then $A - \lambda I$ is positive semidefinite, i.e. all eigenvalues are nonnegative; note that the eigenvectors of $A - \lambda I$ are the same as those of $A$. One of the eigenvectors is $\mathbf{j} := (1, 1, \ldots, 1)$, and we have

$$\mathbf{j}(A - \lambda I) = (d - \lambda)\mathbf{j}.$$

Then with
$$M := A - \lambda I - \frac{d - \lambda}{n} J,$$

we will have $\mathbf{j}M = 0$. Every other eigenvector $\mathbf{e}$ of $A$ may be taken to be orthogonal to $\mathbf{j}$, so $\mathbf{e}J = 0$, and is seen to be an eigenvector of $M$ with nonnegative eigenvalue. That is, $M$ is also positive semidefinite.

Now let $\phi$ be the characteristic vector of a coclique $S$ consisting of $m$ vertices of $G$, i.e. $\phi(x)$ is 1 if $x \in S$ and 0 otherwise. Then $\phi A \phi^{\top} = 0$ and we have

$$0 \leq \phi M \phi^{\top} = -\lambda \phi \phi^{\top} - \frac{d - \lambda}{n} \phi J \phi^{\top} = -\lambda m - \frac{d - \lambda}{n} m^2.$$

The stated inequality follows.  $\square$

There are extensions and variations on Theorem 31.3 for nonregular graphs—see Haemers (1979).

PROBLEM 31B. Apply Theorem 31.3 to the complement of a simple regular graph $G$ to obtain an upper bound on the size of a clique in $G$ in terms of the spectrum. Find several examples where your bound is met.

PROBLEM 31C. The *line graph* $L = L(G)$ of a simple graph $G$ is the simple graph with $V(L) := E(G)$ and where $a, b \in E(G)$ are adjacent as vertices of $L$ if and only if, as edges in $G$, $a$ and $b$ have a common incident vertex. For example, $L_2(m)$ as defined in Chapter 21 is the line graph of $K_{m,m}$, $T(m)$ is the line graph of $K_m$, and the line graph of a pentagon is a pentagon. Show that the Petersen graph is *not* the line graph of any graph. Prove that if $G$ has more edges than vertices, then the minimum eigenvalue of the line graph $L(G)$ is $-2$. (Suggestion: let $N$ be the incidence matrix of $G$ and consider $N^\top N$.)

L. Lovász observed that the bound on the size of a coclique in Theorem 31.3 is also a bound on what is called the Shannon capacity of a graph. This is a concept that arose in information theory. Suppose a set of "letters" is to be used for transmitting messages. Some pairs of letters are assumed to be "confusable" or "confoundable". We say two potential messages (words or strings of $n$ of these letters) are *confoundable* when in each coordinate the letters are either identical or confoundable. We desire a set of words no two of which are confoundable.

To state this in more graph-theoretic terminology, we introduce the graph $G$ whose vertices are the letters and where two vertices are adjacent if and only if the letters are confoundable. Thus a set of letters, no two of which are confoundable, is exactly a coclique in this graph. The *strong product* $G \otimes H$ of simple graphs $G$ and $H$ is the simple graph defined by $V(G \otimes H) := V(G) \times V(H)$ and where distinct vertices $(x_1, y_1)$ and $(x_2, y_2)$ are adjacent in $G \otimes H$ when $x_1$ and $x_2$ are equal or adjacent, and $y_1$ and $y_2$ are equal or adjacent. The vertices of $G^n := G \otimes G \otimes \cdots \otimes G$ ($n$ factors) correspond to words of length $n$, and two are adjacent in $G^n$ just when the words are confoundable. Thus we are interested in the size of a largest coclique in $G^n$.

The *independence number* $\alpha(G)$ of a graph $G$ is the largest cardinality of a coclique in $G$. The *Shannon capacity* of a graph $G$ is

$$\Theta(G) := \lim_{n \to \infty} (\alpha(G^n))^{1/n} = \sup_n (\alpha(G^n))^{1/n}.$$

That the limit exists and is equal to the supremum follows from the observation that the product of cocliques in $G$ and $H$ is a coclique in $G \otimes H$, and thus that $\alpha(G^{k+m}) \geq \alpha(G^k)\alpha(G^m)$. (See Fekete's lemma in Chapter 11.)

EXAMPLE 31.1. The problem of evaluating the Shannon capacity of such simple graphs as the pentagon $P_5$ was open for many years before the relevance of eigenvalue techniques was noticed by Lovász. It is easy to find $2^n$ words of length $n$, no two of which are confoundable, e.g. all words of 1's and 3's. This shows $\Theta(P_5) \geq 2$. This is not so good; we get only four words of length 2 this way, when it is not hard to find a set of five:

$$(1,1), \ (2,3), \ (3,5), \ (4,2), \ (5,4).$$

For even values of $n$, we can take the product of $n/2$ copies of the above five words to get $(\sqrt{5})^n$ pairwise nonconfoundable words of length $n$; this shows $\Theta(P_5) \geq \sqrt{5}$. Can we do better, perhaps for very large values of $n$? No. By the Corollary to Theorem 31.4 below, $\Theta(P_5) \leq \sqrt{5}$.

The following is one of the theorems from Lovász (1979).

THEOREM 31.4. *Let $G$ be a finite graph. Suppose $M$ is a symmetric matrix with rows and columns indexed by the vertices of $G$ such that the entry in row $x$ and column $y$ is 1 whenever $x$ and $y$ are not adjacent (this implies that $M$ has 1's on the diagonal, but the entries where $x$ and $y$ are adjacent are arbitrary). Then the Shannon capacity of $G$ is bounded above by the largest eigenvalue of $M$.*

PROOF: Let $\lambda$ be the largest eigenvalue of such a matrix $M$. Note that the $k$-fold Kronecker product $M^{\otimes k} := M \otimes \cdots \otimes M$ has the property that its rows and columns are indexed by the vertices of $G^k$ and it has a 1 in the entry corresponding to nonadjacent vertices of $G^k$. The eigenvalues of the Kronecker product of two matrices are the products of the eigenvalues of the matrices (see Problem 21E). Thus the largest eigenvalue of $M^{\otimes k}$ is $\lambda^k$.

Let $\phi$ be the characteristic vector of a coclique $S$ of $G^k$ with $|S| = m$. We compute

$$m^2 = \phi(M^{\otimes k})\phi^\top \leq \lambda^k \phi\phi^\top = m\lambda^k.$$

So $m \leq \lambda^k$ and hence $\alpha(G^k)^{1/k} \leq \lambda$.     □

COROLLARY. *Let $G$ be a graph on $n$ vertices which is regular of degree $d$ and let $\lambda_{\min}$ be the least eigenvalue of $G$. Then*

$$\Theta(G) \le \frac{-n\lambda_{\min}}{d - \lambda_{\min}}.$$

PROOF: In the proof of Theorem 31.3, we saw that with $\lambda := \lambda_{\min}$, $A - \lambda I - \frac{d-\lambda}{n} J$ was positive semidefinite. This is equivalent to saying that $J - \frac{n}{d-\lambda} A$ has maximum eigenvalue $\frac{-n\lambda}{d-\lambda}$. This latter matrix may be taken as $M$ in the statement of Theorem 31.4.          □

LEMMA 31.5. *Let $A$ be a symmetric matrix of order $n$ with eigenvalues*

$$\lambda_1 \ge \lambda_2 \ge \cdots \ge \lambda_n.$$

*Suppose $N$ is an $m$ by $n$ real matrix such that $NN^\top = I_m$ so $m \le n$, let $B := NAN^\top$, and*

$$\mu_1 \ge \mu_2 \ge \cdots \ge \mu_m$$

*be the eigenvalues of $B$. Then the eigenvalues of $B$ "interlace" those of $A$; that is,*

$$\lambda_i \ge \mu_i \ge \lambda_{n-m+i}$$

*for $i = 1, 2, \ldots, m$.*

PROOF: Let $\mathbf{e}_1, \ldots, \mathbf{e}_n$ be an orthonormal basis of eigenvectors of $A$ corresponding to $\lambda_1, \ldots, \lambda_n$, and let $\mathbf{f}_1, \ldots, \mathbf{f}_m$ be an orthonormal basis of eigenvectors of $B$ corresponding to $\mu_1, \ldots, \mu_m$.

Fix $i$ and consider $U := \text{span}\{\mathbf{f}_1, \ldots, \mathbf{f}_i\}$. By (31.1),

$$\frac{\mathbf{x} B \mathbf{x}^\top}{\mathbf{x}\mathbf{x}^\top} \ge \mu_i$$

for every nonzero $\mathbf{x} \in U$. Let $W := \{\mathbf{x}N : \mathbf{x} \in U\}$. Then $W$ is an $i$-dimensional subspace and

(31.2)          $$\frac{\mathbf{y} A \mathbf{y}^\top}{\mathbf{y}\mathbf{y}^\top} \ge \mu_i$$

for every $\mathbf{y} \in W$. Choose $\mathbf{y} \neq 0$ in $W \cap \mathrm{span}\{\mathbf{e}_i, \mathbf{e}_{i+1}, \ldots, \mathbf{e}_n\}$. Then in addition to (31.2), we have

$$\frac{\mathbf{y} A \mathbf{y}^\top}{\mathbf{y}\mathbf{y}^\top} \leq \lambda_i$$

from (31.1) and this proves $\lambda_i \geq \mu_i$.

A similar argument with $U := \mathrm{span}\{\mathbf{f}_i, \ldots, \mathbf{f}_m\}$ will prove that $\mu_i \geq \lambda_{n-m+i}$.     □

The term interlacing seems most natural in the case $m = n - 1$ where

$$\lambda_1 \geq \mu_1 \geq \lambda_2 \geq \mu_2 \geq \lambda_3 \geq \cdots \geq \lambda_{n-1} \geq \mu_{n-1} \geq \lambda_n.$$

An important special case of Lemma 31.5 occurs when the rows of $N$ are $m$ distinct "standard basis" vectors of length $n$, i.e. with a single 1 in each row. In this case the matrix $B$ is just an $m$ by $m$ principal submatrix of $A$. The following observation is due to D. M. Cvetković.

THEOREM 31.6. *The size of a coclique in a graph $G$ cannot exceed the number of nonnegative eigenvalues, or the number of nonpositive eigenvalues, of $G$.*

PROOF: If $G$ has a coclique of size $m$, then the adjacency matrix $A$ has an $m$ by $m$ principal submatrix of all zeros. If the eigenvalues of $A$ are $\lambda_1 \geq \lambda_2 \geq \cdots \geq \lambda_n$, then by interlacing, $\lambda_m \geq 0$ and $0 \geq \lambda_{n-m+1}$, so $A$ has at least $m$ eigenvalues which are $\geq 0$ and at least $m$ which are $\leq 0$.     □

The next theorem is a result of A. J. Hoffman on the chromatic number of a graph. We remark that for regular graphs, this would follow from Theorem 31.3.

THEOREM 31.7. *Let $\lambda_1 \geq \cdots \geq \lambda_n$ be the eigenvalues of a graph $G$. Then*

$$\chi(G) \geq 1 + \lambda_1/(-\lambda_n).$$

PROOF: Suppose $G$ can be properly colored with $m$ colors. The color classes induce a partition of the adjacency matrix

$$A = \begin{pmatrix} A_{11} & \cdots & A_{1m} \\ \vdots & & \vdots \\ A_{11} & \cdots & A_{1m} \end{pmatrix}$$

where $A_{ij}$ is the submatrix consiting of the rows indexed by the vertices of color $i$ and columns indexed by the vertices of color $j$. Each of the diagonal blocks $A_{ii}$ is a square zero matrix.

Let $\mathbf{e}$ be an eigenvector of $A$ corresponding to the maximum eigenvalue $\lambda_1$ and write $\mathbf{e} = (\mathbf{e}_1, \dots, \mathbf{e}_m)$ where $\mathbf{e}_i$ has coordinates indexed by the vertices of color $i$. Let

$$
N := \begin{pmatrix}
\frac{1}{\|\mathbf{e}_1\|}\mathbf{e}_1 & 0 & 0 & \cdots \\
0 & \frac{1}{\|\mathbf{e}_2\|}\mathbf{e}_2 & 0 & \cdots \\
0 & 0 & \frac{1}{\|\mathbf{e}_3\|}\mathbf{e}_3 & \cdots \\
\vdots & \vdots & \vdots & \cdots
\end{pmatrix}
$$

(an $m$ by $n$ matrix) and let $B := NAN^\top$. By Lemma 31.5, the eigenvalues $\mu_1, \dots, \mu_m$ of $B$ interlace those of $A$ and hence are between $\lambda_1$ and $\lambda_n$. On the other hand, we have constructed $B$ so that $\lambda_1$ *is* an eigenvalue since

$$(\|\mathbf{e}_1\|, \dots, \|\mathbf{e}_m\|)B = \mathbf{e}AN^\top = \lambda_1 \mathbf{e}N^\top = \lambda_1(\|\mathbf{e}_1\|, \dots, \|\mathbf{e}_m\|).$$

Finally note that the diagonal entries of $B$ are zeros, so

$$0 = \text{trace}(B) = \mu_1 + \cdots + \mu_m \geq \lambda_1 + (m-1)\lambda_n.$$

This proves the theorem. (But perhaps we should have said that if some of the $\mathbf{e}_i$ are zero, those rows should not be included in $N$.) $\qquad\square$

PROBLEM 31D. Show that for any finite graph $G$, $\chi(G) \leq 1 + \lambda_1$ where $\lambda_1$ is the largest eigenvalue of $G$. Hint: consider the degrees of an induced subgraph of $G$ that is minimal with respect to having the same chromatic number as $G$.

In addition to being symmetric, the adjacency matrix of a graph is also a *nonnegative* matrix. There is a useful theory of such matrices, at the center of which is the Perron-Frobenius theorem, Theorem 31.8 below. A proof may be found in Gantmacher (1959).

A square matrix $A$ with rows and columns indexed by a set $X$, say, is called *irreducible* when it is *not* possible to find a proper subset $S$ of $X$ so that $A(x, y) = 0$ whenever $x \in S$ and $y \in X \setminus S$. Equivalently, $A$ is not irreducible if and only if it is possible to apply a simultaneous row and column permutation to obtain a matrix of the form

$$\begin{pmatrix} B & 0 \\ C & D \end{pmatrix}$$

where $B$ and $D$ are square and of order at least 1. It should be clear that if $A$ is the incidence matrix of a graph, then it is irreducible if and only if the graph is connected.

PROBLEM 31E. Let $A$ be the adjacency matrix of a digraph $D$. Show that $A$ is irreducible if and only if $D$ is *strongly connected*, that is, if and only if for any two vertices of $D$, there is a *directed* path from $x$ to $y$.

THEOREM 31.8 (PERRON-FROBENIUS). *Let $A$ be an irreducible $n$ by $n$ nonnegative matrix. There is an eigenvector $\mathbf{a} = (a_1, \ldots, a_n)$ all of whose coordinates $a_i$ are strictly positive. The corresponding eigenvalue $\lambda$ has multiplicity 1 (so in particular, $\mathbf{a}$ is unique up to scalar multiples) and has the property that $\lambda \geq |\mu|$ for any eigenvalue $\mu$ of $A$.*

The simple eigenvalue $\lambda$ corresponding to the positive eigenvector of such an irreducible nonnegative matrix $A$ is called the *dominant* eigenvalue of $A$. It has multiplicity 1 in the strong sense that it is a simple root of the characteristic polynomial.

THEOREM 31.9. *Let $\lambda$ be the dominant eigenvalue of a connected graph $G$. Then $G$ is bipartite if and only if $-\lambda$ is also an eigenvalue of $G$.*

PROOF: First suppose $G$ is bipartite. Then the adjacency matrix $A$ of $G$ is, after renumbering the vertices if necessary so that the color classes consist of the first $k$ and last $n - k$, of the form

$$A = \begin{pmatrix} 0 & B \\ B^\top & 0 \end{pmatrix}$$

where $B$ is $k$ by $n - k$. Let $\mathbf{e}$ be an eigenvector corresponding to an eigenvalue $\mu$ and write $\mathbf{e} = (\mathbf{e}_1, \mathbf{e}_2)$ where $\mathbf{e}_1$ consists of the first $k$ coordinates. It may be checked that $(\mathbf{e}_1, -\mathbf{e}_2)$ is then an eigenvector of eigenvalue $-\mu$. So if $G$ is bipartite (connected or not), its spectrum is in fact symmetric about 0.

Now suppose $\lambda$ is the dominant eigenvalue of a connected graph $G$ and that $-\lambda$ is also an eigenvalue. Let $\mathbf{e} = (\mathbf{e}_1, -\mathbf{e}_2)$ be an eigenvector of unit length corresponding to $-\lambda$, where we have numbered the vertices so that $\mathbf{e}_1$ and $\mathbf{e}_2$ both have all nonnegative coordinates. Partition the adjacency matrix $A$ accordingly:

$$A = \begin{pmatrix} B & C \\ D & E \end{pmatrix}.$$

Then

$$-\lambda = \mathbf{e} A \mathbf{e}^\top = \mathbf{e}_1 B \mathbf{e}_1^\top + \mathbf{e}_2 E \mathbf{e}_2^\top - \mathbf{e}_1 D \mathbf{e}_2^\top - \mathbf{e}_2 C \mathbf{e}_1^\top,$$

and hence

$$(31.3) \qquad\qquad (\mathbf{e}_1, \mathbf{e}_2) A (\mathbf{e}_1, \mathbf{e}_2)^\top \geq \lambda$$

with equality if and only if

$$(31.4) \qquad\qquad \mathbf{e}_1 B \mathbf{e}_1^\top = \mathbf{e}_2 E \mathbf{e}_2^\top = 0.$$

The equation (31.1) and the fact that $\lambda$ is the dominant eigenvalue imply that equality must hold in (31.3) and that $(\mathbf{e}_1, \mathbf{e}_2)$ is an eigenvector corresponding to eigenvalue $\lambda$. Then Theorem 31.8 implies that all coordinates of $(\mathbf{e}_1, \mathbf{e}_2)$ are positive. Finally, (31.4) shows that $B = E = 0$ and this means that $G$ is bipartite. $\qquad\square$

THEOREM 31.10. *Let $G$ be a finite graph and $A$ its adjacency matrix. There exists a polynomial $f(x)$ such that $f(A) = J$ if and only if $G$ is connected and regular.*

PROOF: Suppose $f(A) = J$ for some polynomial $f(x)$. Then, since polynomials in a given matrix commute, $AJ = JA$. The entry in row $x$ and column $y$ of $AJ$ is $\deg_G(x)$; the entry in that position of $JA$ is $\deg_G(y)$. Thus $G$ is regular.

If $f(A) = J$, then, in particular, for any $x$ and $y$, we must have $A^k(x, y) \neq 0$ for some $k$. The entry in row $x$ and column $y$ of $A^k$

is the number of walks of length $k$ from $x$ to $y$ in $G$. Thus $G$ is connected.

Now suppose $G$ is connected and regular of degree $d$. Then $\mathbf{j}$ is an eigenvector of $A$ corresponding to eigenvalue $d$. By Theorem 31.8, the eigenvalue $d$ has multiplicity one. For any symmetric matrix $M$, the matrix of the orthogonal projection onto any of its eigenspaces is a polynomial in $M$; explicitly, if $(x - \mu_1)(x - \mu_2)\ldots(x - \mu_k)$ is the minimal polynomial of $M$, then it is an exercise to show that

$$\frac{1}{(\mu_1 - \mu_2)\ldots(\mu_1 - \mu_k)}(M - \mu_2 I)\ldots(M - \mu_k I)$$

is the orthogonal projection onto the eigenspace $\{\mathbf{a} : \mathbf{a}M = \mu_1\mathbf{a}\}$. Since the orthogonal projection onto the span of $\mathbf{j}$ is $\frac{1}{v}J$, this matrix is a polynomial in $A$.     $\square$

PROBLEM 31F. Let $G$ be a connected regular graph with exactly three distinct eigenvalues. Show that $G$ is strongly regular.

PROBLEM 31G. Let $G$ be a nontrivial strongly regular graph and for $x \in V(G)$, let $\Delta(x)$ denote the subgraph induced by the vertices nonadjacent to $x$. Prove that $\Delta(x)$ is connected by (1) explaining why $k - \mu$ would be an eigenvector of $G$ if $\Delta(x)$ were not connected, and (2) showing by calculation that $k - \mu$ is not an eigenvalue of $G$.

## Notes.

Theorem 31.3 was never published by Hoffman, one of the first "algebraic graph theorists", but caught the fancy of many of those who heard of it, and has become a classic.

Problem 31D is a result of H. Wilf. Problem 31A is an unpublished result of A. J. Schwenk.

## References.

N. Biggs (1974), *Algebraic Graph Theory*, Cambridge University Press.

D. M. Cvetković, M. Doob, and H. Sachs (1979), *Spectra of Graphs, a Monograph*, V. E. B. Deutscher Verlag der Wissenschaften.

F. R. Gantmacher (1959), *The Theory of Matrices*, Chelsea.

W. Haemers (1979), *Eigenvalue Techniques in Design and Graph Theory*, Mathematisch Centrum, Amsterdam.

L. Lovász (1979), On the Shannon capacity of a graph, *IEEE Trans. Information Theory* **25**, 1–7.

# 32

## Graphs: planarity and duality

We begin by introducing the concepts of deletion and contraction, and connectivity. Then planar embeddings of graphs will be considered.

Let $G$ be a graph and $S \subseteq E(G)$. We define two graphs $G'_S$ and $G''_S$ with the same edge set $E(G) \backslash S$. First, let $G : S$ denote the spanning subgraph of $G$ with edge set $S$ (recall that a *spanning subgraph* of $G$ is one that includes all vertices of $G$). The graph $G'_S$ is $G : (E(G) \setminus S)$ with all isolated vertices removed. We speak of the *deletion* of $S$. The vertices of $G''_S$ are to be the (connected) components of $G : S$ and where the endpoints of an edge $e \in E(G) \backslash S$ as an edge of $G''_S$ are the components of $G : S$ that contain the endpoints of $e$ in $G$. This construction is called *contraction* (of $S$) and the graphs arising in this way are called *contractions* of $G$. For example, $K_5$ is a contraction of the Petersen graph (let $S$ be the five "spokes" in the usual drawing). A *minor* of $G$ is a contraction of a subgraph of $G$. Notice that contractions will often have loops and multiple edges even if the original graph has none.

A *subdivision* of a graph $G$ is a graph $H$ obtained by replacing the edges of $G$ by path-graphs of length at least one. Informally, we are inserting extra vertices into some edges. A graph $G$ is a contraction of any of its subdivisions $H$.

PROBLEM 32A. Find a subdivision of $K_{3,3}$ as a subgraph of the Petersen graph.

For the important case $S = \{e\}$, we write simply $G'_e$ and $G''_e$. To repeat our definition in this case, $G''_e$ is obtained by identifying the endpoints of $e$ (and removing $e$ itself). We can think of "reeling in" or "shrinking" an edge in a drawing of a graph as in Fig. 32.1.

Figure 32.1

The *chromatic polynomial* $\chi_G$ of a finite graph $G$ was introduced in Chapter 25: $\chi_G(\lambda)$ is the number of proper colorings of $G$ with $\lambda$ colors. For example, note that

$$\chi_{K_n}(\lambda) = \lambda(\lambda - 1)(\lambda - 2) \cdots (\lambda - n + 1).$$

That $\chi_G$ is a polynomial in $\lambda$ was shown in Chapter 25—also see the remark following (32.1) below.

EXAMPLE 32.1. We claim that for a tree $T$ on $n$ vertices,

$$\chi_T(\lambda) = \lambda(\lambda - 1)^{n-1}.$$

To see this, observe that if $x$ is a monovalent vertex of a tree $T$, there are $\lambda - 1$ colors available for $x$ after the tree on $n - 1$ vertices obtained by removing $x$ is given any proper coloring. The claim follows by induction on the number of vertices.

If $e$ is an edge of $G$, not a loop, we can break the proper colorings of $G'_e$ into two classes: those in which the endpoints of $e$ have different colors (these are proper colorings of $G$) and those in which the endpoints of $e$ have the same color (and these are in one-to-one correspondence with the proper colorings of $G''_e$). This establishes

(32.1)                    $\chi_G(\lambda) = \chi_{G'_e}(\lambda) - \chi_{G''_e}(\lambda).$

The equation (32.1) provides another way to see that $\chi_G(\lambda)$ is a polynomial in $\lambda$. A graph with loops has no proper colorings. For an edgeless graph $H$ on $n$ vertices, $\chi_H(\lambda) = \lambda^n$. These are both polynomials. Now use (32.1) and induction on the number of edges.

The chromatic polynomial of a tree $T$ could also be obtained by induction and (32.1): $T''_e$ is a tree and $T'_e$ has two components that are trees.

PROBLEM 32B. Let $\tau(G)$ denote the number of spanning trees in a graph $G$ (this is sometimes called the *complexity* of $G$). Show that for any nonloop $e \in E(G)$, $\tau(G) = \tau(G'_e) + \tau(G''_e)$.

PROBLEM 32C. Find the chromatic polynomial of the $n$-gon $C_n$. Find the chromatic polynomial of the $n$-wheel $W_n$ (this is the graph obtained from $C_n$ by adding a new vertex and joining it to all vertices of $C_n$).

It makes no sense to ask for the number of proper colorings of a graph with $-1$ colors, but since $\chi_G(\lambda)$ is a polynomial, we can evaluate $\chi_G(-1)$. R. P. Stanley (1973) discovered the combinatorial interpretation of $|\chi_G(-1)|$ that follows. An *orientation* of a graph is one of the directed graphs arising by choosing for each edge one of its endpoints to be the *head*; the other is the *tail*. (So the number of orientations of a graph is $2^m$ where $m$ is the number of edges that are not loops—although for topological purposes, there are reasons why a loop should also be considered to admit two directions.)

THEOREM 32.1. *The number of acyclic orientations, i.e. orientations in which there are no directed circuits, of a graph $G$ is* $(-1)^{|V(G)|}\chi_G(-1)$.

PROOF: If $G$ has loops, then $\chi_G(\lambda) = 0$ and $G$ has no acyclic orientations, so the theorem holds in this case.

Consider an acyclic orientation of $G'_e$ where $e$ is not a loop. There are two ways to direct $e$ to get an orientation of $G$. We claim that one or both of them are acyclic, and that the cases where both are acyclic are in one-to-one correspondence with acyclic orientations of $G''_e$. This is because directing $e$ from one end $x$ to the other $y$ produces a nonacyclic orientation if and only if there is a directed path from $y$ to $x$ in $G'_e$; there cannot be directed paths from $x$ to $y$ and also from $y$ to $x$ in $G'_e$ since we assumed an acyclic orientation; and there is no directed path from $y$ to $x$ in $G'_e$ if and only if identifying $x$ and $y$ produces an acyclic orientation of $G''_e$.

Thus

$$\omega(G) = \omega(G'_e) + \omega(G''_e)$$

when $\omega(H)$ is taken to be the number of acyclic orientations of a graph $H$. But by (32.1), this same recursion is also satisfied when $\omega(H)$ is taken to be $(-1)^{|V(H)|}\chi_H(-1)$. The theorem follows by induction on the number of edges after we check that it is valid for edgeless graphs. □

A *k-separation* of a graph $G$ is a pair $(H, K)$ of nonempty edge-disjoint subgraphs of $G$, each containing at least $k$ edges, such that $H \cup K = G$ and $|V(H) \cap V(K)| = k$. The graph $G$ is said to be *k-connected* when it has no $\ell$-separations for $\ell < k$. Thus a graph is 1-connected if and only if it is connected in the usual sense. Another term for 2-connected is *nonseparable*, which we tend to use. Note that a nonseparable graph cannot contain a loop unless it is the loop-graph itself (with one vertex and one loop) because taking $H$ as the loop-graph and $K$ as the spanning subgraph with all other edges would give a 1-separation if there are other edges.

The definition of $k$-connected we have given is that of W. T. Tutte (1984). A strongly related concept is *vertex connectivity*. A graph $G$ is said to be *k-vertex connected* when $|V(G)| \geq k + 1$ and the removal of any $k - 1$ vertices (and any incident edges) from $G$ does not result in a disconnected graph.

PROBLEM 32D. *Let $G$ be a graph with $|V(G)| \geq k + 1$. Then $G$ is $k$-connected if and only if $G$ is $k$-vertex connected and has no polygons with fewer than $k$ edges.*

REMARKS. An important theorem on vertex connectivity is that of H. Whitney (1932) who proved that a graph $G$ with at least $k + 1$ vertices is $k$-vertex connected if and only if, for any two distinct vertices $x$ and $y$ of $G$, there exist $k$ internally disjoint simple paths joining $x$ and $y$, i.e. paths which share no vertices other than $x$ and $y$. Also see Menger's theorem, e.g., in Chartrand and Lesniak (1986). These theorems may also be derived from the maxflow-mincut theorem of Chapter 7—see Ford and Fulkerson (1962).

If $G$ has a 1-separation $(H, K)$, it is clear that an edge of $H$ and an edge of $K$ cannot be contained together in any polygon of $G$.

THEOREM 32.2. *Any two edges in a nonseparable graph $G$ are contained in a circuit (the edge set of a polygon subgraph).*

PROOF: The proof is in two steps.

(1) We claim that a nonseparable graph $G$ with at least two edges arises from any of its maximal proper nonseparable subgraphs $H$ as the union of $H$ with another subgraph which is the graph of a simple path $p$ that joins two distinct vertices of $H$ and that has no terms in $H$ other than those two vertices. (The path is a "handle", and this assertion can be called the "handle lemma" for nonseparable graphs.)

To see this, let $H$ be a maximal proper nonseparable subgraph of a nonseparable graph $G$. It may be that $V(H) = V(G)$. In this case, let $e$ be an edge in $G$ but not $H$ and let $p$ be the path consisting of $e$ and its ends. Since adding $e$ to $H$ produces a nonseparable graph and $H$ is maximal among the proper nonseparable subgraphs, the union of $H$ and $p$ is $G$. If $V(H) \neq V(G)$, the connectivity of $G$ implies that there exists an edge $e$ with one end $x \in V(H)$ and one end $y \notin V(H)$. By the connectivity of $G - x$, there exists a simple path that does not pass through $x$ joining $y$ to some other vertex $w$ of $H$. If $z$ is the first vertex of $H$ on this path (as we proceed from $y$ to $w$), the edge $e$ and the initial part of this path provide a simple path $p$ from $x$ to $z$ that has no edge terms in $H$ and no vertex terms in $H$ other than its ends. The union of $H$ and all terms of $p$ is a nonseparable graph larger than $H$, and so is equal to $G$.

(2) We assert that given distinct vertices $x, y$ and an edge $e$ of a nonseparable graph $G$, there exists a simple path from $x$ to $y$ which traverses $e$. So if $x, y$ are the ends of another edge $e'$, we obtain a circuit containing $e$ and $e'$, and the theorem will follow. We prove this assertion by induction on the total number of edges.

The edge $e$ with its incident ends is a nonseparable subgraph. Let $H$ be a maximal proper nonseparable subgraph containing $e$. If both $x$ and $y$ are in $H$, we are done by the induction hypothesis. Otherwise $G$ is the union of $H$ and the graph of a path $p$ joining vertices $a$ and $b$ of $H$, say, and at least one of $x$ or $y$ is an *internal* vertex of $p$. If both $x$ and $y$ are vertices of $p$, a simple path from

$a$ to $b$ in $H$ that traverses $e$ together with segments of $p$ provides the required simple path. Finally, assume $x$ is internal to $p$ and $y \in V(H)$. Say $y \neq a$. Then a simple path in $H$ from $y$ to $a$ that traverses $e$ together with the segment of $p$ from $x$ to $a$ provides the required simple path.                                    □

COROLLARY. *Any two vertices of a nonseparable graph $G$ (other than the link-graph consisting of one edge joining two distinct vertices) are contained in some polygon in $G$.*

By a *surface*, we shall mean a compact 2-manifold $S$. This is a compact connected Hausdorff space with the property that for each of its points $a$, there is a neighborhood $N_a$ of $a$ that is a 2-cell, i.e. that is homeomorphic to the open disk $\{(x,y) : x^2 + y^2 < 1\}$ in the plane $\mathbb{R}^2$. Examples include the sphere (the boundary of the ball) in 3-space, the torus (boundary of the donut), and the real projective plane (which cannot be embedded in 3-space). While we give definitions for general surfaces and discuss them in the following chapter, it is drawings of graphs on the sphere that we are concerned with in this chapter.

By an *embedding* or *proper drawing* of a graph $G$ on a surface $S$, we mean a drawing where the edges do not cross but meet only at vertices. More precisely, we mean a representation of the graph where the vertices correspond to points in $S$, the edges correspond to Jordan arcs (continuous one-to-one images of the unit interval) in $S$ that join the points of $S$ corresponding to the ends of the edge in $G$, and where no internal point of any of the Jordan arcs is a point of any other Jordan arc or one of the points corresponding to the vertices of $G$. We consider only embeddings of finite graphs. While the plane is not compact, it can be naturally embedded in the sphere (its one-point compactification) and *a graph is planar if and only if it admits an embedding on the sphere.*

Given an embedding of a graph $G$ on a surface $S$, we define the *faces* or *regions* with respect to the embedding to be the topologically connected components that result when the vertices and edges (more precisely, the points corresponding to vertices and sets of points in the Jordan arcs corresponding to edges) are removed from $S$. We denote the set of faces by $F(G)$. We realize that this is *horrible* notation, since the faces depend on the embedding (in

particular, on the surface) and are not determined by $G$ (although
see Theorem 32.9 below). But we will use it anyway.

In general, these faces may or may not be 2-cells. But for an
embedding of a connected graph on the sphere, the regions *will* be
2-cells. This fact is related to the Jordan curve theorem which asserts that if the points of a one-to-one continuous image of the unit
circle in the sphere are removed, the resulting topological space
has exactly two simply connected components that are 2-cells. A
*2-cell embedding* of a graph is one in which all regions are 2-cells.
A surface $S$ together with a 2-cell embedding of a graph $G$ on that
surface is a *map*. The vertices and edges of $G$ that are contained in
the topological closure of a face are *incident* with that face. (More
precisely, we mean that the points and Jordan arcs corresponding to the abstract vertices and edges of $G$ are contained in the
closure of the face. We will occasionally be sloppy and not distinguish between a vertex or edge of $G$ and the point or set of points
representing them in a drawing.)

Figure 32.2

We call attention to the following combinatorial properties of
an embedded graph, without formally discussing the topological
details. See Fig. 32.2. Firstly, a Jordan arc divides a small neighborhood of one of its internal points into two "halves". Thus in
addition to being incident with two vertices (which may coincide),
an edge in a map is incident with two regions (which may coincide). Secondly, there is an induced cyclic ordering of the edges incident with a given vertex (in a small neighborhood of the vertex).
Thirdly, there is an induced cyclic ordering of the edges incident
with any region that is a 2-cell; travelling along the boundary gives
a closed walk in the embedded graph. We need to be a little careful
in considering these cyclic orderings; an edge may be incident twice

with a vertex or a region and so may occur twice in the ordering, and we are not distinguishing a cyclic ordering from its reverse.

The first property above allows us to define the *dual* graph $G^*$ of a graph $G$ with respect to a 2-cell embedding of $G$ on a surface $S$. The vertices $V(G^*)$ are to be the regions with respect to the embedding; the edges $E(G^*)$ are to be the same as the edges of $G$; the ends of an edge $e$ in $G^*$ are to be the regions with which $e$ is incident. Two examples are shown in Fig. 32.3.

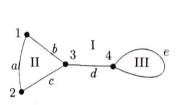

| ends in $G$ | edge | ends in $G^*$ |
|:---:|:---:|:---:|
| 1,2 | $a$ | I,II |
| 1,3 | $b$ | I,II |
| 2,3 | $c$ | I,II |
| 3,4 | $d$ | I,I |
| 4,4 | $e$ | I,III |

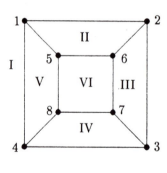

| ends in $G$ | edge | ends in $G^*$ |
|:---:|:---:|:---:|
| 1,2 | $a$ | I,II |
| 2,3 | $b$ | I,III |
| 3,4 | $c$ | I,IV |
| 4,1 | $d$ | I,V |
| 5,6 | $e$ | II,VI |
| 6,7 | $f$ | III,VI |
| 7,8 | $g$ | IV,VI |
| 8,5 | $h$ | V,VI |
| 1,5 | $i$ | II,V |
| 2,6 | $j$ | II,III |
| 3,7 | $k$ | III,IV |
| 4,8 | $l$ | IV,V |

Figure 32.3

We have defined $G^*$ abstractly, but it is important to note that it admits an embedding on the same surface $S$: we can represent a region by a point in its interior; the interior points in two regions sharing an edge $e$ can be joined by a Jordan arc crossing only the

Jordan arc representing $e$ in $G$. The drawing of the dual of the cube in Fig. 32.3 is shown in Fig. 32.4 below.

Examples will convince the reader that $(G^*)^*$ is isomorphic to $G$. Two other facts that we invite the reader to think about but do not attempt to prove formally are: A graph that is not connected cannot have a 2-cell embedding; the dual graph with respect to a 2-cell embedding is always connected.

The planar duals of the series of polygons are the *bond-graphs*, i.e. the graphs with two vertices and any positive number of edges joining them. In particular, the loop-graph and the link-graph are dual.

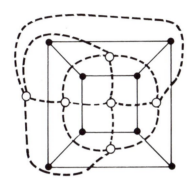

Figure 32.4

In general, a graph may have several planar dual graphs. Fig. 32.5 shows two isomorphic cotrees (graphs with one vertex) and their nonisomorphic planar duals. Fig. 32.6 shows two embeddings of a nonseparable graph $G$ and the dual graphs with respect to each embedding, which are not isomorphic.

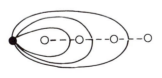 

Figure 32.5

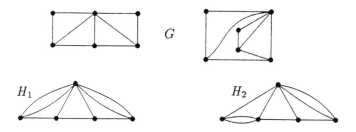

Figure 32.6

If we have a drawing of a graph $G$ on a surface $\mathcal{S}$ and $e \in E(G)$, it is clear that there is a natural drawing of $G'_e$ on $\mathcal{S}$. If $e$ is a nonloop, there is also a natural *induced* drawing of $G''_e$ on $\mathcal{S}$: the ends of the edge can be continuously brought closer on the surface (see Fig. 32.1). (According to our combinatorial definition, contracting a loop is the same as deleting it.) In particular, if $G$ admits an embedding in a surface $\mathcal{S}$, then so does every minor of $G$.

THEOREM 32.3 (EULER'S FORMULA). *Let $C(G)$ denote the set of components of a planar graph $G$. For any embedding of $G$ on the sphere,*

$$(32.2) \qquad 1 - |F(G)| + |E(G)| - |V(G)| + |C(G)| = 0.$$

PROOF: We proceed by induction on the number of edges of $G$. For an edgeless graph, $|C(G)| = |V(G)|$ and $|F(G)| = 1$, so (32.2) holds.

Suppose $G$ has a nonloop $e$. Consider an embedding of $G$ and note that contracting an edge will not affect the number of faces. Apply the induction hypothesis to $G''_e$ to get

$$1 - |F(G''_e)| + |E(G''_e)| - |V(G''_e)| + |C(G''_e)| = 0.$$

We have $|E(G)| = |E(G''_e)| + 1$, $|V(G)| = |V(G''_e)| + 1$, $|C(G)| = |C(G''_e)|$, and also $|F(G)| = |F(G''_e)|$. So (32.2) holds for $G$.

Suppose $G$ has loops; let $e$ be one of these, and apply the induction hypothesis to $G'_e$ to get

$$1 - |F(G'_e)| + |E(G'_e)| - |V(G'_e)| + |C(G'_e)| = 0.$$

Of course, $|E(G)| = |E(G'_e)| + 1$, $|V(G)| = |V(G'_e)|$, and $|C(G)| = |C(G'_e)|$. On the sphere, the Jordan curve theorem implies that the two faces incident with $e$ are distinct, so $|F(G)| = |F(G'_e)| + 1$ and we obtain (32.2) for $G$. □

REMARK. For an embedding of a graph $G$ on any surface,

$$1 - |F(G)| + |E(G)| - |V(G)| + |C(G)| \leq 0$$

will follow by a similar argument. The only difference is that the relation of $|F(G)|$ and $|F(G'_e)|$ depends on whether the two faces incident with $e$ are distinct (in which case $|F(G)| = |F(G'_e)| + 1$) or not (when $|F(G)| = |F(G'_e)|$).

For a drawing of a connected graph $G$ on the sphere or plane, Euler's formula is

$$f - e + v = 2$$

where $f$, $e$, and $v$ are, respectively, the numbers of faces, edges, and vertices.

EXAMPLE 32.2. We show that $K_5$ and $K_{3,3}$ are not planar. We use the fact that the set of edges incident with a face in a planar embedding of a nonseparable graph $G$ (other than the link-graph) is the edge set of a polygon in $G$ (see the Corollary to Theorem 32.8 below). The *degree* of a face is the number of incident edges (i.e. its degree as a vertex of the dual graph). Of course, the sum of the degrees of the faces is twice the number of edges.

Consider a hypothetical embedding of $K_5$ in the plane. By Euler's formula, there are seven faces. But each has degree at least 3, so the sum of the degrees is at least 21, which, of course, is a contradiction to $|E(K_5)| = 10$.

Similarly, there would be five faces in a hypothetical embedding of $K_{3,3}$ in the plane, each of degree at least 4 since $K_{3,3}$ has girth 4. Then the sum of the degrees would be at least 20, contradicting $|E(K_{3,3})| = 9$.

(A similar argument shows that the Petersen graph is not planar. But we already knew this since both $K_5$ and $K_{3,3}$ occur as minors of the Petersen graph and they are not planar.)

PROBLEM 32E. Determine all pairs $(d_1, d_2)$ of integers greater than 1 so that there exists a planar graph (not necessarily simple) that is regular of degree $d_1$ whose dual with respect to some planar drawing is regular of degree $d_2$. (This will include the degrees of the vertices and faces of the five Platonic solids.)

An important theorem that we shall not prove here is the following (see Chartrand and Lesniak (1986), or Tutte (1984), for a proof).

THEOREM 32.4 (KURATOWSKI). *A graph $G$ is planar if and only if it has no subgraphs isomorphic to subdivisions of $K_5$ or $K_{3,3}$.*

PROBLEM 32F. If a trivalent graph $H$ occurs as a minor of a graph $G$, then $G$ contains a subdivision of $H$ (as a subgraph). If $K_5$ is a minor of $G$, then $G$ contains a subdivision of $K_5$ or a subdivision of $K_{3,3}$ (cf. Problem 32A).

Recall that the *incidence matrix* $N = N(G)$ of a graph $G$ is the matrix whose rows are indexed by $V(G)$, whose columns are indexed by $E(G)$, and where

$$N(x, e) = \begin{cases} 1 & \text{if } x \text{ is incident with a nonloop } e, \\ 2 & \text{if } e \text{ is a loop incident with } x \\ 0 & \text{if } x \text{ is not incident with } e. \end{cases}$$

We are going to work over the field $\mathbb{F}_2$ of two elements, so loops essentially correspond to columns of all zeros. But every other column contains two ones and, in particular, the sum modulo 2 of all rows of $N$ is the zero vector.

By the *code $\mathcal{C}(G)$* of a graph $G$, we mean the binary code generated by the incidence matrix $N(G)$. Usually, $\mathcal{C}(G)$ is called the *cutset space* or *coboundary space* of $G$, and the dual code $\mathcal{C}(G)^\perp$ is called the *cycle space* of $G$. See below. (As in Chapter 20, we use the term *support* of a codeword $\mathbf{x}$ to denote the subset of the indexing set for its coordinates that are nonzero in $\mathbf{x}$. For binary codes, we can identify the codewords with their supports.) The binary sum of the rows of the incidence matrix indexed by vertices in $X$ has support equal to the set of edges with one end in $X$ and one end in $Y := V(G) \setminus X$. These supports of words in $\mathcal{C}(G)$ are called *cutsets* of $G$. Supports of words in $\mathcal{C}(G)^\perp$ are sets $S$ of edges so that every vertex has even degree in the spanning subgraph with edge set $S$ and are called *cycles* of $G$.

Note that if $G$ is connected, the cutset of edges with one end in $X$ and one in $Y$, where $(X, Y)$ is a partition of $V(G)$, is nonempty unless $X = \emptyset$ or $X = V(G)$. That is, there is exactly one nontrivial linear relation among the rows of $N(G)$ and so the dimension of $C(G)$ over $\mathbb{F}_2$ is $|V(G)| - 1$. (In general, the dimension is $|V(G)| - |C(G)|$, where $C(G)$ is the set of components of $G$.)

EXAMPLE 32.3. The code of a polygon with $n$ edges consists of all even weight words of length $n$; the code of a bond-graph with $n$ edges consists of the two constant words. The code of a tree consists of all words; the code of a cotree consists of only the zero word.

EXAMPLE 32.4. The cycle spaces of $K_6$ and of the Petersen graph are binary codes of length 15 of dimensions 10 and 6, respectively. The minimum distances are 3 and 5. We remark that the vector of all 1's can be added to a generating set for these two codes and the minimum distances remain 3 and 5 while the dimension goes up by 1 in both cases.

PROPOSITION 32.5. *Let $G$ be a graph.*

*(i) A set $S \subseteq E(G)$ is a minimal nonempty cycle if and only if it is the edge set of a polygon subgraph of $G$.*

*(ii) If $G$ is connected, a set $S \subseteq E(G)$ is a minimal nonempty cutset if and only if it is the cutset of edges with one end in $X$ and one in $Y$ where $(X, Y)$ is a partition of $V(G)$ and the subgraphs of $G$ induced by $X$ and $Y$ are both connected. A subset $S \subseteq E(G)$ is a minimal nonempty cutset of an arbitrary graph $G$ if and only if it is a minimal nonempty cutset of one of the components of $G$.*

PROOF: Clearly the edge set of a polygon is a minimal nonempty cycle. A nonempty cycle is certainly not the edge set of a forest (since every tree has monovalent vertices, say) and so contains the edge set of a polygon; so a minimal nonempty cycle is the edge set of a polygon.

Let $S$ be the cutset consisting of all edges with one end in $X$ and one in $Y$. If the subgraphs induced by $X$ and $Y$ are connected, then $S$ cannot properly contain a nonempty cutset because adding any edge from $S$ to the edges not in $S$ (those with both ends in $X$ or both in $Y$) yields a connected graph. If the subgraph induced on $X$, say, is not connected, there is a partition of $X$ into nonempty subsets $X_1$ and $X_2$ with no edges between them. Then the cutset consisting of all edges with one end in $X_1 \cup Y$ and the other in $X_2$ is contained in $S$, and the connectivity of $G$ implies that it is both nonempty and a proper subset of $S$.

The last statement is left for the reader to check. $\qquad\square$

The terms *circuits* and *bonds* are used, respectively, for minimal nonempty cycles and cutsets. In other words, given a set $S$ of edges of a connected graph $G$, $S$ is a circuit if and only if deleting the edges not in $S$ (and any isolated vertices that result) produces a polygon; $S$ is a bond if and only if contracting the edges not in $S$ produces a bond-graph. The word "circuit" is overused, but we will be careful to use it only in this sense in the remainder of the chapter. We could have defined a bond as a set of edges whose removal increases the number of components, but which is minimal with respect to that property.

PROBLEM 32G. In a $k$-connected graph with at least $2k - 2$ edges, every circuit and every bond contains at least $k$ edges.

PROBLEM 32H. Show that the circuits of a graph span the cycle space, and the bonds span the cutset space. To do this, just verify that the nonzero codewords in a binary linear code $\mathcal{C}$, whose supports are minimal among the supports of all nonzero codewords, generate the code. For extra credit, establish this last assertion for linear codes over any field.

A graph $H$ is a *Whitney dual* of a graph $G$ when the code $\mathcal{C}(G)$ of $G$ is equivalent to the dual code $\mathcal{C}(H)^\perp$ of $H$. Clearly, $H$ is a Whitney dual of $G$ if and only if $G$ is a Whitney dual of $H$. Often it is notationally or conceptually convenient to assume the graphs have the *same* abstract edge set $E$ and then they are Whitney duals when the cycle space of $G$ *is* exactly the cutset space of $H$ and/or the cutset space of $G$ *is* exactly the cycle space of $H$. By Problem 32H, we can say $G$ and $H$ are Whitney duals when the circuits of $G$ are exactly the bonds of $H$ and/or when the bonds of $G$ are exactly the circuits of $H$.

EXAMPLE 32.5. The two graphs described by each table in Fig. 32.3 are Whitney duals. This is because each pair consists of planar duals (see Theorem 32.6 below). For example, $\{a, e, g, c\}$ and $\{a, j, b\}$ are bonds in the cube $G$ and circuits in $G^*$; the set $\{a, b, c, d\}$ is a circuit in $G$ and a bond in $G^*$.

PROBLEM 32I. Let $G$ and $H$ be Whitney duals with $E(G) = E(H) = E$. Explain why $G'_e$ and $H''_e$ are Whitney duals for any $e \in E$.

PROBLEM 32J. Let $G$ and $H$ be connected graphs with $E(G) = E(H) = E$ that are Whitney duals. Show that $S$ is the edge set of a spanning tree in $G$ if and only if $E \setminus S$ is the edge set of a spanning tree in $H$.

THEOREM 32.6. *A graph $G$ is planar if and only if it has a Whitney dual. In particular, a dual $G^*$ of $G$ with respect to any planar drawing of $G$ is a Whitney dual of $G$.*

We need a lemma for the "if" part of Theorem 32.6.

LEMMA 32.7. *Let $G$ and $H$ be graphs with the same edge set $E$. Suppose $G$ is the union of edge-disjoint subgraphs $G_1$ and $G_2$ so that $|V(G_1) \cap V(G_2)| \leq 1$. Let $H_1 = H'_{E \setminus E(G_2)}$ and $H_2 = H'_{E \setminus E(G_1)}$ be the subgraphs of $H$ that have edge sets $E(G_1)$ and $E(G_2)$, respectively.*

*(i) If $G$ and $H$ are Whitney duals, then $G_i$ and $H_i$ are Whitney duals for both $i = 1$ and $i = 2$.*

*(ii) If $G_i$ and $H_i$ are Whitney duals for both $i = 1$ and $i = 2$ and $|V(H_1) \cap V(H_2)| \leq 1$, then $G$ and $H$ are Whitney duals.*

PROOF: From Proposition 32.5, it is easy to see that every bond of $G$ is contained fully in either $E(G_1)$ or $E(G_2)$. Moreover, a subset $S \subseteq E(G_i)$ is a bond of $G_i$ if and only if it is a bond of $G$.

Suppose $G$ and $H$ are Whitney duals. If $S$ is a bond of $G_i$, then $S$ is a circuit of $H$ contained in $E(H_i)$ and so is a circuit of $H_i$. And if $S$ is a circuit of $H_i$, then it is a circuit of $H$, hence a bond of $G$ contained in $E(G_i)$ and so is a bond of $G_i$.

Suppose that $G_i$ and $H_i$ are Whitney duals for $i = 1, 2$ and that $|V(H_1) \cap V(H_2)| \leq 1$. Let $S$ be a bond of $G$. Then $S$ is a bond of $G_1$ or $G_2$ and hence a circuit in either $H_1$ or $H_2$. But then certainly

$S$ is a circuit in $H$. If $S$ is a circuit of $H$, then it is a circuit in either $H_1$ or $H_2$. Hence it is a bond of $G_1$ or $G_2$ and so is a bond of $G$.

PROOF OF THEOREM 32.6: Let $G^*$ be a planar dual of $G$. We first show that $G^*$ is a Whitney dual of $G$.

Let $S$ be a circuit of $G$. So $S$ is the edge set of a polygon subgraph $P$ of $G$ that divides the plane into two regions; let $\mathcal{F}_1$ be the set of regions inside, and $\mathcal{F}_2$ the set of regions outside, $P$. This is a partition of the vertices of $G^*$. Clearly, each edge of $P$ is incident with one region of $\mathcal{F}_1$ and one of $\mathcal{F}_2$. Any other edge of $G$ lies (except possibly for its ends) in one of the regions and so is incident only with members of $\mathcal{F}_1$ or $\mathcal{F}_2$. In summary, $S$ is a cutset in $G^*$. The circuits generate the cycle space of $G$, so we have shown that the cycle space of $G$ is contained in the cutset space of $G^*$.

We could now complete the proof by an appeal to Euler's formula: the dimension of the cycle space of $G$ is $|E(G)| - (|V(G)| - |C(G)|)$ while the dimension of the cutset space of $G^*$ is $|F(G)| - 1$. By Euler's formula, these two spaces have the same dimension, so they coincide and thus $G^*$ is a Whitney dual of $G$.

But it is in fact easy to see that the cutset space of $G^*$ is contained in the cycle space of $G$ for an embedding on *any* surface: if the faces of $G$ are colored red and blue, then the number of edges incident with any vertex $x$ of $G$ that have one red side and one blue side is certainly even; see Fig. 32.7. This is the same as saying that if the vertices of a polygon are colored red and blue, then the number of edges with ends of different colors is even.

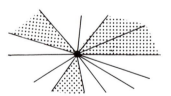

Figure 32.7

For the proof of the converse, we first establish the following assertion. Given a pair $G$, $H$ of Whitney duals with the same edge set $E$ and no isolated vertices, we claim that there exists a possibly different Whitney dual $H^0$ which admits what we will call a *system of walks* for $G$. This is a family $\{w_x : x \in V(G)\}$ of walks $w_x$ in $H^0$, one for each vertex $x$ of $G$, so that $w_x$ traverses in $H^0$ exactly the edges incident with $x$ in $G$, loops being traversed twice and nonloops once. We will use induction on the number of edges. We may assume $G$ has no isolated vertices.

First note that the assertion is easy if $G$ is nonseparable. Of course, it is true for a loop-graph. Otherwise there are no loops and for every vertex $x$, the set $S(x)$ of edges incident with $x$ in $G$ is a bond in $G$ and hence a circuit in $H$. We need only let $w_x$ be a simple closed path in $H$ traversing the edges of the circuit $S(x)$ in $H$.

If $G$ is not nonseparable, there exist edge-disjoint subgraphs $G_1$ and $G_2$, each with at least one edge, and that have no vertices or one vertex in common, such that their union is $G$. Let $H_1 := H'_{E \backslash E(G_2)}$ and $H_2 := H'_{E \backslash E(G_1)}$. By Lemma 32.7, $H_i$ is a Whitney dual of $G_i$, $i = 1, 2$. By the induction hypothesis, there exist graphs $H_i^0$ that are Whitney duals of $G_i$ and that admit systems of walks $\{w_x^{(i)} : x \in V(G_i)\}$ for $G_i$, $i = 1, 2$. We may take $H_1^0$ and $H_2^0$ to be vertex disjoint. If $G_1$ and $G_2$ are disjoint, let $H^0$ be the disjoint union of $H_1^0$ and $H_2^0$; then the union of the two systems of walks provides a system of walks for $G$. If $G_1$ and $G_2$ have one vertex $z$ in common, there are walks $w_z^{(1)}$ and $w_z^{(2)}$ in $H_1^0$ and $H_2^0$ that traverse the edges incident with $z$ in $G_1$ and $G_2$, respectively. Then let $H^0$ be obtained as the union of $H_1^0$ and $H_2^0$ but where we identify some vertex in the walk $w_z^{(1)}$ in $H_1^0$ with some vertex in the walk $w_z^{(2)}$ in $H_2^0$. We combine these two walks into one which traverses the edges incident with $z$ in $G$; this together with the remaining walks in the systems in $H_1$ and $H_2$ provides a system of walks for $G$. In either case, $H^0$ is a Whitney dual of $G$ by Lemma 32.7.

Finally, we prove the following. Let $G$ be a graph with a Whitney dual $H$ which admits a system of walks $\{w_x : x \in V(G)\}$ for $G$. We assert that $G$ has an embedding in the sphere so that the edges incident with $x$ in $G$ occur in same cyclic order as they occur in the walk $w_x$. We proceed by induction on the number of edges.

If $G$ has only loops, then $H$ is a forest. Let $e$ be a loop in $G$

such that $e$ is incident with a monovalent vertex in $H$. Let $z$ be the vertex incident with $e$ in $G$. The walk $w_z$ in $H$ evidently traverses $e$ twice in succession, say

$$w_z = (s_0, e, s_1, e, s_2 = s_0, a_3, s_3, a_4, s_4, \ldots, a_k, s_0)$$

where the $s_i$'s are vertices of $H$. The graphs $G'_e$ and $H''_e$ are Whitney duals and suppressing $e$ in $w_z$ will provide a system of walks in $H''_e$ for $G'_e$, so by the induction hypothesis, $G'_e$ admits a planar drawing such that the edges incident with $z$ in $G'_e$ occur in same cyclic order as they occur in the walk $w_z$ with $e$ suppressed. In a small neighborhood of the point representing $z$ in the embedding, we can insert the loop $e$ to obtain the required embedding of $G$ (see Fig. 32.8).

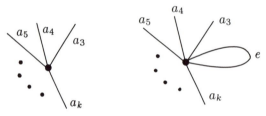

Figure 32.8

Let $e$ be a nonloop in $G$ with ends $y$ and $z$. Then $G''_e$ and $H'_e$ are Whitney duals and we claim that there is a natural system of walks $\{u_x : x \in V(G''_e)\}$ in $H'_e$ for $G''_e$. For $x \neq y, z$, we take $u_x := w_x$. For the new vertex $x_0$ obtained by identifying $y$ and $z$, we combine $w_y$ and $w_z$ omitting $e$ as follows. Say

$$w_y = (s_0, e, s_1, a_2, s_2, \ldots, s_{k-1}, a_k, s_0)$$

and

$$w_z = (t_0, e, t_1, b_2, t_2, \ldots, t_{m-1}, b_m, t_0)$$

where $e, a_2, a_3, \ldots, a_k$ and $e, b_2, b_3, \ldots, b_m$ are the edges incident with $y$ and $z$, respectively, in $G$ (loops appearing twice) and the $s_i$'s and $t_j$'s are vertices of $H$. By reversing one of these walks if necessary, we can assume $s_0 = t_0$ and $s_1 = t_1$. Then take

$$u_{x_0} := (s_1, a_2, s_2, \ldots, s_{k-1}, a_k, s_0, b_m, t_{m-1}, b_{m-1}, \ldots, b_3, t_2, b_2, s_1).$$

By the induction hypothesis, $G_e''$ admits an embedding in the plane
so that the edges incident with $x$ in $G_e''$ occur in same cyclic order
as they occur in the walk $u_x$. In a small neighborhood of the point
representing $x_0$ in the embedding, we can "uncontract" $e$ to obtain
the required embedding of $G$ (see Fig. 32.9).                    □

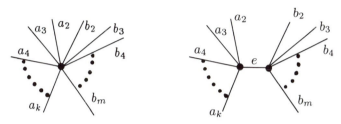

Figure 32.9

We will now prove some results about planar graphs and their
duals using Whitney duality (and not referring to the plane or
sphere at all). To warm up, note that a planar graph $G$ is bipartite
if and only if its dual $G^*$ is Eulerian—since the first property is
equivalent to the entire edge set being a cutset, and the second
property is equivalent to the entire edge set being a cycle.

THEOREM 32.8. *If $G$ and $H$ are Whitney duals (without isolated
vertices) and $G$ is nonseparable, then $H$ is nonseparable.*

PROOF: Theorem 32.2 shows that any two edges of $G$ are contained
in a circuit. Hence any two edges of $H$ are contained in a bond of
$H$. But if two edges $a$ and $b$ are contained in a bond, say in the
set of edges with one end in $X$ and one in $Y$, we can find a simple
path in the subgraph induced by $X$ joining one end of $a$ and one
of $b$, and a simple path in the subgraph induced by $Y$ joining the
other ends, and we find a polygon containing $a$ and $b$. So any two
edges of $H$ are contained in a circuit of $H$. This implies that $H$ is
nonseparable.                    □

COROLLARY. *The set of edges incident with a face in a planar
embedding of a nonseparable graph $G$, not a link-graph, is a circuit
in $G$.*

PROOF: The set $S$ of edges incident with a face is the set of edges incident with a vertex in $G^*$. But $G^*$ is nonseparable by Theorem 32.8; deleting a vertex does not disconnect $G^*$, so $S$ is a bond in $G^*$. Then $S$ is a circuit in $G$. □

PROBLEM 32K. Two edges $a$ and $b$ of a connected graph $G$ that are contained together in a circuit of $G$ are also contained in some bond of $G$.

THEOREM 32.9. *If $G$ and $H$ are Whitney duals (without isolated vertices) and $G$ is 3-connected, then $H$ is 3-connected.*

PROOF: Assume $G$ is 3-connected. For $x \in V(G)$, let $S(x)$ denote the set of edges incident with $x$. Each "star" $S(x)$ is a bond with the property that $G'_{S(x)}$ is nonseparable, and these bonds span the code (cutset space) of $G$ by definition.

Thus in $H$, there exists a family $\mathcal{F}$ of circuits that spans the cycle space of $H$ and where each member $S$ of $\mathcal{F}$ is a circuit with the property that $H''_S$ is nonseparable.

We know $H$ is nonseparable; suppose for contradiction that it is not 3-connected. Then $H$ has a 2-separation: $H$ is the edge-disjoint union of subgraphs $A$ and $B$ of $H$, each with at least two edges, and where $A \cap B$ is a graph consisting of two isolated vertices, say $x$ and $y$. If a polygon contains edges of $H$ contains edges of both $A$ and $B$, it must have $x$ and $y$ among its vertices. If the edges of such a polygon are contracted, we would obtain a separable graph since it would be the union of the graphs obtained from $A$ and $B$ by contracting the edges of the polygon in each (which would identify $x$ and $y$); these contracted subgraphs will have only this vertex resulting from the identification of the vertices of the polygon in common (and the reader can check that there still remains at least one edge in each contracted subgraph).

Thus the circuits in $\mathcal{F}$ are contained completely in either $E(A)$ or $E(B)$. Let $C$ be any circuit of $H$. Since $\mathcal{F}$ spans the cycle space of $H$, $C \cap E(A)$ and $C \cap E(B)$ are, respectively, the modulo 2 sum of members of $\mathcal{F}$ that are contained in $E(A)$ and $E(B)$. Hence each of $C \cap E(A)$ and $C \cap E(B)$ would be cycles, and then the minimality of $C$ would imply that $C \cap E(A)$ or $C \cap E(B)$ is empty. It follows that *every* circuit of $H$ is contained in either $E(A)$ or $E(B)$, which contradicts the nonseparability of $H$. □

PROBLEM 32L.

(i) Let $G$ be a 3-connected graph and $S$ a bond in $G$. Show that $G'_S$ is nonseparable if and only if $S$ is equal to the star $S(x)$ for some vertex $x$ of $G$.

(ii) If $G$ and $H$ are 3-connected graphs with equivalent codes, then $G$ is isomorphic to $H$.

THEOREM 32.10. *Let $G$ be a 3-connected graph embedded in the sphere. A circuit $C$ of $G$ is the set of edges incident with some face in this embedding if and only if $G''_C$ is nonseparable.*

PROOF: The circuit $C$ is the set of edges incident with a face if and only if it is the set of edges incident with a vertex in the dual $H := G^*$. By Problem 32L(i), this is the case if and only if $H'_C$ is nonseparable. This latter graph is a Whitney dual of $G''_C$. □

Thus the faces of a 3-connected planar graph are uniquely determined by the graph. This is not true for nonseparable graphs in general—see Fig. 32.6. If we know the faces, i.e. the circuits that bound faces, the embedding is determined (see the next chapter).

COROLLARY. *A 3-connected planar graph has a unique embedding on the sphere.*

EXAMPLE 32.6. Here is another proof that the 3-connected graphs $K_5$, $K_{3,3}$, and the Petersen graph are not planar. Note that contracting the edges of any 3-gon in $K_5$, any 4-gon in $K_{3,3}$, and any 5-gon in the Petersen graph, always yields a nonseparable graph. So if these graphs were planar, every 3-gon, 4-gon, and 5-gon, respectively, would bound a face in the embedding in view of Theorem 32.10. But every edge belongs to three 3-gons, to four 4-gons, and to four 5-gons in the respective graphs.

E. Steinitz (1922) characterized graphs that arise from a convex polytope in Euclidean 3-space. A *convex polytope* $P$ in $\mathbb{R}^n$ is the convex hull of finitely many points, i.e.

$$P = \{\lambda_1 \mathbf{x}_1 + \cdots + \lambda_k \mathbf{x}_k : \lambda_i \geq 0, \; \lambda_1 + \cdots + \lambda_k = 1\}.$$

The dimension of a polytope is the dimension of its affine span, as usual.

A *supporting hyperplane* of $P$ is a hyperplane

$$H = \{\mathbf{x} : \ell(\mathbf{x}) = c\}$$

(where $\ell$ is a nonzero linear functional $\mathbb{R}^n \to \mathbb{R}$ and $c$ a scalar) with $H \cap P \neq \emptyset$ but such that $P$ lies entirely in one of the closed halfspaces of $H$, say

$$P \subseteq \{\mathbf{x} : \ell(\mathbf{x}) \geq c\}.$$

A *face* of $P$ is the intersection of a supporting hyperplane of $P$ with $P$. A *vertex* of $P$ is a 0-dimensional face; an *edge* of $P$ is a 1-dimensional face. The *graph* $G$ of a polytope $P$ (also called the 1-*skeleton* of $P$) has $V(G)$ and $E(G)$ equal to the vertices and edges, respectively, of $P$ with incidence determined by inclusion.

A face $F$ is a convex polytope in its own right; indeed, it is the convex hull of the vertices of $P$ that it contains (see Grünbaum, 1967). The graph of $F$ is a subgraph of the graph of $P$.

THEOREM 32.11. *A graph $G$ is the graph of a 3-dimensional convex polytope if and only if $G$ is planar, has at least four vertices, and is 3-connected.*

PARTIAL PROOF: We shall prove here only the easier part of this theorem, namely that the graph $G$ of a 3-dimensional convex polytope $P$ is planar and 3-connected. That $G$ is planar is immediate since the surface of $P$ is homeomorphic to the sphere. Since $G$ is clearly a simple graph and has at least four vertices, it remains only to show $G$ is 3-vertex connected.

We require the following fact from the theory of linear programming (see Grünbaum, 1967): If $s$ is a vertex of a polytope $P$ in $\mathbb{R}^n$ and $\ell$ a linear functional, either $\ell(s) \geq \ell(x)$ for all $x \in P$ or there exists a vertex $t$ of $P$ adjacent to $s$ such that $\ell(t) > \ell(s)$.

Let $x, y, a, b$ be four distinct vertices in the graph $G$ of a convex polytope $P$ in $\mathbb{R}^3$. We want to find a path in $G$ from $x$ to $y$ that avoids $a$ and $b$. Consider the affine line $L$ in $\mathbb{R}^3$ spanned by $a$ and $b$. Some plane $H$ on $L$ is such that both $x$ and $y$ are on the same side of $H$, i.e. $H = \{z : \ell(z) = c\}$ and $\ell(x) \geq c$, $\ell(y) \geq c$. Since $P \nsubseteq H$, we can assume $P$ contains points $z$ with $\ell(z) > c$ (otherwise

$x$ and $y$ are in $H$ and we can replace $\ell$ and $c$ by their negatives— i.e. consider the other side of $H$). By the remark of the previous paragraph, there exist paths $p$ and $q$ in $G$ from $x$ and $y$ to vertices $x'$ and $y'$, respectively, at which $\ell$ attains its maximum value $c'$, and such that no vertices of $p$ and $q$ other than their initial vertices lie in $H$. The intersection $P \cap H'$, where $H' := \{z : \ell(z) = c'\}$, is a face of $P$ of dimension 0, 1, or 2 (a vertex, an edge, or a convex polygon) and in particular has a connected graph, so there is a path $r$ in $G$ from $x'$ to $y'$ with vertices in $H'$. We concatenate $p$, $r$, and the reverse of $q$ to find the required path from $x$ to $y$. $\qquad\square$

## Notes.

Hassler Whitney (1907–1989), a pioneer in topology who was also intensely concerned with mathematical education for the last twenty years of his life, was a professor at Harvard and later at the Institute for Advanced Study in Princeton. He was awarded the Wolf Prize in 1982. Whitney had a degree in music from Yale, played the violin, viola and piano, and was concertmaster of the Princeton Community Orchestra.

Ernst Steinitz (1871–1928) was professor of mathematics at Breslau (now Wroclaw) and Kiel. He made important contributions to field theory (with Hilbert).

The Polish mathematician Kazimierz (or Casimir) Kuratowski (1896–1880) taught at Warsaw University (for nearly 40 years) and Lwow Polytech before retiring in 1966. His celebrated theorem characterizing planar graphs was proved in 1930.

The proof of part of Theorem 32.11 generalizes to show that the 1-skeleton of a $d$-dimensional polytope is $d$-vertex connected, a result due to M. Balinski (1961).

We have not explicity mentioned matroids in this chapter, although they are implicit in several arguments. Binary codes are more-or-less equivalent to binary matroids. Matroids were introduced by Hassler Whitney in 1935 in order to abstract certain properties of linear independence, of circuits and bonds in graphs, and of duality. Their theory is "cryptomorphic" to that of geometric lattices. See Welsh (1976) and Crapo and Rota (1971).

**References.**

M. Balinski (1961), On the graph structure of convex polyhedra in *n*-space, *Pacific J. Math.* **11**, 431–434.

G. Chartrand and L. Lesniak (1986), *Graphs and Digraphs*, 2nd edn., Wadsworth.

H. Crapo and G.-C. Rota (1971), *On the Foundations of Combinatorial Theory: Combinatorial Geometries*, M. I. T. Press.

L. R. Ford, Jr. and D. R. Fulkerson (1962), *Flows in Networks*, Princeton University Press.

B. Grünbaum (1967), *Convex Polytopes*, J. Wiley (Interscience).

R. P. Stanley (1973), *Acyclic orientations of graphs*, Discrete Math. **5**, 171–178.

E. Steinitz (1922), Polyeder und Raumeinteilungen, *Enzykl. Math. Wiss.* **3**, 1–139.

W. T. Tutte (1984), *Graph Theory*, Encylodpedia of Math. and its Appl. **21**, Addison-Wesley.

D. J. A. Welsh (1976), *Matroid Theory*, Academic Press.

H. Whitney (1932), Nonseparable and planar graphs, *Trans. Amer. Math. Soc.* **34**, 339–362.

H. Whitney (1935), On the abstract properties of linear dependence, *Amer. J. Math.* **57**, 509–533.

# 33
## Graphs: colorings and embeddings

We begin by coloring the vertices of planar graphs and then we consider other surfaces.

Historically, it was the coloring of the regions of a spherical map that was first considered. Of course, this is equivalent to coloring the vertices of the dual map. The four color problem, "Can the countries of a map on the sphere be colored with four or fewer colors so that adjacent countries are colored differently?", occurs in a letter dated October 23, 1852, from Augustus de Morgan to Sir William Rowan Hamilton. It is likely that the originator of the question was the brother, Francis Guthrie, of a student, Frederick Guthrie, of de Morgan. Four colors are clearly necessary since $K_4$ is planar, but it is possible to find many other planar graphs $G$ with chromatic number $\chi(G) = 4$. This problem has motivated a great deal of graph theory. More on the history of the four color problem may be found in Biggs, Lloyd, and Wilson (1976).

In 1890, P. J. Heawood proved the Five Color Theorem, Theorem 33.2 below: A loopless planar graph $G$ has chromatic number $\chi(G) \leq 5$. Heawood used ideas of A. B. Kempe (in particular, the recoloring idea reviewed below). In 1976, K. Appel and W. Haken finally settled the four color problem with an announcement of a proof of the Four Color Theorem: $\chi(G) \leq 4$. Surprisingly, perhaps, their proof did not use the vast quantity of work and theories developed in this century but went back to the ideas of Kempe and Heawood. Another surprise, perhaps, is that their proof required over 1000 hours of computer time. It was not clear when they began that the computer would ever finish. As an oversimplification, there were many cases to consider but it was not certain that their number would be finite; the computer was programmed to generate cases itself and could conceivably have gone on forever breaking cases into subcases. See Appel, Haken, and Koch (1977), and Appel and Haken (1977). We have no space to discuss this work in any detail, but will prove the Five Color Theorem below.

PROPOSITION 33.1. *Every non-null simple planar graph $G$ has a vertex of degree at most 5.*

PROOF: This follows quickly from Euler's formula. It suffices to prove that a component has such a vertex, so assume $G$ is connected. Suppose a drawing of $G$ has $f$ regions, $e$ edges, and $v$ vertices. If no vertex of degree less than six exists, then the sum of the $v$ degrees of the vertices, which is $2e$, is at least $6v$; that is,

$$v \leq \frac{e}{3}.$$

Since $G$ is simple (and we may assume it is not a link-graph with two vertices joined by a single edge), each region is incident with at least three edges, so the sum of the $f$ degrees of vertices of $G^*$, which is $2e$, is at least $3v$; that is

$$f \leq \frac{2e}{3}.$$

But then we have the contradiction

$$2 = f - e + v \leq \frac{e}{3} - e + \frac{2e}{3} = 0.$$

$\square$

Of course, coloring a loopless graph is equivalent to coloring a simple graph (just suppress the multiple edges). An immediate consequence of the above result is the Six Color Theorem: A loopless planar graph $G$ has $\chi(G) \leq 6$. When a vertex $x$ of degree at most five is deleted from a simple planar graph and the result colored with six colors inductively, there is always a color available for $x$.

We recall the idea of recoloring introduced in Chapter 3. Suppose $G$ is properly colored and let $\alpha$ and $\beta$ be two of the colors. The colors $\alpha$ and $\beta$ can be switched on any *component* of the subgraph of $G$ induced by the vertices of colors $\alpha$ and $\beta$ to obtain another proper coloring. So given two vertices $x$ and $y$ of colors $\alpha$ and $\beta$, respectively, either there is a path (of odd length) with vertex terms $a_0 = x, a_1, \ldots, a_n = y$ that alternate colors $\alpha$ and $\beta$ (a so-called *Kempe chain*), or there is a proper coloring in which both $x$ and $y$ have color $\alpha$, say, while vertices of colors other than $\alpha$ and $\beta$ retain the same color.

THEOREM 33.2. *If $G$ is a loopless planar graph, then $\chi(G) \leq 5$.*

PROOF: We use induction on the number of vertices. The assertion is surely true for graphs with at most five vertices. Consider a planar drawing of a loopless graph $G$ and assume the theorem holds for all graphs with one vertex less. We can assume $G$ is simple.

By Proposition 33.1, there exists a vertex $x \in V(G)$ of degree $\leq 5$. Let $G_x$ denote the graph obtained from $G$ by deleting the vertex $x$ and the incident edges. We have a natural planar drawing of $G_x$ obtained from the drawing of $G$. By the induction hypothesis, $G_x$ has a proper coloring with five colors. There is a color available for $x$, and we are done, unless $x$ has degree exactly five and the vertices $y_1, y_2, \ldots, y_5$ (which we write in the cyclic order given by the drawing) adjacent to $x$ in $G$ have received five distinct colors in the coloring of $G_x$. So we assume that $y_i$ has received color $i$, $i = 1, 2, 3, 4, 5$. See Fig. 33.1.

Consider colors 1 and 3. Either there exists a proper coloring of $G_x$ in which $y_1$ and $y_3$ both have color 1 (while $y_2$, $y_4$, and $y_5$ retain their original colors) or there exists a simple path with vertex terms

(33.1) $$y_1, a_1, b_1, a_2, b_2, \ldots, a_s, b_s, y_3,$$

say, in $G_x$ with vertices alternately colored 1 and 3. In the first case, $x$ may be assigned color 3 and we are done. Similarly, either there exists a proper coloring of $G_x$ in which $y_2$ and $y_4$ both have color 2 (while $y_1$, $y_3$, and $y_5$ retain their original colors) or there exists a simple path with vertex terms

(33.2) $$y_2, c_1, d_1, c_2, d_2, \ldots, c_t, d_t, y_4,$$

say, in $G_x$ with vertices alternately colored 2 and 4. In the first case, $x$ may be assigned color 4 and we are done.

To complete the proof, we note that not both paths above can exist. If the path in (33.1) exists, then

$$(33.3) \qquad x, y_1, a_1, b_1, a_2, b_2, \ldots, a_s, b_s, y_3, x$$

is the vertex sequence of a simple closed path, which divides the plane into two regions. The vertex $y_2$ is in one region while $y_4$ is in the other. Any path from $y_2$ to $y_4$ in $G_x$ must necessarily have a vertex in common with the path in (33.2) and in particular must have a vertex of color 1 or 3; thus a path as in (33.3) cannot exist. $\square$

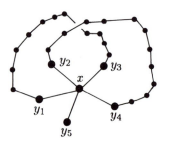

Figure 33.1

We remark that a graph $G$ can be properly colored with four colors if and only if $E(G)$ is the union of two cutsets: If we have two partitions $(X_1, X_2)$ and $(Y_1, Y_2)$ of $V(G)$ so that every edge either has one end in $X_1$ and one in $X_2$, or one end in $Y_1$ and one in $Y_2$, then $X_1 \cap Y_1, X_1 \cap Y_2, X_2 \cap Y_1, X_2 \cap Y_2$ are the color classes of a proper 4-coloring. The converse is also immediate.

A *Tait coloring* (after the English mathematician P. G. Tait) of a trivalent graph is a coloring of the edges with three colors, say $\alpha$, $\beta$, and $\gamma$, so that each color appears at every vertex. Fig. 33.2 shows such a coloring. The existence of a Tait coloring of a trivalent graph is equivalent to partitioning the edges into three perfect matchings (cf. Chapter 5) and so a bipartite trivalent graph has a Tait coloring by Problem 5A.

An *isthmus* of a graph is an edge whose deletion increases the number of components (i.e. disconnects the graph if it was originally connected). This is the dual concept of a loop. An isthmus is an edge $e$ so that the singleton set $\{e\}$ is a cutset; a loop is an edge $e$ so that the singleton set $\{e\}$ is a cycle. (Some authors use *acyclic edge* for an isthmus since it is an edge contained in no circuits. The term *bridge* has also been used.)

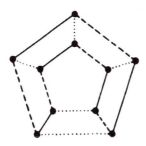

Figure 33.2

PROBLEM 33A. Show that a trivalent graph with an isthmus does not have a Tait coloring.

The Petersen graph is an example of an isthmus-free trivalent graph which has no Tait coloring.

The Four Color Theorem is equivalent to the assertion that every planar isthmus-free trivalent graph has a Tait coloring. To see this, we first claim that to prove the Four Color Theorem it is sufficient to show that the regions of a *trivalent* (isthmus-free) planar graph $G$ can be properly 4-colored. (An isthmus in $G$ would be a loop in the dual which could then not be properly colored.) To this end, given a planar graph $G$, replace each vertex $x$ of degree $\neq 3$ by the configuration indicated for the case of a vertex of degree 5 in Fig. 33.3 below.

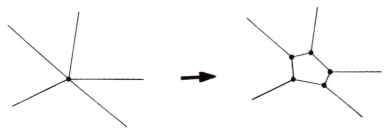

Figure 33.3

This creates more regions (more vertices in the dual $G^*$) but preserves adjacency of the original regions, so if the regions of this larger but trivalent graph can be 4-colored, then we obtain a coloring for the original graph. The equivalence with the Four Color Theorem is completed with the following theorem.

THEOREM 33.3. *The regions of a trivalent planar graph $G$ can be properly 4-colored if and only if the graph admits a Tait coloring.*

PROOF: The regions are the vertices of the dual graph $G^*$. The dual $G^*$ can be properly 4-colored if and only if $E(G)$ can be expressed as the union of subsets $A, B \subseteq E(G)$ each of which is a cutset in $G^*$, or equivalently, each of which is a cycle in $G$.

We claim that $G$ admits a Tait coloring if and only if there exist two cycles $S_1$ and $S_2$ in $G$ so that $E(G) = S_1 \cup S_2$. Given the coloring, take $S_1$ to be the union of the edges of colors $\alpha$ and $\gamma$, and $S_2$ to be the union of the edges of colors $\beta$ and $\gamma$. These are both cycles since every vertex is incident with two edges of $S_1$ and two edges of $S_2$. Conversely, if $E(G)$ is the union of two cycles $S_1$ and $S_2$, color an edge $e$ with color

$$\begin{cases} \alpha & \text{if } e \in S_1, e \notin S_2, \\ \beta & \text{if } e \notin S_1, e \in S_1, \\ \gamma & \text{if } e \in S_1 \cap S_1. \end{cases}$$

Every vertex $x$ has even degree (between 0 and 3) in the spanning subgraphs with edge sets $S_1$ and $S_2$; but $E(G) = S_1 \cup S_2$ implies that both degrees are 2 and that there is one edge of each color incident with $x$.                                                    □

The theorem above can be proved without explicit mention of Whitney duality as follows, but the proof is not really different. Regard $\{0, \alpha, \beta, \gamma\}$ as the elements of the Klein 4-group, so that each of $\alpha, \beta, \gamma$ has order 2 and the sum of any two of them is the third. Given a 4-coloring of the regions, color an edge with the sum of the colors on the two incident regions to get a Tait coloring. Given a Tait coloring, color one region 0 and proceed by assigning color $i + j$ to a region whenever it is joined by an edge of color $i$ to a region already colored $j$; this must be shown to yield a well defined coloring, however.

Tait conjectured in 1880 that every 3-connected planar trivalent graph has a Hamiltonian circuit. This would imply that the graph admits a Tait coloring. W. T. Tutte (1954) found a counterexample but did prove that every 4-connected planar trivalent graph is Hamiltonian.

We now consider surfaces other than the sphere. The classification of surfaces (see Fréchet and Fan, 1967) shows that there are two infinite families of surfaces, up to homeomorphism. We use $\mathcal{T}_g$, $g \geq 0$, to denote the orientable surface of *genus g* (the *g-torus*). This can be realized as a sphere with $g$ "handles" or $g$ "holes"; $\mathcal{T}_0$ is the sphere. We use $\mathcal{N}_n$, $n \geq 1$, to denote the nonorientable surface which can be constructed by inserting $n$ "cross-caps" on the sphere (these cannot be embedded in $\mathbb{R}^3$). We state the extension of Euler's formula. A proof may be found in Fréchet and Fan (1967), or in Chartrand and Lesniak (1986) for the orientable case.

THEOREM 33.4. *For a 2-cell embedding of a graph G on* $\mathcal{T}_g$,

$$|F(G)| - |E(G)| + |V(G)| = 2 - 2g.$$

*For a 2-cell embedding of a connected graph G on* $\mathcal{N}_n$,

$$|F(G)| - |E(G)| + |V(G)| = 2 - n.$$

*For any embedding, not necessarily a 2-cell embedding, the inequalities which result by replacing "=" by "≥" in the equations above will hold.*

EXAMPLE 33.1. Fig. 33.4 shows an embedding of the Petersen graph $P$ on the real projective plane $\mathcal{N}_1$. In this diagram, diametrically opposite points on the boundary are to be identified. There are six faces, each a pentagon. More precisely, the dual graph with respect to this embedding is the complete graph $K_6$.

(This establishes a curious one-to-one correspondence between the 15 edges of $P$ and the 15 edges of $K_6$. We have previously seen that there is a one-to-one correspondence between the 10 vertices of $P$ and the 10 edges of $K_5$.)

(The real projective plane can be thought of as the surface of the sphere where diametrically opposite points have been identified. The Petersen graph can be thought of as the graph of a regular dodecahedron where diametrically opposite points have been identified.)

Figure 33.4

EXAMPLE 33.2. Fig. 33.5 shows an embedding of the complete graph $K_7$ on the torus $\mathcal{T}_1$. We use the usual representation of the torus; here the horizontal and vertical boundary lines are to be identified. There are 14 triangular faces. The dual graph with respect to this embedding has 14 vertices of degree 3 and seven hexagonal faces, any two of which are adjacent. This is the *Heawood graph* which is given as an example of a map on the torus whose faces require seven colors.

The Heawood graph is isomorphic to the bipartite graph whose vertices are the points and lines of the Fano configuration of Example 19.6, and where a point and a line are joined when they are incident.

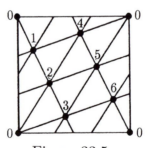

Figure 33.5

THEOREM 33.5. *If a loopless graph $G$ admits an embedding in $\mathcal{T}_g$, $g > 0$, then*

$$\chi(G) \leq \frac{7 + \sqrt{1 + 48g}}{2}.$$

PROOF: Suppose $G$ can be emdedded in $\mathcal{T}_g$. We can assume that $G$ is simple, connected and has at least three vertices. We claim that there exists a vertex $x$ of degree at most $N - 1$ where $N :=$ $(7 + \sqrt{1 + 48g})/2$. Once this is established, the theorem will follow by induction on the number of vertices; for if this vertex $x$ is deleted and the remaining graph properly $\lfloor N \rfloor$-colored, there remains a color free for $x$.

By Theorem 33.4, we have

$$2 - 2g \le f - e + v$$

where $f := |F(G)|$, $e := |E(G)|$, $v := |V(G)|$. Since every face is incident with at least three edges, $3f \le 2e$. Let $d$ be the average degree of the vertices of $G$. We assume for contradiction that $d > N - 1$. This implies that $v > N$. Also $d > 6$ since $N \ge 7$ for $g \ge 1$, and we have

$$2 - 2g \le \frac{2}{3}e - e + v = -\frac{1}{3}e + v = -\frac{1}{6}vd + v = \frac{1}{6}v(6 - d),$$

$$12g - 12 \ge v(d - 6) > N(N - 7) = 12g - 12.$$

This provides the promised contradiction.        □

Here is a small table of bounds from Theorem 33.5:

| genus $g$ | 1 | 2 | 3 | 4 | 5 | 6 | 7 | 8 | 9 | 10 | 11 | 12 | 13 | 14 | 15 |
|---|---|---|---|---|---|---|---|---|---|---|---|---|---|---|---|
| $\chi \le$ | 7 | 8 | 9 | 10 | 11 | 12 | 12 | 13 | 13 | 14 | 15 | 15 | 16 | 16 | 16 |

The assertion that Theorem 33.5 was the best possible result, i.e. that $\lfloor \frac{7 + \sqrt{1 + 48g}}{2} \rfloor$ colors were required for some graph which could be embedded in $\mathcal{T}_g$, was known as the Heawood conjecture until it was proved by Ringel and Youngs in 1968. See Ringel (1974). What they did is to prove the following stronger result.

THEOREM 33.6 (RINGEL-YOUNGS). *Given $g \ge 0$, let*

$$n := \lfloor \frac{7 + \sqrt{1 + 48g}}{2} \rfloor.$$

*Then the complete graph $K_n$ can be embedded in $\mathcal{T}_g$.*

We will prove here only a small part of this result (see Theorem 33.7 below). For example, Theorem 33.6 shows that $\chi(G) \le 19$ for graphs $G$ emdedded in $\mathcal{T}_{20}$, $\mathcal{T}_{21}$, and $\mathcal{T}_{22}$. Theorem 33.7 shows that $K_{19}$ can be embedded in $\mathcal{T}_{20}$ (and hence in $\mathcal{T}_{21}$ and $\mathcal{T}_{22}$), verifying the Heawood conjecture for $g = 20, 21, 22$.

We need to understand 2-cell embeddings combinatorially, in terms of the graph itself. Here is what is important.

Given a 2-cell embedding of a graph $G$ on a surface $\mathcal{S}$, we may traverse the boundary of any face $F$ and we obtain a closed walk $w_F$ in $G$. The starting vertex is not important, nor is the direction. In many cases, this may be a simple closed path in $G$, but, e.g. when $G$ is a tree embedded in the sphere, there is a single face and this walk passes through every edge twice (see Fig. 2.4). In general, every edge will either belong to exactly two walks in $\{w_F : F \in F(G)\}$ or occur twice in one of these walks. A system $\mathcal{M}$ of closed walks in $G$ with this property that every edge of $G$ occurs twice among the walks in $\mathcal{M}$ will be called a *mesh* in $G$.

EXAMPLE 33.3. Here are two meshes in $K_5$. The walks are described by their vertex sequences:

$$\mathcal{M}_1 = \{(1, 2, 3, 4, 5, 1),$$
$$(1, 2, 4, 1),\ (2, 3, 5, 2),\ (3, 4, 1, 3),\ (4, 5, 2, 4),\ (5, 1, 3, 5)\},$$

$$\mathcal{M}_2 = \{(1, 2, 4, 5, 1),\ (1, 3, 2, 5, 3, 4, 2, 3, 5, 4, 1),\ (1, 4, 3, 1, 5, 2, 1)\}.$$

The mesh $\mathcal{M}_1$ consists of simple paths (one pentagon and five triangles), but the second walk in $\mathcal{M}_2$ traverses two edges twice.

A *triangular mesh* is a mesh consisting of simple closed paths of length 3. So a triangular mesh for $K_n$ is equivalent to a 2-design $S_2(2, 3, n)$ with blocks of size 3 and index 2. But these designs rarely correspond to embeddings of $K_n$ on surfaces because there is an important additional condition that must hold.

Given a mesh $\mathcal{M}$ in a connected graph $G$, we may try to construct a 2-cell embedding of $G$ in a surface so that these walks bound the faces. There is really no choice—we must proceed as follows. For each walk $w \in M$ of length $\ell$, we need a corresponding closed disk $F_w$, for example a convex $\ell$-gon and its interior in the plane (even for $\ell = 1$ and 2, although we cannot use straight line segments as the edges in these cases). Label the vertices and edges of $F_w$ with the vertices of the walk $w$ in the order they occur in $w$. (So some vertices and perhaps edges of $F_w$ will receive the same label if $w$ is not a simple closed path.) We will "paste" or "sew" these disks together at their boundaries.

Consider the disjoint union of all disks $F_w$. For each vertex $x$ of $G$, identify all the points labeled with an $x$. For each edge $e$ of $G$, identify the interior points of the edges of the disks labeled with $e$ in a one-to-one continuous manner in the direction indicated by their ends. (It is necessary to agree that loops have two directions.)

Figs. 33.6 and 33.7 show, respectively, polygonal regions in the plane (disks) labeled by the vertices of the walks in the meshes of Example 33.3. In Fig. 33.8, we have started the pasting process for the disks in Fig. 33.6; it remains to identify diametrically opposite points on the boundary (and we obtain the real projective plane with $K_5$ drawn on it).

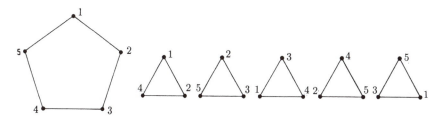

Figure 33.6

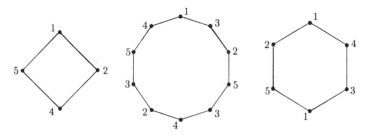

Figure 33.7

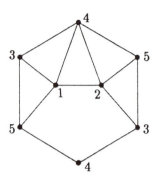

Figure 33.8

In general, the topological space constructed by this pasting has neighborhoods homeomorphic to the open disk (2-cell) about points $x$ in the interior of the disks $F_x$ and also for points in the interior of an edge on the boundary of the disks (because every edge is in two of the walks of the mesh). But there may be trouble at vertices, where many of the disks $F_w$ meet. For example, suppose that the walks that traverse a vertex $x$ in a certain mesh have vertex sequences

$$(\ldots, 1, x, 2, \ldots), (\ldots, 2, x, 3, \ldots), (\ldots, 3, x, 1, \ldots),$$
$$(\ldots, 4, x, 5, \ldots), (\ldots, 5, x, 6, \ldots), \text{ and } (\ldots, 6, x, 4, \ldots).$$

The disks corresponding to the first three walks fit together to make a neighborhood of $x$ which is a 2-cell, but with the other three, a neighborhood of $x$ will look like two cones touching at their apexes. The graph $\mathcal{M}_x$ defined in the next paragraph would consist of two disjoint triangles in this case.

For a vertex $x \in V(G)$, consider the graph $\mathcal{M}_x$ where $V(\mathcal{M}_x)$ is the set of edges of $G$ incident with $x$ and where two members $a, b$ of $V(\mathcal{M}_x)$ are adjacent in $\mathcal{M}_x$ when they occur as consecutive edges of some walk in $\mathcal{M}$ (with the vertex $x$ between them). Of course, join $a$ and $b$ by two edges in $\mathcal{M}_x$ if they occur consecutively twice, which probably will not often be the case. Clearly, $\mathcal{M}_x$ is regular of degree 2. The *vertex condition* requires that $\mathcal{M}_x$ is connected (i.e. a single polygon) for every vertex $x$.

The topological space constructed by this pasting from a mesh in a connected graph is a surface if and only if the mesh satisfies the vertex condition. There is a natural embedding of the graph $G$ in this surface. The meshes in Example 33.3 satisfy the vertex condition. See Example 33.4 and Fig. 33.9.

To discover what surface $S$ has been constructed from a mesh $\mathcal{M}$, we first compute the *Euler characteristic* $h := |\mathcal{M}| - |E(G)| + |V(G)|$. In view of Theorem 33.4, if $h$ is odd, we must have $S = \mathcal{N}_{2-h}$. But if $h$ is even, we must decide whether the surface is orientable or not (unless $h = 2$, when the sphere is the only possibility): $S = \mathcal{T}_{1-h/2}$ if $S$ is orientable and $S = \mathcal{N}_{2-h}$ if $S$ is not orientable. We state without proof that $S$ is orientable if and only if the walks can be directed (one of the two possible directions being chosen for each walk) so that *every edge is traversed exactly once in each direction*, in which case we say the mesh is *orientable*.

It is easy to check whether a mesh satisfying the vertex condition is orientable. We give one walk a direction and that determines the directions of all walks sharing an edge, which in turn determines the directions of further walks. The vertex condition implies that the direction of every walk in the mesh is determined (why?). Then we check each edge to see in what directions it is traversed.

EXAMPLE 33.3 (CONTINUED). The mesh $\mathcal{M}_1$ provides an embedding of $K_5$ on the real projective plane $\mathcal{N}_1$. The mesh $\mathcal{M}_2$ is orientable—the walks may be directed as they have been written down—and we have a 2-cell embedding of $K_5$ on the double torus $\mathcal{T}_2$.

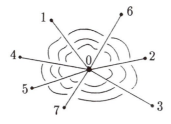

Figure 33.9

In describing walks in simple graphs in the examples in the sequel, we list the vertex sequences in square brackets but suppress the terminal vertex since it is the same as the initial vertex. And it is convenient to refer to paths of length $3, 4, \ldots$ as triangles, quadrilaterals, etc., ignoring the difference between walks in a graph and subgraphs that are polygons.

EXAMPLE 33.4. We show that $K_8$ can be embedded on the double torus $T_2$. This shows that at least eight colors are required for $T_2$ and proves the Heawood conjecture for $g = 2$.

There must be 18 faces (so that $f - e + v = -2$), and their degrees must add to 56. We display a solution with two quadrilaterals and 16 triangles as the faces. We use "difference methods" (cf. Chapter 19).

As the vertices of $K_8$, we take $\mathbb{Z}_8$, the integers modulo 8. The quadrilaterals will have vertex sequences $[0, 2, 4, 6]$ and $[1, 3, 5, 7]$. The triangles are to be the translates modulo 8 of the two initial triangles $[0, 1, 4]$ and $[0, 3, 2]$. The reader should check that every edge appears in exactly two of these walks. Let us check the vertex condition. Since translation modulo 8 preserves the polygons, we need only check the vertex 0. The walks traversing 0 are $[0, 2, 4, 6]$, $[0, 1, 4]$, $[7, 0, 3]$, $[4, 5, 0]$, $[0, 3, 2]$, $[5, 0, 7]$, and $[6, 1, 0]$. They fit together to form a single polygon as indicated in Fig. 33.9.

It remains to check whether the mesh is orientable. It turns out that if the quadrilaterals are directed as $[0, 2, 4, 6]$ and $[7, 5, 3, 1]$, and every other triangle is also reversed (i.e. whenever the initial triangles are translated by an odd element of $\mathbb{Z}_8$), we obtain an oriented mesh.

PROBLEM 33B. Find a mesh of five 4-gons in $K_5$, a mesh of six 5-gons in $K_6$, and a mesh of seven 6-gons in $K_7$, all of which satisfy the vertex condition. What surfaces have you constructed?

PROBLEM 33C. Find a mesh of 16 pentagons in the Clebsch graph (see Example 21.4). Check the vertex condition and also check for orientability. What surface, if any, has been constructed? What is the dual graph with respect to this embedding? (Cf. Problem 33H.)

PROBLEM 33D. (i) Show that if a simple graph $G$ is embedded in the nonorientable surface $\mathcal{N}_g$, $g \geq 1$, then

$$\chi(G) \leq \frac{7 + \sqrt{1 + 24g}}{2}.$$

(ii) Show that if a simple graph is embedded in the Klein bottle $\mathcal{N}_2$, then either it has a vertex of degree $\leq 5$ or it is the graph $K_7$. Show that there is a unique mesh of 14 triangles in $K_7$ that satisfies the vertex condition and that it is orientable.

(Thus $K_7$ does not embed in the Klein bottle after all. We conclude that $\chi(G) \leq 6$ for graphs embedded in $\mathcal{N}_2$.)

PROBLEM 33E. Before we continue with surfaces other than the sphere, we remark that our understanding of meshes allows a quick proof of the fact that if $G$ and $H$ are *nonseparable* Whitney duals, then there exists an embedding of $G$ in the sphere with respect to which the dual $G^*$ is isomorphic to $H$. As in the proof of Theorem 32.6, the set of edges incident with a vertex of $H$ is a circuit in $G$. The system of walks we get in $G$ is a mesh $\mathcal{M}$. Show that $\mathcal{M}$ satisfies the vertex condition. Explain why the surface constructed is the sphere and why the dual $G^*$ is isomorphic to $H$.

PROBLEM 33F. Consider a connected graph $G$ and a mesh $\mathcal{M}$ in $G$ satisfying the vertex condition (i.e. a map). To keep things simple, assume that $G$ is simple and that the walks in $\mathcal{M}$ are simple closed paths—which we identify with polygons and call *faces*. An *automorphism* of this map is a permutation of $V(G)$ which takes edges to edges and faces to faces (i.e. an automorphism of $G$ that preserves $\mathcal{M}$). Show that if an automorphism $\alpha$ fixes a vertex $x$, an edge $e$ incident with $x$, and a face $F$ incident with $e$, then it is the identity.

(It follows, for example, that a simple map as above with graph $G$ has at most $4|E(G)|$ automorphisms. Examples where equality occurs include the Platonic solids and the maps in Examples 33.1 and 33.2.)

THEOREM 33.7. *If $n \equiv 7 \pmod{12}$, then the complete graph $K_n$ admits a triangular embedding on an orientable surface of genus*

$$g := (n-3)(n-4)/12.$$

PROOF: We construct an orientable triangular mesh in $K_n$ satisfying the vertex condition. Write $n = 12s + 7$. As the vertices of $K_n$, we use $\mathbb{Z}_n$.

The construction is encoded in the diagrams of Fig. 33.10. The diagrams are for the cases $n = 7$, $n = 19$, $n = 31$, and $n = 12s + 7$ in general. (In the general case, the vertical edges are directed alternately and are labeled $1, 2, \ldots, 2s$, consecutively.) We could describe this construction simply as an application of difference methods and not give any figures, but that would be a shame since the diagrams were conceptually very helpful in finding this and other embeddings.

Note that the edges are labeled with the integers from 1 to $6s+3$. The "conservation" law, or "Kirchhoff's current law", holds: at every vertex, the sum of values on the incoming edges is equal to the sum of the values on the outgoing edges. (Except that we have no source or sink, this is a flow of strength zero as defined in Chapter 7.)

Actually, it will be useful to think that the reverses of the indicated directed edges are also present; the reverse of the edge shown in Fig. 33.10 will carry a label equal to the negative of the shown edge. Then all nonzero values modulo $n = 12s + 7$ appear on the directed edges, each exactly once. The vertices are of two types: solid (representing "clockwise") and hollow (representing "counterclockwise").

Every vertex in the diagrams of Fig. 33.10 will provide a family of $n$ directed triangles in $K_n$ which are translates of one another modulo $n$. For each solid vertex that is the tail of directed edges with values $a$, $b$, $c$ *in clockwise order*, we take the $n$ directed triangles

$$[x, x + a, x + a + b], \quad x \in \mathbb{Z}_n.$$

Remember that $a + b + c = 0$ in $\mathbb{Z}_n$, so we get the same triangles (but with different initial vertices) if we take $[y, y + b, y + b + c]$, $y \in \mathbb{Z}_n$ or $[z, z + c, z + c + a]$, $z \in \mathbb{Z}_n$. For each hollow vertex that is the tail of directed edges with values $a$, $b$, $c$ *in counter-clockwise order*, we take the $n$ directed triangles

$$[x, x + a, x + a + b], \quad x \in \mathbb{Z}_n.$$

For example, in the third diagram in Fig. 33.10 (when $n = 31$), there is a hollow vertex with incoming edges labeled 7 and 2 and an outgoing edge labeled 9. We regard this as outgoing edges labeled $-7$, $-2$, and 9 (counter-clockwise), and take the 31 directed triangles obtained as the translates of $[0, -7, -9]$ modulo 31 (or translates of $[0, 9, 2]$ or $[0, -2, 7]$). Another vertex in that diagram yields the 31 directed triangles obtained as the translates of $[0, 14, 13]$ modulo 31.

It is not hard to see that the collection of directed triangles obtained from the diagrams form an oriented triangular mesh. For example, when $n = 31$, to find a triangle traversing the edge joining 4 to 23 in that direction, notice that $23 - 4 = -12$ in $\mathbb{Z}_n$. The edge labeled $-12$ leaves a solid vertex with other outgoing edges labeled 15 and $-3$, and the triangle $[4, 4 + (-12), 4 + (-12) + 15] = [4, 23, 7]$ is found. The triangle traversing the edge joining 23 to 4 in that order is found to be $[23, 4, 6]$.

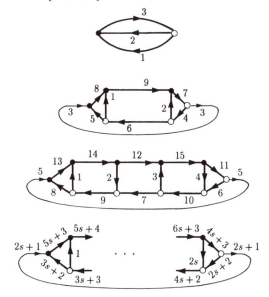

Figure 33.10

It will suffice to check the vertex condition only at 0. Let $e_i$ denote the edge $\{0, i\}$ of $K_n$. Consider the triangle of the mesh that traverses 0 and enters 0 on edge $e_a$. In the diagram, the edge labeled $a$ will enter a vertex that is the tail of edges with labels $-a$, $p$, $q$, say, in clockwise or counter-clockwise order, depending on whether the vertex is solid or hollow. That is, the triangle is

$[a, 0, p]$ and leaves $0$ on the edge $e_p$. To summarize, if a triangle enters $0$ on an edge $e_i$ of $K_n$, the value $j$ such that the triangle leaves $0$ on $e_j$ is found from the diagram as follows: find the edge labeled $i$ in the diagram—turn clockwise or counter-clockwise, as appropriate, at the head of that edge and leave on the next edge—the label on that edge of the diagram will be $j$.

For example, when $n = 31$, we find the sequence of edges $e_{i_\ell}$ of $K_n$ so that the triangle which enters $0$ on $e_{i_\ell}$ will leave on $e_{i_{\ell+1}}$ to be

$$e_1, e_{-13}, e_{-8}, e_{-9}, e_{-7}, e_{-10}, e_{-6}, e_5,$$

$$e_{13}, e_{14}, e_{12}, e_{15}, e_{11}, e_6, e_{-4}, e_{-15}, e_{-3}, e_7, e_{-2}, e_{-14},$$

$$e_{-1}, e_8, e_{-5}, e_{-11}, e_4, e_{10}, e_3, e_{-12}, e_2, e_9, e_1, \ldots.$$

Consecutive pairs above will be adjacent in $\mathcal{M}_0$, which is seen to be a single polygon. The reader should check that all diagrams in Fig. 33.10 will similarly yield under this procedure a sequence traversing each edge of the diagram once in each direction. This is quite amazing. □

PROBLEM 33G. (i) If the complete graph $K_n$ admits an embedding on the orientable surface $\mathcal{T}_g$ so that all faces are bounded by triangles, then $g = (n-3)(n-4)/12$.

(ii) If the complete bipartite graph $K_{n,n}$ admits an embedding on the orientable surface $\mathcal{T}_g$ so that all faces are bounded by quadrilaterals, what is $g$ in terms of $n$? What is the least value of $n > 2$ for which such an embedding exists?

In order to present the proof of Theorem 33.7 as quickly and as simply as possible, we have taken the construction out of its original context. We close this chapter with a brief discussion of some of the ideas that allowed Ringel and Youngs to come up with constructions for all values of $n$.

A technique for embedding graphs on orientable surfaces was described by J. Edmonds (1960). It may be thought of as arising from the mesh in the dual graph with respect to the embedding. If we have a connected simple graph $G$ embedded in an orientable surface $\mathcal{S}$, then the choice of an orientation on $\mathcal{S}$ will induce at every vertex a cyclic permutation of the edges incident with that vertex.

Edmonds pointed out that any such "local cyclic permutations" determine an embedding. Let $G$ be a simple graph. Suppose a cyclic permutation $\sigma_x$ of the set $S(x)$ of edges incident with $x$ is given for each vertex $x$. Then there is a natural oriented mesh $\mathcal{M}$ determined as follows. In the directed walk traversing an edge $e$ directed from one of its ends $y$ to the other $x$, the edge following $e$ is to be $\sigma_x(e)$.

The examples below concern simple graphs, so the local permutations may be abbreviated as permutations of the adjacent vertices and walks may be described by their vertex sequences.

EXAMPLE 33.5. Consider $K_5$ with vertex set $\{1, 2, 3, 4, 5\}$ and the following local permutations:

$$
\begin{aligned}
1 &: \ (2435) \\
2 &: \ (1435) \\
3 &: \ (4125) \\
4 &: \ (1325) \\
5 &: \ (1234)
\end{aligned}
$$

It might be good to think of the resulting walks as a decomposition of the edges of the complete directed graph on 5 vertices. Let $e_{ij}$ denote the edge directed from $i$ to $j$. Here are the edge terms of the walks:

$$(e_{12}, e_{24}, e_{45}, e_{51}),$$
$$(e_{13}, e_{32}, e_{25}, e_{53}, e_{34}, e_{42}, e_{23}, e_{35}, e_{54}, e_{41}),$$
$$(e_{14}, e_{43}, e_{31}, e_{15}, e_{52}, e_{21}).$$

The vertex sequences of these walks are exactly those of $\mathcal{M}_2$ in Example 33.3.

In the family of graphs with $4s + 2$ vertices in Fig. 33.10, we specified a cyclic order at each vertex by the use of a solid vertex to indicate that the local permutation is to be clockwise and a hollow vertex to indicate counter-clockwise. As part of the proof of Theorem 33.7, we had to check that the resulting embeddings have exactly one face.

Given a finite group $\Gamma$ and a subset $S$ of nonidentity elements of $\Gamma$ such that $\alpha \in S$ implies that $\alpha^{-1} \in S$, the *Cayley graph* $G(\Gamma, S)$ is the simple graph with vertex set $\Gamma$ and where vertices $\alpha$ and

$\beta$ are adjacent if and only if $\beta\alpha^{-1} \in S$. A complete graph $K_n$ is a Cayley graph with respect to any group of order $n$, where $S$ consists of all nonidentity elements. Often it is required that $S$ generates the group—this ensures that the corresponding Cayley graph is connected.

W. Gustin (1963) introduced the theory of "quotient manifolds" and developed methods to embed Cayley graphs into orientable surfaces. We have no space to describe this precisely, but give some examples. Consider the case where the local permutations at the vertices of a Cayley graph are all "shifts" of an initial cyclic permutation $(s_1 \; s_2 \; \ldots \; s_k)$ of the set $S$ of vertices adjacent to 0; that is, the local permutation at $\alpha$ is

$$(s_1 + \alpha \quad s_2 + \alpha \quad \ldots \quad s_k + \alpha)$$

(with additive notation).

EXAMPLE 33.6. Consider $K_7$ with vertex set $\{0, 1, 2, 3, 4, 5, 6\}$ and with local permutation $(1 + i \; 3 + i \; 2 + i \; 6 + i \; 4 + i \; 5 + i) \pmod{7}$ at $i$ as below:

$$
\begin{array}{rl}
0: & (132645) \\
1: & (243056) \\
2: & (354160) \\
3: & (465201) \\
4: & (506312) \\
5: & (610423) \\
6: & (021534)
\end{array}
$$

The walks are:

$$[1, 2, 4] \quad [2, 3, 5] \quad [3, 4, 6] \quad [4, 5, 0] \quad [5, 6, 1] \quad [6, 0, 2] \quad [0, 1, 3]$$
$$[3, 5, 6] \quad [4, 6, 0] \quad [5, 0, 1] \quad [6, 1, 2] \quad [0, 2, 3] \quad [1, 3, 4] \quad [2, 4, 5]$$

This is the triangular embedding of $K_7$ on the torus we have seen in Example 33.2.

If we try this method on $K_{31}$, for example, with vertex set $\mathbb{Z}_{31}$ and initial local permutation

$$(1, -13, -8, -9, -7, -10, -6, 5, 13, 14, 12, 15, 11, 6, -4,$$
$$-15, -3, 7, -2, -14, -1, 8, -5, -11, 4, 10, 3, -12, 2, 9),$$

we obtain the same triangular mesh as in the proof of Theorem 33.7. The diagrams of Fig. 33.10 actually represent the "quotient manifolds" of the dual graphs of $K_n$ with respect to the prescribed embeddings. For $n = 12s + 7$, the embedded graph represented by the diagram has $4s + 2$ vertices, $6s + 3$ edges, and 1 face; the resulting embedding of $K_n$ has $(4s+2)n$ faces, $(6s+3)n$ edges, and $(1)n$ vertices.

PROBLEM 33H. Let $\omega$ be a nonzero element in $\mathbb{F}_q$ where $q$ is a power of a prime. Say $\omega$ has order $m$ in the multiplicative group and assume that $-1$ is a power of $\omega$. Consider the Cayley graph $G(\mathbb{F}_q, \langle \omega \rangle)$ where $\langle \omega \rangle = \{1, \omega, \omega^2, \dots\}$. We are using the additive group of $\mathbb{F}_q$ for the Cayley graph, so $a$ and $b$ are adjacent if and only if $a - b \in \langle \omega \rangle$. (This is a complete graph when $\omega$ is a primitive element in $\mathbb{F}_q$. This is the Clebsch graph when $q = 16$ and $m = 5$.)

Let us take the local permutation at 0 to be $(1, \omega, \omega^2, \dots, \omega^{m-1})$ and obtain the other local permutations as shifts. What is the size and number of the faces in the resulting mesh?

## Notes.

P. J. Heawood (1861–1955) spent most of his career at Durham as professor of mathematics, finally to become vice-chancellor.

A. B. Kempe (1849–1922) took a degree in 1872 with special distinction in mathematics at Trinity College, Cambridge where he also gained a musical reputation with his fine counter-tenor voice. He chose law as his profession while not relinquishing his mathematical studies.

## References.

K. Appel and W. Haken (1977), The solution of the four-color map problem, *Scientific American* **237**, 108–121.

K. Appel, W. Haken, and J. Koch (1977), Every planar map is four colorable, *Illinois J. Math.* **21**, 429–567.

N. L. Biggs, E. K. Lloyd, and R. J. Wilson (1976), *Graph Theory 1736–1936*, Oxford University Press.

G. Chartrand and L. Lesniak (1986), *Graphs and Digraphs*, 2nd edn., Wadsworth.

J. Edmonds (1960), A combinatorial representation for polyhedral surfaces, *Notices Amer. Math. Soc.* **7**, 646.

M. Fréchet and K. Fan (1967), *Initiation to Combinatorial Topology*, Prindle, Weber and Schmidt.

W. Gustin (1963), Orientable embedding of Cayley graphs, *Bull. Amer. Math. Soc.* **69**, 272–275.

S. MacLane (1937), A structural characterization of planar combinatorial graphs, *Duke Math. J.* **3**, 460–472.

G. Ringel (1974), *Map Color Theorem*, Springer-Verlag.

G. Ringel and J. W. T. Youngs (1968), Solution of the Heawood map coloring problem, *Proc. Nat. Acad. Sci. U.S.A.* **60**, 438–445.

P. G. Tait (1880), Remarks on the colouring of maps, *Proc. Roy. Soc. Edinburgh* **10**, 501–503.

W. T. Tutte (1956), A theorem on planar graphs, *Trans. Amer. Math. Soc.* **82**, 99–116.

A. T. White (1973), *Graphs, Groups, and Surfaces*, Mathematical Studies **8**, North-Holland.

# 34

# Electrical networks and squared squares

We start with the *matrix-tree* theorem, which expresses the number of spanning trees in a graph as the determinant of an appropriate matrix.

THEOREM 34.1. *The number of spanning trees in a connected graph $G$ on $n$ vertices and without loops is the determinant of any $n-1$ by $n-1$ principal submatrix of the matrix $D-A$, where $A$ is the adjacency matrix of $G$ and $D$ is the diagonal matrix whose diagonal contains the degrees of the corresponding vertices of $G$.*

For multigraphs $G$ without loops we agree that the adjacency matrix $A$ is defined to have $A(x,y)$ equal to the number of edges joining $x$ and $y$, for distinct vertices $x$ and $y$. We postpone the proof until we have developed some tools, but we give some examples now.

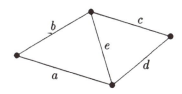

Figure 34.1

EXAMPLE 34.1. Let $G$ be the graph in Fig. 34.1. Then

$$D - A = \begin{pmatrix} 2 & -1 & 0 & -1 \\ -1 & 3 & -1 & -1 \\ 0 & -1 & 2 & -1 \\ -1 & -1 & -1 & 3 \end{pmatrix}$$

and $G$ has 8 spanning trees.

EXAMPLE 34.2. Take $G$ to be the complete graph $K_n$ in Theorem 34.1. In this case, the matrix $D - A$ is $nI - J$, where $I$ is the identity matrix of order $n$ and $J$ is the $n$ by $n$ matrix of all 1's. The calculation of the determinant of a $n - 1$ by $n - 1$ principal submatrix of this simple matrix can be done in several ways (row operations, consideration of eigenvalues) and is left as an exercise. Once this is done, we obtain yet another proof of Cayley's theorem, cf. Chapter 2, that the complete graph $K_n$ has $n^{n-2}$ spanning trees.

PROBLEM 34A. Let $M$ be an $n$ by $n$ matrix all of whose linesums are 0. Then one of the eigenvalues of $M$ is $\lambda_1 = 0$; let $\lambda_2, \lambda_3, \ldots, \lambda_n$ denote the other eigenvalues. Show that all principal $n - 1$ by $n - 1$ submatrices of $M$ have the same determinant, and this value is equal to the product $\frac{1}{n}\lambda_2\lambda_3 \cdots \lambda_n$.

EXAMPLE 34.3. We can use the observation of the above problem to calculate the number of spanning trees in regular graphs whose spectrum is known. For example, let $A$ be the adjacency matrix of the Petersen graph. Then $A$ has eigenvalues

$$3, 1, 1, 1, 1, 1, -2, -2, -2, -2$$

(see Chapter 21). The matrix $M$ is $3I - A$, which has eigenvalues

$$0, 2, 2, 2, 2, 2, 5, 5, 5, 5.$$

We conclude that the Petersen graph has 2000 spanning trees.

The following lemma is known as the Cauchy-Binet theorem. It is more commonly stated and applied with the diagonal matrix $\Delta$ below taken to be the identity matrix.

LEMMA 34.2. *Let $A$ and $B$ be, respectively, $r$ by $m$ and $m$ by $r$ matrices. Let $\Delta$ be the $m$ by $m$ diagonal matrix with entry $e_i$ in row $i$, column $i$. For an $r$-subset $S$ of $\{1, 2, \ldots, m\}$, let $A_S$ and $B^S$ denote, respectively, the $r$ by $r$ submatrices of $A$ and $B$ consisting of the columns of $A$, or the rows of $B$, indexed by elements of $S$. Then*

$$\det(A\Delta B) = \sum_S \det(A_S) \det(B^S) \prod_{i \in S} e_i$$

*where the sum is extended over all $r$-subsets $S$ of $\{1, 2, \ldots, m\}$.*

PROOF: We will prove this assuming that $e_1, \ldots, e_m$ are independent indeterminates. Of course, it will then hold for all values of $e_1, \ldots, e_m$.

The entries of the $r$ by $r$ matrix $A\Delta B$ are linear forms in the indeterminates $e_1, \ldots, e_m$; explicitly, if $A = ((a_{ij}))$ and $B = ((b_{ij}))$, then the $(i, k)$ entry of $A\Delta B$ is $\sum_{j=1}^{m} a_{ij} b_{jk} e_j$. Thus $\det(A\Delta B)$ is a homogeneous polynomial of degree $r$ in $e_1, \ldots, e_m$.

Consider a monomial $e_1^{t_1} e_2^{t_2} \ldots$ where the number of distinct indeterminates $e_i$ that occur, i.e. that have exponent $t_i > 0$, is less than $r$. Substitute 0 for the indeterminates $e_i$ which do not occur in $e_1^{t_1} e_2^{t_2} \ldots$. The monomial $e_1^{t_1} e_2^{t_2} \ldots$ and its coefficient are unaffected by this substitution. But after this substitution, the rank of $\Delta$ is less than $r$ so the polynomial $\det(A\Delta B)$ evaluates to the zero polynomial.

So we see that the coefficient of a monomial in the polynomial $\det(A\Delta B)$ is 0 unless the monomial is a product of $r$ distinct indeterminates $e_i$, i.e. unless it is of the form $\prod_{i \in S} e_i$ for some $r$-subset $S$. The coefficient of a monomial $\prod_{i \in S} e_i$ in $\det(A\Delta B)$ can be obtained by setting $e_i$, $i \in S$, equal to 1 and $e_j$, $j \notin S$, equal to 0. When this substitution is made in $\Delta$, $A\Delta B$ evaluates to $A_S B^S$. So the coefficient of $\prod_{i \in S} e_i$ in $\det(A\Delta B)$ is $\det(A_S) \det(B^S)$. $\qquad\square$

We will deal extensively with directed graphs $H$ in this chapter. All graph-theoretic terms (trees, components, circuits, etc.) will be applied to a directed graph with the same meaning as for the underlying undirected graph we obtain by ignoring the directions of the edges. Thus a path in a digraph will traverse some edges in a "forward" direction, other edges will be traversed "backward". It is really undirected graphs we are interested in, but it is convenient for the theory to choose an orientation and produce a digraph; any orientation will do and produces essentially the same theory.

As we have often done in previous chapters, we use the functional notation $M(i, j)$ to denote the entry in row $i$ and column $j$ of a matrix $M$, and $f(i)$ to denote the $i$-th coordinate of a vector $f$.

The *incidence matrix* $N$ of a directed graph $H$ is the matrix whose rows are indexed by $V(H)$, whose columns are indexed by

$E(H)$, and where

$$N(x,e) = \begin{cases} 0 & \text{if } x \text{ is not incident with } e, \text{ or } e \text{ is a loop,} \\ 1 & \text{if } x \text{ is the head of } e, \\ -1 & \text{if } x \text{ is the tail of } e. \end{cases}$$

We mention that

(34.1) $$\text{rank}(N) = |V(H)| - |C(H)|$$

where $C(H)$ is the set of components of $H$. To see this, suppose that $g$ is a row vector with coordinates indexed by $V(H)$ and that $gN = 0$. This means for every edge $e$, directed from $x$ to $y$, say, that $g(y) - g(x) = 0$. It is clear then, that $gN = 0$ if and only if $g$ is constant on the vertex set of every component of $H$, and hence the dimension of the space of all $g$ such that $gN = 0$ is $|C(H)|$.

We also mention that the determinant of any square matrix that has at most two nonzero entries in each column, at most one 1 and at most one $-1$, is equal to 0 or $\pm 1$. This follows by induction: If every column has a $+1$ and a $-1$, then the sum of all rows is the zero vector, so the matrix is singular. Otherwise, expand the determinant by a column with one nonzero entry to find it is $\pm 1$ times the determinant of a smaller matrix with the same property. So every square submatrix of the incidence matrix of a digraph has determinant 0 or $\pm 1$. (Matrices with this property are called *totally unimodular*.)

PROOF OF THEOREM 34.1: Let $H$ be a connected digraph with $n$ vertices, with incidence matrix $N$. Let $S$ be a set of $n-1$ edges and, with the notation of the Cauchy-Binet theorem, consider the $n$ by $n-1$ submatrix $N_S$, whose columns are indexed by elements of $S$, of the incidence matrix $N$. By (34.1), $N_S$ has rank $n-1$ if and only if the spanning subgraph of $H$ with edge set $S$ is connected, i.e. if and only if $S$ is the edge set of a tree in $H$. Let $N'$ be obtained by dropping any single row from the incidence matrix $N$. Since the sum of all rows of $N$ (or $N_S$) is the zero vector, the rank of $N'_S$ is the same as the rank of $N_S$. The observations of this and the preceding paragraph prove that

(34.2)

$$\det(N'_S) = \begin{cases} \pm 1 & \text{if } S \text{ is the edge set of a spanning tree in } H, \\ 0 & \text{otherwise.} \end{cases}$$

Let a connected loopless graph $G$ on $n$ vertices be given, let $H$ be any orientation of $G$, and let $N$ the incidence matrix of $H$. Then $NN^\top = D - A$ because

$$NN^\top(x,y) = \sum_{e \in E(G)} N(x,e)N(y,e)$$

$$= \begin{cases} \deg(x) & \text{if } x = y, \\ -t & \text{if } x \text{ and } y \text{ are joined by } t \text{ edges in } G. \end{cases}$$

An $n-1$ by $n-1$ principal submatrix of $D - A$ is of the form $N'N'^\top$ where $N'$ is obtained from $N$ by dropping any one row. By Cauchy-Binet,

$$\det(N'N'^\top) = \sum_S \det(N'_S)\det(N'_S{}^\top) = \sum_S \left(\det(N'_S)\right)^2$$

where the sum is extended over all $(n-1)$-subsets $S$ of the edge set. By (34.2), this is the number of spanning trees of $G$. $\square$

REMARK. If we view $E(G)$ as a set of indeterminates and apply Lemma 34.2 to $n-1$ by $n-1$ principal submatrices of $N\Delta N^\top$ where here $\Delta$ is the diagonal matrix whose rows and columns are indexed by the edges and where the diagonal entries are the edges themselves (i.e. $\Delta(e,e) := e$), we find that $\det(N\Delta N^\top)$ is the sum of monomials corresponding to the edge sets of the trees in $G$. For example, for the graph $G$ in Fig. 34.1,

$$N\Delta N^\top = \begin{pmatrix} a+b & -b & 0 & -a \\ -b & b+c+e & -c & -e \\ 0 & -c & c+d & -d \\ -a & -e & -d & a+d+e \end{pmatrix}.$$

Of course, $N\Delta N^\top$ is a singular matrix (all linesums are 0), but the determinant of the 3 by 3 matrix obtained by dropping e.g. the last row and column is

$$(a+b)(b+c+e)(c+d) - b^2(c+d) - c^2(a+b) =$$
$$abc + abd + acd + ace + ade + bcd + bce + bde,$$

representing the eight spanning trees in $G$.

$$* * *$$

Let $N$ be the incidence matrix of a digraph $H$. The rowspace of $N$ is called the *coboundary space* of $H$; the nullspace of $N^{\top}$, i.e. the space of all row vectors $f$ such that $fN^{\top} = 0$, is called the *cycle space* of $H$. These are of importance in the theory of electrical networks and algebraic topology. From (34.1), the dimension of the coboundary space of a digraph $H$ is $|V(H)| - |C(H)|$, and hence the dimension of the cycle space is $|E(H)| - |V(H)| + |C(H)|$.

From the definition, a vector $f$ is a cycle if and only if for each vertex $x$, the sum of the values $f(e)$ on outgoing edges is equal to the sum of the values $f(e)$ on incoming edges. An example of a cycle $f$ (a so-called *elementary* cycle) is obtained from any simple closed path $p$ in $H$ by defining $f(e)$ to be $+1$ on the forward edges of the path, $-1$ on the backward edges, and $0$ on all edges not in the path. A vector is a coboundary if and only if it is orthogonal to all cycles, so in particular, the sum of the values of a coboundary on the forward edges of a simple closed path will equal the sum of the values on the backward edges of the path.

PROBLEM 34B. A vector $g$ with coordinates indexed by $E(H)$ is a coboundary on a digraph $H$ if and only if for every closed walk $w$, the signed sum of the values of $g$ on the edges (the sum of the values of $g$ on the forward edges of $w$ minus the sum of the values of $g$ on the backward edges of $w$) is zero.

Let us agree that an *electrical network* consists of a digraph $H$, together with a function $r$ associating to each edge $e$ a "resistance" $r(e) \geq 0$, and a function $s$ associating to each edge $e$ an "impressed electromotive force" or "voltage source" $s(e)$. That is, each edge is to be thought of as a resistor or a battery or both. If batteries and resistors are connected, a "current flow" $f(e)$ will pass through each edge and a "potential difference" or "voltage" $g(e)$ will be measurable across each edge. All that we need to know here is that the vectors $f$ and $g$ are determined from $r$ and $s$ by the rules called *Kirchhoff's laws* and *Ohm's law*. Kirchhoff's current law and voltage law assert, respectively, that $f$ is a cycle and that $g$ is a coboundary. Ohm's law says that $g(e) = -s(e) + r(e)f(e)$, or, in matrix notation, $g = -s + fR$ where $R$ is the diagonal matrix with the numbers $r(e)$ on the diagonal and where we think of $s$, $g$, and $f$ as row vectors.

PROBLEM 34C. Suppose $\{e : r(e) = 0\}$ contains no circuits, i.e. is the edge set of a forest. Show that the electrical network has a unique solution, by which we mean that there exist unique vectors $f$ and $g$ satisfying the above laws.

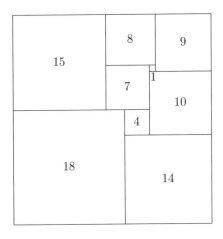

Figure 34.2

By a *squared rectangle* we mean a rectangle which has been partitioned into a finite number of squares. Two examples appear in Figs. 34.2 and 34.3. These rectangles have dimensions 33 by 32, and 177 by 176, respectively. Squared rectangles are thought to be elegant when all internal squares have distinct sizes and when they contain no smaller squared rectangle.

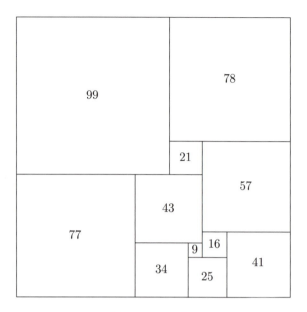

Figure 34.3

PROBLEM 34D. Find integral dimensions for the small squares in Fig. 34.4 so that the diagram represents a squared rectangle.

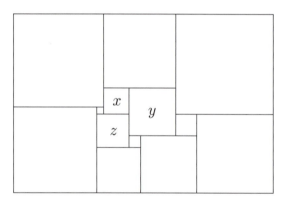

Figure 34.4

For many years, it was not known whether a *square* could be partitioned into a finite number of *unequal* smaller squares. In the late 1930s, such squared squares were finally discovered. One example was shown on the cover of the *Journal of Combinatorial Theory* until a smaller example was found by A. J. W. Duijvestijn (1978);

the cover illustration was updated in January of 1979. This is a
partition into the least possible number (21) of squares. Computers are used for these results, but an important tool in the study of
squared rectangles is the connection with electrical networks which
was discovered by four undergraduates at Cambridge in 1936–1938.
A popular account of some of their work is given by W. T. Tutte
(1961) and (1965).

We give an informal description of this connection with electrical
networks. Let us use the term *special network* for an electrical
network with a distinguished edge $e_0$ so that $s(e) = 0$ for $e \neq e_0$,
and where

$$r(e) = \begin{cases} 1 & \text{if } e \neq e_0, \\ 0 & \text{if } e = e_0. \end{cases}$$

That is, we have a system of unit resistors and a single battery.

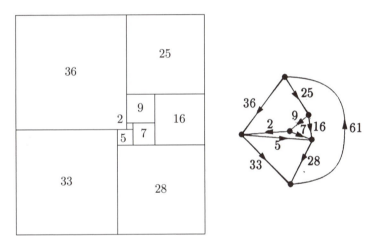

Figure 34.5

A special electrical network arises from a squared rectangle as follows. Define a digraph $H$ whose vertices are the maximal horizontal
line segments in the diagram. The edges of $H$ are to be the small
squares, and we include one special edge $e_0$ for the entire rectangle.
We direct each edge (except $e_0$) downward, i.e. towards the lower
line segment. An example is shown in Fig. 34.5. The digraph $H$ is
clearly planar. There is a real number $f(e)$ naturally associated to
edge $e$, namely the size of the corresponding small square; we take

$f(e_0)$ to be the width of the large rectangle. It is not hard to see that $f$ is a cycle on $H$: for each maximal horizontal line segment, the sum of the sizes of the squares "sitting on" that segment is equal to the sum of the squares "hanging from" it. Moreover, were it not for the special edge, $f$ would also be a coboundary: if for each vertex (horizontal line segment) $x$, we take $h(x)$ to be the distance of the line segment from the top side of the large rectangle, then for $e \neq e_0$, $f(e)$ is equal to $h(\text{head of } e) - h(\text{tail of } e)$. So if we let $s(e_0)$ be the height of the rectangle and take $g$ be the coboundary of $h$, then $f$ and $g$ are the current flow and voltage vectors which solve this special network, i.e. $g = -s + Rf$.

Conversely, a special electrical network where the digraph is *planar* leads to the construction of a squared rectangle. We do not give a proof. The ratio of its width to its height will be the ratio of the current flow $f(e_0)$ to the voltage $-g(e_0)$ in the solution to the network. (Some of the edges may have negative current flow when the network is solved, but we could reverse the direction on these edges to get a positive value. The edges with zero current flow may be deleted—or contracted; they will not give rise to squares.) Thus a squared square corresponds to a planar network of resistors which have the net effect of a resistance of one ohm.

PROBLEM 34E. Solve the special networks in Fig. 34.6 below by any means you wish, and sketch the corresponding squared rectangles. The dual graphs may also be thought of as special networks— what are the squared rectangles that arise from the dual networks?

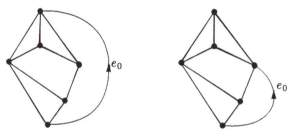

Figure 34.6

Let $N$ be the incidence matrix of a digraph $D$ and $\Delta$ a diagonal matrix indexed by $E(D)$ with $\Delta(e, e) = r(e) > 0$, the resistance of the edge $e$; let $C := N\Delta^{-1}N^{\top}$. If an additional edge $e_0$ is directed from one vertex $y$ to another vertex $x$ of $D$, and a sufficient voltage source is impressed on that single edge so that a current $f(e_0)$ flows through it, then Kirchhoff's laws imply that the potential difference $g(e)$ across an edge $e$ joining vertex $a$ to vertex $b$, say, is given by

$$(34.3) \qquad g(e) = \frac{f(e_0)}{\tau(D)}(xy.ab)$$

where $(xy.ab)$ denotes the cofactor of the element $c_{yb}$ in the cofactor of the element $c_{xa}$ of $C = (c_{ij})$ and $\tau(D)$ denotes the number of spanning trees of $D$ (the complexity of $D$). See Jeans (1908).

A consequence of this is the following theorem.

THEOREM 34.3. *If the voltage source $s(e_0)$ on the distinguished edge $e_0$ of a special network with graph $G$ is taken to be the number of spanning trees on the edge $e_0$, then all values $f(a)$ of the resulting current flow will be integral, and the current flow $f(e_0)$ through the distinguished edge will be equal to the number of spanning trees off the edge $e_0$.*

PROOF: In the discussion preceding the statement of the theorem, take $D$ to be an orientation of $G'_{e_0}$ and $\Delta$ the identity matrix. If the current flow $f(e_0)$ is $\tau(D) = \tau(G'_{e_0})$, then it is clear from (34.3) that all values of $f(a) = g(a)$ are integral. It remains only to show that $-g(e_0)$ is equal to the number of spanning trees *on* the edge $e_0$.

By (34.3), $g(e_0) = (xy.yx) = -(xy.xy)$. By definition, $(xy.xy)$ is the determinant of the matrix $N''N''^{\top}$ where $N''$ consists of $N$ with rows $x$ and $y$ deleted. By the Cauchy-Binet theorem, this is equal to

$$\sum_{S} (\det(N''_S))^2$$

where the sum is extended over all $(n-2)$-subsets $S$ of the edge set and $N''_S$ is the square submatrix of $N''$ whose rows are indexed by elements of $S$.

The proof is completed by the observation that $\det(N''_S)$ is 0 unless $S \cup \{e_0\}$ is the edge set of a spanning tree in $G$, in which case it is $\pm 1$. This is similar to the proof of (34.2) and is left as an exercise. $\qquad \square$

In summary, Theorem 34.3 says that the resistance between the ends of an edge $e_0$ of a graph $G$ when all other edges are thought of as unit resistors, is the ratio $\tau(G''_{e_0})/\tau(G'_{e_0})$. Readers might check that this agrees with their knowledge of electrical networks when $G$ is a polygon or a bond-graph. A squared *square* arises when we find a planar graph $G$ and an edge $e_0$ so that $\tau(G''_{e_0}) = \tau(G'_{e_0})$. This statement alone *does not necessarily help* in the search for squared squares (and electrical engineers will think this would be a waste of resistors anyway). Such graphs appear to be quite rare.

## Notes.

The first appearance in the literature of the problem of using incongruent squares to make a rectangle seems to be M. Dehn (1903).

The first extensive tables of squared squares were published by C. J. Bouwkamp, A. J. W. Duijvestijn, and P. Medema (1960).

W. T. Tutte (1918–) has made numerous significant contributions to graph theory. He is Emeritus Professor of Mathematics in the Department of Combinatorics and Optimization of the University of Waterloo, one of the few academic departments in the world whose name includes the word "combinatorics".

## References.

C. J. Bouwkamp, A. J. W. Duijvestijn, P. Medema (1960), *Tables relating to simple squared rectangles of order nine through fifteen*, T. H. Eindhoven.

A. J. W. Duijvestijn (1978), Simple perfect squared square of lowest order, *J. Combinatorial Theory* (B) **25**, 555–558.

M. Dehn (1903), Zerlegung von Rechtecke in Rechtecken, *Math. Ann.* **57**.

J. H. Jeans (1908), *The Mathematical Theory of Electricity and Magnetism*, Cambridge.

W. T. Tutte (1961), Squaring the square, in: M. Gardner, *The 2nd Scientific American Book of Mathematical Puzzles and Diversions*, Simon and Schuster.

W. T. Tutte (1965), The quest of the perfect square, *Amer. Math. Monthly* **72**, No. 2, 29–35.

# 35
# Pólya theory of counting

We return to counting in this chapter. There are many instances when we are not interested in the number of primitive objects, but rather the number of equivalence classes of objects with respect to an appropriate equivalence relation. Moreover, these equivalence relations are often induced by certain permutation groups in a natural way.

QUESTION 1. What is the number of "essentially different" necklaces which can be made with $n$ beads of two different colors? For $n = 6$, this number is, by inspection, 13.

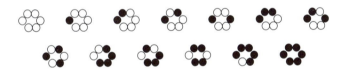

Note that we choose to regard the necklaces

(35.1)                              and

as essentially the same, since one can be reflected through a vertical axis (or flipped over) to get the other. Evidently, underlying our concept of "essentially different" is the dihedral group $D_n$ of automorphisms of the $n$-gon.

QUESTION 2. What is the number of nonisomorphic simple graphs on $n$ vertices? For $n = 4$, there are 11.

QUESTION 3. What is the number of essentially different ways to paint the faces (or edges, or vertices) of the cube with $n$ colors?

QUESTION 4. Given three white balls and one black ball, what is the number of ways to distribute these into two square boxes and one round box?

| | | | |
|---|---|---|---|
| [ooo•] [ ] ( ) | [ooo] [•] ( ) | [oo] [o•] ( ) | [oo•] [o] ( ) |
| [ooo] [ ] (•) | [oo] [o] (•) | [oo•] [ ] (o) | [oo] [•] (o) |
| [o•] [o] (o) | [o•] [ ] (oo) | [o] [•] (oo) | [oo] [ ] (o•) |
| [o] [o] (o•) | [•] [ ] (ooo) | [o] [ ] (oo•) | [ ] [ ] (ooo•) |

Let $A$ and $B$ be finite sets and $G$ a group of permutations of $A$ (or more generally, a finite group acting on $A$). The elements of $B$ will be called *colors*. $G$ acts on the set $B^A$ of mappings $f : A \to B$ when for $\sigma \in G$ and $f \in B^A$, we define $\sigma(f) \in B^A$ by

$$(\sigma(f))(x) := f(\sigma^{-1}(x)).$$

It is the orbits of an appropriate $G$ on the set $B^A$ that we wish to count in the first three questions above.

(The use of $\sigma^{-1}$ rather than $\sigma$ on the righthand side above is not an error. It is required to ensure that $\sigma(\tau(f)) = (\sigma\tau)(f)$, which is necessary in order to have a legitimate "action" of $G$ on $B^A$. Formally, we have a homomorphism of $G$ into the symmetric group on $B^A$.)

In the necklace problem, we would take $A$ to be the vertex set of an $n$-gon, and $G$ as its automorphism group represented as a permutation group on the vertices (i.e. the dihedral group $D_n$ of order $2n$ in its usual representation). A slightly different problem arises when the beads to be used are not round, but, say, flat on one side. Then the necklaces cannot be flipped over and those in (35.1) are regarded as distinct.

Here, we would take $G$ to be the cyclic group of order $n$ in its regular representation; cf. Example 10.5.

Asking for the number of orbits of mappings is the simplest question we can pose. We might further want to know the number of orbits of a given "weight"; for example, the number of necklaces of $n$ beads, $k$ of which are to be white, or the number of graphs on $n$ vertices with $k$ edges. We might introduce a permutation group $H$

on $B$ and, with respect to a more general equivalence relation, ask for the number of "configurations". (This would be necessary for the distribution problem.)

To begin, we recall Burnside's lemma, Theorem 10.5: The number of orbits of a finite group $G$ on a set $X$ is the average number of fixed points:

$$(35.2) \qquad \frac{1}{|G|} \sum_{\sigma \in G} \psi(\sigma),$$

where $\psi(\sigma)$ denotes the number of points of $S$ that are fixed by the permutation $\sigma$. To review, this formula followed from counting ordered pairs $(x, \sigma) \in X \times G$ such that $\sigma(x) = x$.

THEOREM 35.1. *Let $A$ and $B$ be finite sets and let $G$ act on $A$. Denote by $c_k(G)$ the number of permutations in $G$ that have exactly $k$ cycles in their cycle decomposition on $A$. Then the number of orbits of $G$ on the set $B^A$ of all mappings $f : A \to B$ is*

$$\frac{1}{|G|} \sum_{k=1}^{\infty} c_k(G) |B|^k.$$

PROOF: By Burnside's lemma, the number of orbits is given by (35.2) where $\psi(\sigma)$ is here the number of mappings $f : A \to B$ such that $\sigma(f) = f$, i.e. $f(a) = f(\sigma^{-1}(a))$ for all $a \in A$. But a mapping $f$ is fixed by $\sigma$ if and only if $f$ is constant on every cycle of $\sigma$; the reader should check this. Such mappings are obtained by assigning an element of $B$ to each cycle of $\sigma$, and thus if $\sigma$ has $k$ cycles, the number of mappings $f$ fixed by $\sigma$ is $|B|^k$. $\qquad \square$

We encourage the reader to pause at this point and to verify that Theorem 35.1 produces the answer 13 when applied to the necklace counting problem raised as Question 1 above.

Since it is to be used extensively later, it is appropriate to introduce the *cycle index* of a permutation group at this point. First, in this chapter, it will be convenient to use the venerable notation $(1^{k_1} 2^{k_2} \ldots n^{k_n})$ for the partition of the integer $n$ which has $k_i$ parts of size $i$, $i = 1, 2, \ldots, n$. Do not compute powers or multiply; this is just notation.

For a permutation $\sigma$ of a set $A$, let $z_i(\sigma)$ denote the number of cycles of $\sigma$ having length $i$; so $(1^{z_1(\sigma)}2^{z_2(\sigma)}\dots)$ is a partition of $n = |A|$, called the *type* of $\sigma$. Given a group $G$ acting on $A$, we define the cycle index $Z_G$ as a polynomial in $n$ letters $X_1, X_2, \dots, X_n$ by

$$Z_G(X_1, X_2, \dots, X_n) := \frac{1}{|G|}\sum_{\sigma\in G} X_1^{z_1(\sigma)}\cdots X_n^{z_n(\sigma)}.$$

The assertion of Theorem 35.1 is that the number of orbits of $G$ on $B^A$ is

$$Z_G(b, b, \dots, b) = \frac{1}{|G|}\sum_{\sigma\in G} b^{z_1(\sigma)+z_2(\sigma)+\cdots+z_n(\sigma)}$$

where $b := |B|$.

EXAMPLE 35.1. A cyclic group $C_n$ of order $n$ has $\varphi(d)$ elements of order $d$ for each divisor $d$ of $n$. As a permutation in the regular representation of $C_n$, an element of order $d$ has $n/d$ cycles of length $d$. Thus, for the regular representation of $C_n$, we have

$$Z_{C_n}(X_1, \dots, X_n) = \frac{1}{n}\sum_{d|n}\varphi(d)X_d^{n/d}.$$

We can now count the number of "one-sided" necklaces of $n$ beads of two colors. The answer is $Z_{C_n}(2, 2, \dots, 2) = \frac{1}{n}\sum_{d|n}\varphi(d)2^{n/d}$; cf. equation (10.12). For $n = 6$, we get 14, one more than with dihedral equivalence (the two necklaces in (35.1) are now distinguished). For $n = 10$, we get 108.

EXAMPLE 35.2. To find the cycle index of the dihedral group $D_n$ in its usual representation on $n$ points, we must account for the permutations (reflections) outside of its cyclic subgroup of order $n$. There are two cases depending on the parity of $n$:

$$Z_{D_n} = \begin{cases} \frac{1}{2n}(\sum_{d|n}\varphi(d)X_d^{n/d} + nX_1X_2^{\frac{n-1}{2}}) & \text{for } n \text{ odd,} \\ \frac{1}{2n}(\sum_{d|n}\varphi(d)X_d^{n/d} + \frac{n}{2}X_1^2X_2^{\frac{n}{2}-1} + \frac{n}{2}X_2^{\frac{n}{2}}) & \text{for } n \text{ even.} \end{cases}$$

The number of "two-sided" necklaces with $n$ beads of two different colors is $Z_{D_n}(2, 2, \ldots, 2)$. For $n = 6$, we get 13 (verifying the list following Question 1). For $n = 10$, we get 78.

PROBLEM 35A. Compute the cycle index of the group of rotations of the cube represented as permutations of the six faces. (There are 24 rotations—including the identity—that map the cube, as a rigid body in 3-space, onto itself. We are ignoring the 24 reflections of the cube.) What is the number of essentially different ways to paint the faces of the cube in 2 colors? In 3 colors? In $n$ colors?

Before going further, let us write down the cycle index of the symmetric groups. We noted the formula for the number of *partitions* of a given type $(1^{k_1} 2^{k_2} \ldots n^{k_n})$ in equation (13.3). Each block of size $i$ may be equipped with a cyclic permutation in $(i-1)!$ ways, so we have

$$Z_{S_n} = \sum_{(1^{k_1} 2^{k_2} \ldots)} \frac{1}{1^{k_1} 2^{k_2} \ldots n^{k_n} k_1! k_2! \ldots k_n!} X_1^{k_1} X_2^{k_2} \ldots X_n^{k_n}.$$

We list the first several polynomials below:

$$1! Z_{S_1} = X_1,$$
$$2! Z_{S_2} = X_1^2 + X_2,$$
$$3! Z_{S_3} = X_1^3 + 3X_1 X_2 + 2X_3,$$
$$4! Z_{S_4} = X_1^4 + 6X_1^2 X_2 + 3X_2^2 + 8X_1 X_3 + 6X_4,$$
$$5! Z_{S_5} = X_1^5 + 10X_1^3 X_2 + 15X_1 X_2^2 + 20X_1^2 X_3 + 20X_2 X_3$$
$$\qquad + 30X_1 X_4 + 24X_5,$$
$$6! Z_{S_6} = X_1^6 + 15X_1^4 X_2 + 45X_1^2 X_2^2 + 40X_2^3 + 40X_1^3 X_3 + 15X_2^3$$
$$\qquad + 120X_1 X_2 X_3 + 90X_1^2 X_4 + 90X_2 X_4 + 144X_1 X_5 + 120X_6.$$

EXAMPLE 35.3. Theorem 35.1 can be used for the enumeration of graphs. Let $V$ be a fixed set of $n$ vertices and let $E$ consist of all 2-subsets of $V$. Then the simple graphs with vertex set $V$ can be viewed as mappings $f : E \to \{0, 1\}$, the graph corresponding to $f$ having edge set $f^{-1}(1) = \{e : f(e) = 1\}$.

Two graphs on $V$ are isomorphic if and only if there exists a permutation of $V$ taking the edges of one onto the edges of the other. To describe this in a manner suitable for the application of Theorem 35.1, let $S_n$ be the symmetric group on $V$ and let $S_n^{(2)} = \{\sigma^{(2)} : \sigma \in S_n\}$ be the induced group, isomorphic to $S_n$, of permutations of $E$ where

$$\sigma^{(2)} : \{x, y\} \mapsto \{\sigma(x), \sigma(y)\}.$$

Then $S_n^{(2)}$ acts on $\{0, 1\}^E$ and we note that $f, g : E \to \{0, 1\}$ correspond to isomorphic graphs if and only if they lie in the same orbit of $S^{(2)}$ on $\{0, 1\}^E$.

It is not a pleasant task to compute the types of all permutations in $S_n^{(2)}$. For example, if $\sigma \in S_5$ has type $(2^1 3^1)$, then it turns out that $\sigma^{(2)}$ has type $(1^1 3^1 6^1)$ on the 10 edges of $K_5$. The cycle index of $S_5^{(2)}$ is

$$\frac{1}{120}(X_1^{10} + 10X_1^4 X_2^3 + 15X_1^2 X_2^4 + 20X_1 X_3^3$$
$$+ 20X_1 X_3 X_6 + 30X_2 X_4^2 + 24X_5^2).$$

In general, we will have a sum over all partitions of $n$. The number of nonisomorphic graphs on five vertices is found, by substituting 2 for all indeterminates, to be 34.

We now generalize Theorem 35.1 to allow for weights. Let $A$ be an $n$-set, let $G$ act on $A$, and let $B$ be a finite set of colors. Let $R$ be a commutative ring containing the rationals and let $w : B \to R$ assign a *weight* $w(b) \in R$ to each color $b \in B$. For $f : A \to B$, define

$$W(f) := \prod_{a \in A} w(f(a)) \in R.$$

$W(f)$ is the *weight* of the function $f$. Note that two mappings representing the same orbit of $G$ on $B^A$ have the same weight, i.e. $W(\sigma(f)) = W(f)$ for every $\sigma \in G$. The sum

$$\sum_{f \in R} W(f),$$

extended over a system of representatives $R$ for the orbits, is called the *configuration counting series*. (The orbits are often called *configurations*; the term *series* becomes appropriate when the weights are monomials.) If all weights $w(b)$ are 1, then the following theorem reduces to Theorem 35.1.

THEOREM 35.2. *With the terminology as above, the configuration counting series is given by*

$$\sum W(f) = Z_G \left( \sum_{b \in B} w(b), \sum_{b \in B} [w(b)]^2, \dots, \sum_{b \in B} [w(b)]^n \right).$$

PROOF: Let $N = \sum W(f)$, the sum extended over all pairs $(\sigma, f)$ with $\sigma \in G$, $f \in B^A$, and $\sigma(f) = f$. We have

$$N = \sum_{f \in B^A} W(f)|G_f|,$$

where $G_f$ is the stabilizer of $f$. Consider the terms $W(f)|G_f|$ as $f$ ranges over an orbit $\mathcal{O}$ of $G$ on $B^A$: each is equal to $W(f_0)|G|/|\mathcal{O}|$ where $f_0$ is a representative of the orbit $\mathcal{O}$, so these terms sum to $|G|W(f_0)$. It is now clear that $N$ is $|G|$ times the configuration counting series.

On the other hand,

$$N = \sum_{\sigma \in G} \left( \sum_{\sigma(f)=f} W(f) \right).$$

Recalling the definition of $Z_G$, we see that the proof will be complete if we show that

$$\sum_{\sigma(f)=f} W(f) = \left( \sum_{b \in B} w(b) \right)^{k_1} \left( \sum_{b \in B} [w(b)]^2 \right)^{k_2} \cdots \left( \sum_{b \in B} [w(b)]^n \right)^{k_n}$$

whenever $\sigma$ is a permutation of type $(1^{k_1} 2^{k_2} \cdots n^{k_n})$.

A mapping $f : A \to B$ is fixed by $\sigma$ if and only if $f$ is constant on every cycle of $\sigma$ on $A$. Let

$$C_1, C_2, \dots, C_k \quad (k := k_1 + k_2 + \cdots + k_n)$$

be the cycles of $\sigma$. The mappings $f \in B^A$ fixed by $\sigma$ are in one-to-one correspondence with $k$-tuples $(b_1, b_2, \ldots, b_k)$ of elements of $B$ (the corresponding mapping is the one associating $b_i$ to all elements of $C_i$). The weight of the mapping $f$ corresponding to $(b_1, \ldots, b_k)$ is

$$W(f) = \prod_{i=1}^{k} [w(b_i)]^{|C_i|},$$

and summing over all $k$-tuples $(b_1, b_2, \ldots, b_k)$,

$$\sum_{\sigma(f)=f} W(f) = \sum_{b_1, b_2, \ldots, b_k} [w(b_1)]^{|C_1|} [w(b_2)]^{|C_2|} \cdots [w(b_k)]^{|C_k|}$$

$$= \prod_{i=1}^{k} \left( \sum_{b_1} [w(b_1)]^{|C_1|} \right) \left( \sum_{b_2} [w(b_2)]^{|C_2|} \right) \cdots$$

$$= \prod_{c=1}^{n} \left( \sum_{b} [w(b)]^c \right)^{k_c},$$

as required.                                                          □

EXAMPLE 35.4. Consider the necklace problem with cyclic equivalence. Here $G = C_n$, $B = \{black, white\}$. We take $R$ to be the polynomial ring $Q[X]$ and define $w(black) = 1$, $w(white) = X$. Then the weight of a coloring $f : A \to B$ is

$$W(f) = X^k$$

where $k$ is the number of white beads.

Thus the number of essentially different cyclic necklaces with $n$ beads, $k$ of which are white, is the coefficient of $X^k$ in the configuration counting series

$$Z_{C_n}(1 + X, 1 + X^2, 1 + X^3, \ldots, 1 + X^n)$$

$$= \frac{1}{n} \sum_{d|n} \varphi(d)[1 + X^d]^{n/d}$$

$$= \frac{1}{n} \sum_{d|n} \varphi(d) \sum_{r=0}^{n/d} \binom{n/d}{r} x^{rd}.$$

This coefficient is

$$\frac{1}{n} \sum_{d|(k,n)} \varphi(d) \binom{n/d}{k/d}.$$

So if $k$ and $n$ are relatively prime, the number is simply $\frac{1}{n}\binom{n}{k}$, while for $n = 12$, $k = 4$, there are $\frac{1}{12}(\varphi(1)\binom{12}{4} + \varphi(2)\binom{6}{2} + \varphi(4)\binom{3}{1}) = $ 43 necklaces.

PROBLEM 35B. What is the number of essentially different ways to paint the faces of a cube such that one face is red, two are blue, and the remaining three are green?

Let $A$ and $B$ be finite sets, $|A| = n$, and let $G$ and $H$ be finite groups, $G$ acting on $A$ and $H$ on $B$. The direct product $G \times H$ acts on $B^A$ when for $f \in B^A$ and $(\sigma, \tau) \in G \times H$ we define $(\sigma, \tau)(f)$ by

$$((\sigma, \tau)(f))(a) = \tau(f(\sigma^{-1}(a))).$$

THEOREM 35.3. *The number of orbits of $G \times H$ on $B^A$ is*

$$\frac{1}{|H|} \sum_{\tau \in H} Z_G(m_1(\tau), m_2(\tau), \dots, m_n(\tau))$$

*where*

$$m_i(\tau) := \sum_{j|i} j z_j(\tau), \quad i = 1, 2, \dots, n.$$

PROOF: By Burnside's lemma, the number of orbits is

$$\frac{1}{|G||H|} \sum_{\sigma \in G, \tau \in H} \psi(\sigma, \tau)$$

where $\psi(\sigma, \tau)$ is the number of mappings $f \in B^A$ with $(\sigma, \tau)(f) = f$. To complete the proof, it will suffice to show that for each $\tau \in H$,

$$(35.3) \qquad \frac{1}{|G|} \sum_{\sigma \in G} \psi(\sigma, \tau) = Z_G(m_1(\tau), m_2(\tau), \dots, m_n(\tau)).$$

Fix $\sigma \in G$ and $\tau \in H$, and let

$$C_1, C_2, \dots, C_k \quad (k := z_1(\sigma) + z_2(\sigma) + \cdots + z_n(\sigma))$$

be the cycles of $\sigma$ on $A$. A mapping $f : A \to B$ is fixed by $(\sigma, \tau)$ if and only if all the restrictions $f_i := f|C_i$ are fixed, $1 \le i \le k$. So $\psi(\sigma, \tau)$ is the product, over $i$ in the range $1 \le i \le k$, of the numbers of mappings $f_i : C_i \to B$ satisfying

$$f_i(\sigma(a)) = \tau(f_i(a)) \quad \text{for all } a \in C_i.$$

Let $C_i$ have length $\ell$ and fix $a_o \in C_i$. Suppose that $f_i : C_i \to B$ is fixed by $(\sigma|C_i, \tau)$ and that $f_i(a_o) = b$. Then $f_i$ is completely determined: $f_i(\sigma^t(a_o)) = \tau^t(b)$. Moreover, $b = f_i(a_o) = f_i(\sigma^\ell(a_o)) = \tau^\ell(b)$ and so we see that the cycle of $\tau$ containing $b$ must have length $\ell'$, a divisor of $\ell$.

Conversely, if $b$ is an element of $B$ lying in a cycle of $\tau$ having a length dividing $\ell := |C_i|$, then we can define a mapping $f_i$ on $C_i$ by $f_i(\sigma^t(a_o)) := \tau^t(b)$ (check that $f_i$ is well defined) and this $f_i$ is fixed by $(\sigma|C_i, \tau)$.

In summary, the fixed mappings $f_i : C_i \to B$ are equinumerous with the elements of $B$ lying in cycles of $\tau$ of length dividing $|C_i|$. The number of such elements is, of course,

$$m_{|C_i|}(\tau) = \sum_{j \mid |C_i|} j z_j(\tau).$$

Then

$$\psi(\sigma, \tau) = \prod_{i=1}^{k} m_{|C_i|}(\tau) = [m_1(\tau)]^{z_1(\sigma)} [m_2(\tau)]^{z_2(\sigma)} \cdots [m_n(\tau)]^{z_n(\sigma)},$$

from which the desired equation (35.3) is immediate. $\qquad\square$

EXAMPLE 35.5. What is the number of ways to distribute 2 red, 2 yellow, and 4 green balls into 1 round and 3 square boxes? We take

$$A = \{R_1, R_2, Y_1, Y_2, G_1, G_2, G_3, G_4\},$$

$$G = S_2 \times S_2 \times S_4, \quad B = \{r, s_1, s_2, s_3\}, \quad H = S_1 \times S_3.$$

It is easy to see that the cycle index of $G$ is the product of the cycle indices of the symmetric groups of which it is the product, so

$$(35.4) \quad Z_G = Z_{S_2} \cdot Z_{S_2} \cdot Z_{S_4}$$

$$= \frac{1}{2!2!4!}(X_1^2 + X_2)^2(X_1^4 + 6X_1^2 X_2 + 3X_2^2 + 8X_1 X_3 + 6X_4).$$

There are three types of permutations in $H$. If $\tau$ is the identity, then

$$m_1(\tau) = 4, \quad m_2(\tau) = 4, \quad m_3(\tau) = 4, \quad m_4(\tau) = 4;$$

if $\tau$ transposes two of $\{s_2, s_2, s_3\}$, then

$$m_1(\tau) = 2, \quad m_2(\tau) = 4, \quad m_3(\tau) = 2, \quad m_4(\tau) = 4;$$

if $\tau$ fixes only $r$, then

$$m_1(\tau) = 1, \quad m_2(\tau) = 1, \quad m_3(\tau) = 4, \quad m_4(\tau) = 1.$$

By Theorem 35.3, the number of distributions is

$$\frac{1}{3!}\frac{1}{2!2!4!}[(4^2 + 4)^2(4^4 + 6 \cdot 4^2 \cdot 4 + 2 \cdot 4^2 + 8 \cdot 4 \cdot 4 + 6 \cdot 4)$$
$$+ 3(2^2 + 4)^2(2^4 + 6 \cdot 2^2 \cdot 4 + 3 \cdot 4^2 + 8 \cdot 2 \cdot 2 + 6 \cdot 4)$$
$$+ 2(1^2 + 1)^2(1^4 + 6 \cdot 1^2 \cdot 1 + 3 \cdot 1^2 + 8 \cdot 1 \cdot 4 + 6 \cdot 1)]$$
$$= 656 \text{ (if we did the arithmetic correctly)}.$$

PROBLEM 35C. Given $G$ on $A$, $H$ on $B$, find an expression for the number of orbits of injective mappings $f : A \to B$.

We give, without proof, a statement of the extension of Theorem 35.3 to include weights. See De Bruijn (1964) for a proof. If all weights are 1, then this reduces to Theorem 35.3.

THEOREM 35.4. *Let $G$ act on $A$, $H$ on $B$, let $R$ be a commutative ring and $w : B \to R$. Assume that $w$ is constant on each orbit of $H$ on $B$. With $W(f) := \prod_{a \in A} w[f(a)]$, the sum $\sum W(f)$, extended over a system of representation for the orbits of $G \times H$ on $B^A$, is equal to*

$$\frac{1}{|H|} \sum_{\tau \in H} Z_G(M_1(\tau), M_2(\tau), \ldots, M_n(\tau))$$

*where*

$$M_i(\tau) = \sum_{\tau^i(b) = b} [w(b)]^i.$$

EXAMPLE 35.6. We continue Example 13.5. Take

$$w(r) = r, \quad w(s_1) = w(s_2) = w(s_3) = s,$$

where $R = \mathbb{Q}[r, s]$. The number of distributions with $t$ balls in the round box and the remaining $8 - t$ in square boxes is the coefficient of $r^t s^{8-t}$ in

$$\frac{1}{6}\Big(Z_G(r + 3s, r^2 + 3s^2, r^3 + 3s^3, r^4 + 3s^4)$$

$$+ 3Z_G(r + s, r^2 + 3s^2, r^3 + s^3, r^4 + 3s^4) + 2Z_G(r, r^2, r^3 + 3s^3, r^4)\Big)$$

where $Z_G$ is the polynomial in (35.4).

PROBLEM 35D. Identify necklaces of two types of colored beads with their *duals* obtained by switching the colors of the beads. (We can now distinguish between the two colors, but we can't tell which is which.) Now how many of these reduced configurations are there? (For example, with $n = 6$ and dihedral equivalence, there are 8 distinct configurations.) How many necklaces are selfdual?

* * *

The theorems of this chapter are also sources of algebraic identities.

Consider the case of "distribution of like objects into unlike cells". (We are recalling something which was discussed in Chapter 13.) For example, in how many ways can we distribute 23 apples to Fred, Jane, and George? More generally, consider the distributions of a set $A$ of $n$ apples to a set $B$ of $x$ individuals (the elements of $A$ are "like", whereas the elements of $B$ remain distinct, i.e. are "unlike"). Such a distribution amounts to the selection of a family $(k_b : b \in B)$ of nonnegative integers $k_b$ with $\sum_{b \in B} k_b = n$, $k_b$ apples going to individual $b$.

Thus the number of distributions, as shown in Theorem 13.3, is

$$\binom{n + x - 1}{n}.$$

However, the best formal definition of such a distribution is as an orbit of mappings $f : A \to B$ with respect to the symmetric group $S_n$ acting on $A$. (Two mappings $f, g \in B^A$ determine the same distribution if and only if they "differ" by a permutation of the apples.) By Theorem 35.1, the number of distributions is

$$\frac{1}{n!} \sum_{k=0}^{n} c_k(S_n) x^k,$$

where $c_k(S_n)$ is the number of permutations of $n$ letters with exactly $k$ cycles. In Chapter 13, the numbers $c_k(S_n)$ were called the signless Stirling numbers of the first kind, and denoted by $c(n, k)$. By Theorem 35.2, for every nonnegative integer $x$,

$$(x + n - 1)_{(n)} = \sum_{k=0}^{n} c_k(S_n) x^k,$$

and hence this must hold as a polynomial identity. We have therefore re-proved the formula (13.7).

Take $G$ to be $S_n$ on an $n$-set $A$, let $B$ be finite and let $w$ be the insertion of $B$ into the polynomial ring $\mathbb{Q}[B]$. Here the weight of a mapping $f : A \to B$ is a monomial and two mappings are equivalent, that is, represent the same configuration, if and only if they have the same weight. The configuration counting series

$$(35.5) \qquad \sum W(f) = Z_{S_n} \left( \sum_{b \in B} b, \sum_{b \in B} b^2, \ldots, \sum_{b \in B} b^n \right)$$

expresses the so-called homogeneous-product-sum symmetric functions as a polynomial in the power-sum symmetric functions. For example, with $n = 3$ and $B = \{X, Y\}$,

$$6(X^3 + X^2Y + XY^2 + Y^3) =$$
$$(X + Y)^3 + 3(X + Y)(X^2 + Y^2) + 2(X^3 + Y^3).$$

With $n = 4$ and $B = \{X, Y, Z\}$, (35.5) expresses

$$\sum_{i+j+k=4} X^i Y^j Z^k$$

as a polynomial in

$$X + Y + Z, \ X^2 + Y^2 + Z^2, \ X^3 + Y^3 + Z^3, \ \text{and} \ X^4 + Y^4 + Z^4.$$

**Notes.**

We remark that $Z_{S_n}(X_1, \dots, X_n)$ is the coefficient of $Y^n$ in

$$\exp(X_1 Y + \frac{1}{2} X_2 Y^2 + \frac{1}{3} X_3 Y^3 + \cdots)$$

as an element of $(\mathbb{Q}[X_1, X_2, \dots])[[Y]]$.

The answers provided by Pólya theory have the advantage (over inclusion-exclusion, say) that the formula produced is the sum of positive terms rather than an alternating sum.

G. Pólya (1887–1985) was a Hungarian mathematician perhaps best known for his book *Problems and Theorems in Analysis* written in 1924 with G. Szegő. It is still a classic. His book *How to Solve It* sold more than one million copies. He wrote papers on number theory, complex analysis, combinatorics, probability theory, geometry and mathematical physics.

**References.**

N. G. de Bruijn (1964), Pólya's theory of counting, in: E. F. Beckenbach (ed.), *Applied Combinatorial Mathematics*, Wiley.

F. Harary and E. D. Pulver (1966), The power group enumeration theorem, *J. Combinatorial Theory* **1**.

# 36
## Baranyai's theorem

In this chapter we shall give an elegant application of the integrality theorem on flows to a problem in "combinatorial design".

EXAMPLE 36.1. Suppose we have been entrusted to draw up a schedule for the "Big Ten" football teams. Each weekend they are to divide into 5 pairs and play. At the end of 9 weeks, we want every possible pair of teams to have played exactly once.

Here is one solution. Put 9 teams on the vertices of a regular 9-gon and one in the center. Start with the pairing indicated below and obtain the others by rotating the figure by multiples of $2\pi/9$.

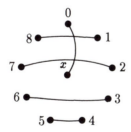

Explicitly, here is the schedule:

| $x$•—•0 | $x$•—•1 | $x$•—•2 | $x$•—•3 | $x$•—•4 | $x$•—•5 | $x$•—•6 | $x$•—•7 | $x$•—•8 |
|---|---|---|---|---|---|---|---|---|
| 1•—•8 | 2•—•0 | 3•—•1 | 4•—•2 | 5•—•3 | 6•—•4 | 7•—•5 | 8•—•6 | 0•—•7 |
| 2•—•7 | 3•—•8 | 4•—•0 | 5•—•1 | 6•—•2 | 7•—•3 | 8•—•4 | 0•—•5 | 1•—•6 |
| 3•—•6 | 4•—•7 | 5•—•8 | 6•—•0 | 7•—•1 | 8•—•2 | 0•—•3 | 1•—•4 | 2•—•5 |
| 4•—•5 | 5•—•6 | 6•—•7 | 7•—•8 | 8•—•0 | 0•—•1 | 1•—•2 | 2•—•3 | 3•—•4 |

There are many other ways to do this. For example, start with the initial pairing shown below.

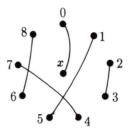

A perfect matching in a graph is also called a *1-factor*. A partition of the edge set of a graph into 1-factors is a *1-factorization*. In the above example, we constructed 1-factorizations of $K_{10}$. There are 396 nonisomorphic 1-factorizations of $K_{10}$, a computer result of E. N. Gelling (1973). See also Mendelsohn and Rosa (1985).

EXAMPLE 36.2. Let $\Gamma$ be any abelian group of odd order. Consider the complete graph on vertex set $\Gamma \cup \{\infty\}$. For $g \in \Gamma$, let

$$\mathcal{M}_g := \{\{g, \infty\}\} \cup \{\{a, b\} : a + b = 2g, a \neq b\}.$$

Then $\{\mathcal{M}_g : g \in \Gamma\}$ is a 1-factorization of the complete graph.

We consider the following generalization of this problem: let us use the term *parallel class* of $k$-subsets of an $n$-set to mean a set of $n/k$ $k$-subsets which partition the $n$-set. Can the set of all $k$-subsets be partitioned into parallel classes of $k$-subsets? Of course, this necessitates that $k$ divide $n$. The number of parallel classes required would be $\frac{k}{n}\binom{n}{k} = \binom{n-1}{k-1}$.

This was not hard for $k = 2$. It is much harder for $k = 3$, but was done by R. Peltesohn in 1936, and for $k = 4$ by J.-C. Bermond (unpublished). The reader might try to find appropriate parallel classes for $n = 9$, $k = 3$. It is certainly not clear that our generalization always admits a solution and so it was surprising when Zs. Baranyai proved Theorem 36.1 below in 1973. All known proofs use a form or consequence of Theorem 7.2. The proof we give is due to A. E. Brouwer and A. Schrijver (1979).

THEOREM 36.1. *If $k$ divides $n$, the set of all $\binom{n}{k}$ $k$-subsets of an $n$-set may be partitioned into disjoint parallel classes $\mathcal{A}_i$, $i = 1, 2, \ldots, \binom{n-1}{k-1}$.*

PROOF: In this proof, we will use the term *m-partition* of a set $X$ for a multiset $\mathcal{A}$ of $m$ pairwise disjoint subsets, some of which may be empty, of $X$ whose union is $X$. (The normal use of "partition" forbids the empty set, but it is important here to allow it to occur, perhaps with a multiplicity, so that the total number of subsets is $m$.)

In order to get an inductive proof to work, we prove something seemingly stronger than the original statement. Let $n$ and $k$ be given, assume $k$ divides $n$, and let $m := n/k$, $M := \binom{n-1}{k-1}$. We assert that for any integer $\ell$, $0 \le \ell \le n$, there exists a set

$$\mathcal{A}_1, \mathcal{A}_2, \ldots, \mathcal{A}_M$$

of $m$-partitions of $\{1, 2, \ldots, \ell\}$ with the property that each subset $S \subseteq \{1, 2, \ldots, \ell\}$ occurs in exactly

$$(36.1) \qquad \binom{n-\ell}{k-|S|}$$

of the $m$-partitions $\mathcal{A}_i$. (The binomial coefficient above is interpreted as zero if $|S| > k$, of course, and for $S = \emptyset$, the $m$-partitions containing $\emptyset$ are to be counted with multiplicity equal to the number of times the empty set appears.)

Our assertion will be proved by induction on $\ell$. Notice that it is trivially true for $\ell = 0$ where each $\mathcal{A}_i$ will consist of $m$ copies of the empty set. Also notice that the case $\ell = n$ will prove Theorem 36.1, since the binomial coefficient in (36.1) is then

$$\binom{0}{k-|S|} = \begin{cases} 1 & \text{if } |S| = k, \\ 0 & \text{otherwise.} \end{cases}$$

REMARK. This somewhat technical statement is not really more general, but would follow easily from Theorem 36.1. If $M$ parallel classes exist as in the statement of Theorem 36.1, then for any set $L$ of $\ell$ points of $X$, the intersections of the members of the parallel classes with $L$ will provide $m$-partitions of $L$ with the above property.

Assume for some value of $\ell < n$ that $m$-partitions $\mathcal{A}_1, \ldots, \mathcal{A}_M$ exist with the required property. We form a transportation network as follows. There is to be a source vertex $\sigma$, another named $\mathcal{A}_i$

for each $i = 1, 2, \ldots, M$, another vertex named $S$ for every subset $S \subseteq \{1, 2, \ldots, \ell\}$, and a sink vertex $\tau$. There is to be a directed edge from $\sigma$ to each $\mathcal{A}_i$ with capacity 1. There are to be directed edges from $\mathcal{A}_i$ to the vertices corresponding to members of $\mathcal{A}_i$ (use $j$ edges to $\emptyset$, if $\emptyset$ occurs $j$ times in $\mathcal{A}_i$); these may have any integral capacity $\geq 1$. There is to be a directed edge from the vertex corresponding to a subset $S$ to $\tau$ of capacity

$$\binom{n - \ell - 1}{k - |S| - 1}.$$

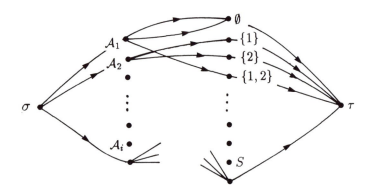

We exhibit a flow in this network: assign a flow value of 1 to the edges leaving $\sigma$, a flow value of $(k - |S|)/(n - \ell)$ to the edges from $\mathcal{A}_i$ to each of its members $S$, and a flow value of $\binom{n-\ell-1}{k-|S|-1}$ to the edge from $S$ to $\tau$. This is easily checked to be a flow: the sum of the values on edges leaving a vertex $\mathcal{A}_i$ is

$$\sum_{S \in \mathcal{A}_i} \frac{k - |S|}{n - \ell} = \frac{1}{n - \ell}\left(mk - \sum_{S \in \mathcal{A}_i} |S|\right) = \frac{1}{n - \ell}(mk - \ell) = 1.$$

The sum of the values on the edges into a vertex $S$ is

$$\sum_{i : S \in \mathcal{A}_i} \frac{k - |S|}{n - \ell} = \frac{k - |S|}{n - \ell}\binom{n - \ell}{k - |S|} = \binom{n - \ell - 1}{k - |S| - 1}.$$

Since all the edges leaving $\sigma$ are saturated, this is a maximum flow and has strength $M$. The edges into $\tau$ are also saturated in this, and hence in any, maximum flow.

By Theorem 7.2, this network admits an *integral*-valued maximum flow $f$. All edges leaving $\sigma$ will be saturated, so it is clear that for each $i$, $f$ assigns the value 1 to one of the edges leaving $\mathcal{A}_i$ and 0 to all others. Say $f$ assigns 1 to the edge from $\mathcal{A}_i$ to its member $S_i$. For each subset $S$, the number of values of $i$ such that $S_i = S$ is $\binom{n-\ell-1}{k-|S|-1}$.

Finally, we obtain a set of $m$-partitions $\mathcal{A}_1', \ldots, \mathcal{A}_M'$ of the set $\{1, 2, \ldots, \ell+1\}$ by letting $\mathcal{A}_i'$ be obtained from $\mathcal{A}_i$ by replacing the distinguished member $S_i$ by $S_i \cup \{\ell+1\}$, $i = 1, \ldots, M$. Readers should check that each subset $T$ of $\{1, 2, \ldots, \ell+1\}$ occurs exactly

$$\binom{n-(\ell+1)}{k-|T|}$$

times among $\mathcal{A}_1', \ldots, \mathcal{A}_M'$. This completes the induction step. $\square$

PROBLEM 36A. Let $A = ((a_{ij}))$ be an $n$ by $m$ matrix of real numbers such that the rowsums and columnsums of $A$ are all integers. Prove that there exists an *integral* $n$ by $m$ matrix $B = ((b_{ij}))$ with the *same* rowsums and columnsums and such that for each $i, j$, either $b_{ij} = \lfloor a_{ij} \rfloor$ or $b_{ij} = \lceil a_{ij} \rceil$.

PROBLEM 36B. Let $v$ and $u$ be integers with $v \geq 2u$ and $v$ even, and consider the complete graph $K_u$ as a subgraph of $K_v$. Suppose the edges of $K_u$ are colored with $v-1$ colors so that distinct edges of the same color are disjoint. Show that this coloring can be extended to a coloring of $E(K_v)$ with $v-1$ colors so that edges of the same color are disjoint. (Such a coloring of $E(K_v)$ is equivalent to a 1-factorization of $K_v$, with the edges of any given color forming a 1-factor.)

## Notes.

Zsolt Baranyai (1948–1978) was a Hungarian mathematician who was also a professional recorder player. He toured Hungary with the *Barkfark Consort* giving concerts and died in a car accident on a country road after one of them. His mathematical work included many other wonderful results on "complete uniform hypergraphs".

One-factorizations of complete graphs are related to symmetric Latin squares—see Chapter 17. The result of Problem 36B is due to Cruse (1974). Generalizations may be found in Baranyai and Brouwer (1977).

**References.**

Zs. Baranyai and A. E. Brouwer (1977), Extension of colourings of the edges of a complete (uniform hyper) graph, *Math. Centrum Dep. Pure Math. ZW.* **91**, 10 pp.

A. E. Brouwer and A. Schrijver (1979), Uniform hypergraphs, in: A. Schrijver (ed.), *Packing and Covering in Combinatorics*, Mathematical Centre Tracts **106**, Amsterdam.

A. Cruse (1974), On embedding incomplete symmetric Latin squares, *J. Combinatorial Theory* (A) **16**, 18–22.

E. N. Gelling (1973), On 1-factorizations of the complete graph and the relationship to round robin schedules, M.Sc. Thesis, University of Victoria.

E. Mendelsohn and A. Rosa (1985), One-factorizations of the complete graph—a survey, *Journal of Graph Theory* **9**.

R. Peltesohn (1936), Das Turnierproblem für Spiele zu je dreien, Dissertation Berlin, August Pries, Leipzig.

# Appendix 1

## Hints and comments on problems

PROBLEM 1A. Show that the $\binom{5}{2}$ pairs from $\{1, \ldots, 5\}$ can be used to label the vertices in such a way that a simple rule determines when there is an edge. To find the full automorphism group, consider the subgroup that fixes a vertex and its three neighbors. This graph is known as the *Petersen graph*. It was first studied by the Danish mathematician J. P. C. Petersen (1839–1910). See also Chapter 21.

PROBLEM 1B. Let the vertex set $V$ be the disjoint union of $V_1$ and $V_2$, with no edges from $V_1$ to $V_2$. What is the maximum number of edges possible?

PROBLEM 1C. (i) Consider a vertex of degree 1; use induction. (ii) A circuit has as many edges as vertices; if the graph is connected, every vertex has degree $\geq 1$.

PROBLEM 1D. Call the vertices $a_1, a_2, a_3$ and $b_1, b_2, b_3$. First, omit $a_3$ and show that there is only one way to draw the graph with the remaining six edges in the plane. A better way is to use Euler's formula (see Chapter 32).

PROBLEM 1E. Each color occurs an even number of times in a circuit. See J. A. Bondy, Induced subsets, *J. Combinatorial Theory* (B) **12** (1972), 201–202.

PROBLEM 1F. Consider the vertices at distance 2 from a fixed vertex. See Fig. 1.4.

PROBLEM 2A. By "backwards induction" on $k$, show that the set $\{\{b_i, a_i\} : i = k, \ldots, n-1\}$ is the edge set of a tree.

PROBLEM 2B. See Chapter 9 of J. H. van Lint, *Combinatorial Theory Seminar Eindhoven University of Technology*, Lecture Notes in Mathematics **382**, Springer-Verlag, 1974.

PROBLEM 2C. One approach is as follows. Show that $T_n$ has the property that for each of its edges $a$, $a$ has ends in different components of $G : \{e \in E(G) : c(e) < c(a)\}$. Then show that any spanning tree $T$ with this property is a cheapest spanning tree.

PROBLEM 3A. It is possible to use the same type of argument as given for $K_6$ (below this problem). Show that there must be a vertex that is in two monochromatic triangles. Then, argue on the remaining six vertices. The argument will be long and tricky. The reader who has cheated and has read on, knows the Corollary to Theorem 3.2.

PROBLEM 3B. Call the terms, both even, on the righthand side $n_1$ and $n_2$. Suppose equality holds and consider a graph with $n_1 + n_2 - 1$ vertices. If there is no red $K_p$ and no blue $K_q$, what is the red-degree of any vertex? How many red edges are there in this graph?

PROBLEM 3C. In $\mathbb{Z}_{17}$ consider $\pm 2^i$, with $i = 0, 1, 2, 3$. For the other problem use $\mathbb{Z}_{13}$. Apply (3.4).

PROBLEM 3D. (a) Fix a vertex and consider all edges to, respectively from, this vertex. Pick the larger set and use induction.

(b) Use the probabilistic method. A solution can be found in Theorem 1.1 of P. Erdős and J. Spencer (1974), where this problem is used to illustrate the method.

PROBLEM 3E. Take as vertices $\{1, 2, \ldots, n\}$. Color $\{i, j\}$ with the color of $|i - j|$. To avoid a triple for two colors, if 1 is red, then 2 must be blue and hence 4 must be red, etc. $N(2) = 5$. Separate two such 2-colored configurations by four numbers with a third color to estimate $N(13)$. See I. Schur, Über die Kongruenz $x^m + y^m \equiv z^m$ (mod $p$), *Jber. Deutsche Math. Ver.* **25** (1916), 114–116.

PROBLEM 3F. Color with four "colors" as follows:

$$\{i, j\} \mapsto (a_{ij}, a_{ji}) \quad (i < j).$$

PROBLEM 4A. (i) Color $K_{10}$ in such a way that red corresponds to an edge of $G$, blue to a nonedge. There are triangles with 3 red edges, 2 red and 1 blue, 2 blue and 1 red, and finally with 3 blue edges. Let $a_i$ ($i = 1, 2, 3, 4$) be the numbers of these. Set up a system of equations and inequalities for these numbers, expressed in the degrees of the vertices of $G$. These should show that there are at least four triangles in $G$ and equality can then be excluded, again by looking at the equations.

(ii) A second solution is as follows. Show there is a triangle. Then consider a triangle, the complementary 7-set, and edges in between. This gives a number of new triangles of two types; estimate it.

PROBLEM 4B. Use induction. First review the proof of Theorem 4.1.

PROBLEM 4C. The number of triangles on an edge $a$ with ends $x$ and $y$ is at least $\deg(x) + \deg(y) - n$. Sum this over all edges $a$.

PROBLEM 4D. Fix any vertex $x$ and let $D_i$ denote the set of vertices at distance $i$ from $x$ in $G$.

For $g = 2t + 1$, $|V(G)| \geq 1 + r + r(r-1) + \cdots + r(r-1)^{t-1}$ since it is easy to see that $|D_i| = r(r-1)^{i-1}$ for $i = 1, 2, \ldots, t$.

For $g = 2t$, we still have $|D_i| = r(r-1)^{i-1}$ for $i = 1, 2, \ldots, t-1$. We can also get a lower bound $|D_t| \geq \frac{r-1}{r} D_{t-1}$ since each vertex $y \in D_{t-1}$ is adjacent to $r - 1$ vertices in $D_t$ while a vertex $y \in D_t$ can of course be adjacent to at most $r$ vertices in $D_{t-1}$.

PROBLEM 4E. The graph is bipartite.

PROBLEM 4F. See L. Lovász (1979).

PROBLEM 5A. (i) Given $A \subseteq X$, one side of a bipartition, count the number of edges that have one end in $A$ and the other in $\Gamma(A)$.

(ii) A trivalent graph with no perfect matching is the graph on four vertices and six edges with one vertex incident with three non-loops and loops incident with the other three vertices. Replace the loops by appropriate simple graphs.

PROBLEM 5B. Construct a sequence of sets $A_i$ of size $m_i$ using new elements only when it is necessary to keep condition $H$ satisfied. A solution is given on p. 41 of Van Lint (1974).

PROBLEM 5C. Use induction on the number of nonzero entries of the matrix. The argument of Theorem 5.5 can be used similarly.

PROBLEM 5D. Theorem 5.5.

PROBLEM 6A. Define a poset using both the index and the size of the integers $a_i$.

PROBLEM 6B. (i) Consider $A$ and $\overline{A}$. (ii) Let $x$ be in all the sets.

PROBLEM 6C. Show that large sets can be replaced by their complements, and apply Theorem 6.5.

PROBLEM 6D. (i) No two of the subsets are in the same chain. (ii) Use induction.

PROBLEM 6E. See C. Greene and D. Kleitman, Strong versions of Sperner's Theorem, *J. Combinatorial Theory* (A) **20** (1976), 80–88.

PROBLEM 7A. This is a straightforward application of the algorithm. The minimum cut capacity is 20.

PROBLEM 7B. Given a flow $f$ with nonzero strength, show that the edges $e$ with $f(e) \neq 0$ contain the edges of a simple directed path from $s$ to $t$. Subtract a scalar multiple of the corresponding elementary flow to obtain a flow $f'$ which vanishes on more edges than $f$ and proceed by induction.

Four vertices suffice for the required example.

PROBLEM 7C. Consider a maximum flow $f$. From Equation (7.1), edges from $X_i$ to $Y_i$ are saturated and edges from $Y_i$ to $X_i$ have zero flow value.

PROBLEM 7D. Given the bipartite graph $G$ with vertices $X \cup Y$, construct a network by adding two vertices $s$ and $t$, edges directed from $s$ to elements of $X$ of capacity 1, and edges directed from elements of $Y$ to $t$ of capacity 1. Direct all original edges from $X$ to $Y$; it is convenient to assign them a large capacity, say $|X| + 1$, so that no such edges are present in a minimum cut. Explain why there exists a complete matching from $X$ to $Y$ in $G$ if and only if this network has maximum flow strength $X$.

PROBLEM 8A. Assume that some subsequence of $n$ consecutive elements occurs twice. Show that $\alpha$ satisfies an equation of degree less than $n$.

PROBLEM 8B. The vertices of the digraph are the ordered pairs from $\{0, 1, 2\}$; the edges are the 27 ordered triples.

PROBLEM 9A. Apply the algorithm.

PROBLEM 9B. A trivial addressing of length $n$ as in Theorem 9.2 is easily found. What is the diameter of the graph?

PROBLEM 9C. Use Theorems 9.1 and 9.5. Somewhere in the calculation $\sum k \cos(kx)$ will appear. This is the derivative of $\sum \sin(kx)$ and that sum is determined by multiplying by $\sin(\frac{1}{2}x)$.

PROBLEM 10A. Use Theorem 10.1, where $E_i$ $(i = 1, 2, 3, 4)$ are the integers $\leq 1000$ divisible by 2,3,5,7, respectively.

PROBLEM 10B. Use inclusion-exclusion and the fact that if $f(i) = 0$, then $(x - i)$ divides $f(x)$.

PROBLEM 10C. Use $\lfloor x \rfloor = \sum_{k \leq x} 1$ and Theorem 10.4.

PROBLEM 10D. Multiply $\sum a_n n^{-s}$ and $\sum b_m m^{-s}$. Determine the coefficient of $k^{-s}$, and use Theorem 10.3.

PROBLEM 10E. Determine $\sum_{d|n} \log f_d(z)$; use Theorem 10.4.

PROBLEM 10F. There are clearly $2n + 1$ colorings. Use inclusion-exclusion where a set $E_i$ consists of colorings with $i$ red and $i - 1$ blue. To determine $N_j$, use Example 10.6.
　　For a direct solution, calculate

$$\sum_{n=0}^{\infty} \sum_{k=0}^{n} (-1)^k \binom{2n - k}{k} 2^{2n-2k} x^{2n}.$$

To do this, use (10.6) and the fact that $\sum (2n + 1) x^{2n} = (\frac{x}{1-x^2})'$.

PROBLEM 10G. Form a permanent and expand by the first row. The Fibonacci sequence has occurred before.

PROBLEM 11A. Consider an all-one matrix of size $n$ by $k$. Adjoin a column of 0's and then complete to a square matrix by adding rows of type $(0, \dots, 0, 1)$.

PROBLEM 11B. (i) Theorem 5.3; (ii) Theorem 11.5; (iii) use a direct product of matrices $J_k$.

PROBLEM 11C. Determine $A = (a_{ij})$ with $a_{ij} = |A_i \cap B_j|$. To find the permanent of this matrix, compare with the *problème des ménages*.

PROBLEM 12A. Use Theorem 12.1.

PROBLEM 12B. Suppose $AA^\top$ is decomposable. Explain the block of 0 entries by considering some row of $A$, its nonzero entries, and the inner products with other rows.

PROBLEM 12C. Do four averaging operations on pairs of rows and similarly for columns. Note that the element $a_{55}$ does not change during these operations. Determine the permanent and minimize. This shows that the matrix was $\frac{1}{5}J$ from the start.

PROBLEM 13A. A linear sequence can be made circular except if it has $a_1 = a_n = 1$.

PROBLEM 13B. (i) Multiply two powers of $(1 - x)$. (ii) Let $k$ be in position $a + 1$; what can precede this term?

PROBLEM 13C. (i) Choose $A_1$ and then choose $A_2$ as a subset of its complement; determine the sum of products of binomial coefficients. (ii) Consider a 2 by $n$ matrix for which the rows are the characteristic functions of $A_1$, respectively $A_2$. Translate the condition on $A_1$ and $A_2$ to a condition on this matrix.

PROBLEM 13D. First, fix the union $U$ of the sets in $\mathcal{A}$. Apply Example 13.6 and then sum over all choices of $U$. The answer shows that we could have found a better solution. To find this better solution, again depict $\mathcal{A}$ by a (0,1)-matrix of size $k$ by $n$. To this matrix adjoin a "special" row, namely the characteristic function of the union of the sets of $\mathcal{A}$. Calculate the number of these (0,1)-matrices that have a 1 in a specified position in their "special" row.

PROBLEM 13E. In each path, consider the *last* step of type $(x, y) \rightarrow (x + 1, y + 1)$. Use Theorem 13.1. For the last question, insert $k$

balls into $n+k$ possible positions; let $x_k$ be the length of the interval preceding the $k$-th ball.

PROBLEM 13F. (i) Prove by induction. If a permutation of 1 to $n-1$ is written as a product of cycles, there are $n$ ways to adjoin the element $n$, one of which has one more cycle. (ii) In Theorem 13.7, sum both sides over $k$; use (13.5).

PROBLEM 13G. Insert (13.13) into (13.11) and change the order of summation.

PROBLEM 14A. The sequences end in 10 or in 1. Establish the Fibonacci recursion and solve by trying $a_n = t^n$ as a solution (use linearity). The method of Example 14.3 also can be used.

PROBLEM 14B. See p. 25. of J. H. van Lint *Combinatorial Theory Seminar, Eindhoven University of Technology*, Lecture Notes in Mathematics **382**, Springer–Verlag, 1974.

PROBLEM 14C. Fix one edge; number others cyclicly: $a, b, c, \dots$ . Give the chords the obvious names; be careful about the order in which these names are given. See also p. 25. of Van Lint (1974), *ibid*.

PROBLEM 14D. Use the method of Fig. 2.2. One fixed point produces an arborescence; the remainder is a mapping with no fixed points. See (14.15).

PROBLEM 14E. Fix an edge and then consider the quadrilateral that it belongs to. The other three sides divide the $(n+1)$-gon into possibly empty polygons; see (14.10). Solve $f = x + f^3$ by applying Theorem 14.3.

A combinatorial proof that the number of ways of dividing an $(n+1)$-gon into quadrangles by nonintersecting diagonals is 0 if $n$ is even, and $\frac{1}{3k+1}\binom{3k+1}{k}$ if $n = 2k+1$, can be given as follows. Since the case of even $n$ is trivial, we assume $n = 2k+1$. There are $k-1$ diagonals and $k$ quadrangles in the dissection. Put a vertex in each quadrangle and on each edge; fix one edge-vertex as a root. Vertices of quadrangles are joined by an edge if the quadrangles share a diagonal; an edge-vertex is joined to the quadrangle-vertex of its quadrangle. Describe a walk around the tree, that was just defined, by a sequence of symbols $x$ and $y$, where we use an $x$ every time we reach a quadrangle vertex not visited before, and similarly a $y$ for edge-vertices. We find a sequence of $k$ symbols $x$ and $2k+1$ symbols $y$. We know, see Example 10.6, that there are $\frac{1}{3k+1}\binom{3k+1}{k}$ *circular* sequences of this type. It remains to show a one-one correspondence between these circular sequences and the sequences describing the trees. Describe the $x, y$ sequence as follows: an $x$ corresponds to a step in the $X$-$Y$ plane one unit to the right and one unit upwards; a $y$ to a step one unit to the right and $k/(2k+1)$ steps downward. The walk starts in $(0,0)$ and ends in $(3k+1,0)$, never meeting the $X$-axis in between. If we describe a circular sequence of the given type in the same way, it will have a *unique* minimum because $(k, 2k+1) = 1$. This will give us the unique starting point to make it correspond to one of the trees.

PROBLEM 14F. See Appendix 2 on formal operations. Take logarithms and use Theorem 10.3.

PROBLEM 14G. Show equivalence with Example 14.11.

PROBLEM 14H. Make a path as in Example 14.8 and find a lowest point.

PROBLEM 14I. (i) See (14.10). (ii) Put a vertex on the circle, say between 1 and $2n$. This will be the root of a tree. Put a vertex in each region and join vertices in the obvious way. Show equivalence with one of the Catalan problems. See Van Lint (1974), p.26.

PROBLEM 14J. A labeled regular graph of valency 2 is a union of (labeled) polygons. Apply Theorem 14.2.

PROBLEM 15A. Distinguish between $x_k = 1$ and $x_k > 1$, in which case take $y_i = x_i - 1$. Use induction.

PROBLEM 15B. Let $E$, $I$, and $S$ denote the number of equilateral, isosceles, and scalene triangles, respectively. Calculate $E$, $I + E$, and express $\binom{n}{3}$ in $E$, $I$, and $S$.
   (This proof is due to J. S. Frame, *Amer. Math. Monthly* **47**, (1940), 664.)

PROBLEM 15C. The denominator of the generating function factors into linear terms of the form $(1 - \zeta x)$ where $\zeta$ is a root of unity.

Use partial fractions. The constant $c$ is the coefficient of $(1-x)^{-t}$.

PROBLEM 15D. See the argument in the paragraph preceding the problem. For any three consecutive blue balls with red balls on either side, i.e. for a fixed triple, corresponding to one of the 3's to be counted, count the number of ways of inserting more red balls on either side to form a composition. Sum over all possible choices of the three consecutive blue balls; all the terms in this sum, except those corresponding to the initial and last three balls, are the same.

A second solution by induction is possible. Consider the case $n=3$ first.

PROBLEM 15E. Leave out the first column.

PROBLEM 15F. For the unequal odd parts, use "hook" shaped figures to form a Ferrers diagram and read this in the usual way.

PROBLEM 15G. Show a correspondence with Example 14.8.

PROBLEM 16A. Use induction; compare two initial segments of $\mathbf{r}$ and $\mathbf{s}$.

PROBLEM 16B. Use Problem 16A and switch edges. Another possibility is to describe the graph by its adjacency matrix and use the same idea as in the proof of Theorem 16.2.

PROBLEM 16C. Estimate the number of solutions of $s_1 + \cdots + s_n = \frac{1}{2}n^2$ by just using the fact that each $s_i$ is at most $n$.

PROBLEM 16D. $A(5,3) = A(5,2)$. See Theorem 16.3, i.e. (16.2). Counting directly is possible as follows: (i) Fix the first row as $(1,1,0,0,0)$; this means that the final result must be multiplied by 10. (ii) Put a 1 in position (2,1); this means that we should also multiply by 4. (iii) Split into two cases: a 1 in position (2,2), etc. or a 1 in position (2,3), in which case a factor 3 is introduced, etc.

Another solution is obtained by observing that $\mathcal{A}^*(5,2)$ has cardinality $\frac{1}{2}5!4! = 1440$, and that there are $(\binom{5}{2})^2 \cdot 3! = 600$ decomposable matrices in $\mathcal{A}(5,2)$.

PROBLEM 17A. Index the first two rows by $x$ and $y$, the first two columns by $u$ and $v$, all under the assumption that we have the multiplication table of a group of order 5. Show that $(v^{-1}u)^2 = 1$.

PROBLEM 17B. This is already difficult, despite the small size of the square! Take $a = 1$. Note that the 1 in the upper lefthand corner is in subsquares of type $\begin{pmatrix} 1 & i \\ i & 1 \end{pmatrix}$ for $i = 1, 2, 3, 4$. Consider any Latin square with this property. If the (2,3) entry is not 4, then permuting rows 4 and 5, columns 4 and 5, and symbols 4 and 5, changes the square into an equivalent one with a 4 in position (2,3). At this point, there is only one way to complete the square. So, if a Latin square has any cell with a property like the one we mentioned for the upper lefthand 1, then it is equivalent to this one.

Using the same kind of argument one can show that there are only a few possibly inequivalent squares of order 5. For some, one then has to establish the equivalence, a tedious task.

PROBLEM 17C. At this point, trial and error is prescribed. However, see Chapter 22.

PROBLEM 17D. Sufficiency is shown by an example (you can, but do not have to, use Theorem 17.1). For necessity, observe that just appending one more column to a Latin square of order $m$ requires $m$ new symbols.

PROBLEM 17E. Apply the algorithm.

PROBLEM 17F. For part (i), describe an algorithm that recovers the original Latin square of order $n$ from the constructed Latin square of order $n+1$ with constant back diagonal. For (ii), explain why the number of Latin squares of order $n$ with $n$'s on the back diagonal is equal to $N(n)/n!$.

PROBLEM 17G. (a) For $n = 6$, an example is $6, 1, 5, 2, 4, 3$.
  (b) Add the differences.
  (c) Assume two pairs are equal and solve.

PROBLEM 18A. See Problem 19D.

PROBLEM 18B. Use the method of Theorem 18.1. Note that a permutation of rows and columns, that preserves the 0-diagonal, does not help us to make a matrix symmetric or antisymmetric. So, only multiplications by $-1$ will do the job. Normalize the first row and make the first column (except for the 0th entry) all 1 or all $-1$. It is now sufficient to show that the (2,3) entry and the (3,2) entry are the same, respectively opposites, whenever $n \equiv 2$ (mod 4), respectively $n \equiv 0$ (mod 4).

PROBLEM 18C. A trivial exercise!

PROBLEM 18D. We have $28 = 1+9+9+9$ and $28 = 1+1+1+25$. Use the first of these. Note that each $W_i$ has 1 on the diagonal and if it has $w_{ij} = 2$, then $w_{7-i,j}$ is also 2; similarly for $-2$. The observation about occurrences of the $U^i$ in the matrices $W_j$ then completely forces the choice of the four matrices $W_i$. They prove to work.

PROBLEM 18E. See the reference E. C. Posner (1968).

PROBLEM 18F. For both questions use induction.

PROBLEM 18G. For (2), show that the number of rows with first coordinate 0 (or 1) is at most $n$.

PROBLEM 19A. For the isomorphism in part (i): Label the vertices of $K_6$ with 0000, 1000, 0100, 0010, 0001, 1111 in $\mathbb{F}_2^4$ and label an edge with the sum in $\mathbb{F}_2^4$ of the labels on its ends. This gives a one-to-one correspondence between the edges of $K_6$ and the 15 nonzero vectors in $\mathbb{F}_2^4$.

For part (ii), consider configurations consisting of three edges (there are four types). The construction is due to E. S. Kramer (1990), *Discrete Math.* **81**, 223–224.

PROBLEM 19B. Calculate $b_i$, $i = 3, 2, 1, 0$; $b_1 \notin \mathbb{Z}$.

PROBLEM 19C. This is Theorem 19.4 .

PROBLEM 19D. (i) Let $N$ have columnsums $c_i$. Then $\sum c_i = 66$. The sum of all inner products of rows is $\sum \binom{c_i}{2}$. Use the Cauchy-Schwarz inequality. Conclude that all columnsums are 6 and all inner products are 3. (ii) To show uniqueness, consider the complement: a 2-(11,5,2) design. Without loss of generality, the first row is (11111000000); the remaining ten rows start with the $\binom{5}{2}$ pairs (110...), (1010...), etc. Then show that the rows 2 to 5 can be completed in one way. This leaves a 6 by 6 square to be completed. At first sight, there seem to be two ways to do this. Find a permutation that maps one into the other.

PROBLEM 19E. See p. 4 of P. J. Cameron and J. H. van Lint (1980), *Graphs, Codes and Designs*, London Math. Soc. Lecture Note Series **43**, Cambridge University Press.

PROBLEM 19F. The mapping $x \mapsto 2x$ has six orbits, two of which are too long to use (e.g. (1,2,4,8,16,11)). See Example 28.2.

PROBLEM 19G. Use the corresponding orthogonal array. Use the common $\{0, 1\}$ to $\{1, -1\}$ mapping.

PROBLEM 19H. See p. 9 of the Cameron and Van Lint reference given above.

PROBLEM 19I. (i) Fix a point $x$ on $O$. Count pairs $(y, L)$, $y \in O$, $L$ a line through $x$ and $y$. (ii) From (i) we see that if $|O| = n + 2$, then every line that meets $O$, meets it twice. Take a point $z \notin O$ and count the lines through $z$ that meet $O$. (iii) Take any four points, no three on a line. These determine six lines, that together contain 19 points of the plane.

PROBLEM 19J. See D. Jungnickel and S. A. Vanstone, Hyperfactorizations of graphs and 5-designs, *J. Univ. Kuwait (Sci.)* **14** (1987), 213–223.

PROBLEM 19K. Let $B_1 \sim B_2$. How many blocks meet $B_1$? Take $x \in B_1$, $y \in B_2$; how many blocks meet both $B_1$ and $B_2$? There are $n + 1$ equivalence classes, each with $n$ lines (blocks). To each class append a point that is to be added as an extra point to each line of that class, etc.

PROBLEM 19L. Use Theorem 19.11. Consider the equation modulo 3.

PROBLEM 19M. For $v$ even, use quadratic forms, as in the proof of Theorem 19.11. See J. H. van Lint and J. J. Seidel (1966), Equilateral point sets in elliptic geometry, *Proc. Kon. Nederl. Akad. Wetensch.* **69**, 335–348.

PROBLEM 19N. (a) The method of Example 19.14 also works if $|V| = 1$; $(85 = 7(13 - 1) + 1)$. (b) $15 = 7(3 - 1) + 1$.

PROBLEM 20A. Use (20.5).

PROBLEM 20B. For each of the three ways in which the conditions for a codeword can be violated, change a suitable coordinate to remedy the situation.

PROBLEM 20C. Apply the pigeonhole principle to the last two bits of the words. A binary code with three words is easily handled!

PROBLEM 20D. See Van Lint (1982), p. 53 and p. 150.

PROBLEM 20E. (i) See Problem 20A. (ii) Count words of weight 5 "covered" by codewords of weight 7; the other words of weight 5 are covered by codewords of weight 8. Using this idea, set up a system of linear equations for the $A_i$.

PROBLEM 20F. This should not be difficult if the statement is understood.

PROBLEM 20G. The codewords in the dual of the binary Hamming code of length $2^r - 1$ may be identified with linear functionals on $\mathbb{F}_2^r$ (cf. the geometric description of Reed-Muller codes in Chapter 18); each nonzero codeword has weight $2^{r-1}$.

PROBLEM 20H. Use (19.6) and the fact that $G_{24}$ is linear and has minimum weight 8; note that there are $\binom{21}{2}$ pairs of lines. If $\alpha = 1$, consider the 7-point configuration $B^*$ in the plane. Any line that meets $B^*$ must meet it in 1 point or in 3 points. How many meet $B^*$ in 3 points?

PROBLEM 20I. (i) Find the standard generator matrix for $C$. (ii) The conditions (1) and (2) clearly define a linear code. In (1) we have 2 choices for parity; in (2) we have $4^3$ choices for the codeword; in five of the columns of $A$ we then have 2 possibilities (not in the last one!). For the choice "even", it is clear that a nonzero codeword in $C$ forces weight $\geq 8$; $\mathbf{0}$ has to be treated separately. For the choice "odd", it is clear that we obtain a weight of at least 6; condition (1) ensures that equality cannot hold. For more on this wonderful description of the Golay code and many consequences see: J. H. Conway, The Golay codes and the Mathieu groups, Chapter 11 in J. H. Conway and N. J. A. Sloane, *Sphere Packings, Lattices and Groups*, Springer-Verlag, 1988.

PROBLEM 20J. See Van Lint (1982), p. 44.

PROBLEM 21A. The argument is the same as for (21.4), now with inequalities.

PROBLEM 21B. Use the integrality condition. Also see the reference A. J. Hoffman and R. R. Singleton (1960) from Chapter 4.

PROBLEM 21C. Use (19.2); see Problem 20F. If two blocks are disjoint, count the number of blocks that meet both of them, each in two points. How many meet only one of them? Similar counting arguments (distinguish a few cases!) are used to show that $\mu = 6$.

PROBLEM 21D. Suppose such a graph exists. Pick a vertex $x$ and consider the sets $\Gamma(x)$ and $\Delta(x)$. We consider these as the points and blocks of a design. Show that this is a 2-(9,4,3) design. Next, fix a block of this design and let $a_i$ be the number of blocks that meet this block in $i$ points. Show that $a_0 \leq 4$ but that $a_0$ should be at least 5.

PROBLEM 21E. The first two questions are straightforward. For the third we refer to textbooks on matrix theory (see "interlacing") or to Chapter 31. The theorem we are alluding to states that the eigenvalues of a principal submatrix $B$ of a symmetric matrix $A$ are bounded by the largest and the smallest eigenvalue of the matrix $A$. To prove this, just consider any vector $\mathbf{x}$ and calculate $\mathbf{x}^\top A \mathbf{x} / \mathbf{x}^\top \mathbf{x}$. Specialize by letting $\mathbf{x}$ be an eigenvector of $B$ with 0's appended.

PROBLEM 21F. As in the proof of Theorem 21.5, first show that $\overline{\Gamma(x)}$ is bipartite. Define a *grand clique* to be a clique of size $n$ and show that any edge is in exactly one grand clique. Show that there are exactly $2n$ grand cliques, which fall into two classes of $n$ so that grand cliques in the same class are disjoint. See S. S. Shrikhande, The uniqueness of the $L_2$ association scheme, *Ann. Math. Stat.* **30** (1959), 781–798.

PROBLEM 21G. Count flags and other configurations, e.g. two intersecting lines and a point on one of them.

PROBLEM 21H. See Problem 21E for $v$; $k$ is obvious. Take two points on line $L$. Points joined to both of them are on $L$ or on one of the other $R - 1$ lines through one of them. A similar argument yields $\mu$. For $r$ and $s$ see (21.6).

PROBLEM 21I. On $\Gamma(x)$ use the fact that $\mu = 1$. Then consider $\{x\} \cup \Gamma(x)$.

PROBLEM 22A. Show that the complement of the graph of an $(n, n)$-net is the graph of an $(n, 1)$-net by computing the parameters of the complement. More generally, an $(n, r)$-net may be completed to an $(n, n + 1)$-net if and only if the complement of the graph of the $(n, r)$-net is the graph of an $(n, n + 1 - r)$-net.

PROBLEM 22B. Use the construction from fields described above the problem. Take $S(x, y) := (x + y)/2$ for $q$ odd, for example.

For even $q$, we can use $A$ and $S$ to schedule mixed-doubles tennis matches. Suppose we have $q$ couples—Mrs. $i$ and Mr. $i$ for $i \in \mathbb{F}_q$. Let Mrs. $i$ and Mrs. $j$ play against each other with respective partners Mr. $A(i, j)$ and Mr. $A(j, i)$ during round $S(i, j)$. There will be $q - 1$ rounds, labeled by the off-diagonal symbols in $S$, during which everyone is playing in exactly one match. It will be found that no one meets her or his spouse as partner or opponent the entire tournament, but meets everyone else exactly once as an opponent, and has everyone of the opposite sex exactly once as a partner. For $q$ odd, we can have $q$ rounds in which one couple sits out each round.

PROBLEM 22C. Show that $m, m+1, t$ and $u$ each have the property that for each prime $p < x$, they are either prime to $p$ or divisible by $p^x$.

PROBLEM 22D. Review the connection between sets of pairwise orthogonal Latin squares and transversal designs.

PROBLEM 23A. Note that the closure of a set $S$ of edges consists of all edges that have both ends in the same connected component of the spanning subgraph of $G$ with edge set $S$.

PROBLEM 23B. The lines of $AG_r(2)$ have size 2.

PROBLEM 23C. Dispose first of the cases when the union of two lines is the whole point set. Then show how to set up a one-to-one correspondence between the points of any two lines.

PROBLEM 23D. The hardest part is showing that $\mathcal{F} = \{F_1 \cup F_2 : F_1 \in \mathcal{F}_1, F_2 \in \mathcal{F}_2\}$. Prove that $\overline{F_1 \cup F_2} = F_1 \cup F_2$ using the fact that the rank of $\overline{F_1 \cup F_2}$ is the sum of the ranks of $F_1$ and $F_2$.

See H. Crapo and G.-C. Rota (1970) for a complete discussion of connectedness of combinatorial geometries and irreducibility of geometric lattices.

PROBLEM 23E. Suppose $A \cap F = \emptyset$ and $\text{rank}(A) + \text{rank}(F) = \text{rank}(\overline{A \cup F})$. Show that these equations remain valid if $A$ is replaced by $A' := \overline{A \cup \{x\}}$ where $x$ is any point not in $\overline{A \cup F}$.

PROBLEM 23F. Use (23.4).

PROBLEM 23G. We may work in $PG_2(\mathbb{F})$ and assume without loss of generality that $L_1 = [1,0,0]$ and $L_2 = [0,1,0]$. Then (explain why) $a_1 = \langle 0, \alpha_1, 1 \rangle$, $b_1 = \langle 0, \beta_1, 1 \rangle$, and $c_1 = \langle 0, \gamma_1, 1 \rangle$, while $a_2 = \langle \alpha_2, 0, 1 \rangle$, $b_2 = \langle \beta_2, 0, 1 \rangle$, and $c_2 = \langle \gamma_2, 0, 1 \rangle$.

PROBLEM 24A. One answer is $e_i = i(m - k + i)$.

PROBLEM 24B. Check that $x$ and $y$ are contained in a coset of a subspace $U$ if and only if $x - y$ is contained in $U$.

PROBLEM 25A. For a subspace $U$ of the $r$-dimensional space $W$, let $f(U)$ be the number of $k$-subspaces which intersect $W$ in $U$. Use (25.5).

PROBLEM 25B. Use (25.2) to calculate $\mu(0, x)$ for $x$ of rank 1 or 2.

PROBLEM 25C. There are only $n - 1$ partitions $x \neq 0_L$ (each consisting of one block of size 2 and $n - 2$ singletons) that satisfy $x \wedge a = 0_L$.

PROBLEM 26A. The number of lines on a point $x$ not in $A$ which meet $A$ is a constant, i.e. independent of $x$.

PROBLEM 26B. Any line disjoint from the arc can be used to obtain a partition of the blocks into parallel classes, one parallel class for each point of the line.

PROBLEM 26C. The point of intersection of the line though $\langle 1, 0, 0 \rangle$ and $\langle 0, 1, 0 \rangle$ and the line though $\langle 0, 0, 1 \rangle$ and $\langle 1, 1, 1 \rangle$, for example, is $\langle 1, 1, 0 \rangle$.

PROBLEM 26D. Let $\mu_i$ be the cardinality of the intersection of the $i$-th line with $S$. The inequality is easy if some $\mu_i \geq \sqrt{n} + 1$, so assume otherwise and consider $\sum_i (\mu_i - 1)(\mu_i - \sqrt{n} - 1)$.

PROBLEM 26E. Whether the characteristic is even or odd, $f$ is degenerate if and only if $\mathbf{x}(C + C^\top) = 0$ for some $\mathbf{x}$ with $f(\mathbf{x}) = 0$. To find such an $\mathbf{x}$ when $f$ is degenerate, take the last row of a nonsingular matrix $A$ such that $ACA^\top$ has a last row and column of all zeros, which exists because $x_n$, say, does not occur in some projectively equivalent form. Also see J. W. P. Hirschfeld (1979).

PROBLEM 26F. It is relatively straightforward to show that $Q'$ is a nondegenerate quadric in $W$. There is nothing more to do if $n$ is odd. Theorem 26.5 can be used to complete the proof for $n$ even. It is easy to see that if $Q'$ is hyperbolic, then so is $Q$. Next use equation (26.4) to show that every point $p$ on a hyperbolic quadric $Q$ in $PG_{n-1}(q)$ is contained in a flat $F \subseteq Q$ of projective dimension $n/2 - 1$. Check that such a flat is contained in $T_p$ so that the flat $F \cap W \subseteq Q'$ has dimension one less than that of $F$.

PROBLEM 27A. It is not difficult to see that the even numbered items are equivalent to each other, and similarly for the odd numbered items. To show (3) and (4) are equivalent, for example, introduce an incidence matrix $N$ and show that one is equivalent to the matrix equation $NN^\top = (k - \lambda)I + \lambda J$, and the other to $N^\top N = (k - \lambda)I + \lambda J$.

PROBLEM 27B. View $(\mathbb{Z}_2)^4$ as a vector space over the field $\mathbb{F}_2$. We may assume the difference set contains the zero vector. It will span $(\mathbb{Z}_2)^4$ and so contains a basis.

PROBLEM 27C. Count separately the number of times a nonzero element of $G \times H$ can be written as a difference of two elements of $A \times (H \setminus B)$, a difference of two elements of $(G \setminus A) \times B$, and as a difference of one element of each of these sets.

PROBLEM 27D. Multiply the sum of the elements, that we want to show is 0, by any square.

PROBLEM 27E. We already said that this problem is similar to the illustrated calculation of $\lambda_2$.

PROBLEM 27F. Use Theorem 27.5. (A table of cyclic difference sets may be found in L. D. Baumert, *Cyclic Difference Sets*, Lecture Notes in Math. **182**, Springer-Verlag, 1971.)

PROBLEM 27G. A zero $\omega$ of the polynomial $y^3+3y+2$ (coefficients in $\mathbb{F}_7$) is a primitive element of $\mathbb{F}_{7^3}$.

PROBLEM 27H. The cardinality of the intersection of a translate $D + g$ of $D$ with the set. $-D$ is the number of times $g$ arises as a *sum* from $D$.

PROBLEM 27I. For distinct $i$ and $j$, the differences $\mathbf{x} - \mathbf{y}$, $\mathbf{x} \in U_i, \mathbf{y} \in U_j$, comprise all vectors in $V$, each exactly once.

PROBLEM 28A. The Frobenius automorphism of $\mathbb{F}_{p^{t(n+1)}}$ is an automorphism of the symmetric design of 1-dimensional and $n$-dimensional subspaces of $\mathbb{F}_{p^{s(n+1)}}$ over $\mathbb{F}_{p^t}$.

PROBLEM 28B. We may take $S(x)$ as in Lemma 28.4 with $D = \{1, 2, 4\}$ but where $\alpha$ is not a multiplier.

PROBLEM 28C. If $n \equiv 0 \pmod{10}$, say, then for any $x \in D$, $\{x, 2x, 4x, 5x\} \subseteq D$ and the difference $x$ is seen to occur twice unless $3x = 0$. If $3x = 0$ for all $x \in D$, then $3x = 0$ for all $x$ in the group, and then $n^2 + n + 1$ must be a power of 3. Show that $n^2 + n + 1$ is never divisible by 9.

PROBLEM 28D. The numbers of normalized difference sets with these parameters are, respectively: 2, 2, 4, 2, 0, 4.

PROBLEM 28E. There are two normalized $(15, 7, 3)$ difference sets; they are equivalent since one consists of the negatives of the elements of the other. There are no normalized difference sets for any of the other parameters. To prove this in each case, first find a multiplier by Theorem 28.7.

For e.g. $(v, k, \lambda) = (25, 9, 3)$, Example 28.3 shows 2 is a multiplier of a hypothetical normalized difference set $D$. If the group is cyclic, $D$ is the union of cycles of $x \mapsto 2x$ on $\mathbb{Z}_{25}$ which are

$$\{0\}, \{5, 10, 20, 15\} \text{ and } \mathbb{Z}_{25} \setminus \{0, 5, 10, 15, 20\};$$

but no such union has nine elements. If the group is elementary abelian (exponent 5), then $D$ must consist of 0 and two cycles of the form $\{x, 2x, 4x, 8x = 3x\}$; but the difference $x$ occurs five times already among 0 and the elements of such a cycle.

PROBLEM 28F. Assume $D(x^{-1}) = D(x)$ and use Lemma 28.2 with $p = 2$.

PROBLEM 28G. Recall the construction of a symmetric design with those parameters from a Latin square of order 6 in Problem 19G.

PROBLEM 28H. Assume $D(x^q) = D(x)$. Note that $q^3 \equiv -1$ (mod $|H|$). Show $D(x^\alpha) \equiv 0$ (mod $q, H$) and conclude that one of the coefficients of $D(x^\alpha)$ is $q + 1$ and all others are 1.

PROBLEM 28I. The first part is similar to Problem 28H. Then show that if, in $PG_3(q)$, a line meets a set $S$ of size $q^2 + 1$ in $k \geq 3$ points, then one of the $q + 1$ planes on that line contains more than $q + 1$ points of $S$.

PROBLEM 29A. Review the proof of Theorem 20.6.

PROBLEM 29B. Review the latter part of the proof of Proposition 29.5.

PROBLEM 29C. The answer is $(v, k, \lambda) = (4n - 1, 2n - 1, n - 1)$ or $(4n - 1, 2n, n)$, i.e. Hadamard or the complement of Hadamard.

PROBLEM 30A. $A_1 A_2 = 4A_1 + 4A_2 + 9A_3$ and $A_1^3 = 90A_0 + 83A_1 + 56A_2 + 36A_3$.

PROBLEM 30B. By (30.3), $A_i A_j = \sum_{\alpha=0}^{k} P_i(\alpha) P_j(\alpha) E_\alpha$. Then by (30.4),

$$N A_i A_j = \sum_{\alpha=0}^{k} P_i(\alpha) P_j(\alpha) \sum_{\beta=0}^{k} Q_\alpha(\beta) A_\beta,$$

so evidently $p_{ij}^\ell = \frac{1}{N} \sum_{\alpha=0}^{k} P_i(\alpha) P_j(\alpha) Q_\alpha(\ell)$. The second eigenmatrix $Q$ can be calculated by inverting $P$ (also see Theorem 30.2).

PROBLEM 30C. For the Latin square graphs,

$$P = \begin{pmatrix} 1 & (n-1)r & (n-1)(n-r+1) \\ 1 & n-r & -n+r-1 \\ 1 & -r & r-1 \end{pmatrix}.$$

PROBLEM 30D. To evaluate the entries $P_\ell(0)$ of the top row of $P$, take the inner product (as introduced in the proof of Theorem 30.2) of both sides of (30.3) with $E_0$.

PROBLEM 30E. This is the dual to Problem 30B.

PROBLEM 30F. There is a trivalent simple graph on eight vertices with the required property.

PROBLEM 30G. Sum $|A \cap \sigma(B)|$ over all automorphisms $\sigma$ in two ways.

PROBLEM 30H. Given a metric scheme arising from a distance regular graph $G$, first note that $A_1^i$ is a linear combination of $A_0, A_1, \ldots, A_i$ where the coefficient of $A_i$ is positive. Given a $P$-polynomial scheme, define a graph $G$ by saying that $x$ and $y$ are adjacent in $G$ if and only if they are 1st associates in the scheme, and show that two points are $i$-th associates if and only if they are at distance $i$ in $G$.

PROBLEM 31A. Consider the intersection of the eigenspaces of $A_1$ and $A_2$ corresponding to the eigenvalue 1.

PROBLEM 31B. Since the adjacency matrix of the complement of a graph $G$ with adjacency matrix $A$ is $J - I - A$, the eigenvalues of $\overline{G}$ are easily determined from those of $G$. The first examples in Chapter 21 are instances in which equality can occur.

PROBLEM 31C. The Petersen graph has 3-*claws*, i.e. induced subgraphs on four vertices consisting of three edges incident with a common vertex (and no other edges). Explain why a line graph cannot contain a 3-claw. The matrix $N^\top N$ is positive semidefinite and is equal to $2I + A$, where $A$ is the adjacency matrix.

PROBLEM 31D. From the equation following (31.1) with $\mathbf{x}$ the vector of all 1's, it follows that the largest eigenvalue of a graph is bounded below by the minimum degree of the graph. By Lemma 31.5, the largest eigenvalue of a graph $G$ is bounded below by the minimum degree of any induced subgraph $H$ of $G$. If $\chi(H) = \chi(G)$ and the deletion of any vertex $x$ of $H$ decreases the chromatic number, then $\deg(x) \geq \chi(H) - 1$.

PROBLEM 31E. If there are no directed paths from $x$ to $y$ in a digraph $G$ with adjacency matrix $A$, let $S$ be the set of vertices $z$ such that there does exist a directed path in $G$ from $x$ to $z$. Then $A(x, y) = 0$ whenever $a \in S$ and $b \in V(G) \setminus S$, so $A$ is not irreducible. The converse is even easier.

PROBLEM 31F. Most of the work has been done in the proof of Theorem 31.10: If the distinct eigenvalues of $G$ are its degree $d$ and $\mu_i$, $i = 1, 2$, then $(A - \mu_1 I)(A - \mu_2 I) = \frac{1}{v}(d - \mu_1)(d - \mu_2)J$ where $A$ is the adjacency matrix of $G$. This means that $A^2$ is in the algebra generated by $I$, $J$, and $A$. So $G$ is strongly regular.

PROBLEM 31G. Let $G$ be an $srg(v, k, \lambda, \mu)$. The eigenvalues of $\Delta(x)$ interlace the eigenvalues

$$k, r, r, \ldots, r, s, s, \ldots, s \quad (r > s, \text{ say})$$

of $G$. Any sequence that interlaces this can have at most one term which is greater than $r$.

Except in the half-case, $r$ and $s$ are integers with $rs = \mu - k$; in particular, $r \le k - \mu$ (even in the half-case). The graph $\Delta(x)$ is regular of degree $k - \mu$. Were it not connected, $k - \mu$ would be an eigenvalue of $\Delta(x)$ of multiplicity greater than one.

PROBLEM 32A. More strongly, a subdivision of $K_{3,3}$ (where edges are replaced by paths) occurs as a subgraph of the Petersen graph.

PROBLEM 32B. Briefly explain why spanning trees of $G$ that contain the edge $e$ are in one-to-one correspondence with the spanning trees of $G''_e$ and why spanning trees of $G$ that do not contain the edge $e$ are in one-to-one correspondence with the spanning trees of $G'_e$.

PROBLEM 32C. For the polygons, use induction, Equation (32.1), and Example 32.1.

PROBLEM 32D. If $(H, K)$ is an $\ell$-separation and the deletion of the $\ell$ vertices $S := V(H) \cap V(K)$ does not disconnect the graph, then at least one of $H$ or $K$ has vertex set contained in $S$.

PROBLEM 32E. Use Euler's formula. The only pairs $(d_1, d_2)$ that arise other than those $(3, 3)$, $(3, 4)$, $(3, 5)$, $(4, 3)$, $(5, 3)$ that correspond to the five Platonic solids are those $(2, n)$ and $(n, 2)$, $n \geq 2$, that correspond to the polygons and bond-graphs.

PROBLEM 32F. Suppose a vertex of $G_S''$ is incident with three distinct edges $e_1, e_2, e_3$. Such a vertex is a connected component $C$ of $G : S$. So $e_i$ has an end $x_i$ in $C$, $i = 1, 2, 3$ (these ends are not necessarily distinct). Explain why the connectivity of $C$ implies that there exists a vertex $x$ and possibly degenerate but internally disjoint paths in $C$ from $x$ to each of $x_1$, $x_2$, and $x_3$. That is, we have a subgraph of $C$ isomorphic to a subdivision of the letter $Y$ with the $x_i$'s at the ends of the arms if the paths all have length $\geq 1$, but otherwise some of the arms may be degenerate. This observation immediately leads to a solution of the first part of the problem.

Given four vertices $x_i$, $i = 1, 2, 3, 4$, of a connected graph $C$, show that there is a subgraph isomorphic to a subdivision of either the letter $X$ or the letter $I$, though with possibly degenerate arms, and with the $x_i$'s at the ends of the arms.

Check that if a vertex, say 1, of $K_5$ is replaced by two vertices $1_L$ and $1_R$ joined by an edge, and the edges joining 1 to the other vertices are split, say $1_L$ is joined to 2 and 3, and $1_R$ is joined to 4 and 5, then the resulting graph has $K_{3,3}$ as a subgraph. Combine this with the observation of the previous paragraph to finish the problem.

PROBLEM 32G. This is easy for circuits. For the case of bonds, it is helpful to first note that every vertex of such a $k$-connected graph has degree at least $k$.

PROBLEM 32H. Use induction on the cardinality of the support of codewords.

PROBLEM 32I. The circuits of $G_e'$ are exactly those circuits of $G$ that do not contain $e$. What are the bonds of $H_e''$?

PROBLEM 32J. One way to do this would be to first prove two things about subsets $S$ of the edge set of a connected graph $K$: (1) $S$ is the edge set of a spanning tree in $K$ if and only if no circuit of $K$ is contained in $S$ and $S$ is maximal with respect to this property and (2) $S$ is the edge set of a spanning tree in $K$ if and only if every bond of $K$ meets $S$ nontrivially and $S$ is minimal with respect to this property.

PROBLEM 32K. Here is an idea to construct bonds. Deleting any edge of a spanning tree of $G$ leaves a subgraph with two components; the set of edges with one end in each component is a bond of $G$.

PROBLEM 32L. This should not be difficult if all definitions are well understood.

PROBLEM 33A. Show first that the two components that result from deleting an isthmus from a connected trivalent graph each have an odd number of vertices.

PROBLEM 33B. Take the vertices of $K_n$ to be $\mathbb{Z}_n$. For $n = 5$, use all cyclic shifts (translates) of $[1, 2, 4, 3]$ (mod 5).

PROBLEM 33C. Let us drop the last coordinate in the description of Example 21.4 and describe the Clebsch graph as having vertex set $\mathbb{F}_2^4$ and where $\mathbf{x}$ and $\mathbf{y}$ are adjacent when $\mathbf{x} + \mathbf{y}$ has weight 1 or 4. Try to choose one walk $w$ and obtain the others as translates of $w$ by all elements of $\mathbb{F}_2^4$.

PROBLEM 33D. Part (i) is similar to Theorem 33.5. For the proof of uniqueness of the mesh, we may take $V(K_7) = \{0, 1, \ldots, 6\}$ and, without loss of generality, the walks traversing 0 are

$$[6, 0, 1], \quad [1, 0, 2], \quad [2, 0, 3], \quad [3, 0, 4], \quad [4, 0, 5], \quad [5, 0, 6].$$

The walk traversing the edge $(6, 1)$ has third vertex 3 or 4 (otherwise the vertex condition is violated at 1 or 6) and in either case the other walks are uniquely determined.

PROBLEM 33E. The vertices of the graph $\mathcal{M}_x$ are the edges incident with $x$. These edges form the edge set of a bond in $G$ and hence are the edges of a circuit in $H$.

PROBLEM 33F. Let us use the term *flag* for a triple $(x, e, F)$ consisting of a vertex $x$, an edge $e$ incident with $x$, and a face $F$ incident with $e$. There is exactly one other vertex $x'$ so that $(x', e, F)$ is a flag, exactly one other edge $e'$ so that $(x, e', F)$ is a flag, and exactly one other face $F'$ so that $(x, e, F')$ is a flag; these objects must hence be fixed by any automorphism $\alpha$ that fixes the terms of $(x, e, F)$.

PROBLEM 33G. Use Euler's formula, Theorem 32.3.

PROBLEM 33H. All faces will have the same size. Consider the walk traversing the edge $(0, 1)$. The next several edges will be $(1, 1 - \omega)$, $(1 - \omega, 1 - \omega + \omega^2)$, $(1 - \omega + \omega^2, 1 - \omega + \omega^2 - \omega^3)$, etc. If $m \equiv 2 \pmod 4$, for example, then the walk has length $m/2$.

PROBLEM 34A. The sum of the determinants of all principal $n - 1$ by $n - 1$ submatrices of $M$ is an appropriate sign times the coefficient of $x$ in the characteristic polynomial $\det(xI - M)$ of $M$.

PROBLEM 34B. If a vector $g$ with coordinates indexed by $E(H)$ has the property that for every closed walk $w$, the signed sum of the values of $g$ on the edges of $w$ is zero, then define a vector $h$ indexed by $V(H)$ as follows. (We assume $H$ is connected for convenience; otherwise work separately with each component.) Fix a vertex $x$, define $h(x) := 0$, and for any vertex $y$, let $h(y)$ be the signed sum of the values of $g$ on the edges of any walk from $x$ to $y$. Check that $h$ is well defined and that $g$ is the coboundary of $h$.

PROBLEM 34C. Let $\mathcal{Z}$ denote the cycle space and $\mathcal{B}$ the coboundary space. These are orthogonal complements, so their (direct) sum is $\mathbb{R}^m$ (the space of all vectors with coordinates indexed by $E(D)$). Show that the linear transformation $R$ is one-to-one on $\mathcal{Z}$ and that $\mathcal{Z}R \cap \mathcal{B} = \{0\}$, so that $\mathbb{R}^m$ is the direct sum of these subspaces.

PROBLEM 34D. The sizes of all squares in Fig. 34.4 may be expressed as integral linear combinations of the sizes $x, y, z$ of the three indicated squares. For example, the smallest square has size $z - x$ and the next smallest has size $x + z - y$. It will be found that the sizes of some squares can be so expressed in more than one way, and we thus have linear relations between $x, y, z$. Eventually, we can express the size of any square as a rational scalar multiple of $x$, and the least common denominator of these rational numbers is the value we should take for $x$.

PROBLEM 34E. It may be quickest to procede as in Problem 34D; let $x, y, z$, for example, denote the values of the current on three edges, etc. But one can evaluate determinants as in equation (34.3) to find the solutions.

PROBLEM 35A. The rotations include: the identity (1), rotations of 90, 180, and 270 degrees about an axis through the centers of two opposite faces $(3 + 3 + 3)$, rotations of 180 degrees about an axis through the centers of two opposite edges (6), and rotations of 120 and 240 degrees about an axis through two opposite vertices $(2 + 2 + 2 + 2)$. The latter eight rotations, for example, contribute $8X_3^2$ to the cycle index.

PROBLEM 35B. Use Theorem 35.2 and Problem 35A.

PROBLEM 35C. Explain why the number of injections $f$ fixed by $(\sigma, \tau)$ depends only on the cycle structure of $\sigma$ and $\tau$ (i.e. on the numbers $z_i$ of cycles of length $i$ for each $i$) and find an expression for the number of injections in terms of the $z_i(\sigma)$'s and the $z_i(\tau)$'s.

PROBLEM 35D. In Theorem 35.3, take $G$ to be dihedral and $H$ to be the symmetric group $S_2$.

PROBLEM 36A. It suffices to prove the statement for matrices with entries $0 \leq a_{ij} \leq 1$. Construct a network with $1+m+n+1$ vertices: a source $s$, one vertex $x_i$ for each of the $m$ rows and one vertex $y_j$ for each column $y_j$ and a sink $t$. There are to be edges directed from $s$ to all $x_i$, an edge from $x_i$ to $y_j$ of capacity 1 if $a_{ij} > 0$, and edges from all $y_j$ to $t$. We omit the remainder of the description. Use Theorem 7.2.

PROBLEM 36B. Let $m := v/2$. For each of the $M := v - 1$ colors $i$, we obtain an $m$-partition of $V(K_v)$ by taking the edges of color $i$, all singletons $\{x\}$ for $x$ not incident with an edge of color $i$, and enough copies of the empty set to make an $m$-partition. Check that the condition (36.1) holds. Review the proof of Theorem 36.1.

# Appendix 2
## Formal power series

Before introducing the subject of this appendix, we observe that many of the assertions that we shall make will not be proved. In these cases the proof will be an easy exercise for the reader.

Consider the set

$$\mathbf{C}^{\mathbf{Z}_+} := \left\{ (a_0, a_1, \dots) \ : \ \forall_{i \in \mathbf{Z}_+} [a_i \in \mathbf{C}] \right\}.$$

On this set we introduce an addition operation and a multiplication as follows:

$$(a_0, a_1, \dots) + (b_0, b_1, \dots) := (a_0 + b_0, a_1 + b_1, \dots),$$

$$(a_0, a_1, \dots)(b_0, b_1, \dots) := (c_0, c_1, \dots),$$

where $c_n := \sum_{i=0}^{n} a_i b_{n-i}$.

This definition produces a *ring* that we denote as $\mathbf{C}[[z]]$ and call the ring of *formal power series*. This name is explained in the following way. Let $z := (0, 1, 0, 0, \dots)$. Then $z^n$ is the sequence $(0, \dots, 0, 1, 0, \dots)$ with a 1 in the $n$-th position. So, formally we have

$$\mathbf{a} = (a_0, a_1, \dots) = \sum_{n=0}^{\infty} a_n z^n =: a(z).$$

We shall use both notations, i.e. $\mathbf{a}$ and $a(z)$, for the sequence. We say that $a_n$ is the *coefficient of $z^n$* in $\mathbf{a}$ (or $a(z)$). Notice that $\mathbf{C}[z]$, the ring of polynomials with coefficients in $\mathbf{C}$, is a subring of $\mathbf{C}[[z]]$. Some of the power series will be convergent in the sense of analysis. For these, we can use results that we know from analysis. Quite often these results can be proved in the formal sense, i.e. not using convergence or other tools from analysis.

EXAMPLE 1. Let $f := (1, 1, 1, \dots)$. Using the definition of multiplication and $1 - z = (1, -1, 0, \dots)$, we find $(1-z)f = (1, 0, 0, \dots) = 1$. So, in $\mathbb{C}[[z]]$, we have $f = (1 - z)^{-1}$, i.e.

$$\frac{1}{1 - z} = \sum_{n=0}^{\infty} z^n,$$

a result that we knew from analysis.

The regular elements in $\mathbb{C}[[z]]$ are the power series with $a_0 \neq 0$. This can be seen immediately from the definition of multiplication. From the relation $a(z)b(z) = 1$ we can calculate the coefficients $b_n$, since $b_0 = a_0^{-1}$ and $b_n = -a_0^{-1} \sum_{i=1}^{n} a_i b_{n-i}$. From this we see that the quotient field of $\mathbb{C}[[z]]$ can be identified with the set of so-called *Laurent series* $\sum_{n=k}^{\infty} a_n z^n$, where $k \in \mathbb{Z}$. Since it plays a special role further on, we give the coefficient of $z^{-1}$ a name familiar from analysis. If $a(z)$ is the series, we say that $a_{-1}$ is the *residue* of $a(z)$. This will be written as Res $a(z)$.

Let $f_n(z) = \sum_{i=0}^{\infty} c_{ni} z^i$ $(n = 0, 1, 2, \dots)$ be elements of $\mathbb{C}[[z]]$ with the property

$$\forall_i \exists_{n_i} [n > n_i \Rightarrow c_{ni} = 0].$$

Then we can formally define

(1)
$$\sum_{n=0}^{\infty} f_n(z) = \sum_{i=0}^{\infty} \left( \sum_{n=0}^{n_i} c_{ni} \right) z^i.$$

This definition allows us to introduce *substitution* of a power series $b(z)$ for the "variable" $z$ of a power series $a(z)$. If $b_0 = 0$, which we also write as $b(0) = 0$, then the powers $b^n(z) := (b(z))^n$ satisfy the condition for formal addition, i.e.

$$a(b(z)) := \sum_{n=0}^{\infty} a_n b^n(z)$$

makes sense.

EXAMPLE 2. Let $f(z) := (1-z)^{-1}$, $g(z) = 2z - z^2$. Then formally

$$h(z) := f(g(z)) = 1 + (2z - z^2) + (2z - z^2)^2 + \cdots$$
$$= 1 + 2z + 3z^2 + 4z^3 + \cdots .$$

From calculus we know that this is the power series expansion of $(1-z)^{-2}$, so this must also be true in $\mathbb{C}[[z]]$. Indeed we have

$$(1-z)h(z) = \sum_{n=0}^{\infty} z^n = (1-z)^{-1}.$$

This also follows from (legitimate) algebraic manipulation:

$$f(g(z)) = \left(1 - (2z - z^2)\right)^{-1} = (1-z)^{-2}.$$

Many power series that we often use represent well known functions, and in many cases the inverse function is also represented by a power series. This can also be interpreted formally for series $f(z)$ with $f_0 = 0$ and $f_1 \neq 0$. We "solve" the equation $f(g(z)) = z$ by substitution. This yields $f_1 g_1 = 1$, $f_1 g_2 + g_1^2 = 0$, and in general an expression for the coefficient of $z^n$ starting with $f_1 g_n$ for which the other terms only involve coefficients $f_i$ and coefficients $g_k$ with $k < n$. Setting this 0 allows us to calculate $g_n$.

EXAMPLE 3. The reader who likes a combinatorial challenge can give a proof by counting of the formula

$$\sum_{k=0}^{n} \binom{2k}{k} \binom{2n - 2k}{n - k} = 4^n.$$

This would be equivalent to proving that the formal power series $f(z) := \sum_{n=0}^{\infty} \binom{2n}{n}(z/4)^n$ satisfies the relation $f^2(z) = (1-z)^{-1}$. By similar arguments or by algebraic manipulation one can then find the formal power series that deserves the name $(1+z)^{\frac{1}{2}}$.

Now consider the formal power series $f(z) := 2z + z^2$. The "inverse function" procedure that we described above will yield a power series $g(z)$ that satisfies $2g(z) + g^2(z) = z$, i.e. $(1 + g(z))^2 = 1 + z$. So this should be the series that we just called $(1+z)^{\frac{1}{2}}$. It should no longer be surprising that algebraic relations that hold for convergent power series are also true within the theory of formal power series.

REMARK. We point out that substitutions that make perfect sense in calculus could be forbidden within the present theory. The power series $\sum_{n=0}^{\infty} \frac{z^n}{n!}$ will of course be given the name $\exp(z)$. In calculus we could substitute $z = 1+x$ and find the power series for $\exp(1+x)$. We do not allow that under formal addition.

We now introduce *formal derivation* of power series.

DEFINITION. If $f(z) \in \mathbb{C}[[z]]$, then we define the *derivative*

$$(Df)(z) = f'(z)$$

to be the power series $\sum_{n=1}^{\infty} n f_n z^{n-1}$.

The reader should have no difficulty proving the following rules by using the definitions in this appendix:

(D1) $(f(z) + g(z))' = f'(z) + g'(z)$;
(D2) $(f(z)g(z))' = f'(z)g(z) + f(z)g'(z)$;
(D3) $(f^k(z))' = k f^{k-1}(z) f'(z)$;
(D4) $(f(g(z)))' = f'(g(z)) g'(z)$.

The chain rule (D4) is an example of a more general statement, namely that

$$D \left( \sum_{n=0}^{\infty} f_n(z) \right) = \sum_{n=0}^{\infty} D(f_n(z)).$$

If convergence plays a role, this is a difficult theorem with extra conditions, but for formal power series it is trivial!

The familiar rule for differentiation of a quotient is also easily proved. So we can carry the theory over to Laurent series. We shall need the following two facts. Again we leave the proof as an exercise.

If $w(z)$ is a Laurent series, then

(R1) $\mathrm{Res}(w'(z)) = 0$;
(R2) the residue of $w'(z)/w(z)$ is the least integer $\ell$ such that the coefficient of $z^\ell$ in $w(z)$ is nonzero.

We have already mentioned the idea of an "inverse function" and have shown how to calculate its coefficients recursively. The next theorem gives an expression for the coefficients.

THEOREM 1. *Let* $W(z) = w_1 z + w_2 z^2 + \cdots$ *be a power series with* $w_1 \neq 0$. *Let* $Z(w) = c_1 w + c_2 w^2 + \cdots$ *be a power series in* $w$, *such that* $Z(W(z)) = z$. *Then*

$$c_n = \mathrm{Res}\ \left( \frac{1}{nW^n(z)} \right).$$

PROOF: Observe that $c_1 = w_1^{-1}$. We now use formal derivation and apply it to the relation $Z(W(z)) = z$. This yields

$$(2) \qquad 1 = \sum_{k=1}^{\infty} k c_k W^{k-1}(z) W'(z).$$

Consider the series obtained by dividing the righthand side of (2) by $nW^n(z)$. If $n \neq k$, then the term $W^{k-1-n}(z)W'(z)$ is a derivative by (D3) and hence has residue 0 by (R1). By applying (R2) to the term with $n = k$ we find the assertion of the theorem. □

This theorem makes it possible for us to give a proof of the Lagrange inversion formula (see Theorem 14.3) within the theory of formal power series. Let $f(z)$ be a power series with $f_0 \neq 0$. Then $W(z) := z/f(z)$ is a power series with $w_1 \neq 0$. Now apply Theorem 1. We find that $z = \sum_{n=1}^{\infty} c_n w^n$ with

$$c_n = \mathrm{Res}\ \left( \frac{f^n(z)}{nz^n} \right) = \frac{1}{n!}(D^{n-1}f^n)(0),$$

which is (14.19). This approach is based on P. Henrici, An algebraic proof of the Lagrange-Bürmann formula, *J. Math. Anal. and Appl.* **8** (1964), 218–224. The elegant simplification is due to J. W. Nienhuys.

As was observed above, we define

$$(3) \qquad \exp(z) := \sum_{n=0}^{\infty} \frac{z^n}{n!}.$$

From previous knowledge we expect that replacing $z$ by $-z$ will yield the inverse element in the ring $\mathbb{C}[[z]]$. Formal multiplication and the fact that $\sum_{k=0}^{n}(-1)^k \binom{n}{k} = 0$ for $n > 0$ shows that this is true.

It is now completely natural for us to also define

$$
(4) \qquad\qquad \log(1+z) := \sum_{n=1}^{\infty}(-1)^n \frac{z^n}{n}.
$$

Again, calculus makes us expect a relation between the "functions" log and exp. By our substitution rule, it makes sense to consider the power series $\log(\exp(z))$. From (D4) we then find

$$
D(\log(\exp(z))) = \frac{\exp(z)}{\exp(z)} = 1,
$$

i.e. $\log(\exp(z)) = z$. (Here we have used the fact that the formal derivative of $\log(1+z)$ is $(1+z)^{-1}$.)

Of course, much more can be said about formal power series. One could explore how many familiar results from analysis can be carried over or given a formal proof. We hope that this sketchy treatment will suffice to make things clearer.

# Name Index

# Subject Index